Analysis and Design of Dynamic Systems

ANALYSIS AND DESIGN OF DYNAMIC SYSTEMS

Second Edition

IRA COCHIN
New Jersey Institute of Technology

HAROLD J. PLASS, JR.
University of Miami

HarperCollins*Publishers*

Project Coordination: Science Typographers, Inc.
Cover Coordinator: Heather Zeigler
Cover Design: Wanda Lubelska Design
Production: Beth Maglione

ANALYSIS AND DESIGN OF DYNAMIC SYSTEMS, Second Edition

Library of Congress Cataloging-in-Publication Data

Cochin, Ira.
    Analysis and design of dynamic systems / Ira Cochin, Harold J.
Plass, Jr.—2nd ed.
        p.   cm.
    Includes bibliographical references.
    ISBN 0-06-041314-X
    1. Engineering design.   2. System analysis.   3. Dynamics.
I. Plass, Harold J.   II. Title
TA174.C57   1990
620'.0042—dc20   89-26782
                                                          CIP

        90   91   92   9   8   7   6   5   4   3   2

Contents

Preface

This text introduces fundamental systems techniques for the analysis and design of dynamic systems. The text may be used for undergraduate courses in vibrations, dynamics, systems, automatic controls, and senior design. The present text was actually developed in the process of teaching these four subjects. The text has undergone a number of changes. Many applications were added or modified, and new methods were adapted in a continuing effort to make the text more suitable for study of a difficult subject by an undergraduate audience. The present form of the text, its depth, order, treatment of many aspects of a single system, and policy of practical involvement are the logical conclusions of its inception and subsequent growth.

The literature abounds with formulation oriented techniques, in the face of a dearth of design. In response to this dilemma, this text devotes equal time to each of these two areas. This, in turn, required a search of every possible source of design methods. Nevertheless, this left a number of windows in the design spectrum, requiring the authors to create a number of original techniques. At the same time, this also required the creation of many new problems never before aired in the literature. Also, since half the text is design oriented, it was natural to include a number of real-world problems involving the latest practices.

Chapters 1 and 2 introduce the reader to the basic elements to be used in systems of various disciplines.

Chapter 3 introduces modeling techniques and elementary design.

Chapter 4 introduces a number of solution techniques, the meaning of transient and steady state response, the use of differential equations, Laplace transforms, and TUTSIM dynamic modeling.

Chapter 5 delves into first order systems, employing electrical, mechanical, and fluidic elements. Some examples and problems include the design of a capacitor flashlight, a water tower, and an annealing oven.

Chapter 6 delves into second order systems, employing various disciplines. Classical methods as well as short cuts are featured in order to treat initial conditions and impulse, step, and ramp inputs. Some examples and problems cover the following devices: lunar landing system; spring scale with various types of damping; suspension for motorcycle.

Chapter 7 is concerned with frequency response using traditional methods. This covers forced vibration, transmissibility, self-excitation, and seismic instruments. Some examples and problems cover the following: phono pickup; motorcycle suspension; accelerometer; design of a voice actuated circuit; design of an inductive pickup for telephone; design of a proximity switch; design of a metal detector; design of an intruder alarm; and design of a carrier current transmission system.

Chapter 8 treats the same subject as that in Chapter 7, but using short cuts such as log-log (Bode) plots. Some examples and problems cover the following: a stepper motor; design of an electrostatic speaker; design of the suspension system for a motorcycle; design of a vibrating reed tachometer.

Chapter 9 concentrates upon inputs described by either a Fourier series or a random distribution. Some examples and problems cover the following: design of an electrical circuit with a triangular input; design of a vibrator pump using frequency doubling; Mechanical low pass filter of mechanical noise; and design of vibration powered tools.

Chapter 10 provides an in-depth view of the new field of microelectromechanical devices; Valves, nozzles, vibrator pumps, accelometers, pressure sensors, torsion mirrors for optical communication, ink jet printers, and motors that are smaller than the period at the end of this sentence. The subject requires the integration of all the material in the book up to this point (which dictated where to locate this chapter). This small size is required for the unique means of fabrication. No cutting, fastening, or assembly tools are employed. The entire device is manufactured by photolithographic means, followed by chemical etching. Thus, it is possible to produce one thousand items for the price of one. Needless to say, this field has a potential future. Some examples and problems include the design of microdevices such as the following: check valve, pressure sensor, control valve, heart attack alert, insulin dispenser, ink jet printer, gas chromatograph, and various surgical instruments fastened to a catheter.

Chapter 11 discusses all of the transient properties of dynamic systems. In addition to the usual calculus approach, a novel method is introduced that makes use of what the authors refer to as the scatter function. Some examples and problems cover the following: quick acting circuit, pneumatic control valve, pulse width modulation, and rotary conveyor.

Chapter 12 presents various means by which the set of differential equations for high order systems may be formulated. Some examples and problems cover the following devices: motor and pump; coupled electrical circuit; suspension for motorcycle; damped double compound pendulum;

superheterodyne interstage amplifier; Houdaille damper; the gyroscope; and transistor amplifier.

Chapter 13 considers systems that are described by either block or signal flow diagrams.

Chapter 14 covers various means with which to solve the set of simultaneous differential equations including damping. In addition to the conventional means, two original methods are presented: the "folding" of any order of symmetric systems that cuts the order in half, and the compatible system that provides a closed form solution for as many as eleven unknowns. Some examples and problems include design of the following: undamped dynamic absorber, two mass low pass filter, resonating electrical circuit, vibration absorber for hair clipper, broadband meter movement, band eliminator for alarm system and broadband amplifier with built-in noise suppressor.

Chapter 15 presents Evans' root locus method. Material covered: means to construct the root locus, transient response, stability, design, synthesis, and an in-depth study of a third order system.

Chapter 16 deals with the analysis of automatic control systems, errors, disturbances, stability, transient response, and time delay. The chapter makes use of the following techniques: calculus, differential equations, Laplace transforms, block diagrams, and root locus. Some examples and problems cover the following: rolling mill; hydraulic servo; electromechanical servo; roll stabilization for aircraft; servo positioner for valve; pressure regulator; water level regulator; temperature control; and robot hand.

Chapter 17 is concerned with the design and compensation of automatic control systems. Some examples and problems involving design cover the following: hydraulic servo to meet conflicting requirements with and without a time delay; electromechanical servo with lead-lag control; roll stabilization with integral plus proportional control; need and use of derivative control and derivative feedback; means to meet the several simultaneous (usually mutually exclusive) specifications.

A solution manual for instructors is available from the publisher.

In conclusion, one might regard this text as a blend of many of the most up-to-date methods available today. Each of the techniques may have some limited area where it could stand alone, but for the most part, several techniques are used in conjunction, each one reinforcing and supporting the other. The material is seasoned with experience, and is spiced with practical illustrative examples. The text was designed to bridge the gap between theory and practice and yet remain within the corridor of time-proven methodology. Perhaps there is nothing new under the sun, but we can have a field day with innovation in the communication of these not-so-new ideas.

We gratefully acknowledge the suggestions provided by reviewers of our text: Professor Gray Costello, Cooper Union College, Professor James H. Williams, Jr., Massachusetts Institute of Technology, and Professor E. F. Obert, University of Wisconsin-Madison.

Ira Cochin
Harold J. Plass, Jr.

Chapter 1

Introduction to System Concepts

1.1 INTRODUCTION TO SYSTEMS

The word "system" has become a household word. We hear of a plumbing system, a transportation system, an electrical system, our digestive system, our judiciary system, and our economic system. In more technical circles we hear of an ignition system, a braking system, the XYZ Broadcasting System, a communications system, a system of particles, and a navigation system. What do all of these "systems" have in common?

1.1.1 System Classification

Definition of System A *system* is a collection of interconnected components in which there is a specified set of dynamic variables called *inputs* (or excitations) and a dependent set called *outputs* (or responses). Considering a hierarchy among systems, a system can be expressed as a collection of collections.

Classification of Systems The classification of systems depends upon the degree of interconnection of the *events* from none to total, where each event consists of a collection of interconnected components. It should be pointed out that even for no connection, there must be some relation. If the events are unrelated altogether, there is no system. Systems will be divided into three classes (for analytical and design purposes) according to the degree of interconnection of the events. The classes are: independent, cascaded, and coupled. These are discussed in turn.

1

Independent   If the events have no effect upon one another, then the system is classified as *independent*. The events may be studied individually.

Cascaded   If the effects of the events are unilateral (that is, part A affects part B, B affects C, C affects D, and so forth but not vice-versa) the system is classified as *cascaded*. For such a system, part A is studied first. Then part B is studied next, but where the effect of part A is included. Then, part C is studied next, but where the effects of parts A and B are included. And so on. Thus the events may be studied one at a time, but in the sequence dictated by the "flow" from one cascade to the next. This sequential process is referred to as the *cascade rule*.

Coupled   If the events mutually affect each other, the system is classified as *coupled*. In this case, the events cannot be studied one at a time, but must be studied simultaneously. If one were to draw a "flow" line from one part to another, there will be at least one closed loop. That is, at least one sequence of events will close upon itself. Thus, if one were to study the system starting at any point in the sequence, it would be seen that the overall process is an iterative one.

1.1.2 Survey of Some Common Systems

Independent Systems

Lathe Feed   The cross feed and longitudinal feed of a machine lathe illustrate an example of related yet independent events. If the cross feed is truly perpendicular to the spindle axis, and if the longitudinal feed is truly parallel to the spindle axis, then the operator can perform facing and turning operations anywhere on the machine. If these feeds are out of line, then the operator finds that the feeds are not independent. That is, facing may be good at one end of the (longitudinal feed) machine but not at the other.

Stereo Tape Player   A stereo tape has two separate tracks on the same tape. The pacing between the two tracks is sufficient to keep the tracks from interfering with each other. Separate magnetic pickup heads, amplifiers, and speakers maintain the individuality of each channel. The two events are related since the final result must be heard simultaneously. If the system is constructed properly, the events are independent.

Cascaded Systems

Waterway   Consider a river that flows over a waterfall into a lake. Overflow from the lake is controlled by a dam that provides irrigation. The flow is unidirectional and related, hence this is a system and it is cascaded. To verify that the system is cascaded, note that any foreign matter that enters the system will only affect the downstream areas.

If the water in the irrigated area should somehow rise higher than the dam, this would interfere with the flow, and thus affect the upstream events.

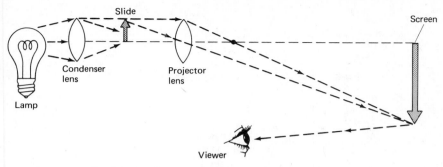

Figure 1.1.2-1   Cascaded optical projection system.

Also this would permit contamination to enter the upstream area. Though unlikely, if this did occur, the system would no longer be cascaded.

Phonograph   Perhaps we have not thought of this familiar device as a cascaded system. A motor turns the record and the needle follows the groove. This produces an electrical signal that is amplified and finally operates a speaker. Whether the speaker operates or not, the motor turns at the same speed. This is a cascaded system

If the speaker can resonate and thus produce vibration of the record or of the needle, or if the mechanical separation of the speaker and turntable is inadequate, the vibration will also be amplified and passed on to the speaker. This may reinforce the vibration, which will self-perpetuate the process. Thus, the system could become coupled, an undesirable effect for a phonograph.

Slide Projector   A lamp is illuminated by an electric current. The light from the lamp is concentrated by a condenser lens. The condensed or concentrated light passes through the transparent slide. An image of the illuminated slide is projected by means of a second lens and comes to focus on a screen of highly reflective material. The focused image is reflected and is seen by the viewer. This cascaded system is shown in Fig. 1.1.2-1.

Formation of Chlorine   HCl is admitted to the system through a thistle tube. The end of the tube lies below the liquid level in the flask (to prevent escape of gases). MnO_2 in the flask reacts with the acid to produce $MnCl_2$, water and chlorine gas. The gas, which is heavier than air, fills a second flask by upward displacement. This cascade system is shown in Fig. 1.1.2-2.

Coupled Systems

Production of Sulfuric Acid   Sulfuric acid is formed by firing sulfates, such as pyrites. A commercial process, shown in Fig. 1.1.2-3, burns pyrites in a furnace. Sulfuric acid, formed upstream, is pumped to the containers at the top of the Glover tower. This trickles down as the SO_2 from the furnace rises. A weak acid is formed and is washed in the steam chambers. Some of this is

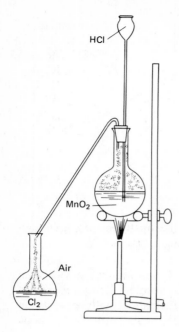

Figure 1.1.2-2 Cascaded chlorine forming system.

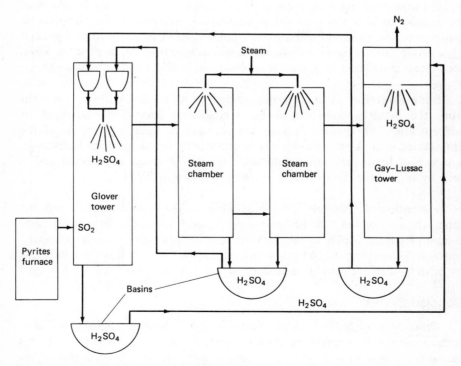

Figure 1.1.2-3 Coupled sulfuric acid manufacturing system.

fed back to the Glover tower. The residual gases rise in the Gay-Lussac tower as sulfuric acid trickles down, forming more sulfuric acid. The process keeps recycling until a high quality acid collects in the basins at the bottoms of the towers. The system is obviously coupled, and the coupling has been put to the most effective use.

1.1.3 Change of Classification

Consider a city where two companies receive shipments of raw materials from a large supplier (infinite source). These companies have constructed cooling facilities for this equipment using water from the local river (infinite sink). The two events are related by commonality (using the same source and sink) and yet they may be independent companies. This is an independent system.

Now suppose that the supplier cannot provide sufficient material for both companies, and suppose company A is the preferred customer. The amount of material received by company B will definitely be affected by the amount company A orders and by the amount that the supplier has. On the other hand, the amount that company B requires will not affect the shipment to company A. Hence the system has changed its classification from independent to cascaded.

Next, suppose that the reduction in material to company B forces it to change its process somewhat, and that the result is that company B must use excessive cooling to the point of raising the temperature of the commonly shared river. As a result, company A cannot process all the material that it receives. Hence, company A must reduce its order for material. (Note that under these conditions, company B affects company A.) Now, company B can order more and consequently it need not raise the water temperature as much as it did before. This allows company A to order more, and the cycle repeats. Here we see the iteration process and dynamic reciprocating effects, which characterize the behavior of a coupled system. If the system reaches equilibrium, it is stable. If the changes get larger and larger, the system is unstable and the two companies will destroy each other.

1.1.4 Further Classification of Systems

Within each of the classifications discussed in the previous sections there may be further classifications. These classifications may affect the type of analytical tools required. Such classes are listed below:

1. Continuous time.
2. Discrete time.
3. Continuous systems.
4. Discrete systems.
5. Static.
6. Dynamic.
7. Constrained.
8. No constraints.

9. Linear.
10. Nonlinear.
11. Approximate.
12. Exact.
13. Optimal.
14. Minimal, maximal, extremal.
15. And so forth.

Considering the large number of system classifications, it may be difficult to discuss the entire field of systems with any one method. Further difficulties are encountered in the design of the various systems. Hence, the classification of systems is not merely a matter of semantics, but instead it has the responsibility of providing meaningful direction in the analysis and the design of systems. Keeping the ultimate goal in mind, some methods of modeling, designing, and analyzing are covered in the subsequent sections.

1.2 ANALYSIS OF SYSTEMS

1.2.1 Analysis

Analysis is the separation of a subject into its constituent parts or elements in an effort to either clarify or to enhance the understanding of the subject. If the subject is a system, this dictionary definition may be extended. System analysis includes the determination of the performance, behavior, or response of a system to a given set of inputs that are applied to a given configuration of defined parts or elements.

In the pursuit of analysis, a suitable model of the system and its inputs must be formulated. For reasons given in Section 1.2.3, the model must be a suitable representation of the actual system. *The model must be sufficiently sophisticated to demonstrate the significant outputs without becoming too cumbersome to make use of available methods of analysis.*

It should be pointed out that the determination of the system output is not the end of the problem. One must proceed beyond the solution of a set of equations or the laboratory testing of a model. The analysis process also demands the interpretation of these results and the establishment of valuable conclusions. Very often the answer to a problem is not as important as the *reason* for that particular answer.

1.2.2 Synthesis

Synthesis is the composition or combination of parts or elements so that the system performs, behaves, or responds according to a given set of specifications.

While synthesis may appear, offhand, to be the opposite of analysis, synthesis has several distinct aspects for which there is no parallel in analysis. For example, in analysis the only unknowns are system outputs, which usually are few in number. In synthesis, the outputs are known (these are usually given in the form of a specification), while most of the system elements are unknown.

While this appears to be a handicap, it is self-offsetting since the lack of such definition enables the engineer to select the configurations that are easily and accurately modeled. In analysis, the configuration is given, leaving little flexibility in choice of geometry. Thus, there are shortcuts and unlimited use of convenient relationships available to synthesis that are excluded from analysis. Consequently, the subject of synthesis proceeds with almost its own set of techniques that are totally different from those available to the subject of analysis.

1.2.3 Modeling

The purpose of modeling a system is to expose its internal workings and to present it in a form useful to engineering study. Within the broad field of modeling there are certain specialties whose aim is to highlight or isolate certain specifics of a given system. The temptation to limit one's horizon in the modeling process is engendered by the natural desire to concentrate upon one specific area at a time. However, there is a danger in such a practice and much risk resides in the hidden coupled effect which may easily be overlooked in the process.

There are a number of means by which modeling may be achieved. A few examples are: scale models (for laboratory testing), field models (for controlled testing by consumers), mathematical models (for analytical studies), and approximate models (for rough estimates of system behavior), to name a few. Within each model category, there are subdivisions. For example, in a mathematical model, there may be a number of subdivided models, such as: computer model, analytical model, linear model, nonlinear model, and others.

In Section 1.1 use was made of one common form of system modeling, the flow diagram. In this type of diagram, the system is modeled by a network of discrete blocks, each having an input and the corresponding output. Whether the system is an actual piece of equipment or a diagram on paper, it is necessary to understand and to correctly model the inputs and the system elements. If these are modeled correctly, then the resulting outputs will be correct. Here we have the immortal problem of modeling; *include enough of the system activity to develop a model of suitable accuracy, yet keep the model simple enough to permit a feasible means to determine the outputs.* Needless to say, a great amount of engineering judgment is required to reach a satisfactory compromise.

In some cases, the escalator approach proves successful. In this approach, the first model is very rough, designed for a "once over lightly" view. Such an early model is often necessary to provide the insight and overview before these become obscured by the subsequent details. The next model includes a few more effects than the first, and in addition, outputs other than overall may be included in the results. Subsequent models could become increasingly more sophisticated, each one expanding upon the previous model. In this way, the engineer can grow with the system, having it within reach at each plateau. In addition, there is a model of some level that is available at any time, and each of these can provide the criteria and data for the next plateau. The highest plateau of this escalating modeling technique may be limited by time or cost allotment or by the technical acuity of the investigator.

1.2.4 General Sequence of Steps

The following suggested list of operations may serve as a guide in either the analysis or the design of a dynamic system.

1. Define the system and its components.
2. List all assumptions.
3. Select the significant inputs and outputs.
4. Model the system elements.
5. Model the system.
6. Solve for the appropriate outputs.
7. Check for violation of assumptions.
8. Interpret the system response
9. If necessary,suggest requirements for redesign.

1.3 DESIGN OF SYSTEMS

1.3.1 Distinction Between Synthesis and Design

Some engineers distinguish between synthesis and design. They refer to synthesis as the formulation of the system configuration, while design is the assignment of numerical parameters to an existing configuration. Other engineers include the choice of materials and the selection of existing components or elements as part of design. Yet others include routine procedures, such as "company practice," design sheets, and standards books as legitimate design methods. Certainly, in practice, a combination of all of these prevails.

1.3.2 Workable and Optimal Designs

There are many possible designs that can meet the required specifications. However, they may not all have equal merit. Here we distinguish between a *workable* (or acceptable) and an *optimal* (or best) design. The workable design will comply with all of the above requirements. The optimal design also satisfies these, but at the same time, it minimizes a particular criterion.

Keep in mind that a workable design is acceptable, and that the discovery of an optimal one may be regarded as "service beyond the call of duty." In many cases, a workable design is accepted because the search for an optimal one may be considered out of proportion to the possible benefits to be gained.

1.3.3 Design by Search Method (Trial and Error Analysis)

When there is a profusion of system constraints, or where a design approach is not apparent, it is sometimes more expedient to design by an ordered sequence of analyses. This can be accomplished by assuming a completely defined configuration, and by analyzing the result. Many configurations are investigated by varying one parameter in predetermined steps until the full range of that quantity is spanned. Obviously this approach is effective only when there are so many system constraints that there are only one or two parameters that can be varied The method may be compared to a scanning process where all the possible designs are examined until a suitable one is found.

Admittedly, this method is not much of a tax upon technical acuity and imagination, but it responds to automated methods. However, it is vulnerable to the obvious risk of overlooking a possibly better solution, and certainly is ineffective if either the search pattern or the sizing of steps is not chosen propitiously.

1.3.4 Direct Design Technique

It was pointed out in Section 1.2.2 that synthesis (and design) is not restricted by existing limitations upon configuration. Hence it is often possible to generate a cause-effect relationship that is both direct and conclusive. When the various system unknowns can be expressed in terms of the given specifications, this is referred to as "direct design."

1.3.5 Design with Constraints

A system constraint implies limits on certain parameters. In many cases, constraints represent discontinuities in the otherwise continuous range of system parameters. Constraints have many sources: for example, they may be specified by the customer; they may be self-imposed; they may be due to windows in the spectrum of design know-how; or they may be imposed by natural laws. Some examples of constraints are: no dimension can exceed 2 m (customer); all sheet stock to be cold rolled standard (self); all relationships must be linear (lack of know-how); gears must have an integer number of teeth (natural law).

SUGGESTED REFERENCE READING FOR CHAPTER 1*

Systems [24], p. 1; [8], Chap. 1; [47], Chap. 1.
Design approach [35], p. 3; [17], p. 4.
Optimal design [36], p. 12, Chap. 6.
Cascaded subsystems [17], p. 511–514.

---

*Numbers in brackets refer to entries in the References section, which is to be found at the back of the book, following Chapter 17.

Chapter 2

Modeling of Lumped Elements

The elements of a system are the individual ingredients. It will be seen that if the system elements can be lumped, the differential equations that will subsequently ensue will be ordinary differential equations. If the elements are distributed, partial differential equations will be required. If the equations are to be solved analytically, it is best that they be simple and linear. Hence, the major task of this chapter is to formulate, approximate, and model the elements in the most useful form. The guidelines will seldom be straightforward. On the one hand, the model must be sufficiently sophisticated for an accurate representation of the actual system. On the other hand, the model must be suitably simplified to eliminate the entanglement of the subsequent mathematical development.

Unfortunately, no simple rules or guidelines can be formulated in the construction of suitable models from actual systems. In most cases, individual judgment is required. As a result, modeling from real machines is an art.

2.1 EQUIVALENT LUMPED MASSES FOR MECHANICAL, ELASTIC, AND FLUID ELEMENTS

Mass is a fundamental property which may be thought of as the amount of matter within a body. Although mass is a fundamental property, it is difficult to observe this quantity in its pure form due to the coexistence of friction and the gravitational field. For this reason, mass is often confused with weight. *Weight* is *not* a fundamental property since it *varies* over the earth and on different celestial bodies, while *mass* is *constant*.

A precise scale does not weigh, but instead, it compares the *mass* of the unknown body to the *mass* of a known one. If the balance is centered, the two masses are equal. The balance will remain centered at the poles of the earth, at the equator, or anywhere on the earth. It will remain centered even at points off the earth, at any altitude, even on the moon.

Although it is commonly done, it is not a good practice to express mass in terms of its weight. It should be pointed out that the *weight* varies from one point on the earth to another, and this variation can be as much as $\frac{1}{2}\%$ from the equator to the poles. This variation is even greater with altitude. For this reason, a spring scale (which is not too accurate anyway) is accurate only at one point on the earth. Hence, weight is a secondary property, while mass, which is constant (at velocities well below the speed of light), is a primary property.

2.1.1 Newton's Second Law

Mass enters the dynamic systems scene through *Newton's second law*, which can be stated in simple terms as "A mass is accelerated by an external force in the direction of the force and by an amount that is proportional to the magnitude of the force." In symbols

$$F = M\ddot{x} \qquad [2.1.1-1]$$

or for the torsional system

$$T = J\ddot{\theta} \qquad [2.1.1-2]$$

The computation of acceleration due to a force that is applied to a mass will be performed in either of two metric systems, the centimeter gram second (cgs) and the meter kilogram second (mks) systems. Occasionally, the results will be expressed in English units for comparison only. The two metric systems are described as follows.

cgs SYSTEM	
Distance, displacement, or length	centimeters (cm)
Mass	grams (g)
Time	seconds (s)
Force	dynes (dyn)

In the cgs system, a force of 1 dyn, applied to a mass of 1 g, will produce an acceleration of 1 cm/s².

mks SYSTEM	
Distance, displacement, or length	meter (m) = 100 cm
Mass	kilogram (kg) = 10^3 g
Time	seconds (s)
Force	newton (N) = 10^5 dyn

In the mks system, a force of 1 N, applied to a mass of 1 kg, will produce an acceleration of 1 m/s².

2.1.2 Lumped Mass

A *lumped mass* may be condensed into a single point located at the center of mass of the body. This assumes that the body is rigid. Hence, all points translate together, permitting the condensation into the single point. The mass M is given by eq. 2.1.2-1. The polar moment of inertia J is given by eq. 2.1.2-2.

$$M = \int \rho \, dV \qquad [2.1.2\text{-}1]$$

$$J = \int r^2 \, dm \qquad [2.1.2\text{-}2]$$

where

ρ = mass density

V = volume   [2.1.2-3]

r = distance from polar axis to mass dm

For a number of common configurations, masses and moments of inertia are listed in Table 2.1.3-1, Appendix 2.

2.1.3 Kinetic Energy due to Moving Mass

The energy E_k associated with a mass M moving at velocity $\dot{X}$ is given by

$$E_k = \tfrac{1}{2} M \dot{X}^2 \qquad [2.1.3\text{-}1]$$

The analogous case for a rotating flywheel whose moment of inertia is J is given by

$$E_k = \tfrac{1}{2} J \dot{\theta}^2 \qquad [2.1.3\text{-}2]$$

2.1.4 Equivalent Lumped Mass

Not all masses are rigid bodies. Consequently, certain approximations and equivalents must be determined. Two basic techniques will be used, use of equivalent masses, and the energy method.

■ **EXAMPLE 2.1.4-1** Lumped Mass for a Spring
The spring is an elastic body. If the spring is continuous, the kinetic energy of the equivalent lumped system (having mass M_{eq} and velocity V_{eq}) can be expressed as the integral of the distributed mass and distributed velocity, as

$$\tfrac{1}{2} M_{eq} V_{eq}^2 = \tfrac{1}{2} \int V^2 \, dm \qquad (1)$$

In a coiled spring, assume that one end is fixed while the other end translates at velocity V_m. There will be a linear velocity distribution, where the velocity of any point on the spring is proportional to its distance from the fixed end.

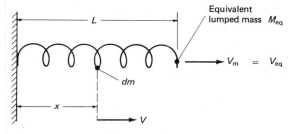

Figure 2.1.4-1   Lumped mass for spring.

Then

$$V = \frac{V_m x}{L} \qquad (2)$$

See Fig. 2.1.4-1.

Assuming uniform density, if the total mass of the actual spring is M, then the differential mass is given by

$$dm = \frac{M\,dx}{L} \qquad (3)$$

Substitute eqs. (2) and (3) into eq. (1), and solve for M_{eq}.

$$M_{eq} = \frac{\int_0^L (x^2/L^2) V_m^2 (M/L)\,dx}{V_m^2} = \frac{M}{3} \qquad (4)*$$

■

2.1.5 Groups of Masses

(a) **Rigidly Connected Masses**   For this case, the masses are rigidly connected, implying that their displacements and velocities are identical. Then

$$V_{eq} = V_1 = V_2 \qquad [2.1.5\text{-}1]$$

and

$$M_{eq} = M_1 + M_2 \qquad [2.1.5\text{-}2]$$

Similarly, for two rotating masses J_1 and J_2

$$J_{eq} = J_1 + J_2 \qquad [2.1.5\text{-}3]$$

(b) **Masses Connected by a Lever (Translation)**   Assume the two masses are fastened to a lever that pivots at one end. Let the distance L_i locate mass M_i from the pivot. (See Fig. 2.1.5-1.)

---

*This result is tabulated in Table 2.1.4-1, Appendix 2.

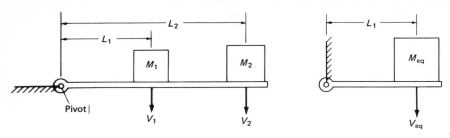

Figure 2.1.5-1 Masses connected by a lever.

We may conveniently locate the equivalent at any point, say, at the site of mass M_1. Thus, all velocities may be obtained in terms of V_1. Then, for small angles of travel,*

$$V_{eq} = V_1 \qquad [2.1.5\text{-}4]$$

$$V_2 = \frac{V_1 L_2}{L_1} \qquad [2.1.5\text{-}5]$$

Apply eq. 2.1.5-5 to eq. 2.1.3-1

$$M_{eq} = M_1 + M_2 \left(\frac{L_2}{L_1}\right)^2 \qquad [2.1.5\text{-}6]$$

(c) Masses on Geared Shafts (Rotation) Consider two masses J_1 and J_2 each rotating on shafts connected by gearing, where mass J_2 is on the *high* speed shaft (it, therefore, has the smaller gear or pinion). Let the number of teeth of the respective gears be N_1 and N_2. (See Fig. 2.1.5-2.)

The equivalent mass J_{eq} may be located at any convenient shaft, say, the site of mass J_1. Then

$$\dot{\theta}_{eq} = \dot{\theta}_1 \qquad [2.1.5\text{-}7]$$

$$\dot{\theta}_2 = \frac{\dot{\theta}_1 N_1}{N_2} \qquad [2.1.5\text{-}8]$$

Note that the subscripts for the above equation are opposite to those in eq. 2.1.5-5. Apply eq. 2.1.5-8 to eq. 2.1.3-1,

$$J_{eq} = J_1 + J_2 \left(\frac{N_1}{N_2}\right)^2 \qquad [2.1.5\text{-}9]$$

(d) Masses in Series Masses do not have two ends (terminals or ports). Consequently, they *cannot* be connected in *series*.

---

*The actual motion is curvilinear and we are approximating this by straight line motion.

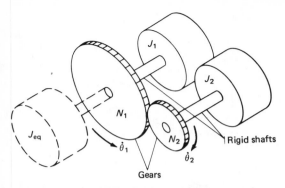

Figure 2.1.5.2   Masses on geared shafts.

(e) Coupled Rotating and Translating Masses   Consider two masses, one rotating about an axis, while the other translates. (See Fig. 2.1.5-3.) The object is to combine the two either as a single equivalent translating mass, or as a single equivalent rotating mass.

Single translating mass M_{eq} (Moment of inertia J is "reflected" to the translating axis): The kinetic energy is given by

$$E = \tfrac{1}{2}M_{eq}V_{eq}^2 = \tfrac{1}{2}MV^2 + \tfrac{1}{2}J\dot{\theta}^2 \qquad [2.1.5\text{-}9a]$$

Since

$$V_{eq} = V \quad \text{and} \quad \dot{\theta} = \frac{V}{R}$$

then

$$\frac{1}{2}M_{eq}V^2 = \frac{1}{2}MV^2 + \frac{1}{2}J\left(\frac{V}{R}\right)^2$$

or

$$M_{eq} = M + \frac{J}{R^2} \qquad [2.1.5\text{-}9b]$$

Single rotating mass J_{eq} (Mass M is "reflected" to the θ axis): In this case, the equivalent mass moment of inertia J_{eq} will rotate at a velocity equal to that of J. Then

$$\dot{\theta}_{eq} = \dot{\theta}$$

$$V = \dot{\theta}R$$

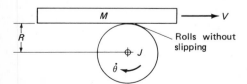

Rolls without slipping

Figure 2.1.5-3 Coupled rotating and translating masses.

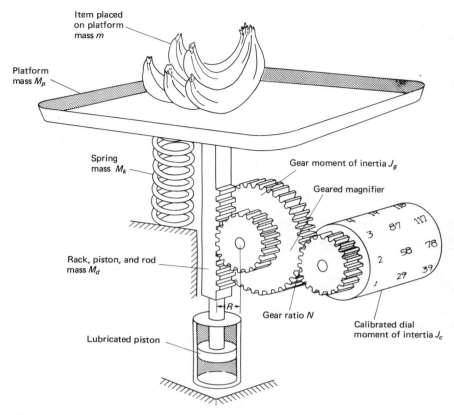

Item placed
on platform
mass m

Platform
mass M_p

Spring
mass M_k

Gear moment of inertia J_g

Geared magnifier

Rack, piston, and rod
mass M_d

Gear ratio N

R

Gear ratio N

Calibrated dial
moment of intertia J_c

Lubricated piston

Figure 2.1.5-4   Spring scale with geared dial.

Apply to eq. 2.1.5-9a

$$\tfrac{1}{2}J_{eq}\dot{\theta}^2 = \tfrac{1}{2}J\dot{\theta}^2 + \tfrac{1}{2}M(\dot{\theta}R)^2$$

or

$$J_{eq} = J + MR^2 \qquad\qquad [2.1.5\text{-}9c]$$

■ **EXAMPLE 2.1.5-1** Spring Scale*

Consider a spring scale, in which the weighing platform is supported on a spring and lubricated piston. The translation of the platform is magnified by appropriate gearing to a calibrated dial. (See Fig. 2.1.5-4.) Various items are placed on the platform. Determine the equivalent translating mass M of the system.

Refer to Ex. 2.1.4-2. The equivalent mass of the spring is

$$M_k' = \frac{M_k}{3} \qquad\qquad (1)$$

---

*Note: The Bureau of Standards does not consider the use of a spring as a legal means to measure weight.

where M_k = total mass of the spring. Referring to Section 2.1.5, the reflected mass M_r' of the calibrated dial is

$$M_r' = \frac{J_c N^2}{R^2} + \frac{J_g}{R^2} \qquad (2)$$

where J_c is the moment of inertia of the dial. Neglecting the mass of the fluid, the equivalent lumped mass M of the spring scale mechanism is

$$M_{eq} = m + M_p + M_d + M_k' + M_r'$$

$$= m + M_p + M_d + \frac{M_k}{3} + \frac{J_c N^2}{R^2} + \frac{J_g}{R^2} \qquad (3) \blacksquare$$

2.2 EQUIVALENT LUMPED SPRINGS FOR MECHANICAL, ELASTIC, AND FLUID ELEMENTS

2.2.1 Linear Springs

Coiled Spring (Translatory Spring) The coiled spring is essentially a helically wound wire that is highly elastic along its axis. If one end of the spring is held fixed, while the other end is displaced a distance X along the axis of the

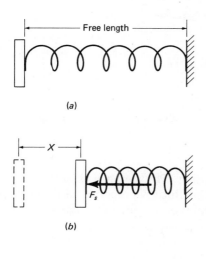

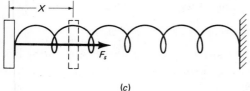

Figure 2.2.1-1 Coiled (translatory) spring. (a) Spring in neutral position. (b) Spring is compressed, force opposes compression. (c) Spring is extended, force opposes extension.

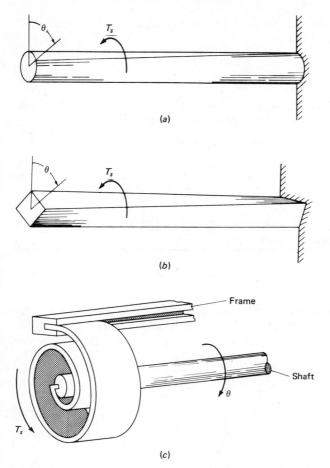

Figure 2.2.1-2 Torsion springs. (a) Circular torsion bar. (b) Square torsion bar. (c) Clock spring.

spring, this will compress the spring. The spring will exert a force F_s in a direction opposing the motion. This force is proportional to the displacement where the constant of proportionality K is called the *spring rate*.

$$F_s = KX \qquad [2.2.1\text{-}1]$$

(See Fig. 2.2.1-1.) The length of a spring in its neutral (zero force) position is referred to as its "free length."

Rotary or Torsion Spring   Like the translatory spring, the torsion spring has two ends, but for the torsional case, the moving end is a shaft that rotates. If the shaft rotates through an angle θ, then this winds up the spring. The spring will exert a torque T_s, which will oppose the motion. This torque is proportional to the *angle* of twist, where the constant of proportionality K is called the *torsional spring rate*.

$$T_s = K\theta \qquad\qquad [2.2.1\text{-}2]$$

(See Fig. 2.2.1-2.)

■ **EXAMPLE 2.2.1-1** Equivalent Spring Using the Definition

A translatory spring K is located at the end of an arm having length L. The arm can pivot about point O. Using the definitions of translatory and torsion springs, determine the equivalent torsion spring K_t in Fig. 2.2.1-3.

Solution.   From the definition of a torsion spring, eq. 2.2.1-2, we have

$$K_t = \frac{T_s}{\theta} \qquad\qquad (1)$$

Assume a small translatory displacement X. This will elongate the spring K and produce a corresponding force F_s according to eq. 2.2.1-1, or alternately,

$$K = \frac{F_s}{X} \qquad\qquad (2)$$

There will be a corresponding torque T_s due to the force F_s at the end of the lever arm. Then,

$$T_s = F_s L \qquad\qquad (3)$$

If the displacement X is very small, then

$$\theta = \frac{X}{L} \qquad (X \ll L) \qquad\qquad (4)$$

Substitute eqs. (3) and (4) into eq. (1) and apply eq. (2)

$$K_t = \frac{F_s L}{X/L} = \frac{L^2 F_s}{X} = L^2 K \qquad\qquad (5) \ ■$$

2.2.2 Nonlinear Springs

The linear relationships expressed by eqs. (2.2.1-1 and 2.2.1-2 represent ideal springs. Actual springs may be nonlinear. To demonstrate the behavior of real springs, consider the testing device shown in Fig. 2.2.2-1. Various weights can

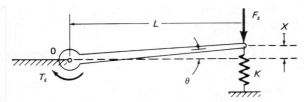

Figure 2.2.1-3   Equivalent spring using the definition.

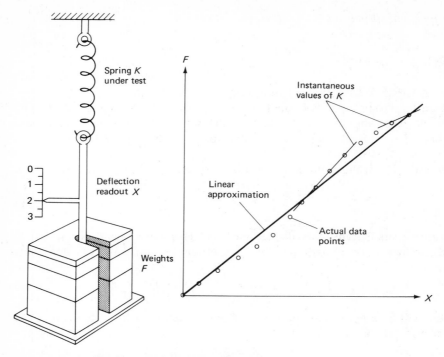

Figure 2.2.2-1   Testing a real (nonlinear) spring.

be applied to the end of the spring under test, and the resulting deflection of the spring can be measured. A plot of the data for different weights for the sample shown is not a straight line, indicating a nonlinearity. A linear approximation to this nonlinear spring can be made by drawing the best straight line.* The slope of the line is approximately the linear equivalent. Needless to say, the procedure for finding the best straight line determines the accuracy of the approximation.

For nonlinear springs, the instantaneous value of the spring rate is found by using the derivative. Then

$$K_{eq} = \frac{dF}{dX} \qquad [2.2.2\text{-}1]$$

2.2.3 Elastic Potential Energy Stored in a Spring

The energy in a spring (elastic structure) is reversible. The energy E_p stored in a spring is equal to the work done to compress or extend the spring. We have

$$E_p = \int_0^{X_{max}} F(X)\, dX \qquad [2.2.3\text{-}1]$$

*The best being defined for a specific range.

where $F(X)$ is the force for displacement X. For a linear spring $F = KX$ and the energy becomes

$$E_p = \int_0^{X_{max}} KX\,dX = \tfrac{1}{2}KX_{max}^2 \qquad [2.2.3\text{-}2]$$

For a linear torsional spring the energy is

$$E_p = \tfrac{1}{2}K\theta_{max}^2 \qquad [2.2.3\text{-}3]$$

For nonlinear springs the energy must be computed using eq. 2.2.3-1 with $F(X)$ given as a function of X.

2.2.4 Equivalent Springs

Not all springs have configurations like those in Figs. 2.2.1-1 and 2.2.1-2. Consequently, many practical springs require approximations and equivalent methods. To accomplish this, two basic techniques will be used, the definition of the spring (eqs. 2.2.1-1, 2.2.1-2, and 2.2.2-1), and the energy method (eqs. 2.2.3-2 and 2.2.3-3).

Mechanical Springs

(a) **Coiled Springs**   The coiled spring was discussed in Section 2.2.1. See Item (1) in Table 2.2.3-1.* It can also be used as a torsion spring, as shown by Items (3) and (4) in Table 2.2.3-4.*

(b) **Rubber Springs**[†]   Rubber is a nonlinear spring in compression, as shown by Item (3) in Table 2.2.3-1,* but it is linear in double shear [Item (4)].

(c) **Pendulum**   The pendulum can be considered a spring [Item (5)] and is linear for small angular travel. A horizontal force F will cause the pendulum to deflect through an angle θ, where

$$F = Mg\sin\theta \qquad [2.2.4\text{-}1]$$

For small angles

$$\theta = \sin\theta = \tan\theta = \frac{X}{L} \qquad [2.2.4\text{-}2]$$

Substitute eq. 2.2.4-2 into eq. 2.2.4-1.

$$F = \frac{MgX}{L} \qquad [2.2.4\text{-}3]$$

---

*Appendix 2.

[†]Rubber also acts like a damper. See Table 2.3.6-1 in Appendix 2.

For a translatory spring,

$$K_{eq} = \frac{F}{X} = \frac{Mg}{L}$$ [2.2.4-4]

(d) Beams   A beam is a linear elastic member. That is, its deflection is proportional to the load.* The constant of proportionality depends upon how the beam is supported and upon where the load is applied. Refer to Table 2.2.3-3, Appendix 2.

(e) Fluid Springs   An air cushion is an example of a fluid spring. For low pressures and for small changes in volume, the air acts as a perfect gas. If the exponent for the perfect gas process is unity, then pressure and volume are linearly related. If the area of the cushion is constant, then the resulting force (pressure times area) is linearly related to volume, and since the cross sectional area is constant, the volume is linearly related to length or change in length. Thus there would be a linear relation between the force exerted and the change in length. This satisfies the definition of a linear spring. Of course, few if any of the above linearizing effects govern the behavior of fluids in even the simplest cases. Consequently, fluids do not behave like linear springs. However, undeꜰ limiting cases, a fair approximation may be found.

2.2.5 Groups of Springs

(a) Springs Connected by Levers   Given two springs K_1 and K_2 connected by levers as shown in Fig. 2.2.5-1(a), determine a single equivalent translatory spring K_{eq} located at the site of K_1. [Spring K_2 has been "reflected" to the locale of spring K_1.]

The equivalent spring will be found using the energy method. The energy stored in a spring is given by eq. 2.2.3-2. For the two springs, this becomes

$$E = \tfrac{1}{2}K_1 X_1^2 + \tfrac{1}{2}K_2 X_2^2$$ [2.2.5-1]

For the single equivalent spring being sought

$$E_{eq} = \tfrac{1}{2}K_{eq} X_{eq}^2$$ [2.2.5-1a]

Equate eqs. 2.2.5-1 and 2.2.5-1a and solve for K_{eq}

$$K_{eq} = \frac{K_1 X_1^2 + K_2 X_2^2}{X_{eq}^2}$$ [2.2.5-1b]

Since we are reflecting to the locale of K_1, and if we assume that the angular travel of the lever is so small that we may approximate the curvilinear motion

---

*For small deflections.

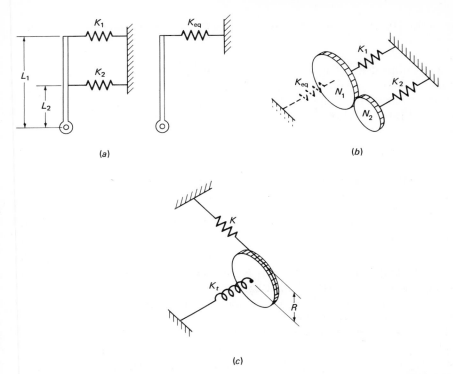

Figure 2.2.5-1 Groups of springs. (a) Springs connected by levers. (b) Springs on geared shafts. (c) Coupled rotating and translating springs.

as straight line motion, we have

$$X_{eq} \approx X_1 \qquad\qquad [2.2.5\text{-}1c]$$

From the physics of levers, we have

$$X_2 = \frac{X_1 L_2}{L_1} \qquad\qquad [2.2.5\text{-}1d]$$

Apply eqs. 2.2.5-1c and 2.2.5-1d to eq. 2.2.5-1b

$$K_{eq} = K_1 + K_2 \frac{(X_1 L_2 / L_1)^2}{X_1^2}$$

$$= K_1 + K_2 \left(\frac{L_2}{L_1}\right)^2 \qquad\qquad [2.2.5\text{-}2]$$

(b) Springs on Geared Shafts Following the approach in Section 2.1.5(c), the result is

$$K_{eq} = K_1 + K_2 \left(\frac{N_1}{N_2}\right)^2 \qquad\qquad [2.2.5\text{-}3]$$

[Torsional spring K_2 has been "reflected" to the locale of torsional spring K_1. See Fig. 2.2.5-1(b).]

(c) **Coupled Rotating and Translating Springs**   Using the energy method similar to that in Section 2.1.5(e), a torsional spring K_t rotating through angle θ and a translating spring K being displaced through a distance X can be combined into a single equivalent translating spring K_{eq} as follows:

$$K_{eq} = K + \frac{K_t}{R^2}$$
  [2.2.5-3a]

They can alternately be combined into a single equivalent torsional spring K_{eqt} as follows:

$$K_{eqt} = K_t + KR^2$$
  [2.2.5-3b]

See Fig. 2.2.5-1(c).

(d) **Springs with Two Movable Boundaries**   A spring has two ends or boundaries. Given a spring K that had zero initial displacement, the question is to determine the resulting force when both ends displace by different amounts. This problem will be treated by using superposition. First, hold one end fixed and determine the magnitude and direction of the force due to displacing the other end. Next, hold the opposite end and find the force due to displacing the end that was previously fixed. Use the sign convention, positive to the right.

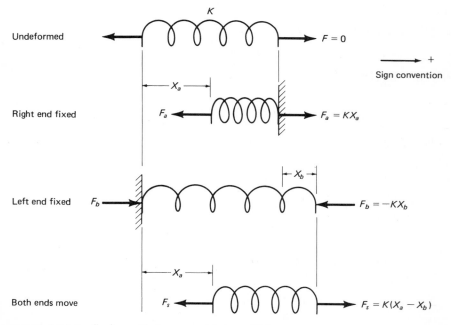

Figure 2.2.5-2   Spring with two movable boundaries.

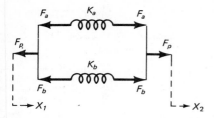

Figure 2.2.5-3   Springs in parallel.

With the right-hand end fixed, and left end movable

$$F_a = KX_a$$

Left-hand end fixed, and right end movable

$$F_b = -KX_b$$

See Fig. 2.2.5-2.

For both ends movable, superpose the above results:

$$F_s = F_a + F_b = KX_a - KX_b = K(X_a - X_b) \qquad [2.2.5\text{-}4]^*$$

Hence, the force exerted by a spring, both of whose ends move, is given by the spring constant multiplied by the *difference* of the end displacements.

(e) **Springs in Parallel** Two springs are in *parallel* if *both* of their respective ends share common boundaries. These boundaries may move or they may be fixed. The boundaries may be walls or masses. Given two springs K_a and K_b in parallel, determine a single equivalent spring K_p. See Fig. 2.2.5-3.

The total force F_p that is exerted on both boundaries is given by the sum of the forces exerted by each individual spring, where these forces are given by eq. 2.2.5-4. Thus,

$$F_p = F_a + F_b \qquad\qquad\qquad [2.2.5\text{-}5]$$

$$F_a = K_a(X_1 - X_2)$$

$$F_b = K_b(X_1 - X_2) \qquad\qquad [2.2.5\text{-}6]$$

and for the equivalent spring

$$F_p = K_p(X_1 - X_2)$$

Substitute eqs. 2.2.5-6 into eq. 2.2.5-5.

$$K_p(X_1 - X_2) = K_a(X_1 - X_2) + K_b(X_1 - X_2)$$

---

*Note that the boundary terms appear in the order indicated by the sign conventions, thus, suggesting an algorithm.

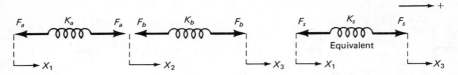

Figure 2.2.5-4   Springs in series.

Or by dividing by $(X_1 - X_2) \neq 0$,

$$K_p = K_a + K_b \qquad [2.2.5-7]$$

(f) Springs in Series   Two springs are in series if the end of one is fastened to one end of the other, and if this interface does not affect anything else, nor does anything else affect it. Springs in series appear like a chain. Given two springs K_a and K_b in series, determine a single equivalent spring K_s as shown in Fig. 2.2.5-4.

For equilibrium at boundary X_2,

$$F_a = F_b$$

For the two systems to be equivalent,

$$F_s = F_a = F_b = F \qquad [2.2.5-8]$$

Using the inverse of eq. 2.2.5-4 on each of the above springs

$$X_1 - X_2 = \frac{F_a}{K_a} = \frac{F}{K_a}$$

$$X_2 - X_3 = \frac{F_b}{K_b} = \frac{F}{K_b} \qquad [2.2.5-9]$$

$$X_1 - X_3 = \frac{F_s}{K_s} = \frac{F}{K_s}$$

Apply eqs. (2.2.5-9 to eq. 2.2.5-8.

$$\frac{F}{K_a} + \frac{F}{K_b} = \frac{F}{K_s}$$

or

$$\frac{1}{K_s} = \frac{1}{K_a} + \frac{1}{K_b} \qquad [2.2.5-10]$$

or

$$K_s = \frac{K_a K_b}{K_a + K_b} \qquad [2.2.5-10a]$$

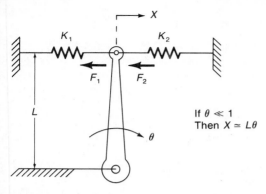

Figure 2.2.5-5   Centering springs.

■ **EXAMPLE 2.2.5-1** Centering Springs

Two springs are used to keep the lever in Fig. 2.2.5-5 in its central position. Determine the equivalent spring for the pair.

One end of each spring is fastened to the lever; the other end of each spring is fastened to the ground. The springs have *both* boundaries in common. Hence, they are in parallel. This can be verified by noting that if the lever turns through a small angle θ (thus translating approximately a small distance X) to the right, spring K_1 is extended and resists by exerting a force F_1 to the left. At the same time, spring K_2 is compressed and resists by exerting a force F_2, also to the left. Thus, the two springs help each other, adding forces, agreeing with eq. 2.2.5-5. This verifies that the two spring *are* in parallel. Equation 2.2.5-7 applies.

$$K_p = K_1 + K_2 \tag{1}$$

■

2.3 EQUIVALENT LUMPED DAMPERS FOR MECHANICAL, ELASTIC, AND FLUID ELEMENTS

2.3.1 Linear Dampers

A damper can be made using two parallel plates with a viscous fluid between them. If one plate is fixed, while the other translates *in its plane* at a velocity $\dot{X}$, this will produce a shearing action upon the fluid. This will result in a horizontal force F_d upon the moving plate in a direction that opposes the motion. The force is proportional to the *velocity* $\dot{X}$ where the constant of proportionality D is called the *damping constant*. See Figs. 2.3.1-1 and 2.3.1-2.

$$F_d = D\dot{X} = Dv \tag{2.3.1-1}$$

Another viscous damper can be constructed using a cylinder with a loose

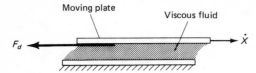

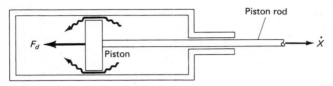

Figure 2.3.1-1   Parallel plate damper.

Figure 2.3.1-2   Piston type damper.

fitting piston. If the piston rod is translated at a velocity $\dot{X}$, this will require that the fluid trapped on one side of the piston squeeze through the annular space between the piston and the cylinder. As in the case of the parallel plate damper, the fluid action will oppose the motion with a magnitude given by eq. 2.3.1-1. For further details, see Sections 2.6.1 and 2.6.2.

Rotary Damper (Paddle Wheel)   Inspired by Joule's experiment, a simple rotary damper can be made using a paddle wheel in a viscous fluid. Let the trough containing the fluid represent the fixed member, while the shaft will be the moving (rotating) member. If the shaft turns at a velocity $\dot{\theta}$, the paddle wheel will produce a torque that will oppose the motion. The torque is proportional to *angular velocity*, and the constant of proportionality D is called the *torsional damping constant*. See Fig. 2.3.1-3.

$$T_d = D\dot{\theta}$$ [2.3.1-2]

A practical torsional damper is that considered in Problem 2-9.

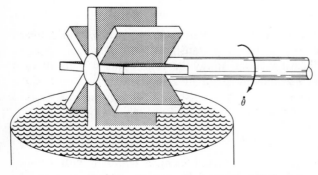

Figure 2.3.1-3   Paddle wheel (rotary or torsional) damper.

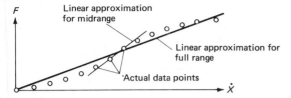

Figure 2.3.2-1   Nonlinear dampers.

2.3.2 Nonlinear Dampers

Following logic similar to that in Section 2.2.2, real dampers may be nonlinear. Using the same testing device described in Section 2.2.2, but measuring the velocity, a damper may be tested. Typical data for a nonlinear sample is shown in Fig 2.3.2-1.

For a nonlinear damper, a plot of force F versus velocity $\dot{X}$ would not be a straight line. The equivalent damper D_{eq} can be approximated by the best straight line within a given range. Note in Fig. 2.3.2-1 that the best straight line for the full range is not the same as that for, say, the midrange.

Over a very small range, we may refer to the "instantaneous" value for the damping constant. This is found by taking the derivative of the force F with respect to velocity $\dot{X}$.

$$D_{inst} = \frac{dF}{d\dot{X}}$$ [2.3.2-1]

2.3.3 Equivalent Dampers in Terms of Fluid Properties

For equivalent damping due to viscous shear, see Section 2.6.1. For equivalent damping due to flow through a fluid restriction, see Section 2.6.2.

2.3.4 Power Dissipated by a Damper

Unlike the mass and the spring, the energy supplied to a damper is not returnable. The power (energy per unit time) dissipated in a damper is given by

$$P_d = F_d V$$ [2.3.4-1]

where

$$F_d = \text{force in the damper} = DV$$ [2.3.4-2]

$$V = \text{velocity} = \dot{X}$$ [2.3.4-3]

Upon substituting eqs. 2.3.4-2 and 2.3.4-3 into eq. 2.3.4-1, we have

$$P_d = D\dot{X}^2$$ [2.3.4-4]

Similarly, for a torsional damper we have

$$P_d = D_t\dot{\theta}^2$$ [2.3.4-5]

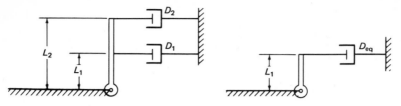

Figure 2.3.5-1   Dampers connected by levers.

The energy dissipated in the time interval $0 \le t \le t_1$ is given by

$$E_d = \int_0^{t_1} P_d \, dt = \int_0^{t_1} F_d \dot{X} \, dt \qquad [2.3.4\text{-}6]$$

2.3.5 Groups of Dampers

(a) Dampers Connected by Levers Analogous to Section 2.1.5(b), we have

$$D_{eq} = D_1 + D_2 \left(\frac{L_2}{L_1} \right)^2 \qquad [2.3.5\text{-}1]$$

(See Fig. 2.3.5-1.)

(b) Dampers on Geared Shafts Analogous to the method in Section 2.1.5(c), the result is

$$D_{eq} = D_1 + D_2 \left(\frac{N_1}{N_2} \right)^2 \qquad [2.3.5\text{-}2]$$

(See Fig. 2.3.5-2.)

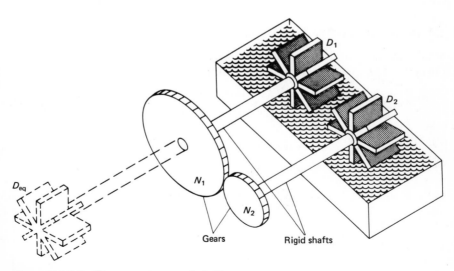

Figure 2.3.5-2   Dampers on geared shafts.

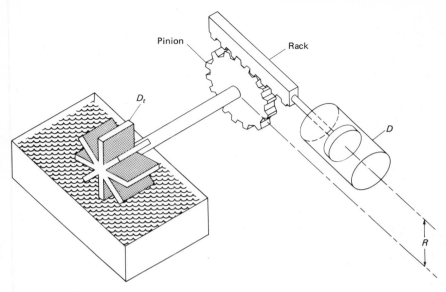

Figure 2.3.5-3   Coupled rotating and translating dampers.

(c) **Coupled Rotating and Translating Dampers**   Using the energy method, similar to that in Section 2.1.5(e), a torsional damper D_t rotating at angular velocity $\dot{\theta}$ and a translating damper D moving at velocity $\dot{X}$, may be combined into a single equivalent translating damper D_{eq}, where

$$D_{eq} = D + \frac{D_t}{R^2}$$ [2.3.5-2a]

Alternately, they may be combined into a single equivalent torsional damper D_{eqt}, where

$$D_{eqt} = D_t + DR^2$$ [2.3.5-2b]

(See Fig. 2.3.5-3.)

(d) **Dampers with Two Movable Boundaries**   Analogous to the method in Section 2.2.5(c), the result is

$$F_d = D\left(\dot{X}_a - \dot{X}_b\right)$$ [2.3.5-3]

(See Fig. 2.3.5-4.)

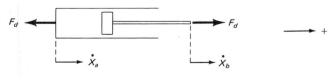

Figure 2.3.5-4   Dampers with two movable boundaries.

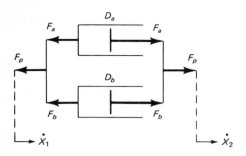
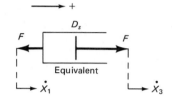

Figure 2.3.5-5   Dampers in parallel.

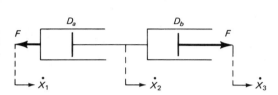

Figure 2.3.5-6   Dampers in series.

(e) Dampers in Parallel   Similar to Section 2.2.5(e) (where velocities $\dot{X}_a$ and $\dot{X}_b$ are used instead of displacements X_a and X_b), the equivalent of two dampers in parallel is given by their sum:

$$D_p = D_a + D_b \qquad [2.3.5\text{-}4]$$

See Fig. 2.3.5-5.

(f) Dampers in Series   Similar to Section 2.2.5(e) (where velocities are used instead of displacements), the equivalent of two dampers in series is given by their product divided by their sum:

$$D_s = \frac{D_a D_b}{D_a + D_b} \qquad [2.3.5\text{-}5]$$

See Fig. 2.3.5-6.

2.4 MODELING OF ELECTRICAL ELEMENTS

It will be assumed that in electrical elements, the current I will flow uniformly throughout the cross section. Current is the time rate of change of charge q.

$$I = \frac{dq}{dt} = \dot{q} \qquad [2.4.1]$$

where

$$\begin{aligned} I &= \text{electrical current} \\ q &= \text{electrical charge} \\ t &= \text{time} \end{aligned} \qquad [2.4.2]$$

2.4.1 Resistance

Resistance is the property of an electrical element which tends to hinder the flow of electric current and at the same time it converts electric energy into heat. Like the damper, this process is irreversible.

The resistance R of a homogeneous body of uniform cross section is given by

$$R = \frac{\rho l}{A} \qquad \text{[2.4.1-1]}$$

where R = resistance (Ω)
       ρ = resistivity ($\Omega \cdot$ m)
       l = length of the path in the direction of current flow (m)
       A = area of the cross section (perpendicular to flow) (m^2)

Values of ρ for various materials are listed in Appendix 6.

Ohm's Law   The voltage drop V across a resistor is proportional to the current I flowing through the resistor, where the constant of proportionality is resistance R. Thus

$$V = IR \qquad \text{[2.4.1-2]}$$

Note that Ohm's law provides an alternate definition for the resistor. Materials that do not obey Ohm's law are referred to as nonlinear resistors. Most metals obey Ohm's law for voltage gradients up to $V/l = 10^8$ V/m. A typical nonlinear resistor is shown in Fig. 2.4.1-1.

Power Consumed by Resistor   The rate of energy dissipated (power consumed) by a resistor is

$$P = I^2 R = \frac{V^2}{R} \text{ (W or J/s)} \qquad \text{[2.4.1-3]}$$

Resistors in Series   Electrical elements are in *series* if they are strung end to end. As a result, the current I must pass through each element. (See Fig. 2.4.1-2.)

The total voltage drop V is given by the sum of the voltage drops across each element.

$$V = V_1 + V_2 \qquad \text{[2.4.1-4]}$$

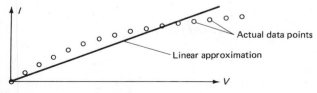

Figure 2.4.1-1   Nonlinear resistance.

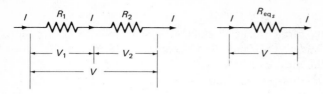

Figure 2.4.1-2 Resistors in series.

Apply Ohm's law to each resistor, and to the equivalent one, and apply to eq. 2.4.1-2

$$IR_{eq_s} = IR_1 + IR_2 \qquad [2.4.1\text{-}4a]$$

Divide through by $I \neq 0$

$$R_{eq_s} = R_1 + R_2 \qquad [2.4.1\text{-}5]$$

The voltage drops across each resistor is given by

$$V_1 = \left[\frac{R_1}{R_1 + R_2} \right] V \qquad [2.4.1\text{-}5a]$$

$$V_2 = \left[\frac{R_2}{R_1 + R_2} \right] V \qquad [2.4.1\text{-}5b]$$

Resistors in Parallel   Electrical elements are in *parallel* if they are connected to the same electrical boundaries. As a result, they all have the same voltage applied across them, and their individual currents (which may be different for each element in parallel) add.

$$I = I_1 + I_2 \qquad [2.4.1\text{-}6]$$

(See Fig. 2.4.1-3.)

Apply Ohm's law to each resistor, including the equivalent one, and apply the results to eq. 2.4.1-6.

$$\frac{V}{R_{eq_p}} = \frac{V}{R_1} + \frac{V}{R_2} \qquad [2.4.1\text{-}6a]$$

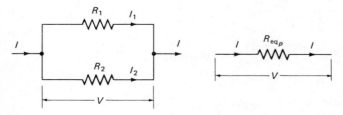

Figure 2.4.1-3 Resistors in parallel.

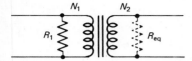

N_1 N_2

R_1 R_{eq}

Figure 2.4.1-4   Resistor reflected through transformer.

Divide through by $V \neq 0$, and solve for R_{eq_p}

$$R_{eq_p} = \frac{R_1 R_2}{R_1 + R_2}$$ [2.4.1-7]

Solve algebraically for the current through each resistor

$$I_1 = \left[\frac{R_2}{R_1 + R_2} \right] I$$ [2.4.1-7a]

$$I_2 = \left[\frac{R_1}{R_1 + R_2} \right] I$$ [2.4.1-7b]

Note that the rules for series and parallel resistors are *opposite* to those for springs and dampers.

Resistor "Reflected" Through a Transformer   A transformer is analogous to a gear train in mechanical systems. (See Fig. 2.4.1-4.)

$$R_{eq} = R_1 \left(\frac{N_2}{N_1} \right)^2$$ [2.4.1-8]

2.4.2 Inductance

Consider a magnetic circuit where a magnetic material provides the path for magnetic flux ϕ. A coil, having N turns and current I, is wound around the magnetic material, thus, requiring that the flux ϕ pass through this coil. (See Fig. 2.4.2-1.)

For such a system, the magnetic inductance L is defined as

$$L = \frac{N \, d\phi}{dI}$$ [2.4.2-1]

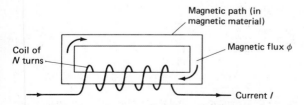

Magnetic path (in magnetic material)

Coil of
N turns

Magnetic flux ϕ

Current I

Figure 2.4.2-1   Flux through magnetic material.

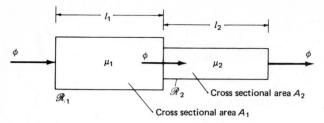

Figure 2.4.2-2   Reluctances in series.

or upon integrating by separation of variables, solving for L

$$L = \frac{N\phi}{I}$$

[2.4.2-2]

where L = inductance (H)
$\quad\quad I$ = current (A)

For simple magnetic paths, the flux may be expressed in a form similar to Ohm's law

$$\phi = \frac{NI}{\mathscr{R}} = \text{magnetic flux (Wb)}$$

[2.4.2-3]

where

$$\mathscr{R} = \frac{1}{\mu A} = \text{magnetic reluctance (1/H)}$$

l = length of path (m)   [2.4.2-3a]

A = cross sectional area of path (m^2)

μ = permeability of the magnetic material* (H/m)

Then

$$L = \frac{\mu A N^2}{l}$$

[2.4.2-4]

Reluctances in Series   Reluctances are in series if the magnetic flux ϕ must pass through one and then the other. Similar to the rule for resistors, reluctances in series are added. (See Fig. 2.4.2-2.) Then

$$\mathscr{R}_{eq_s} = \mathscr{R}_1 + \mathscr{R}_r$$

[2.4.2-5]

or

$$\mathscr{R}_{eq_s} = \frac{l_1}{\mu_1 A_1} + \frac{l_2}{\mu_2 A_2}$$

[2.4.2-5a]

---

*Values of μ for various materials are listed in Appendix 6. Inductance for various configurations can be found in Table 2.4.1-1, Appendix 2.

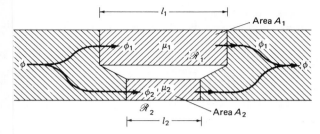

Figure 2.4.2-3   Reluctances in parallel.

Reluctances in Parallel   Reluctances in parallel share common magnetic boundaries, and each reluctance provides an independent path for magnetic flux. (See Fig. 2.4.2-3.)

The equivalent reluctance for two reluctances in parallel is similar to that for two resistors in parallel. Then

$$\mathscr{R}_{eq_p} = \frac{\mathscr{R}_1 \mathscr{R}_2}{\mathscr{R}_1 + \mathscr{R}_2} \qquad [2.4.2\text{-}6]$$

Energy Stored in an Inductance

$$E = \tfrac{1}{2}LI^2 = \tfrac{1}{2}L(\dot{q})^2 \qquad [2.4.2\text{-}7]$$

Voltage Drop in an Inductance

$$V_i = L\dot{I} = L\ddot{q} \qquad [2.4.2\text{-}7a]$$

Inductances in Series   See Fig. 2.4.2-4. Since the voltage drop through an inductance (eq. 2.4.2-7a) is proportional to the inductance L, then the rule for series inductances is the same as that for resistors. Then

$$L_{eq_s} = L_1 + L_2 \qquad [2.4.2\text{-}8]$$

Inductances in Parallel   See Fig. 2.4.2-5. The rule for inductances in parallel is the same as that for resistors in parallel. Then

$$L_{eq_p} = \frac{L_1 L_2}{L_1 + L_2} \qquad [2.4.2\text{-}8a]$$

Figure 2.4.2-4   Inductances in series.

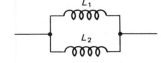

Figure 2.4.2-5   Inductances in parallel.

Note that the rules for series and parallel inductances are the same as those for resistors and reluctances but are opposite to those for springs and dampers.

Flux Linkage Flux linkage λ is defined as follows

$$\Delta V = \frac{d(\Delta \lambda)}{dt} \qquad [2.4.2\text{-}9]$$

Then eq. 2.4.2-7a becomes

$$\frac{d(\Delta \lambda)}{dt} = L \frac{di}{dt}$$

or

$$\Delta \lambda = Li \qquad [2.4.2\text{-}9a]$$

2.4.3 Capacitance

Capacitance C is given by the derivative of charge q with respect to voltage V

$$C = \frac{dq}{dV} \qquad [2.4.3\text{-}1]$$

If the capacitance is independent of voltage, then eq. 2.4.3-1 may be integrated:

$$q = C \Delta V \qquad [2.4.3\text{-}2]$$

or by taking derivatives,

$$\dot{q} = i = C \frac{dV}{dt} \qquad [2.4.3\text{-}2a]$$

The capacitance for a pair of parallel plates (neglecting edge effects) is

$$C = \frac{\epsilon A}{l} \qquad [2.4.3\text{-}3]$$

where

$$C = \text{capacitance (F)}$$

$$A = \text{area of each plate (m}^2)$$

$$l = \text{spacing between plates (m)} \qquad [2.4.3\text{-}4]$$

$$\epsilon = \text{dielectric constant (F/m)}$$

Values of ϵ for various materials are listed in Appendix 6. Capacitances for several common configurations are listed in Table 2.4.2-1, Appendix 2.

Maximum Voltage Gradient If the voltage gradient V/l is too large, the dielectric will break down. This may be evidenced by small holes or by actual burns in the dielectric material. Allowable voltage gradients are listed in Appendix 6.

Stacking of Plates If the plates of a capacitor are stacked (alternate polarity), then both sides of each plate becomes effective, thus doubling the capacity (except for outside plates). For N plates

$$C = C_0(N - 1) \qquad [2.4.3-5]$$

where C_0 = capacitance of one pair of plates

Energy Stored in a Capacitor

$$E = \tfrac{1}{2}CV^2 \qquad [2.4.3-6]$$

Voltage Drop Across Capacitor

$$V_c = \frac{q}{C} = \int \frac{I}{C}\, dt \qquad [2.4.3\text{-}6a]$$

Note that the voltage drop across a capacitor is inversely proportional to the capacitance. Hence, we can expect that the rules for series and parallel capacitors will be opposite to those for resistors.

Capacitors in Series Capacitors in series follow the same rule as that for series springs and dampers. Then

$$C_{eq_s} = \frac{C_1 C_2}{C_1 + C_2} \qquad [2.4.3\text{-}7]$$

(See Fig. 2.4.3-1.)

Capacitors in Parallel Capacitors in parallel follow the same rule as that for springs and dampers. Then

$$C_{eq_p} = C_1 + C_2 \qquad [2.4.3\text{-}8]$$

(See Fig. 2.4.3-2.)

Capacitor "Reflected" Through a Transformer

$$C_{eq} = C_1 \left(\frac{N_1}{N_2} \right)^2 \qquad [2.4.3\text{-}9]$$

Figure 2.4.3-1 Capacitors in series.

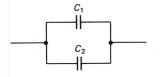

Figure 2.4.3-2 Capacitors in parallel.

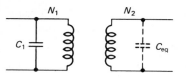

Figure 2.4.3-3 Capacitor reflected through a transformer.

Note that the rules for series and parallel capacitors are opposite to those for resistors and inductors, but are the same as those for springs and dampers. (See Fig. 2.4.3-3.)

2.4.4 Mechanical Equivalents for Magnetic Elements

Magnetic Force in an Inductor   Consider an electromechanical device similar to that in item 4, Table 2.4.1-1, Appendix 2. The current I_0 supplied to the coil is assumed to be constant. The magnetic circuit consists of an air gap of thickness X in series with a loop of magnetic material. The reluctance $\mathscr{R}$ is dominated by the air gap. That is,

$$\mathscr{R} \cong \frac{X}{\mu_0 A}$$

where μ_0 is the permeability of air and A is the pole face area. The magnetic flux Φ is, from eq. 2.4.2-3,

$$\Phi = \frac{NI}{\mathscr{R}} = \frac{\mu_0 NIA}{X}$$

The inductance L is, from eq. 2.4.2-4,

$$L = \frac{N\Phi}{I} = \frac{\mu_0 N^2 A}{X}$$

The energy stored in this device is

$$E_1 = \tfrac{1}{2} L I^2 \qquad [2.4.4\text{-}1]$$

The stored energy E_1 changes as the result of two work inputs, mechanical and electrical. That is,

$$F\,dX + VI\,dt = dE_1 \qquad [2.4.4\text{-}2]$$

We have

$$dE_1 = \frac{dE_1}{dX}\,dX$$

Since $I = I_0 = $ constant,

$$\frac{dE_1}{dX} = \frac{1}{2} I_0^2 \frac{dL}{dX} = -\frac{1}{2} I_0^2 \frac{\mu_0 N^2 A}{X^2} \qquad [2.4.4\text{-}3]$$

Also,

$$V = \frac{d}{dt}(LI_0) = I_0 \frac{dL}{dt}$$

or

$$V\,dt = I_0\,dL = I_0 \frac{dL}{dX}\,dX \qquad [2.4.4\text{-}4]$$

The work-energy equation becomes

$$F\,dX + I_0 \frac{dL}{dX}\,dX = -\frac{1}{2} I_0^2 \frac{\mu_0 N^2 A}{X^2}\,dX \qquad [2.4.4\text{-}5]$$

or

$$F\,dX - I_0^2 \frac{\mu_0 N^2 A}{X^2} = -\frac{1}{2} I_0^2 \frac{\mu_0 N^2 A}{X^2}\,dX$$

Canceling dX, we get

$$F = \frac{1}{2} I_0^2 \frac{\mu_0 N^2 A}{X^2} \qquad [2.4.4\text{-}6]$$

The equivalent magnetic spring constant K_{eq} at $X = X_0$ is the derivative of F with respect to X at $X = X_0$ (linearized spring constant). That is,

$$K_{eq} = K_m = -I_0^2 \frac{\mu_0 N^2 A}{X^3} \qquad [2.4.4\text{-}7]$$

Note that the magnetic spring constant is negative.

Magnetic Damping   The induced voltage V across the inductor L is given by Faraday's law

$$V = \frac{d(LI)}{dt} \qquad [2.4.4\text{-}8]$$

But, since current is constant, then,

$$V = I_0 \frac{dL}{dt} \qquad [2.4.4\text{-}8a]$$

Since inductance L is a function of X, use the chain rule

$$V = I_0 \frac{dL}{dX}\frac{dX}{dt} \qquad [2.4.4\text{-}8b]$$

If the electrical circuit of the coil is closed, and if it has electrical resistance R, then the energy dissipated during a differential time interval is

$$dE_r = \frac{V^2}{R}\,dt \qquad [2.4.4\text{-}9]$$

As the external mechanical system displaces from X_1 to X_2, a voltage V is generated as indicated by eq. 2.4.4-8b. This voltage, in passing through the resistance R will dissipate energy, as indicated by eq. 2.4.4-9. The work required for this is done by the mechanical system. Then,

$$\int_{X_1}^{X_2} F\, dX = \int_{t_1}^{t_2} \frac{dE_r}{dt}\, dt = \int_{t_1}^{t_2} \frac{V^2}{R}\, dt \qquad [2.4.4\text{-}9a]$$

Apply eq. 2.4.4-8b. Note that the terms will be squared, and that in particular, there will be a term $(dX/dt)^2$. Consider this in two parts, one part is dX/dt, which is velocity $\dot{X}$. The other part, also dX/dt, can be used to change the variable of integration from time to displacement. Thus,

$$\int_{X_1}^{X_2} F\, dX = \int_{X_1}^{X_2} \frac{I_0^2}{R}\left[\frac{dL}{dX}\right]^2 \dot{X}\, dX \qquad [2.4.4\text{-}9b]$$

Now, both terms are integrated with respect to the same variable, and over the same limits, and may be taken under one integral.

$$\int_{X_1}^{X_2}\left[F - \frac{I_0^2}{R}\left(\frac{dL}{dX}\right)^2 \dot{X}\right] dX = 0 \qquad [2.4.4\text{-}9c]$$

Since this must be true for arbitrary limits of integration, and since the system does not change sign in the interval, then the integrand itself must be zero. Then

$$F = \frac{I_0^2}{R}\left(\frac{dL}{dX}\right)^2 \dot{X} \qquad [2.4.4\text{-}10]$$

The equivalent damping is obtained by means of eq. 2.3.1-1.

$$D_{eq} = \frac{F}{\dot{X}} = \frac{I_0^2}{R}\left(\frac{dL}{dX}\right)^2 \qquad [2.4.4\text{-}11]$$

The inductance L, the magnetic force F, the magnetic spring K, and the magnetic damping D, for several common configurations appear in Table 2.4.1-1, Appendix 2.

2.4.5 Mechanical Equivalents for Electrostatic Elements

The energy in a capacitor is given by eq. 2.4.3-6, repeated below.

$$E_c = \tfrac{1}{2}CV^2 \qquad [2.4.5\text{-}1]$$

Assume that the voltage V can be held constant while the gap or spacing between the plates changes. Then

$$V = V_0 \qquad [2.4.5\text{-}2]$$

Referring to Table 2.4.2-1, Appendix 2, it reveals that the capacitance C is a function of the spacing X between the plates. Hence, the change in energy due to a change in spacing becomes

$$\frac{dE_c}{dX} = \frac{1}{2} V_0^2 \frac{dC}{dX} \qquad [2.4.5-3]$$

Electrostatic Force in a Capacitor   Consider the parallel plate capacitor of item 1, Table 2.4.2-1. The energy equation, including mechanical and electrical sources, is

$$F\, dX + V\, dQ = dE_c \qquad [2.4.5-4]$$

Assume $V = \text{constant} = V_0$. From eq. 2.4.5-3 we have

$$dE_c = \frac{1}{2} V_0^2 \frac{dC}{dX}\, dX = -\frac{1}{2} V_0^2 \frac{\epsilon A}{X^2}\, dX \qquad [2.4.5-5]$$

Also,

$$dQ = dC \cdot V_0 = \frac{dC}{dX} V_0\, dX = -V_0 \frac{\epsilon A}{X^2}\, dX \qquad [2.4.5-6]$$

After some manipulations, we get

$$F\, dX = \frac{V_0^2 \epsilon A}{2 X^2}\, dX \qquad [2.4.5-7]$$

or

$$F = \frac{V_0^2 \epsilon A}{2 X^2} \qquad [2.4.5-8]$$

The equivalent spring constant at $X = X_0$ is

$$K_{eq} = \frac{dF}{dX}\bigg|_{X=X_0} = -\frac{V_0^2 \epsilon A}{X_0^3} \qquad [2.4.5-9]$$

Note that the equivalent spring constant is negative.

Electrostatic Damping Using a Series Resistor   The ideal capacitor has zero resistance. Hence, no energy would be dissipated in such a device. In order to introduce controlled damping, (not left to the incidental resistance in a real capacitor) let us introduce a resistor R in series with the capacitor. We will assume that the proper circuitry has been introduced to maintain the constant voltage V_0 and that there is provision to keep the power supply voltage V_0 from being applied across this resistor (by means of a blocking capacitor) so that only the current due to the motion of charge in the capacitor must pass through this resistor. Hence, the electrical energy dissipated in this resistor will be equivalent to mechanical damping.

Similar to the procedure in Section 2.4.4, the change of variable, an ultimate step in the derivation, leads to the integral

$$\int_{x_1}^{x_2} \left[F' - V_0^2 \left(\frac{dC}{dX} \right)^2 R\dot{X} \right] dX = 0 \qquad [2.4.5-10]$$

In order that this be true for arbitrary displacements, then

$$F' = V_0^2 \left(\frac{dC}{dX} \right)^2 R\dot{X}$$
[2.4.5-11]

Thus the equivalent damping becomes $\partial F'/\partial \dot{X}$

$$D_{eq} = V_0^2 R \left(\frac{dC}{dX} \right)^2$$
[2.4.5-12]

Note that the resistance used for damping in the electrostatic device appears in the numerator, while that for the magnetic device appeared in the denominator (see eq. 2.4.4-11).

The capacitance C, the electrostatic force F, the electrostatic spring K, and the equivalent damper D, are given for several common configurations in Table 2.4.2-1, Appendix 2.

2.5 MODELING OF THERMAL ELEMENTS

The two variables associated with thermal elements are temperature T and heat stored B,

where $T =$ temperature (°C or °F)
$B =$ heat stored (kg-cal or Btu)

The rate of heat flow q is given by the derivative of B

$$q = \dot{B} = \frac{dB}{dt}$$
[2.5.1]

2.5.1 Thermal Capacitance

Heat stored in a body by virtue of its change in temperature is ascribed to thermal capacitance C,

$$B = C\Delta T$$
[2.5.1-1]

where

$$C = \text{thermal capacitance} = mc$$

$$m = \text{mass of body}$$
[2.5.1-2]

$$c = \text{specific heat of body}$$

Upon taking the derivative of eq. 2.5.1-1

$$q = C\frac{dT}{dt}$$
[2.5.1-3]

The temperature of a body is a measure of the average kinetic energy of its molecules. Hence, thermal capacity stores energy in the form of kinetic energy.

2.5.2 Thermal Capacitances in Series and Parallel

The relationship for thermal capacitance C, eq. 2.5.1-1, is analogous to that for electrical capacitance C, eq. 2.4.3-2. Hence the rules for series and parallel thermal capacitances will follow those for electrical capacitances, eqs. 2.4.3-7 and 2.4.3-8, respectively.

$$C_{ser} = \frac{C_a C_b}{C_a + C_b} \qquad [2.5.2\text{-}1]$$

or

$$\frac{1}{C_{ser}} = \frac{1}{C_a} + \frac{1}{C_b} \qquad [2.5.2\text{-}2]$$

$$C_{par} = C_a + C_b \qquad [2.5.2\text{-}3]$$

2.5.3 Heat Conductor or Dissipator

As heat is transferred through a heat conducting medium, there is a temperature drop. This is analogous to current passing through an electrical conductor with the corresponding voltage drop. The situation may be regarded from either of two points of view.

1. The heat conducting medium is a thermal *resistor* R_t (analogous to an electrical resistor R) which resists heat flow; in the process, it dissipates heat.
2. The heat conducting medium is a thermal *conductor* (analogous to reciprocal electrical resistor $1/R$) which dissipates some of the heat it is conducting.

The thermal element may be called a thermal resistor, a thermal conductor, or a *heat dissipator* D, where

$$D = \frac{1}{R_t} = \text{reciprocal thermal resistance} \qquad [2.5.3\text{-}1]$$

Heat may be transferred by conduction, convection, or by radiation. Only the first two will be discussed in this text.

   (a) **Heat Transferred by Conduction**   Given a rectangular solid body (whose material is homogeneous) with a temperature difference ΔT across two opposite (parallel) faces, heat q will be *conducted* through the body from the hotter to the colder face.

$$q = kA\frac{\Delta T}{L} = D\,\Delta T \qquad [2.5.3\text{-}2]$$

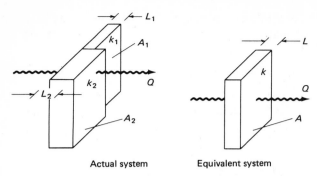

Actual system   Equivalent system

Figure 2.5.4-1   Heat conductors in parallel.

where

q = heat transferred per unit time

k = conductive heat coefficient of the material*

A = area through which heat flows   [2.5.3-3]

ΔT = temperature difference

L = length of the conductive path

$$D = \text{heat conductance or dissipation} = \frac{kA}{L} \qquad [2.5.3\text{-}4]$$

(b) Heat Transferred by Convection   If two parallel surfaces are separated by a medium that can circulate, heat will be transferred from the hotter to the colder surface by *convection*.

$$q = D_0 \, \Delta T \qquad [2.5.3\text{-}5]$$

where

$$D_0 = \text{convective heat dissipation} = hA \qquad [2.5.3\text{-}6]$$

$$h = \text{convective heat transfer coefficient} \qquad [2.5.3\text{-}7]$$

2.5.4 Heat Conductors or Dissipators in Parallel and Series

Refer to Figs. 2.5.4-1 and 2.5.4-2. The expression for a heat dissipator, eq. 2.5.3-2, is analogous to that for a damper, eq. 2.3.1-1. Hence, the rules for parallel and series follow analogously to those expressed by eqs. 2.3.5-4 and 2.3.5-5.

$$D_{\text{par}} = D_a + D_b \qquad [2.5.4\text{-}1]$$

$$D_{\text{ser}} = \frac{D_a D_b}{D_a + D_b} \qquad [2.5.4\text{-}2]$$

---

*Numerical values of k are listed in Appendix 6.

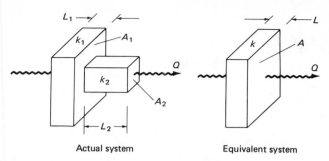

Actual system   Equivalent system

Figure 2.5.4-2   Heat conductors in series.

or

$$\frac{1}{D_{\text{ser}}} = \frac{1}{D_a} + \frac{1}{D_b}$$ [2.5.4-3]

2.6 MODELING OF FLUID ELEMENTS

2.6.1 Viscous Shear in Fluids

Given two parallel plates with a viscous fluid between them, Fig. 2.6.1-1, neglecting edge effects, the shearing force (resisting the relative motion at velocity V) is given by

$$F = \frac{\mu A V}{h}$$ [2.6.1-1]

where

μ = absolute (or dynamic) viscosity of fluid*

A = area of plates (each one)

V = relative velocity between plates   [2.6.1-2]

h = spacing between plates

Apply eq. 2.6.1-1 to the definition of a translatory damper, eq. 2.3.1-1. Then

$$D_{\text{eq}} = \frac{F}{V} = \frac{\mu A}{h}$$ [2.6.1-3]

For two concentric cylinders, Fig. 2.6.1-2, the corresponding torque is given by

$$T = \mu \pi d^3 \frac{L\omega}{4h}$$ [2.6.1-4]

---

*Tabulated in Appendix 6.

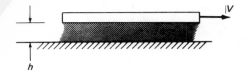

Figure 2.6.1-1 Force for fluid between parallel plates.

where

$$\omega = \dot{\theta} = \text{angular velocity between cylinders}$$
$$d = \text{diameter of inside cylinder} \qquad [2.6.1\text{-}5]$$
$$L = \text{length of cylinders}$$

Apply eq. 2.6.1-4 to eq. 2.3.1-2 (rotary damper)

$$D_{eq} = \frac{T}{\omega} = \pi \mu d^3 \frac{L}{4h} \qquad [2.6.1\text{-}6]$$

2.6.2 Fluid Resistance

Fluid resistance is analogous to electrical resistance. (See Fig. 2.6.2.-1.) The pressure drop is given by

$$\Delta P = P_1 - P_2 = g\rho R Q = R'Q \qquad [2.6.2\text{-}1]$$

We have, by virtue of continuity,

$$Q_{in} = Q_{out} = Q \qquad [2.6.2\text{-}2]$$

or

$$A_1 V_1 = A_2 V_2 = Q \qquad [2.6.2\text{-}3]$$

where

$$R = \text{coefficient of fluid resistance*}$$
$$Q = \text{volumetric flow rate of fluid} = \dot{W}$$
$$V_i = \text{velocity of fluid at point } i$$
$$A_i = \text{cross sectional area at point } i \qquad [2.6.2\text{-}4]$$
$$P_i = \text{pressure at point } i$$
$$\rho = \text{mass density of fluid}$$
$$g = \text{acceleration due to gravity}$$

A damper can be fabricated using fluid resistance. (See Fig. 2.6.2-2.) The reaction force F on the piston is

$$F_1 = (P_i - P_2)A_1 \qquad [2.6.2\text{-}5]$$

---

*Values of R are listed in Table 2.3.4-1, Appendix 2.

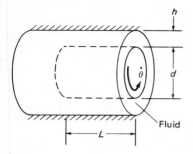

Figure 2.6.1-2   Torque for fluid between concentric cylinders.

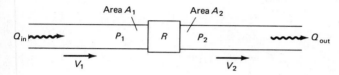

Figure 2.6.2-1   Fluid resistance.

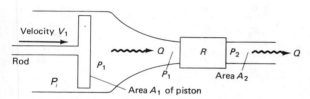

Figure 2.6.2-2   Damping due to flow through a restriction.

Substitute eqs. 2.6.2-1 and 2.6.2-3 into eq. 2.6.2-5

$$F_1 = g\rho R A_1^2 V_1 \qquad [2.6.2\text{-}6]$$

Using the definition of a damper, eq. 2.3.1-1

$$D_{eq} = \frac{F_1}{V_1} = g\rho R A_1^2 \qquad [2.6.2\text{-}7]$$

2.6.3 Fluid Resistances in Series and Parallel

Series   See Fig. 2.6.3-1. To prove that the elements are in series, note the following:

(a) The entire flow rate Q must pass through each element.
(b) Pressure drops P_1 and P_2 add.
(c) No inputs or outputs act at the junction between the two elements.

The equivalent fluid resistance R_s for the series pair is shown in Fig.

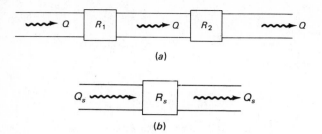

Figure 2.6.3-1   Fluid resistances in series. (a) Actual. (b) Equivalent.

2.6.3-1(b). Statement (a) becomes

$$Q_s = Q_1 = Q_2 = Q \qquad [2.6.3\text{-}1]$$

Statement (b) becomes

$$\Delta P_s = \Delta P_1 + \Delta P_2 \qquad [2.6.3\text{-}2]$$

Apply eq. 2.6.2-1 to each fluid resistance

$$\Delta P_s = R'_s Q_s \qquad [2.6.3\text{-}3]$$

$$\Delta P_1 = R'_1 Q_1 \qquad [2.6.3\text{-}4]$$

$$\Delta P_2 = R'_2 Q_2 \qquad [2.6.3\text{-}5]$$

Substitute the appropriate equations into eq. 2.6.3-2

$$R'_s Q = R'_1 Q + R'_2 Q$$

Divide through by $Q \neq 0$

$$R'_s = R'_1 + R'_2 \qquad [2.6.3\text{-}6]$$

Parallel   See Fig. 2.6.3-2(a). To prove that the elements are in parallel, note the following:

(d) Flow rate Q divides between the elements.
(e) The pressure drop is the same for each element.
(f) The elements share common boundaries at *both* ends.

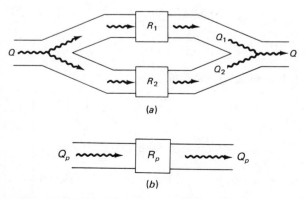

Figure 2.6.3-2   Fluid resistances in parallel. (a) Actual. (b) Equivalent.

The equivalent fluid resistance R_p for the parallel pair is shown in Fig. 2.6.3-2(b).

Statement (d) becomes

$$Q_p = Q_1 + Q_2 \qquad [2.6.3\text{-}7]$$

Statement (e) becomes

$$\Delta P_p = \Delta P_1 = \Delta P_2 = \Delta P \qquad [2.6.3\text{-}8]$$

Apply eq. 2.6.2-1 to the equivalent parallel resistance

$$\Delta P_p = R'_p Q_p \qquad [2.6.3\text{-}9]$$

Solve eqs. 2.6.3-4, 2.6.3-5, and 2.6.3-9 for their respective flow rates:

$$Q_1 = \frac{\Delta P_1}{R'_1} \qquad [2.6.3\text{-}10]$$

$$Q_2 = \frac{\Delta P_2}{R'_2} \qquad [2.6.3\text{-}11]$$

$$Q_p = \frac{\Delta P_p}{R'_p} \qquad [2.6.3\text{-}12]$$

Substitute the appropriate equations into eq. 2.6.3-7 and make use of eq. 2.6.3-8

$$\frac{\Delta P}{R'_p} = \frac{\Delta P}{R'_1} + \frac{\Delta P}{R'_2} \qquad [2.6.3\text{-}13]$$

Divide through by $\Delta P \neq 0$

$$\frac{1}{R'_p} = \frac{1}{R'_1} + \frac{1}{R'_2} = \frac{R'_2 + R'_1}{R'_1 R'_2} \qquad [2.6.3\text{-}14]$$

Take the reciprocal of each side of the equation

$$R'_p = \frac{R'_1 R'_2}{R'_1 + R'_2} \qquad [2.6.3\text{-}15]$$

Note that the rules for series and parallel fluid resistances are the same as those for electrical resistances and inductances, and are the opposite to those for springs and dampers.

2.6.4 Fluid Capacitance

An open tank that is filled through a pipe inserted in its bottom surface is shown in Fig. 2.6.4-1(a). The ratio of fluid flow rate Q to the corresponding rate of pressure rise $\dot{P}$ is defined as *fluid capacitance C*.

$$C = \frac{Q}{\dot{P}} \qquad [2.6.4\text{-}1]$$

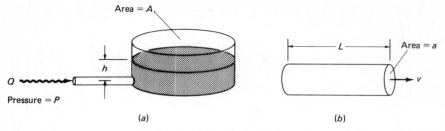

Figure 2.6.4-1   Fluid capacitance (a) and inertance (b).

Use subscript c and note that $Q = \dot{W}$

$$C = \frac{Q_c}{\dot{P}_c} = \frac{\dot{W}_c}{\dot{P}_c}$$

or

$$\dot{P}_c = \frac{\dot{W}_c}{C} \qquad [2.6.4\text{-}2]$$

Integrate both sides

$$\Delta P_c = \frac{W_c}{C} \qquad [2.6.4\text{-}3]$$

where

$$C = \text{fluid capacitance} = \frac{A}{\rho g} \qquad [2.6.4\text{-}4]$$

$A = $ cross sectional area of tank

$\rho = $ mass density of fluid $\qquad [2.6.4\text{-}5]$

$P = $ fluid pressure

Energy stored is in the form of potential energy by virtue of the liquid level or pressure. Then the energy E becomes

$$E = \tfrac{1}{2}C(\Delta P)^2 \qquad [2.6.4\text{-}6]$$

Head   See Fig. 2.6.4-1(a). The pressure P in a tank that is filled to height h is given by

$$P = \frac{F}{A} = \frac{\rho g h A}{A} = \rho g h \qquad [2.6.4\text{-}7]$$

where $F = $ weight of the fluid in the tank
$\qquad A = $ cross sectional area of the tank

The height h is referred to as the *head*.

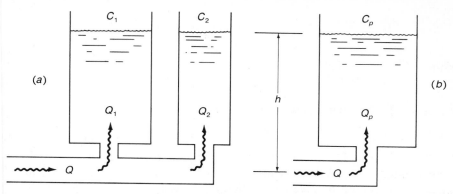

Figure 2.6.4-2   Capacitances in parallel. (a) Individual elements. (b) Equivalent parallel capacitance.

Fluid Capacitances in Series and Parallel

Parallel   See Fig. 2.6.4-2. In order to prove that the elements are in parallel, note the following:

(g) Flow rate Q divides between the elements.
(h) The head h (or pressure P) is the same for each element.
(i) The elements share common boundaries at *both* ends. (This requires some explanation. One boundary is the common feedpipe. The other is the top surface of the fluid in each tank; both are at the same height h and both are at atmospheric pressure.)

Since there are no sources or sinks, statement (g) becomes

$$Q_p = Q_1 + Q_2 \qquad [2.6.4\text{-}8]$$

Solve eq. 2.6.4-1 for Q

$$Q = C\dot{P} \qquad [2.6.4\text{-}9]$$

Apply eq. 2.4.4-9 to each element in parallel and to the equivalent one

$$Q_1 = C_1\dot{P}_1 \qquad [2.6.4\text{-}10]$$

$$Q_2 = C_2\dot{P}_2 \qquad [2.6.4\text{-}11]$$

$$Q_p = C_p\dot{P}_p \qquad [2.6.4\text{-}12]$$

Substitute the appropriate equations into eq. 2.6.4-8

$$C_p\dot{P}_p = C_1\dot{P}_1 + C_2\dot{P}_2 \qquad [2.6.4\text{-}13]$$

Statement (h) becomes

$$h_p = h_1 = h_2 = h \qquad [2.6.4\text{-}14]$$

And equally true are the derivatives

$$\dot{h}_p = \dot{h}_1 = \dot{h}_2 = \dot{h} \qquad [2.6.4\text{-}15]$$

Take the derivative of eq. 2.6.4-7

$$\dot{P} = g\rho\dot{h} \qquad [2.6.4\text{-}16]$$

Apply this to eq. 2.6.4-15

$$\dot{P}_p = \dot{P}_1 = \dot{P}_2 = \dot{P} \neq 0 \qquad [2.6.4\text{-}17]$$

Equation 2.6.4-17 permits us to divide eq. 2.6.4-13 by $\dot{P}$

$$C_p = C_1 + C_2 \qquad [2.6.4\text{-}18]$$

Series   It is difficult to show two capacitances in series. It is possible to concoct some imaginative schemes, but usually these are too contrived to be practical. It should be pointed out that the systems shown in Fig. 6.5.1-1 on page 212 are NOT in series. (The student is cautioned not to judge by appearances.) Consequently, the discussion of capacitances in series will not be covered here.

2.6.5 Fluid Inertance

A mass of fluid is flowing through a pipe in Fig. 2.6.4-1(b). Assume that every particle of fluid is flowing at the same speed v. In order to accelerate the fluid, a force is required. This is analogous to a force that is applied to a rigid body.

$$F = a\,\Delta P = M\frac{dv}{dt} = \rho a L\frac{dv}{dt} \qquad [2.6.5\text{-}1]$$

where

$$\begin{aligned} L &= \text{length of the mass of fluid} \\ a &= \text{cross sectional area of pipe} \end{aligned} \qquad [2.6.5\text{-}2]$$

The fluid velocity v is given by

$$v = \frac{Q}{a} \qquad [2.6.5\text{-}3]$$

Equation 2.6.5-1 becomes

$$\Delta P = \frac{\rho L}{a}\frac{dQ}{dt} = I\frac{dQ}{dt} \qquad [2.6.5\text{-}4]$$

where

$$I = \text{fluid inertance} = \frac{\rho L}{a} \qquad [2.6.5\text{-}5]$$

The energy stored is in the form of kinetic energy by virtue of the fluid velocity v. Then

$$E = \frac{1}{2}Mv^2 = \frac{1}{2}\rho a L\left(\frac{Q}{a}\right)^2 = \frac{1}{2}\left(\frac{\rho L}{a}\right)Q^2 = \frac{1}{2}IQ^2 \qquad [2.6.5\text{-}6]$$

2.6.6 Inertances in Series and Parallel

The rule for series inertances can be found heuristically by letting the areas and densities be equal. Consequently, this becomes equivalent to one continuous inertance whose length is given by the sum of the individual lengths. Then the rule becomes,

$$I_{ser} = I_a + I_b \qquad\qquad [2.6.6\text{-}1]$$

To derive the rule for parallel inertances, heuristically, let the lengths and densities be equal. Consequently, this becomes equivalent to a single inertance whose area is given by the sum of the individual areas. Since area is in the denominator, then the rule becomes

$$\frac{1}{I_p} = \frac{1}{I_a} + \frac{1}{I_b} \qquad\qquad [2.6.6\text{-}2]$$

SUGGESTED REFERENCE READING FOR CHAPTER 2

Equivalent mass [1], p. 50–54; [5], p. 191; [9], p. 430.
Equivalent springs [1], p. 48; [2], p. 104; [3], p. 157, 172, 181; [5], p. 8; [10]; [11], p. 181; [12], p. 48; 62.
Equivalent dampers [5], p. 70–73; [12], p. 59, 177, 183.
Electromechanical equivalents [12], p. 126; [15], p. 298; [17], p. 79–89.
Fluid and thermal elements [8], chap. 3; [12], p. 59, 177, 183; [13]; [14]; [17], chap. 3; [45].
Analogies [1], p. 188–191; [8], p. 47–49; [17], chap. 4; [26], p. 19.
Meaning of lumped parameters [17], p. 14.

PROBLEMS FOR CHAPTER 2

2.1. A nonlinear spring has the following F (N) versus x (mm) relation: $F = 2420x - 21x^3$.
    (a) Find the linearized spring constant in N/m for displacements near $x = 4.5$ mm.
    (b) How much energy (joules) is required to change the stretch x from 2.0 to 6.0 mm?

2.2. For the device shown the distance X (m) varies with time t (s) according to $X = X_0 e^{-(K/D)t}$, where $X_0 = $ initial displacement.
    (a) Find the velocity V ($+$ to the right).
    (b) Find the power dissipated as a function of t.
    (c) Show that the total energy dissipated when the slider moves from $X = X_0$ to the equilibrium position $X = 0$ is given by $E_t = \frac{1}{2}KX_0^2$.

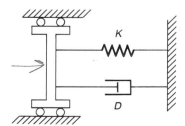

Figure P-2.2

2.3. Find the fluid capacitance of the device shown. The spring is uncompressed when the piston is in contact with the left wall of the cylinder.

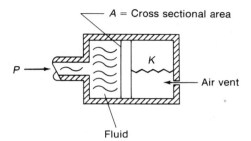

A = Cross sectional area

P

K

Air vent

Fluid

Figure P-2.3

2.4. The wall of a building is made of concrete 8 in. thick. The inside face is covered by 2.0 in. of hair felt and then a layer of wood $\frac{3}{4}$ in. thick. The dimensions of the wall are 20 ft long by 10 ft high. If the inside temperature is 75°F and the outside temperature is 45°F, find the heat transferred (Btu) by the wall in 15 min. Assume conduction transfer only; $k_{\text{wood}} = 0.06$, $k_{\text{hair}} = 0.02$, $k_{\text{conc}} = 0.4$ Btu/h · ft · °F.

2.5. Find the equivalent mass at the point of application of the force f. The stepped pulley is made of two homogeneous disks of equal mass M_0 fastened together.

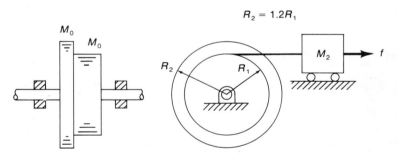

$R_2 = 1.2R_1$

M_0

M_0

R_2

R_1

M_2

f

Figure P-2.5

2.6. The beam AB, with length $L = 20$ in., elastic modulus $E = 15 \times 10^6$ lb/in.², and moment of inertia $I = 7.2 \times 10^{-3}$ in.⁴, has a spring of stiffness $K = 450$ lb/in. under it at the midpoint. Find the equivalent spring constant.

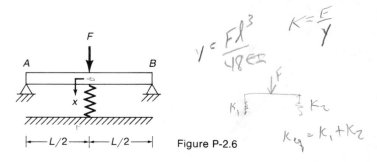

F

A

B

x

$$y = \frac{F\ell^3}{48EI} \qquad K = \frac{F}{y}$$

$$K_1 \qquad K_2$$

$$k_y = k_1 + k_2$$

$\leftarrow L/2 \rightarrow\leftarrow L/2 \rightarrow$ Figure P-2.6

2.7. For the system shown, $K_1 = 40$ N/m, $K_2 = 60$ N/m, and $K_3 = 30$ N/m.
   (a) Find the equivalent spring constant.
   (b) Find the energy stored in the spring system if $f = 108$ N.

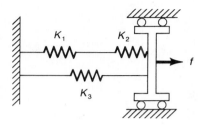

Figure P-2.7

2.8. Find the equivalent damping constant for the stepped pulley system shown.

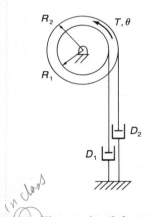

Figure P-2.8

2.9. The rotational damper shown consists of a rotating disk B inside a fixed housing
   C. A viscous fluid occupies the space between B and C. Given $d_1 = 0.300$ m,
   $d_2 = 0.304$ m, $a_1 = 0.050$ m, $a_2 = 0.051$ m, $\mu = 0.24$ N · s/m², and $\rho = 850$
   kg/m³, find the torsional damping constant.

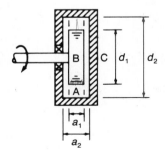

Figure P-2.9

2.10. (a) Draw the necessary free body diagrams and determine the equivalent torsional damping constant.
   (b) What influence does the rotational inertia J have on the damping? Explain.

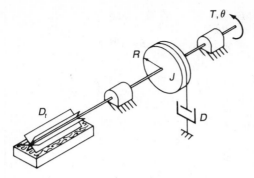

Figure P-2.10

2.11. Given $M = 0.6$ kg, $D = 4.0$ N · s/m, and v (m/s) = $0.2t + 0.12t^2$ [t (s)]:
(a) Find the force f at $t = 0.5$ s.
(b) Find the damper force f_D at $t = 0.5$ s.

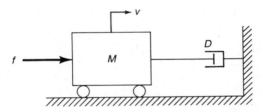

Figure P-2.11

2.12. A wooden box with $\frac{1}{2}$-in. thick walls and outside dimensions as shown rests on a pad of thermal insulation. Find the heat flux q (Btu/hr) due to conduction if the temperatures outside and inside the box are 95 and 75°F, respectively.

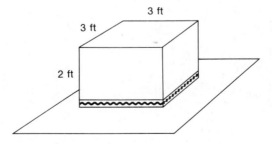

Figure P-2.12

2.13. Consider the geared system shown, with $L = 15$ in., $d = 0.2$ in., and $G = 12 \times 10^6$ lb/in.^{2}. Torque is applied to the shaft of A.

(a) Find the equivalent spring constant at the shaft of A.

(b) Explain why the rotational inertias of gears A and B may be neglected.

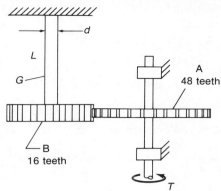

Figure P-2.13

2.14. A rotating disk of diameter d supported on an oil film of thickness h, as shown in item 3, Table 2.3.3-1, is rotated so that θ (rad) $= 30t + 0.5t^2$ [t (s)]. Given $\mu = 0.26$ N $\cdot$ s/m^2, $d = 0.06$ m, $h = 0.0005$ m, mass of the disk $= 0.22$ kg, and $\rho = 850$ kg/m^3, find the torque required at $t = 10.0$ s.

2.15. Given $D_1 = 20$ N $\cdot$ s/m, $D_2 = 80$ N $\cdot$ s/m, and $f = 32 \sin 4t$ [f (N), t (s)]:

(a) Find D_{eq};

(b) Find the energy dissipated by D_1 in one cycle.

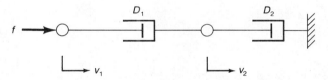

Figure P-2.15

2.16. Find the heat transferred by convection and conduction from air inside a room at 75°F to air outside at 40°F through one wall made of brick 8 in. thick. The wall is 14 ft long by 8 ft high. Assume h for convection (inside and outside) to be $h = 30$ Btu/ft$^2 \cdot$ F.

2.17. Find the equivalent rotational inertia J_{eq} for the system shown. The pulley is a uniform disk of mass M and radius R.

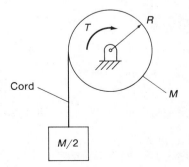

Figure P-2.17

2.18. Given $X = A \sin 2t$ and $A = 0.06$ m [t (s)]:
   (a) Find the damping coefficient D so that 90 joules are dissipated in one cycle.
   (b) Find the peak value of the force required to provide the given displacement.

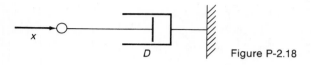

Figure P-2.18

2.19. (a) Assume constant angular velocity, and find the equivalent damping D_{eq} for torque applied to shaft 1.
   (b) Using free body diagrams, find the friction force transmitted between the disks if $T = 20$ N · m.

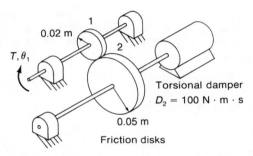

Figure P-2.19

2.20. For the system of springs shown in Fig. 2.2.5-1(a), assume $K_1 = 0.6$ N/m, $K_2 = 180$ N/m, $L_1 = 0.6$ m, and $L_2 = 0.4$ m.
   (a) Find K_{eq}.
   (b) Find the energy stored if the vertical rigid rod is rotated 5°. Assume the two springs to be unstressed when the rod is vertical.

2.21. Find the equivalent torsional spring constant for torque applied to the shaft of the pulley.

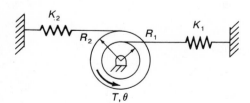

Figure P-2.21

2.22. The friction disks shown do not slip. Consider them to be uniform disks with $M_A = 0.5$ kg and $M_B = 0.8$ kg.
   (a) Find J_{eq} for torque applied to the shaft of B.
   (b) Find the friction force between the disks if $T = 5.76$ N · m.

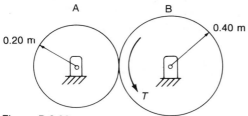

Figure P-2.22

2.23. For the right angle bracket shown, $AB = 0.3$ m, $BC = 0.5$ m, and θ is small.
  (a) Find the spring constant K such that the equivalent torsional spring constant is 20.4 N · m · s.
  (b) Find the spring force at C if $T = 2.0$ N · m.

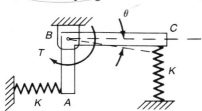

Figure P-2.23

2.24. Consider the system of Fig. P-2.24 with $G = 0.2$ m and $H = 0.4$ m. Given $K_1 = K_2 = 50$ N/m, $K_3 = 20$ N/m, and the angle of rotation of the lever $= 0.05$ rad clockwise, find the external force required at the point where M is attached.

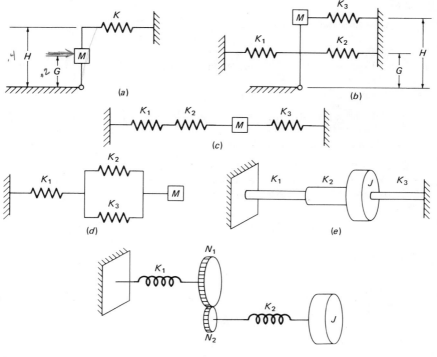

Figure P-2.24   (f)

2.25. The two tubes shown carry a fluid of viscosity $\mu = 3.45 \times 10^{-5}$ N · s/m². Given $d_1 = 0.006$ m, $d_2 = 0.008$ m, $L_1 = 0.5$ m, $L_2 = 1.2$ m, and $P_1 = 40$ N/m², find the flow rate Q in m³/s.

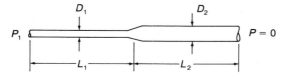

Figure P-2.25

2.26. For the geared system shown, find the equivalent torsional spring constant at the shaft where T is applied.

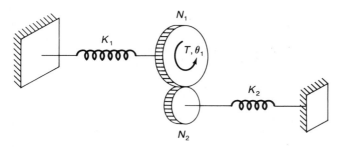

Figure P-2.26

2.27. A rotating damper like that in item 4, Table 2.3.3-1, has the following properties: $\mu = 0.26$ N · s/m², $d = 45.0$ mm, $h = 0.7$ mm, and $L = 120.0$ mm.
(a) Find the torsional damping constant D_{eq} (in N · m · s).
(b) Find the torque (in N · m) required to rotate the damper at a constant angular velocity of 3600 rpm.
(c) Find the power dissipated (in Watts).

2.28. 47.0 in.³/s of oil flow through a tube with 0.50-in. diameter, 120-in. length, $\mu = 4.0 \times 10^{-6}$ lb · s/in.², and $\rho = 6.0 \times 10^{-3}$ lb · s²/in.⁴. Find the pressure drop through the tube (in lb/in.²).

2.29. Given the spring supported bellows shown:
(a) Find the force F to cause a deflection $y_1 = 0.10$ m;
(b) Find the compression y_2 of the spring.

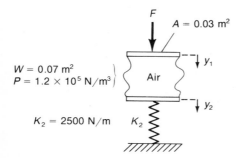

Figure P-2.29

2.30. Using formulas for equivalent mass and equivalent spring constants given in Appendix 2, find the approximate angular frequency ω_n for a cantilever beam made of steel. Assume $E = 30 \times 10^6$ lb/in.^{2} and specific weight = 0.28 lb/in.^{3}. The beam is 20 in. long and has a square cross section $\frac{1}{2} \times \frac{1}{2}$ in.

2.31. For the system shown, $M = 3.7$ kg, $EI = 4.2 \times 10^3$ N $\cdot$ m^2, $L = 0.65$ m, and $K = 7600$ N/m; find the approximate natural angular frequency ω_n.

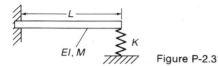

Figure P-2.31

2.32. For the arrangement shown, $K_1 = 12,800$ N/m, $K_2 = 8000$ N/m, $K_3 = 6200$ N/m, and $f = 3500$ N.
(a) Find y for equilibrium.
(b) Find the total stored energy.

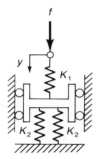

Figure P-2.32

2.33. The system shown is in equilibrium. Given $R = 0.12$ m and $K = 2500$ N/m, find T if $\theta = 0.10$ rad.

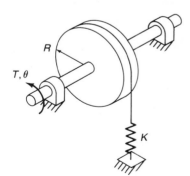

Figure P-2.33

2.34. Given $D_1 = 40$ N $\cdot$ s/m, $D_2 = 60$ N $\cdot$ s/m, and x_1 as given in the graph, find the energy dissipated in the time period from $t = 0$ to 0.3 s.

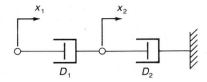

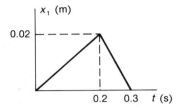

Figure P-2.34

2.35. An 80-lb sack of sand falls from a truck traveling at 35 mph. The sack does not bounce, but slides to a stop with a coefficient of friction = 0.7.
(a) Find the time to stop.
(b) Find the distance to stop.
(c) Find the temperature rise of the sand bag after it has come to rest. The specific heat of sand is 0.12 Btu/lb$_m$ · °F. *Note*: 1 Btu = 778 ft-lb$_f$.

2.36. Consider the fluid capacitor with the performance data shown:

t (s)	P (N / m²)	Q (m³ / s)
0	0	0.1
1	100,000	0.094
2	200,000	0.076
3	300,000	0.046
4	400,000	0.004

(a) Estimate the linearized fluid capacitance in the vicinity of P = 250,000 N/m².
(b) Estimate the volume stored when P = 250,000 N/m².

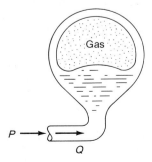

Figure P-2.36

2.37. For the damper loaded as indicated by the graph, find the energy dissipated for $t = 0$ to $t = 0.6$ s.

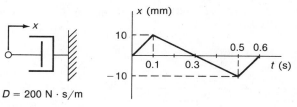

$D = 200$ N · s/m

Figure P-2.37

2.38. Using a material with density $\rho = 8$ g/cm^3, design a solid cylindrical mass such that

$$M = 400 \text{ g}, \qquad J_z \text{ (moment of inertia about axis)} = 1250 \text{ g} \cdot \text{cm}^2$$

2.39. Rework Problem 2.38 using a tubular mass whose inside diameter is equal to 3.0 cm. (Ans. $D = 4.0$ cm, $h = 9.1$ cm.)

2.40. Using a steel plate whose thickness is equal to 0.318 cm design a cantilever beam whose equivalent mass (reflected to the end of the beam) $M = 19.1$ g, and whose equivalent spring rate $K = 9.53 \times 10^6$ dyn/cm. (Ans. $L = 15.3$ cm, $b = 2.17$ cm.)

2.41. Design a coil spring so that its spring rate $K = 9.44 \times 10^4$ N/m, and determine its actual mass M. Use eight active coils and a mean radius $R = 2.5 \times 10^{-2}$ m. (Ans. $d = 9.34 \times 10^{-3}$ m.)

2.42. Design a clock spring so that its torsional spring rate $K = 7.62 \times 10^5$ dyn · cm/rad. Let the spacing between adjacent coils be equal to the thickness of the ribbon, and let the outside diameter be equal to 7.0 cm.

2.43. Design an air spring so that $K = 3.14 \times 10^4$ N/m, where the container is a cube each of whose sides is L, and where the diameter of the piston is equal to $L/2$.

2.44. Design a torsional damper using two concentric cylinders so that $D = 970$ dyn · cm · s. Let the diameter of the inner cylinder be equal to 1.414 cm and let its length be equal to 2.8 cm.

   (a) Let air at 70°F be used as the viscous fluid.
   (b) Is this feasible to manufacture?
   (c) Redesign using SAE10 oil at 70°F.
   (d) Comment on feasibility of manufacture.

2.45. Produce a workable design for a dashpot with $D = 3648$ dyn · s/cm. Neglect the effect of the piston rod.

2.46. A piston having a diameter equal to 0.07 m actuates a bellows-type pump which forces air through a smooth wall pipe having an inside diameter equal to 0.015 m. See Fig. P-2.46. Design a damping device such that $D = 113$ N · s/m considering the following:

   (a) Neglecting the fluid resistance in the pipe and using each of the following values of resistance for a porous material that is to be inserted in the pipe

$$C_1 = 100, 200, 400, 800$$

(b) Using only the fluid resistance of the pipe (without a porous material) where the length of the pipe L_2 is unknown, and where the viscosity of the air is equal to $\mu = 1.76 \times 10^{-5}$ N $\cdot$ s/m^2.

(c) Using both the porous material described in part (a) plus the pipe resistance described in part (b). Let $L_2 = 10$ m.

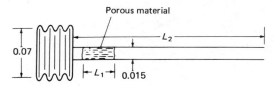

Figure P-2.46

2.47. *Series-Parallel Elements.* For each of the following, perform the following: establish whether the elements are in series, parallel, or combinations; determine the equivalent element in terms of the original elements; determine the equivalent element in terms of the geometry of the components.

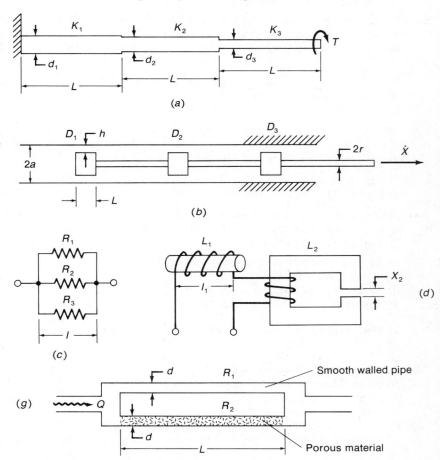

Figure P-2.47

(a) Stepped shaft. See Fig. P-2.47(a).
(b) In-line dampers. All have same dimensions. See Fig. P-2.47(b). Fluid viscosity is μ.
(c) Three electrical resistors have the same length l and the same resistivity ρ but different diameters d_1, d_2, and d_3, respectively. See Fig. P-2.47(c).
(d) Inductances. One is open core with N_1 turns and cross-sectional area A_1. The other has a small air gap with cross-sectional area A_2, path length l_2, and N_2 turns. Both have the same permeability μ. See Fig. P-2.47(d).
(e) Five capacitor plates, each having area A, are stacked a distance X_0 apart. They are numbered 1 through 5. Connections are made, one to plate 1, the other to plate 5. The dielectric constant is ϵ.
(f) Similar to part (e), but one wire connects plates 1, 3, and 5; the other connects plates 2 and 4.
(g) Fluid resistances. One is a smooth-walled pipe. The other is filled with a porous material. Both have same diameter d and length L. The fluid has density ρ and viscosity μ. See Fig. P-2.47(g).
(h) A set of three fluid capacitances with respective areas A_1, A_2, and A_3, sharing a common supply pipe, each open to the atmosphere.
(i) A long resistance wire of diameter d and length l with resistivity ρ is cut into four equal lengths and these are wired with common boundaries.

Chapter **3**

Modeling of Lumped Systems

The subject of modeling as applied to systems may be considered in either of two veins; physical modeling, or mathematical modeling. The purpose of modeling is to gain insight or to produce numerical quantities concerning the behavior or response of the system. In physical modeling, either a full sized or a scale model is built and tested. The dimensions, the environment, and the test procedure must be carefully selected in order to properly represent the actual system. In mathematical modeling, the equation (or equations) of motion is constructed in order to present the same data that would have been produced by the actual system under the identical conditions. Like the physical model, the mathematical model must refer to properly chosen dimensions, inputs, and analytically observable outputs.

Whether the modeling is physical or mathematical, it is often necessary to simplify the model. This is done in order to permit a feasible means to determine the system outputs. On the other hand, a simplified model may eliminate an important property of the system. Thus, the most important guideline for modeling is to make the model sufficiently sophisticated for an accurate representation of the actual system without becoming entangled in either experimental or mathematical processes which might tend to obscure the most significant results.

The formulation of such guidelines is no small task. To assist in this matter, analogies among various systems are derived and are tabulated. In addition to this, a number of hints and suggestions are catalogued and tabulated. These will assist in the evaluation of inputs and initial conditions.

The modeling procedure will be accomplished in four steps. First, the block, circuit, or free body diagram will be drawn. Next, the system elements will be identified. Third, the system will be simplified. Finally, the equations of motion will be found.

3.1 INITIAL CONDITIONS AND INPUTS

The process of modeling dynamic systems involves the formulation of the differential equation including the initial conditions and inputs. The proper identification of initial conditions and inputs in real systems is a difficult task requiring a considerable amount of technical decision making.

One of the difficulties arises when an input force accompanies the corresponding input displacement. These usually are interrelated, one being the dependent quantity, while the other is independent. In this respect, the problem is cascaded. One event is the prime quantity, while the other can be

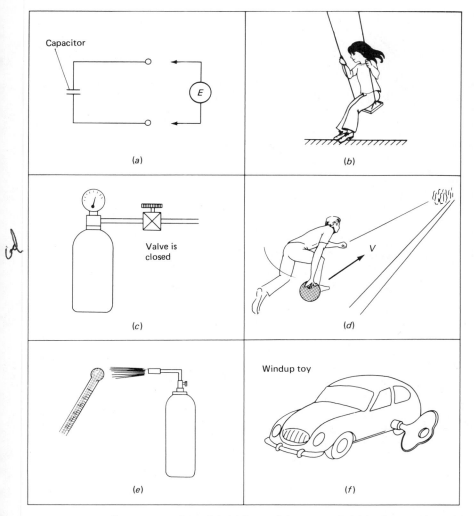

Figure 3.1.1-1 Some examples of initial conditions. (a) Initial charge. (b) Initial displacement. (c) Initial pressure. (d) Initial velocity. (e) Initial temperature. (f) Initial angular displacement.

considered determinable. In such cases, the decision can be based upon identifying the independent quantity.

When the independent quantity is not clearly identifiable, a suggestion is to examine which one influences the other by a greater amount. For example, in Fig. 3.1.1-1(b), a swing is displaced. However, accompanying this, is the restoring torque due to gravity. The restoring torque depends upon the displacement, while the displacement would not change even if the gravity field of the earth changed. Hence, the decision is that there is an *initial displacement* (the restoring torque being dependent, which could be found by solving the problem). Refer to Fig. 3.1.3-1(f), application of the brake pedal.

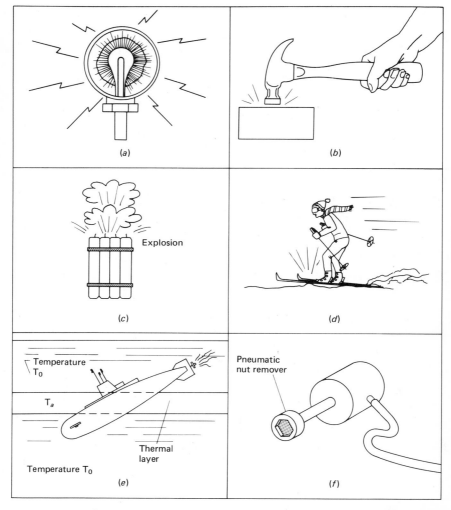

Figure 3.1.2-1   Some examples of impulse inputs. (a) Electronic flash gun. (b) Impulse force. (c) Impulse pressure. (d) Translatory impulse displacement. (e) Thermal shock. (f) Impulse torque.

This is another example requiring some thought. Surely, as the pedal is operated, a force is required. At the same time, there is a sudden displacement of the pedal. Is this a force or a displacement input? Is it the force on the pedal, or is it the angular motion of the lever that actuates the brakes? In power brakes, a hydraulic system actuates the brakes. In order to operate the hydraulic system, a special valve is opened or closed. Very little force or torque is required. Hence, the decision is that this is a *displacement* input.

Another difficulty in choosing the appropriate input resides in the non-compliance of real systems with simple geometric configurations for the inputs. In Fig. 3.1.2-1(e), as the submarine passes quickly through a sharp

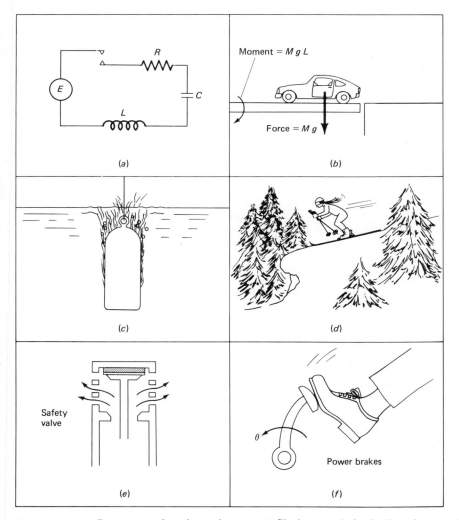

Figure 3.1.3-1   Some examples of step inputs. (a) Closing a switch. (b) Step force and step moment. (c) Quenching hot steel. (d) Translatory step displacement. (e) Step pressure. (f) Rotary step displacement.

temperature change, how close is this to an impulse thermal input? In Fig.
3.1.4-1(e), how uniformly does the temperature rise?

In many cases, the input is not clearly one of the few types listed in the
following sections In addition to that, some inputs are not one, but a
combination of inputs. Hence, a certain amount of approximating, arranging,
and formulating will be necessary.

To assist in the task of identifying and evaluating initial conditions, a
figure is shown for each of the most common types, in the subsequent pages.
Each figure has a number of examples. They are as follows: initial conditions,
impulse input, step input, ramp input, and sinusoidal input.

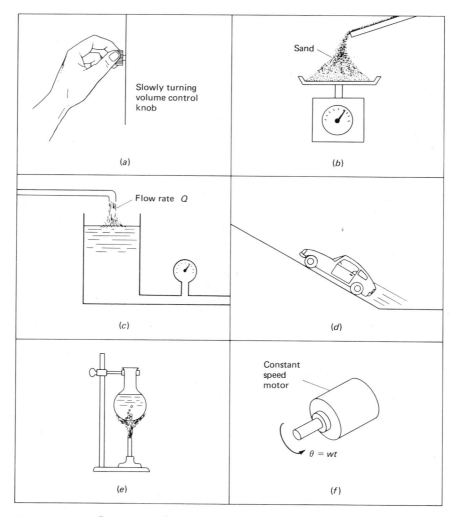

Figure 3.1.4-1   Some examples of ramp inputs. (a) Ramp voltage. (b) Ramp force.
(c) Ramp pressure. (d) Translatory ramp displacement. (e) Constant temperature rise.
(f) Rotary ramp displacement.

3.1.1 Initial Conditions

The initial status of the variables of a system represent the initial conditions, taken at the instant of time when the dynamic problem starts. Some examples are shown in Fig. 3.1.1-1.

3.1.2 Impulse Input

Consider an input of magnitude F and time duration T, where the product

$$FT = 1 \qquad\qquad [3.1.2\text{-}1]$$

Let the magnitude grow larger and larger while the duration grows correspondingly smaller and smaller, maintaining the product of unity. In the limit, the magnitude will approach infinity while the duration will approach zero and the result is called an impulse input. Some practical examples are shown in Fig. 3.1.2-1.

3.1.3 Step Input

A step input is one that is suddenly applied and then held at a constant level. Some examples are shown in Fig. 3.1.3-1.

3.1.4 Ramp Input

A ramp input is one which increases linearly with time. In many practical cases, the problem ends before the input gets too large. Some examples are shown in Fig. 3.1.4-1.

3.2 MODELING OF MECHANICAL SYSTEMS

Free body diagrams are fundamental to an understanding of the physics of the problem. The governing differential equations can be properly formulated with a thought process which isolates bodies and identifies the forces acting on them.

3.2.1 Newton's Laws

In the modeling of physical systems, using the laws of mechanics, the traditional technique is to isolate a body or portion of the system, by an imaginary cut, and by replacing the rest of the system (on the other side of the cut) by equivalent forces and moments.

To assist in visualizing the various forces involved, the force vectors will be shown emanating from the element that is producing the force, and the mass will be shown in physical form. The reader will understand that the mass is to be represented as a point for concurrent forces and that the elements that produce the forces will be omitted. For example, in Fig. 3.2.1-1, the force F_2 is

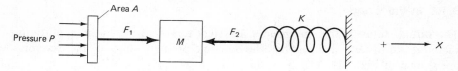

Figure 3.2.1-1   Free body diagram showing sources of forces.

exerted by a spring, and force F_1 is exerted by pressure, where

$$F_2 = KX \quad \text{and} \quad F_1 = PA \qquad [3.2.1\text{-}1]$$

Summing the forces (assuming the mass M can be represented as a point) according to Newton's law, we have

$$\sum F = M\ddot{X} = F_1 - F_2 = PA - KX \qquad [3.2.1\text{-}2]$$

■ **EXAMPLE 3.2.1-1** Vehicle with Spring Bumper Hits Wall
For the system in Fig. 3.2.1-2(a), determine the initial conditions, inputs, draw the free body diagram, and write the differential equation. Assume that the bumper acts as a spring K and a damper D.

Solution.   At first glance, one might be tempted to consider this a problem involving an impulse input. However, for that, the vehicle would be stationary,

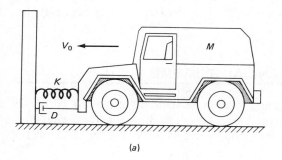

(a)

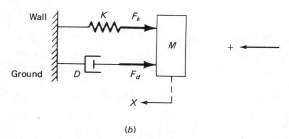

(b)

Figure 3.2.1-2   Vehicle with spring bumper hits wall. (a) Actual system. (b) Free body diagram.

the wall would hit the car, and the wall would retract. Just before impact (let impact with the wall correspond to time = 0) the vehicle has initial velocity V_0. At this time, the springs are at free length, and the bumper end is unrestrained. Let the problem begin just as the bumper touches the wall. At this time, the bumper end of the springs is grounded, and there is no input. The free body diagram is shown in Fig. 3.2.1-2(b).

Taking a positive convention to the left, the vehicle displacement X will also be taken positive to the left. The springs (each equal to $\frac{1}{2}K$) K will oppose the motion with force F_K. At the same time, the braking action, equivalent damper D will also oppose the motion (by applying force F_d). Then, using Newton's law, we have

$$\sum F = M\ddot{X} = -KX - D\dot{X} \tag{1}$$

or

$$M\ddot{X} + D\dot{X} + KX = 0$$
$$\text{IC} \quad X(0) = 0, \qquad \dot{X}(0) = V_0 \tag{2}$$

Note that the differential equation is accompanied by the initial conditions (IC). It is a good habit to get into.   ∎

■ EXAMPLE 3.2.1-2 Spring Scale with Suddenly Applied Input

Consider a spring* scale, in which the weighing platform is supported on a spring and lubricated piston (which provides damping). The translation of the platform is magnified by appropriate gearing to a calibrated dial. See Fig. 3.2.1-3(a). Various items are placed suddenly on the platform, producing a forcing function $F(t)$. Write the equation of motion.

System Modeling

In modeling the system, it is suggested that the mass first be identified. Then determine the location of the remaining system elements and establish the forces these exert upon the mass. The equation of motion can be formulated using Newton's law.

The equivalent mass M_{eq} for this system was modeled in Ex. 2.1.5-1, where it was found that the equivalent mass M_{eq} was given by the sum of two masses, a fixed mass M (associated with the mechanism of the scale) and a variable mass m (associated with the item placed upon the scale). The result is given by eq. (3) of that example, and is repeated below.

$$M_{eq} = m + M \tag{1}$$

$$= m + M_p + M_d + \frac{M_k}{3} + \frac{J_c N^2}{R^2} + \frac{J_g}{R^2} \tag{2}$$

Before the item m is placed upon the scale, the weight F_g of the components

---

*See footnote for Ex. 2.1.5-1, p. 16.

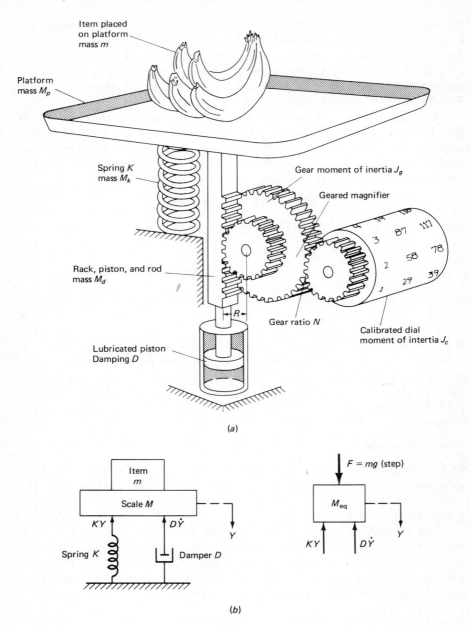

(a)

(b)

Figure 3.2.1-3   Spring scale with suddenly applied load. (a) Actual system. (b) Free body diagram.

of the scale that have vertical movement (having mass M') is given by

$$F_g = M'g \qquad (3)$$

This force is initially in equilibrium with the compressive force $-F_g$ of the spring. This force is given by

$$F_g = KY_0 \qquad (4)$$

where Y_0 is the deflection of the spring from its free length. For a static analysis, or for a maximum stress consideration, this force and its opposite force would be included in the free body diagram. However, for dynamic considerations, these two forces remain in equilibrium, and do not affect the dynamic motion. Hence, *if the displacement in the dynamic problem were measured from the static equilibrium point Y_0, then both the gravitational force and its equal and opposite spring force may be omitted from the problem.* Since all measurements are made from point Y_0, the initial displacement in the equivalent system is zero.

Now, when the item is placed upon the platform, there is a gravitational force on the item. However, this represents a change, for which the system did *not* start in equilibrium. There was no initial compression of the spring to statically oppose this. Hence, the input force F is equal to the force created by the sudden application of item m to the platform. This force is

$$F = mg \quad \text{(step input)} \qquad (5)$$

The scale plus the item represents the total mass M_{eq} as given by eq. (2).

This equivalent mass is supported by a spring K and a damper D, both of which connect the mass to the ground. The free body diagram is shown in Fig. 3.2.1-3(b).

Sum the forces according to Newton's law, and note the convention. The input force mg is directed in the positive direction, while the spring force KY and the damping force $D\dot{Y}$ both opposite the motion and, hence, are negatively directed. Then,

$$\sum F = M_{eq}\ddot{Y} \qquad (6)$$

or

$$mg - KY - D\dot{Y} = (M + m)\ddot{Y} \qquad (7)$$

Upon separating the variables,

$$(M + m)\ddot{Y} + D\dot{Y} + KY = mg \quad \text{(step input)}$$
$$\text{IC} \quad Y(0) = \dot{Y}(0) = 0 \qquad (8)$$

∎

■ **EXAMPLE 3.2.1-3** Phono Pickup with Sinusoidal Input

A phono pickup is a mechanical device that follows a groove cut in a record and actuates an electrical sensor to produce signals in accordance with

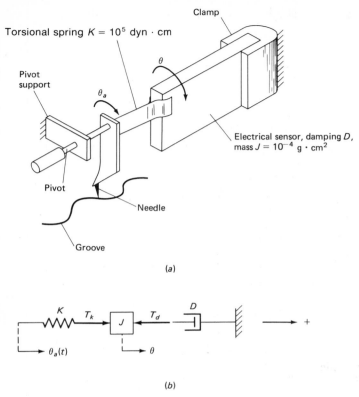

Figure 3.2.1-4   Phono pickup with sinusoidal input. (a) Actual system. (b) Free body diagram.

vibrations cut in the groove. A sharp needle follows the groove, which wavers from side to side, appearing as a displacement input θ_a. See Fig. 3.2.1-4(a). The needle is connected through an elastic rod K to the electrical sensor. Write the equation of motion (differential equation).

Solution. The groove represents sound, which for this problem will be approximated as a sinusoid. Since the needle is pivoted, the action of the needle following the groove will produce an angular displacement input. Although there will also be a torque associated with this input, it is clear that the motion of the needle will correspond to that of the groove. Hence, such torque is dependent upon the operation of the system, and is *not* an input. For convenience, we may start the problem (time = 0) with zero initial conditions.

The input displacement θ_a is transmitted through a spring (elastic rod) K that twists the electrical sensor. The sensor has mass J and structural damping D. Since the rectangular block is the sensor, and since it twists about the pivot, there may be some difficulty in defining the location of the mass J and damper D. We may arbitrarily locate the equivalent mass J_{eq} at point θ, in which case,

$$J_{eq} = \frac{J}{3} \tag{1}$$

The damping coefficient D is usually found experimentally, and it is usually measured at the pivot. As such, the value of D represents a damper that connects a mass J_{eq} at point θ to the ground. The free body diagram is shown in Fig. 3.2.1-4(b). Note that

$$T_k = K(\theta_a - \theta) \tag{2}$$

$$T_d = D\dot{\theta} \tag{3}$$

$$\theta_a = B \sin wt \tag{4}$$

Summing the torques on mass J, according to Newton's law, we have

$$\sum T = J\ddot{\theta} = K(\theta_a - \theta) - D\dot{\theta} \tag{5}$$

Apply eq. (4) and separate the variables.

$$J\ddot{\theta} + D\dot{\theta} + K\theta = KB \sin wt$$
$$\text{IC} \quad \theta(0) = \dot{\theta}(0) = 0 \tag{6} \blacksquare$$

■ **EXAMPLE 3.2.1-4** Recoil Landing System
Rocket transport vehicles, which usually land on terrestrial bodies, use a three-legged landing system. Each leg consists of a spring and damper within a tube. Refer to Fig. 3.2.1-5. Both the spring and damper are connected to the foot (foundation) of the device which has a large enough area not to sink into the ground as the vehicle compresses the landing device. At the instant of landing, the vehicle (mass M) has an initial velocity V_0. The objective of the landing device is to cushion the blow of landing without bouncing. As the vehicle lands, the spring will compress, and at the same time, the damper will

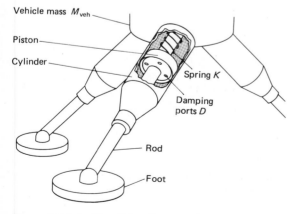

Figure 3.2.1-5   Recoil landing system.

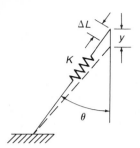

Figure 3.2.1-6 Spring compression for one leg.

absorb the energy. If the vehicle mass is M_{veh}, and if each of the three legs makes an angle θ with the vertical, model the elements of this system.

System Modeling

This system may be described as a single mass M supported on three springs and three dampers which are not aligned. In modeling the system, the three springs and the three dampers will be considered in groups, or clusters.

Assumptions

(a) Each foot remains in place during the landing.
(b) The deflections are small compared to the size of the elements.
(c) By virtue of symmetry, the elements will deform equally, and each will support the same load.
(d) The elements (springs and dampers) are linear.

It is convenient to model the system as a single mass supported by a single vertical spring and a single vertical damper in parallel. The effective spring constant K_e is easily obtained by means of energy methods. Refer to Fig. 3.2.1-6. If the angle θ changes only slightly during descent of the vehicle, the compression ΔL of each spring is related approximately to the decent y of the vehicle by

$$\Delta L = y \cos \theta \tag{1}$$

The energy stored in the three identical springs is

$$E_s = \tfrac{3}{2}[K \Delta L^2] = \left(\tfrac{3}{2}K \cos^2 \theta\right) y^2 \tag{2}$$

Thus, the effective spring constant is

$$K_e = 3K \cos^2 \theta \tag{3}$$

A similar argument using power dissipated by the three dampers leads to the effective damping constant

$$D_e = 3D \cos^2 \theta \tag{4}$$

The model for the system is described by the following differential equation and initial conditions. Summing forces (positive downward), we get:

$$M_e g - D_e \dot{y} - K_e y = M_e \ddot{y} \tag{5}$$

or

$$M_e \ddot{y} + D_e \dot{y} + K_e y = M_e g \quad \text{(a step input)} \tag{6}$$

$$\text{IC} \quad y(0) = 0, \qquad \dot{y}(0) = V_0$$

where M_e is the effective mass of the descending parts. The input is a step since, after the feet of the lander have touched (and do not bounce), the only force acting on the system is $M_e g$. A numerical solution for this system is presented later in Ex. 6.3.3-2. ∎

3.3 MODELING OF ELECTRICAL SYSTEMS

3.3.1 Kirchhoff's Laws

1. The algebraic sum of all the potential drops around a closed loop (or closed circuit) is zero (loop analysis).
2. The algebraic sum of all the currents flowing into a junction (or node) is zero (node analysis).

Loop Analysis   Refer to Fig. 3.3.1-1. Start at any point in the loop (it may be convenient to start at the voltage source V) and sum the voltage drops. These may be expressed either in terms of charge q or current I (noting that current I is equal to the time rate of change of charge), where

$$I = \dot{q} \tag{3.3.1-1}$$

In terms of charge q

$$V - L\frac{d^2q}{dt^2} - R\frac{dq}{dt} - \frac{1}{C}q = 0 \tag{3.3.1-2}$$

or

$$V - L\ddot{q} - R\dot{q} - \frac{q}{C} = 0 \tag{3.3.1-3}$$

In terms of current I

$$V - L\dot{I} - RI - \int \frac{I}{C}\, dt = 0 \tag{3.3.1-4}$$

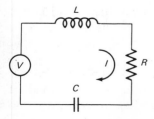

Figure 3.3.1-1 Single loop.

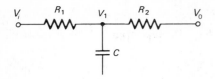

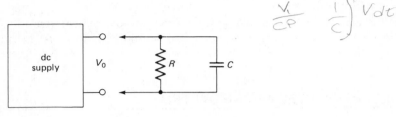

Figure 3.3.1-2   Summing of currents at a junction.

Figure 3.3.1-3   Capacitor discharge.

Node Analysis   The previous paragraph made use of the first of Kirchhoff's laws. When the loops are elaborately intertwined, it sometimes pays to make use of the summing of currents at a junction. Refer to Fig. 3.3.1-2. Summing the currents at the junction, or node,

$$\frac{V_1 - V_i}{R_1} + \frac{V_1 - V_0}{R_2} + \int C\frac{dV_1}{dt}\, dt = 0 \qquad [3.3.1\text{-}5]$$

■ **EXAMPLE 3.3.1-1** Capacitor Discharge
A capacitor is charged by placing its leads across a dc supply (voltage V_0). It is removed from the supply and it is discharged through a resistor R. Draw the circuit diagram and write the differential equation (see Fig. 3.3.1-3).

The dc supply provides an initial charge q_0 in the capacitor C. Since the supply is removed, it no longer serves as a voltage source. Hence, the system has zero input. Once the supply is removed, the capacitor C and resistor R form a closed loop. Using loop analysis, we have upon summing the voltage drops,

$$-\dot{q}R - \frac{q}{C} = 0 \tag{1}$$

$$\text{IC}\quad q(0) = q_0$$

3.3.2 Potentiometer

The potentiometer (also referred to as a "pot" or as a voltage divider) is a mechanical to electrical transducer. The device can be used either as a constant or variable multiplier depending upon whether the mechanical input is fixed or variable.

The operation of the device is shown in Fig. 3.3.2-1. A resistance wire is wound on an insulated core, and the two ends are brought out to terminals A and C. A wiper makes contact with the wire, thus dividing the resistance into

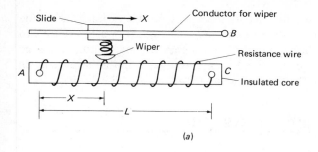

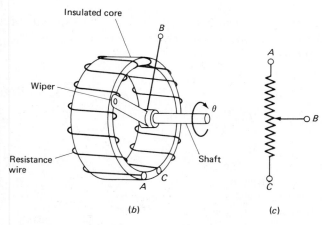

Figure 3.3.2-1   Potentiometer. (a) Translatory pot. (b) Rotary pot. (c) Pot notation.

two parts, whose resistances depend upon the location of the wiper. The wiper (which has negligible resistance) is brought to the wiper terminal B.

3.3.3 Integration and Differentiation

The integrator is designed to take the time integral of its input (multiplied by a preselected constant) whether the input is constant or variable. The electrical analog device makes use of the natural integrating property of the capacitor. Until the capacitor becomes saturated, the total charge accumulated is given by the time integral of the rate of charge into the capacitor. But the rate of

Figure 3.3.3-1   Electrical integration and differentiation. (a) Integrator. (b) Differentiator. (Both operations are approximate.)

charge is the current. The product RC is the time constant or integrator scale factor. By adjusting the value of the resistor R, the constant or scale factor can be preset. Selection of the capacitance C determines the magnitude of charge before saturation occurs (see Fig. 3.3.3-1(a). Electrical differentiation can be obtained by passing the signal through a capacitor [see Fig. 3.3.3-1(b).]

3.4 MODELING OF THERMAL SYSTEMS

3.4.1 Energy Balance

The principal relation governing transient heat transfer is that the time rate of change of the internal energy of a body is equal to the sum of the heat inputs. This relation can be written as:

$$C \frac{dT}{dt} = D \Delta T + D_v \Delta T \qquad [3.4.1\text{-}1]$$

where C = thermal capacity = mc
   m = mass of body
   c = specific heat
   D = heat dissipation (or conductance) = kA/L
   k = conductive heat coefficient
   A = surface area through which heat is transferred
   D_v = convective heat dissipation = hA
   h = convective heat transfer coefficient

3.4.2 Transient Heat Transfer

Equation 3.4.1-1 considers heat transferred by both conduction and convection. In many real cases, the actual heat transfer is a combination of both. However, for simple problems, it is often convenient to approximate the condition as purely one or the other.

■ **EXAMPLE 3.4.2-1** Transient Heat Conduction
A wooden cube (4 in. on each side) at temperature $T = 100°F$ is placed upon a huge steel slab at temperature $T_a = 20°F$. Model the system.

Solution.   Assume that the wooden cube is the body and that the only means of heat transfer is by conduction (neglect convection of the air). Also assume that the mass of and the conduction within the steel slab are both sufficiently high to consider the slab as an infinite sink. Then eq. 3.4.1-1 becomes

$$C \frac{dT}{dt} = D \Delta T \qquad (1)$$

or

$$mc \frac{dT}{dt} = kA \frac{\Delta T}{(L/2)} \qquad (2)$$

($L/2$ is the average path length for heat flux), where

$$m = \rho L^3 \tag{3}$$

$$A = L^2 = (4/12)^2 \text{ ft}^2 \tag{4}$$

$$\Delta T = T_a - T = 20°\text{F} - T \tag{5}$$

$$T(0) = 100°\text{F} \tag{6}$$

Using Appendix 6, we find that

$$\rho = \text{specific weight of wood} = 0.22 \text{ lb/in.}^3 \tag{7}$$

$$c = \text{specific heat of wood} = 0.62 \text{ Btu/lb} \cdot °\text{F} \tag{8}$$

$$k = \text{conductivity of wood} = 0.06 \text{ Btu/(h} \cdot \text{ft} \cdot °\text{F)} \tag{9}$$

Applying numerical values to eq. (2) and simplifying, we have

$$0.7936\dot{T} = 0.04 \, (20°\text{F} - T) \qquad T(0) = 100°\text{F} \tag{10}$$

The cube will always be warmer than the slab. Hence, the rate $\dot{T}$ will always be negative. This is as expected. It implies that the wooden cube is cooling. ∎

■ **EXAMPLE 3.4.2-2** Quenching in Infinite Sink
To harden a steel shaft, it is heated to a temperature $T(0) = 1300°\text{F}$ and then it is quenched in a huge water bath that is maintained at a temperature of $T_a = 75°\text{F}$. See Fig. 3.4.2-1. Assume that tests have been performed which determined that the convective heat transfer coefficient $h = 300 \text{ Btu/(h} \cdot \text{ft}^2 \cdot °\text{F)}$. Model the system.

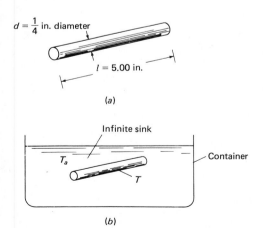

(a)

(b)

Figure 3.4.2-1   Quenching in infinite sink. (a) Shaft before quenching. (b) Quenching operation.

Solution.   Assume that the bath is large enough to be an infinite sink, and assume that the heat transfer is purely convective. Then

$$hA(T_a - T) = mc\dot{T} \tag{1}$$

$$T(0) = 1300°\text{F} \tag{2}$$

where

$$h = 300 \text{ Btu}/(\text{h} \cdot \text{ft}^2 \cdot °\text{F}) \tag{3}$$

$$A = \pi dl + 2\pi\left(\frac{d}{2}\right)^2 = 0.03 \text{ ft}^2 \tag{4}$$

$$m = \rho\pi\left(\frac{d}{2}\right)^2 l = 0.069 \text{ lb} \tag{5}$$

$$c = 0.11 \text{ Btu}/\text{lb} \cdot °\text{F} \tag{6}$$

Applying the numerical values to eq. (1),

$$9(75° - T) = 0.00759\dot{T} \tag{7}$$

$$T(0) = 1300°\text{F} \tag{8}$$

∎

■ EXAMPLE 3.4.2-3 Quenching in a Finite Bath (Coupled System)

Suppose that the shaft used in the previous problem is quenched in a small can containing 1 lb water rather than in a large vat. Naturally the bath temperature will be increased as a result of the quenching process. See Fig. 3.4.2-2.

Note that the system is coupled (shaft and bath affect each other). The quenching is a two-mass heat transfer system, therefore two heat balances are required, one for each mass.

The heat balance for the shaft is the same as that presented in the preceding problem, except that the bath temperature is now a function of time. Therefore:

$$\frac{m_s c_s}{hA}\dot{T} + T - T_b = 0 \tag{1}$$

$$T(0) = 1300°\text{F} \tag{2}$$

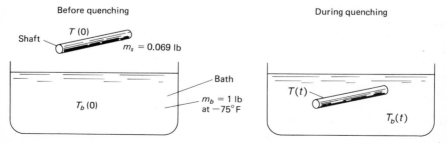

Figure 3.4.2-2   Quenching in finite bath.

And similarly for the water bath:

$$\frac{m_b c_b}{hA} \dot{T}_b + T_b - T = 0 \qquad (3)$$

$$T_b(0) = 75°F \qquad (4)$$

Equations (1) and (3) are simultaneous differential equations that model the coupled system.   ∎

3.5 MODELING OF FLUID SYSTEMS

3.5.1 Laminar Flow

In laminar, or streamline, flow,* the fluid moves in layers, each fluid particle following a smooth and continuous path. The fluid particles in each layer remain in an orderly sequence without passing one another. Soldiers on parade provide a somewhat crude analogy to laminar flow. They march along well-defined lines, one behind the other, and maintain their order even when they turn a corner or pass an obstacle.

3.5.2 Flow Rate in Terms of Head

Consider a reservoir or tank as shown in Fig. 3.5.2-1. The flow rate Q is given by eq. 2.6.2-1, repeated below.

$$Q = \frac{\Delta P}{g\rho R} \qquad [3.5.2\text{-}1]$$

The pressure P and "head"† h of fluid are related by

$$P = g\rho h \qquad [3.5.2\text{-}2]$$

Then

$$Q_{\text{out}} = \frac{h}{R} \qquad [3.5.2\text{-}3]$$

Note that eq. 3.5.2-3 is analogous to Ohm's law.

By virtue of continuity, the difference between the flow in and the flow out will cause the fluid level h to change. If the flow out is greater than the flow in, the level will drop, or the rate of level change will be negative. Then,

$$Q_{\text{out}} - Q_{\text{in}} = -A\frac{dh}{dt} \qquad [3.5.2\text{-}4]$$

Substitute eq. 3.5.2-3 into eq. 3.5.2-4 and arrange terms

$$\frac{h}{R} - Q_{\text{in}} = -A\frac{dh}{dt} \qquad [3.5.2\text{-}5]$$

---

*Reference [45], p. 310

†See eq. 2.6.4-7 for explanation of head.

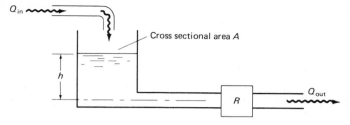

Figure 3.5.2-1   Laminar flow in fluid system.

Separate the variables and normalize

$$\frac{dh}{dt} + \frac{h}{AR} = \frac{Q_{in}}{A}$$

[3.5.2-6]

3.5.3 Fluid Circuit

See Fig. 3.6.2-1(e), Series fluid circuit (with pressure source, initial head h_0, and initial flow rate Q_0). In order to prove that the elements are in series, note the following:

(a) Flow rate Q must pass through each element.
(b) Pressure drops ΔP_r, ΔP_i, and ΔP_c add. Note that the capacitance is located at the end of the line in order to avoid the question of a capacitance in series.
(c) No inputs or outputs act at the junction between any of the elements.

Statement (a) becomes

$$Q_r = Q_i = Q_c = Q$$

[3.5.3-1]

Since $Q = \dot{W}$, it is possible to take either derivatives or integrals, providing various orders of W.

Statement (b) becomes

$$\Delta P = \Delta P_r + \Delta P_i + \Delta P_c$$

[3.5.3-2]

Substitute eqs. 2.6.2-1, 2.6.4-2, and 2.6.5-4 into eq. 3.5.3-2. Make use of eq. 3.5.3-1 and all of the derivatives of W

$$\Delta P = R'\dot{W} + I\ddot{W} + W/C$$

[3.5.3-3]

Now, consider the initial conditions. The initial height h_0 in the accumulator is known. However, the IC must be applied to the variable W. But volume is given by

$$W = hA$$

[3.5.3-4]

At time $t = 0$, this becomes

$$W(0) = h_0 A$$

[3.5.3-5]

In general,

$$\dot{W} = Q$$

At time $t = 0$, this becomes

$$\dot{W}(0) = Q_0 \qquad\qquad [3.5.3\text{-}6]$$

The complete answer is given by eqs. 3.5.3-3, 3.5.3-5 and 3.5.3-6.

A centrifugal pump is used for two reasons. First, the differential equation requires a pressure source. This, in turn, requires that the pressure ΔP is known no matter what the flow rate Q becomes. From experience we know this to be true of a centrifugal pump. Second, since this is a series circuit, if any element should become blocked, or if a valve should be turned off (closed tightly), the flow would stop. From experience, it is known that a centrifugal pump will continue to turn and it will continue to develop the pressure ΔP with no problems. This is not true for positive displacement pumps, such as those in Fig. 3.5.5-1(b) and (c), which would attempt to develop infinite pressure, resulting in damage to the system.

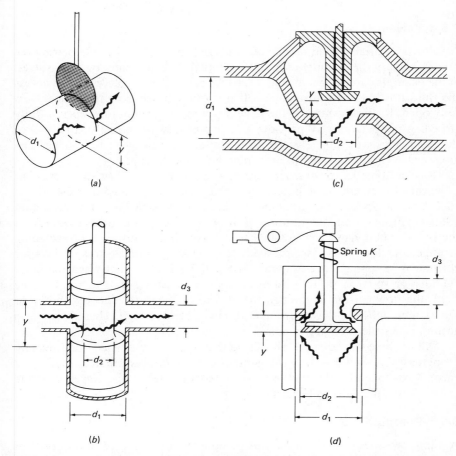

Figure 3.5.4-1   Valves. (a) Gate valve. (b) Spool valve. (c) Globe valve. (d) Poppet valve.

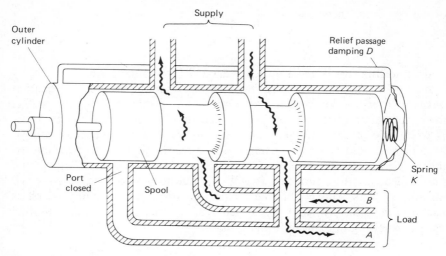

Figure 3.5.4-2   Reversing spool valve.

3.5.4 Valves

A valve is a variable fluid resistance. It is used to regulate the volumetric flow rate Q, to vary the pressure drop ΔP, or both. There are many types of valves. Some examples are shown in Fig. 3.5.4-1. A gate valve (a) places a disk across the fluid flow path. Its advantage is that it offers nearly zero resistance when fully open. Its disadvantage is the large travel required from fully open to fully closed. A poppet valve (d) provides a large passageway upon a small travel of the stem. Its disadvantage is the finite resistance even when fully open. The globe valve (c) is employed in most plumbing installations.

A spool-type valve (b) makes use of a cylinder or spool (with grooves cut in it) that fits snugly in an outer cylinder. In many cases the inner cylinder is solid. Such a device is relatively inexpensive to build, since it can be turned on a lathe. The shape and distribution of the grooves as well as ports to and from the outer cylinder make it possible to achieve almost any type of switching arrangement for fluids. This device is the counterpart of electrical relays for current. A reversing valve is shown in Fig. 3.5.4-2. Note that when the inner cylinder or spool is shifted (by means of the actuating rod) to the extreme right, the flow is from A through the load to B. When the spool is shifted to the extreme left, the flow is from B through the load to A. Hence, the valve accomplishes the reversing action.

All of these valves can be shaped so that as they are opened, the flow rate Q increases in such a way as to maintain a constant pressure drop across the valve. Consequently, the flow rate Q is proportional to the travel y of the valve. (Thus, the valve is linear.)

3.5.5 Pumps

A pump provides power to a fluid. A pump may be considered the opposite of a valve since the pump increases flow Q, pressure P, or both (see Fig. 3.5.5-1). A centrifugal pump (a) produces fluid pressure by centrifugal action. If the

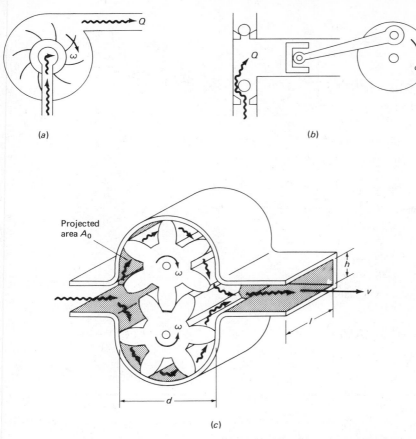

Figure 3.5.5-1 Pumps. (a) Centrifugal pump. (b) Reciprocating pump. (c) Gear pump.

fluid flow is blocked, the pump will turn, maintaining the pressure, without harm to either the fluid line or to the pump itself. By appropriately shaping and curving the blades, the device can sacrifice flow rate Q for pressure P. A reciprocating pump may use a piston (b), a diaphragm, or a bellows (Figs. 7.5.1-2 and 9.3.1-1). Note that all of these must employ check valves. The reciprocating pumps are called positive displacement pumps since they will not permit the fluid to slip by. They have a potential danger in that a fluid blockage will cause the pressure to rise without bound. Another positive displacement pump is the gear type (c). In this device, a pair of gears in a suitably shaped and fitted casing acts as a positive displacement pump. The casing of the device is shaped so that once the fluid has entered the chamber, it must pass around the outside section with the teeth of the gears. The gears fit tightly enough so that the fluid cannot pass through the center of the device. The teeth fit closely enough to the casing that the fluid cannot escape the cavity it occupies between teeth. Hence, it must be carried along until it can exit at the opposite side. A centrifugal pump may be considered as a pressure supply while a positive displacement pump may be considered as a flow rate supply.

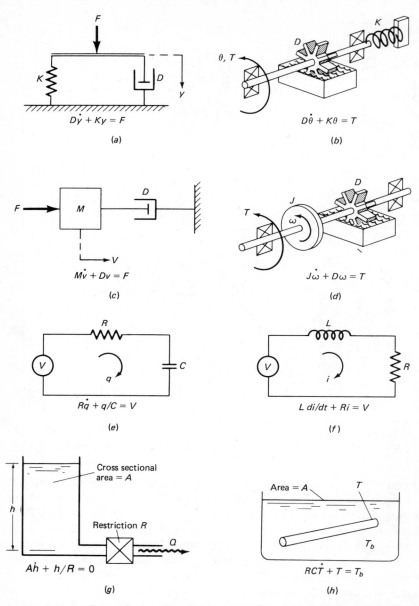

Figure 3.6.1-1   Some first order systems. (a) Translatory displacement. (b) Rotary displacement. (c) Translatory velocity. (d) Rotary velocity. (e) Electrical charge. (f) Electrical current. (g) Fluid flow through restriction. (h) Quenching in infinite bath.

3.6 ANALOGIES AMONG VARIOUS LUMPED SYSTEMS

3.6.1 First Order Systems

The generalized system equation for first order systems is given in general terms by

$$A\dot{\theta} + B\theta = T \qquad [3.6.1\text{-}1]$$

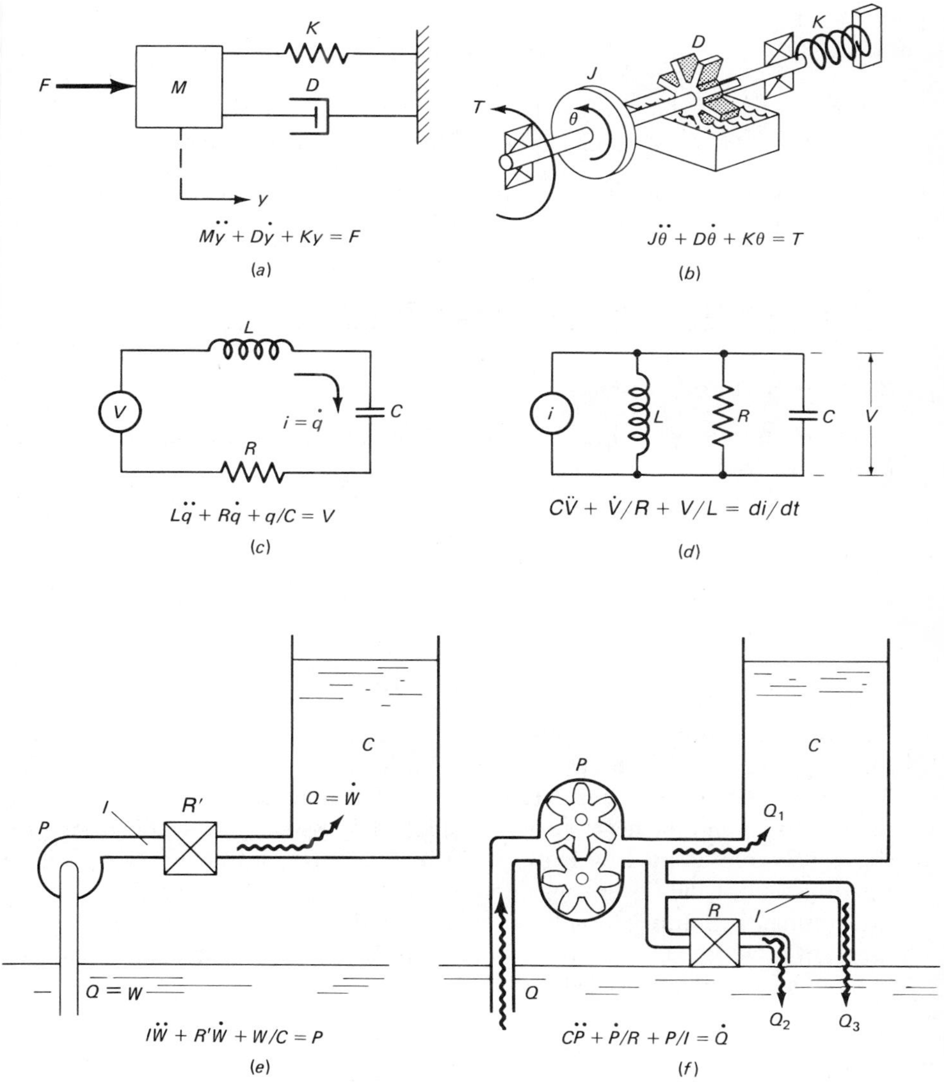
K
F
M
D
y
$M\ddot{y} + D\dot{y} + Ky = F$
(a)
K
D
J
T
θ
$J\ddot{\theta} + D\dot{\theta} + K\theta = T$
(b)
L
V
$i = \dot{q}$
C
R
$L\ddot{q} + R\dot{q} + q/C = V$
(c)
i
L
R
C
V
$C\ddot{V} + \dot{V}/R + V/L = di/dt$
(d)
C
P
I
R'
$Q = \dot{W}$
$Q = W$
$I\ddot{W} + R'\dot{W} + W/C = P$
(e)
P
C
Q_1
R
I
Q
Q_2
Q_3
$C\ddot{P} + \dot{P}/R + P/I = \dot{Q}$
(f)

Figure 3.6.2-1 Some second order systems. (a) Translating system. (b) Rotating system. (c) Series circuit. (d) Parallel circuit. (e) Series fluid circuit. (f) Parallel fluid circuit.

where

$$A, B = \text{constant coefficients} = \text{lumped elements} \qquad [3.6.1\text{-}2]$$

The configurations and the system equations for a number of lumped first order systems are tabulated in Fig. 3.6.1-1.

3.6.2 Second Order Systems

The generalized system equation for second order systems is given in general terms by

$$A\ddot{\theta} + B\dot{\theta} + C\theta = T \qquad [3.6.2\text{-}1]$$

where

$$A, B, C = \text{constant coefficients} = \text{lumped elements} \qquad [3.6.2\text{-}2]$$

The configurations and the system equations for a number of lumped second order systems are tabulated in Fig. 3.6.2-1.

3.7 BLOCK DIAGRAMS OF SYSTEM MODELS

3.7.1 First Order Systems

Consider the system in Fig. 3.6.1-1(a), for which the differential equation and initial condition are

$$D\dot{y} + Ky = F$$
$$\text{IC} \quad y(0) = y. \qquad [3.7.1\text{-}1]$$

Solving this for $\dot{y}$, we get

$$\dot{y} = \frac{F}{D} - \frac{K}{D}y \qquad [3.7.1\text{-}2]$$

Note that y is obtained by integrating $\dot{y}$, subject to the initial condition (IC) $y(0) = y_0$. The relation between y and $\dot{y}$ is given by the block diagram of Fig. 3.7.1-1(a). The value of $\dot{y}$ is determined by a combination of the input force F divided by D and the negative of y multiplied by K/D. In block diagram form this can be represented by Fig. 3.7.1-1(b). The arrows in the diagram indicate the direction of the flow of information, the boxes indicate operations such as integration or multiplication by a constant, and the circle indicates a summing junction in which signals are combined with the algebraic signs indicated. The block diagram is read as follows:

$$y = \int_0^t \dot{y}\, dt + y_0 \qquad [3.7.1\text{-}3]$$

$$\dot{y} = \frac{F}{D} - \frac{K}{D}y \qquad [3.7.1\text{-}4]$$

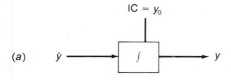

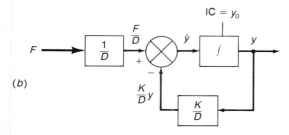

Figure 3.7.1-1   (a) Block diagram relating y to $\dot{y}$. (b) Complete block diagram.

or

$$y = \int_0^t \left[\frac{F}{D} - \frac{K}{D}y \right] dt + y_0 \qquad [3.7.1\text{-}5]$$

The latter equation is an integral equation whose solution is the same as that of the differential equation with its associated initial condition.

3.7.2 Second Order Systems

Consider the system of Fig. 3.6.2-1(a), for which the differential equation and initial conditions are

$$M\ddot{y} + D\dot{y} + Ky = F$$
$$\text{IC} \quad y(0) = y_0, \qquad \dot{y}(0) = v_0 \qquad [3.7.2\text{-}1]$$

Solving the equation for the highest derivative y, we get

$$\ddot{y} = \frac{F}{M} - \frac{D}{M}\dot{y} - \frac{K}{M}y \qquad [3.7.2\text{-}2]$$

The block diagram for the output y and input F is shown in Fig. 3.7.2-1.

Block diagrams such as these are very useful in computer simulation methods for solving the system differential equations that model the systems. Computer simulation will be discussed in Chapter 4.

3.7.3 Alternative Forms for Block Diagrams

The differential equation, eq. 3.7.1-1, can be written in terms of a differential operator $P = d/dt$ as follows:

$$(DP + K)y = F$$
$$\text{IC} \quad y(0) = y_0 \qquad [3.7.3\text{-}1]$$

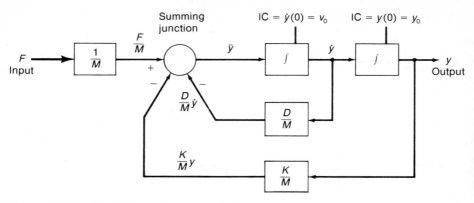

Figure 3.7.2-1   Block diagram for second order system.

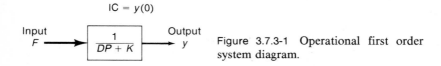

Figure 3.7.3-1   Operational first order system diagram.

Solving for y, we get

$$y = \left[\frac{1}{DP + K} \right] F \qquad [3.7.3\text{-}2]$$

The operator $1/(DP + K)$ is called the first order operator. The block diagram using this operator is shown in Fig. 3.7.3-1.

The second order system of Section 3.7.2 may be diagrammed in a similar manner. The operator form for the differential equation is

$$(MP^2 + DP + K)y = F \qquad [3.7.3\text{-}3]$$

The corresponding block diagram is shown in Fig. 3.7.3-2.

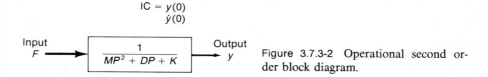

Figure 3.7.3-2   Operational second order block diagram.

SUGGESTED REFERENCE READING FOR CHAPTER 3

Initial conditions [8], p. 114; [17], p. 200–204.
Inputs [8], p. 88.
Free body diagrams [8], p. 17, 31; [17], p. 40.
Modeling of mechanical systems [8], p. 15–35; [12]; [17], p. 53–57; [47], chap. 2.
Modeling of electrical systems [8], p. 35–49; [12]; [17], p. 57–79; [47], chap. 3.
Modeling of fluid systems [8], p. 59–70; [12]; [17], p. 102–120; [47], chaps. 4, 5.
Modeling of thermal systems [8], p. 70–76; [12]; [17], p. 94–101; [36].
System analogies [1], p. 189; [8], p. 232, 233.
Modeling from actual system [17], p. 10–19.

PROBLEMS FOR CHAPTER 3

3.1. (a) Obtain the differential equation for H in Fig. P-3.1.
   (b) If $H(0) = H_0$, find an expression for the initial rate of change of H.

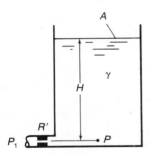

Figure P-3.1

3.2. Draw correct free body diagrams and derive the differential equation in Fig. 3.6.2-1(c). Show that your result becomes that given in Fig. 3.6.1-1(c) when $K = 0$.

3.3. The flow Q (m^3/s) through the orifice of Fig. P-3.3 in terms of pressures P_i (N/m^2) and P (N/m^2) is $Q = 0.008\sqrt{P_i - P}$.
   (a) Derive the differential equation for P.
   (b) Use eq. 2.6.4-1 to find C when $P = 2.0 \times 10^5$ and $P = 4.0 \times 10^5$.

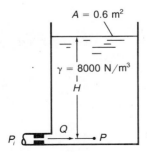

$A = 0.6$ m^2

$\gamma = 8000$ N/m^3

Figure P-3.3

3.4. Derive the differential equation for the output voltage e_0 in Fig. P-3.4. The input is a given current i_i.

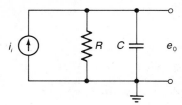

Figure P-3.4

3.5. For the thermal system in Fig. 3.4.2-1, the rod is made of aluminum with $L = 6.0$ in. and $d = 0.30$ in. Assume that heat is transferred by convection with $h = 300$ Btu/h · ft² · °F, the initial temperature is 800°F, and the bath temperature is 75°F, constant. Obtain the differential equation for the rod temperature T (°F).

3.6. Find the differential equation for P in the fluid system shown.

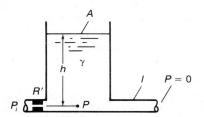

Figure P-3.6

3.7. A solid uniform circular disk of mass 4.0 kg and radius 0.24 m spins on its shaft at $\omega_0 = 50$ rad/s. At $t = 0$, a clutch engages a rotational damper with a damping constant $D_t = 0.28$ N · m · s.
(a) Obtain the differential equation for the angular velocity.
(b) Determine the angular acceleration α of the disk at $t = 0$.

3.8. Obtain the differential equation for e_0 for the circuit shown. The input is a current source i_i. *Hint*: first find the differential equation for i_R.

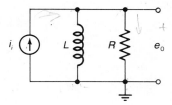

Figure P-3.8

3.9. In Fig. P-3.9, the current i supplied by the source varies with time according to i (A) = 0.2 + 0.3t [t (s)]. Given $R = 20$ Ω and $L = 5.0$ H:

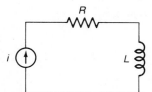

Figure P-3.9

(a) Find the source voltage at $t = 2$ s.

(b) Find the energy stored in the inductor L at $t = 2$ s.

3.10. For the system shown, derive the differential equation for the displacement x of the mass. Assume $D_2 = 4D_1$.

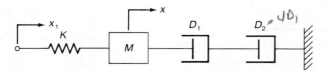

Figure P-3.10

3.11. Derive the differential equation for the output voltage e_0 for the circuit shown. The input i_i is a given current.

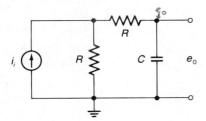

Figure P-3.11

3.12. For the system shown, K, D, and L are given, and AB is a rigid rod of negligible mass. Derive the differential equation for x_B, assuming small angles of rotation of AB. The input is x_1.

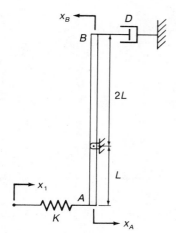

Figure P-3.12

3.13. Derive the differential equation for P, the pressure at the bottom of the tank in Fig. P-3.13. The fluid inertance is I, and the pipe resistance is R', lumped at the right end.

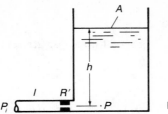

Figure P-3.13

3.14. Draw the appropriate free body diagrams for Fig. 3.6.1-1(a), and derive the differential equation. The rigid member with displacement y remains horizontal and has negligible mass.

3.15. Given $M = 0.6$ kg, $D = 4.0$ N $\cdot$ s/m, and x (m) $= 0.1t^2 + 0.04t^3$ [t (s)]. Find f (N) at $t = 0.5$ s. (See Fig. P-3.15.)

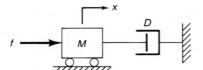

Figure P-3.15

3.16. Derive the differential equation for the angular velocity ω of the flywheel in Fig. P-3.16. The input is the angular velocity ω_i.

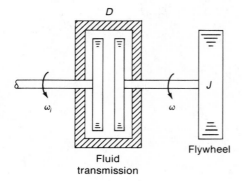

Fluid
transmission

Flywheel

Figure P-3.16

3.17. Draw the necessary free body diagrams and derive the differential equation for the rack displacement x, assuming f, K, D, M, J, and R to be given. (See Fig. P-3.17.)

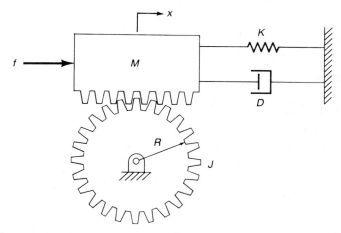

Figure P-3.17

3.18. Derive the differential equation for x in Fig. P-3.18. The input is a given displacement x_i. Correct free body diagrams are required.

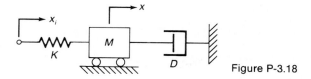

Figure P-3.18

3.19. Consider the power mower blade system shown. The blade has a rotational inertia J, and the air drag between the blade and the housing is a rotational damping D. The shaft has a torsional spring constant K. Derive the differential equation for θ.

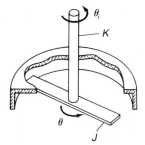

Figure P-3.19

3.20. The tank shown, containing fluid of specific weight γ, with depth h, is drained through an orifice for which the volume flow rate q is given by $q = K\sqrt{p/\rho}$, where K is a constant and ρ is the mass density.

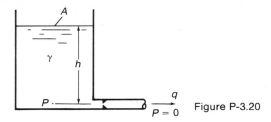

Figure P-3.20

(a) Obtain the (nonlinear) differential equation for h.
(b) Set up an integral from which one could obtain the time for the level of the liquid to drop from an initial depth h_0 to another depth.

3.21. Model the vehicle with the spring-damper bumper of Ex. 3.2.1-1 assuming that the damper is nonlinear, with the damper force $F_d = cV^2$.

3.22. A metal block of mass $M = 4.0$ lb$_m$ and specific heat $c = 0.20$ Btu/lb$_m \cdot$ °F is initially at temperature $T(0) = 40$°F. (See Fig. P-3.22.) The block is insulated except where it contacts the thermal resistance $R_t = 0.25$°F $\cdot$ h/Btu. The input is a given temperature T_a. Derive the differential equation for T.

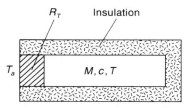

Figure P-3.22

3.23. The system shown is started from rest with a displacement x_0.
(a) Derive the differential equation for x, the displacement of the mass.
(b) From the differential equation find the initial acceleration of the mass.
(c) If instead of starting from rest with displacement x_0, assume the mass is given an initial velocity v_0 to the right when $x = 0$. Find the initial acceleration of the mass.

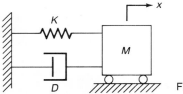

Figure P-3.23

3.24. *Quenching of Hot Metal Parts.* Refer to Ex. 3.4.2-3. A cooling bath contains water initially at a temperature of 75°F. Hot metal parts are quenched in various ways. Model each of the following systems and determine the steady state temperature:
(a) Bath same as that in the example. Rod is aluminum, diameter $= \frac{1}{2}$ in., length $= 5$ in., initial temperature $= 1000$°F.
(b) Bath has 2 lb of water. Rod same as that in part (a).
(c) Bath and rod same as those in part (b). Container is made of 0.2 lb of aluminum.

(d) Bath same as that in part (a). Nine aluminum rods each like that in part (a) and two steel rods each like that in Ex. 3.4.2-3 are dropped into the bath simultaneously.

(e) Bath and container of part (c) with the nine aluminum rods and two steel rods of part (d).

3.25. For each of the following problems define the system by drawing free body diagrams, identify the system elements, define the response variables, write the differential equation of motion, and determine order of system: (unless otherwise noted IC = 0 and shafts are rigid).

(a) A vehicle of mass M, moving horizontally on a frictionless plane with a velocity $V(t)$, experiences an air drag force DV. Input force = 0, $V(0) = V_0$, response = V.

(b) A bomb of mass M is dropped vertically and experiences a drag force DV^2. Input force = Mg, $V(0) = 0$, response is V.

(c) A bomb of mass M is dropped vertically and experiences an air drag of DV^2. Input force = Mg, $V(0) = V_0$, $x(0) = 0$, response is x.

(d) An industrial sewing machine (M) is mounted on springs (k) and dampers (D) to decrease the force transmitted to the floor. The input force = $F(t)$, $\dot{y}(0) = 0$, response is y. Write two differential equations, one for $y(0) = 0$ when the spring force is zero and the other for $y(0) = 0$ when the spring force equals Mg.

(e) A computer (M) is mounted on springs (k) and dampers (D) to reduce its motion (y) caused by the floor motion $y_a(t)$. Input motion = $y_a(t)$, $y(0) = \dot{y}(0) = 0$, response is y. (Define $y(0) = 0$ when spring force = Mg.)

(f) A motor with a gear attached (total inertia J_1) drives a pinion and a flywheel (total inertia J_2). The gear has N_1 teeth and the pinion has N_2 teeth. The flywheel experiences a viscous drag torque of $D\omega_2$. Input torque = $T(t)$, $\omega_2(0) = 0$, response is ω_2.

(g) A motor with a pinion (total rotary inertia J) drives a rack (mass M) that actuates a spool valve (shown in Fig. 3.5.4-2) that is restrained by a spring K. See Fig. P-3.25(g). The valve acts as an equivalent damper D. The pinion radius is R. Input torque from motor = $T(t)$, $x(0) = \dot{x}(0) = 0$, response is rack motion x.

(h) A simple pendulum (mass M and length L) is subjected to large deflections, θ, and to a drag force acting on the mass of DV^2, where V is the tangential velocity. Input force = Mg, $\theta(0) = \theta_0$, $\dot{\theta}(0) = 0$, response = θ.

(i) A cantilever beam is to be modeled with lumped masses, M_1 and M_2, and massless springs, k_1 and k_2. Input forces = F_1 and F_2, $x_1(0) = \dot{x}_1(0) = x_2(0) = \dot{x}_2(0) = 0$, response = x_1 and x_2. x_1 and x_2 are measured from static equilibrium positions of M_1 and M_2.

(j) A U-tube manometer, cross-section area A, contains a fluid of density ρ. A pressure P is suddenly applied to one leg of the tube. The response is the displacement y of the top of the column of fluid from its equilibrium condition. See Fig. P-3.25(j).
(1) Consider inertial effects only.
(2) Consider viscosity effects only (μ is fluid absolute viscosity).
(3) Consider both inertia and viscous effects.

(k) A fluid transmission can be represented by a damper D that drives a load (moment of inertia J) and a spring K. The response is angular displacement θ. See Fig. P-3.25(k).

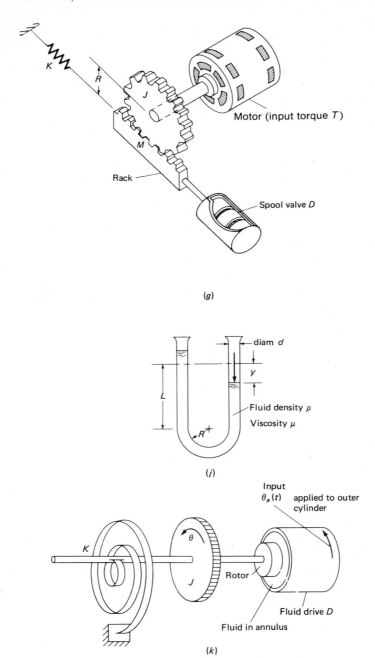

(g)

(j)

(k)

Figure P-3.25 (g) Motor-actuated spool valve. (j) "U" tube manometer. (k) Fluid transmission.

3.26. *Aircraft Landing Gear*. (See Fig. P-3.26.) An aircraft landing gear absorbs the shock of landing by means of a spring and damper. The damper consists of two pistons that are fastened to a long rod that is attached to the landing wheel. The cylinder of the damper is attached to the aircraft frame. Assume that the vertical displacement of the aircraft frame, y, is zero when the landing gear is fully extended (spring force is zero) and the wheel has just touched the runway.
  (a) Draw the free body diagram.
  (b) Discuss each component.
  (c) Write the differential equation.

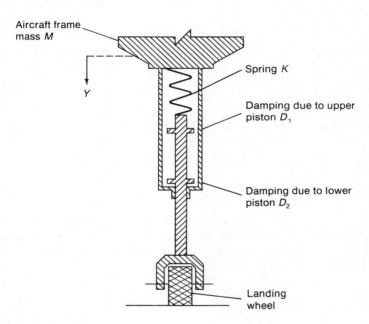

Figure P-3.26

3.27. *Stepped Propeller Shaft*. (See Fig. P-3.27. A long propeller shaft is stepped. The elasticity of each portion can be expressed as springs K_1 and K_2. Assume that the propeller is torsionally rigid and that it has a moment of inertia J. Let

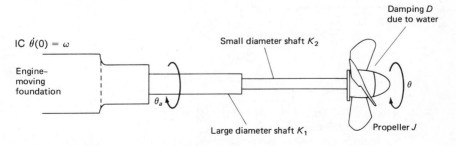

Figure P-3.27

damping due to the water be D. Assume that the engine vibration can be expressed as angular displacement input θ_a.
(a) Draw the free body diagram.
(b) Discuss each component.
(c) Show that the two springs K_1 and K_2 are in series.
(d) Write the differential equation.

3.28. For the fluid system shown, write the differential equation with IC in terms of system geometry as given in the figure. Initially, the tank is filled to level h and the fluid has velocity v in the pipe. Neglect inertance in the porous material, but include it in the pipe.

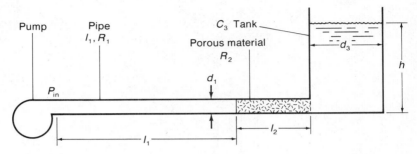

Figure P-3.28

3.29. For the mechanical system shown, write the differential equation with IC in terms of system geometry as given in the figure. Initially, the rotor has been turned through an angle ϕ and is turning at rate ψ: Neglect mass of shaft and torsion spring; include damping and inertia of fluid.

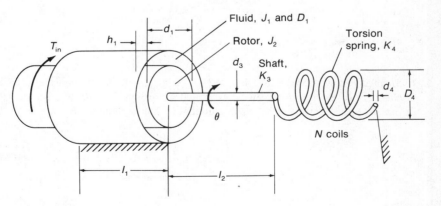

Figure P-3.29

3.30. Refer to Fig. 3.6.1-1(a). Perform the following:
(a) Derive the equation.
(b) Include the initial condition and explain its physical significance.
(c) Add motion of the ground or foundation $y_a(t)$ to the figure, let $F = 0$, and model the system.

3.31. Refer to Fig. 3.6.1-1(b). Perform the following:
  (a) Derive the equation.
  (b) Include the initial condition an explain its physical significance.
  (c) Add rotation $\theta_a(t)$ of the fluid container (in which the paddle wheel operates) to the figure and model the resulting system. (Let $T = 0$.)

3.32. Refer to Fig. 3.6.1-1(c). Perform the following:
  (a) Derive the equation.
  (b) Explain why this is a first order system even though the force applied to the mass requires a second derivative.
  (c) Let $F = 0$, and let the ground or foundation rotate at a constant rate. Model the resulting system.

3.33. Refer to Fig. 3.6.1-1(d). Perform the following:
  (a) Derive the equation.
  (b) Indicate the initial condition and suggest how it might be applied to the physical system.
  (c) Using the pot shown in Fig. 3.3.2-1, show how the displacement θ might be measured.
  (d) Same as Problem 3.32(b).
  (e) Show how to measure $\theta_a - \theta$ in Problem 3.31(c).

3.34. Refer to Fig. 3.6.1-1(e). Perform the following:
  (a) Derive the equation.
  (b) Include the initial condition and explain its physical significance.
  (c) If the massless plate to which force F is applied in Fig. 3.6.1-1(a) and the ground form the plates of capacitor C in Fig. 3.6.1-1(e), and if current flow does not affect the motion, model the system.

3.35. Refer to Fig. 3.6.1-1(f). Perform the following:
  (a) Model the system including the initial condition.
  (b) If the inductance L is due to the configuration shown in entry (1) of Table 2.4.1-1, Appendix 2, and if the horseshoe-like magnetic structure can bend (spring rate K) and if it has equivalent mass M, model the system with an input force F.
  (c) Draw a mechanical system that is an exact analog of the system shown in Fig. 3.6.1-1(f).

3.36. Refer to Fig. 3.6.1-1(g). Perform the following:
  (a) Including the initial condition, model the system.
  (b) Model the system in terms of capacitance C and pressure P.
  (c) Let there be a paddle wheel damper on a vertical axis that uses the liquid in the tank as a viscous damper. As the liquid level changes, the damping coefficient D will change proportionately. Place the system shown in Fig. 3.6.1-1(d) into the tank as indicated and model the system using ω as the response variable.

3.37. Refer to Fig. 3.6.1-1(h). Perform the following:
  (a) Including the initial condition, model the system.
  (b) For certain electrical resistance wire, the resistivity ρ used in eq. 2.4.1-1 is proportional to temperature. Show how this could be used to measure the temperature of the bath.

3.38. Refer to Fig. 3.6.2-1(a). Perform the following:
  (a) Show that the three elements are in parallel.

(b) Indicate how to use the pot shown in Fig. 3.3.2-1 to measure y.

(c) If $F = 0$ and the ground or foundation moves $y_a(t)$, model the resulting system.

(d) Show how the circuit in Fig. 3.6.2-1(d) can be used to study the mechanical system in Fig. 3.6.2-1(a).

3.39. Refer to Fig. 3.6.2-1(b). Perform the following:

(a) Model the system including initial conditions.

(b) Let $T = 0$, and let the paddle wheel container be driven by a motor whose instantaneous displacement is θ_a. Model the resulting system.

3.40. Refer to Fig. 3.6.2-1(c). Perform the following:

(a) Derive the equation and include the initial conditions.

(b) Model the system in terms of current i.

3.41. Refer to Fig. 3.6.2-2(d). Perform the following:

(a) Model the system including initial conditions.

(b) Model the system in terms of current i.

3.42. Refer to Fig. 3.6.2-1(e). Perform the following:

(a) Model the system including initial conditions.

(b) Why is a centrifugal pump used in this model?

(c) Model the system including inertance of the fluid in the tank.

(d) Why is the fluid circuit not closed?

(e) Employ the "ground" symbol in Figs. 3.6.2-1(c) and (e) and list all of the analogous terms. How can the electrical circuit be used to study the fluid system?

3.43. Refer to Fig. 3.6.2-1(f). Perform the following:

(a) Model the system including initial conditions.

(b) Why is a fluid gear pump shown instead of a centrifugal pump?

(c) Show that the systems shown in Figs. 3.6.2-1(a), (b), (d), and (f) are exact analogs.

Chapter **4**

Introduction to Solution Methods

The previous chapters were dedicated to setting up the equations of motion. This chapter introduces some of the methods for solving these equations.

The subject of Laplace transform is a powerful tool in the treatment of linear lumped systems. Following a formal but brief definition, some examples are worked out, having been selected to formulate a useful table of Laplace transforms. This table is then used to solve several problems involving differential equations. The approach is deliberately set in a low key, motivating the reader to explore further. As part of the implanted curiosity, some familiar objects are expressed in Laplace notation. In this way, the reader does not merely learn the mechanics, but begins to develop a comfortable feeling about Laplace transforms.

Some simple differential equations are solved. Then a short cut to determine the steady state response is presented. For the more sophisticated student or for the more advanced course, the method of partial fractions is shown.

4.1 DYNAMIC SYSTEMS

4.1.1 Static and Dynamic Systems

The word *dynamic* indicates change, and a dynamic system is one whose behavior changes. The system output is one quantity that may be associated with system behavior. Thus, a dynamic system has an output that changes. This text is primarily concerned with dynamic systems. Usually, dynamic systems involve differential equations.

Dynamic System of Order *N* A dynamic system may be modeled by either a single or a set of differential equations. If the system can be modeled by a *single* differential equation, then the order of the system is given by the order of the highest derivative (provided there are no integral terms). When the system is modeled by a *set* of equations, then the order of the system is higher than the order of the highest derivative. For many linear systems, the system order is given by the sum of the highest derivative in each differential equation.

Linear System A system is linear if its set of equations is linear. The equations in turn are linear if all the operations upon the *dependent* variable are linear. One simple means to test for linearity involves the distributive property. If

$$f(x + y) = f(x) + f(y)$$

then the function f is linear. The derivative, integral, and multiplication by a constant are linear, while roots, powers, and trigonometric and exponential terms are nonlinear.

Nonlinear Systems and Computer Solutions Nonlinear systems, particularly those containing derivatives or integrals, are usually cumbersome to solve. In most cases, it requires the computer for solution.

Some exercises and problems involving the use of computers are presented in the Problems section of this chapter.

4.1.2 Linear Differential Equation with Constant Coefficients

Of great importance in many engineering problems is the linear differential equation with constant coefficients

$$\frac{d^n x}{dt^n} + a_1 \frac{d^{n-1} x}{dt^{n-1}} + \cdots + a_{n-1} \frac{dx}{dt} + a_n x = f(t)$$

where a_i = constant coefficient

$f(t)$ = forcing function (function of the independent variable)

n = order of the equation

The linear differential equation is important because it represents many practical systems, and it is relatively simple to solve.

Normalizing A differential equation is "normalized" if the coefficient of its highest derivative is unity. Normalizing can be accomplished by dividing through by the actual coefficient if it is not unity. In addition to normalizing by division by the coefficient of the highest derivative, it is often useful to use dimensionless ratios where possible.

4.2 SINUSOIDAL MOTION

4.2.1 Sum of Sinusoids of Same Frequency but Not Same Phase

One method by which to sum two sinusoids, each having some phase angle with respect to a common datum, makes use of a mechanical analogy for the construction of sinusoids. Imagine a peg mounted on a disk that is rotating at an angular velocity ω. Now view the peg by looking in the plane of (or along the surface of) the disk. The peg would appear to oscillate along a straight line. This is the projection of the actual circular motion of the peg. If the peg is set *on* the horizontal axis, as shown on Fig. 4.2.1-1(a), then the motion will describe a *sine* curve for which the amplitude, $X_{max} = A$, is equal to the radius arm A and the frequency is ω. If the peg is *offset* by an angle ϕ, as shown on Fig. 4.2.1-1(b), the motion will *lead* the sine curve by a phase angle ϕ. The radius of the disk (or the radial distance from the center of the disk) to the peg provides the amplitude of the sinusoid. Thus, the polar coordinates of the peg and the angular velocity of the disk represent all the information concerning the sinusoid.

Now we make use of the rotating disk analogy. The various sinusoidal curves are represented by vectors on the rotating disk. Each has its magnitude and phase identified, but all vectors rotate together like a rigid body. Since there is no relative motion between these vectors, the rule for summing out-of-phase sinusoids is readily obtained by taking the vector sum of their rotating disk analogies. Note that the vector resultant also lies on the rotating disk. Hence it represents the resulting sinusoid. Determine its magnitude and direction in order to obtain the amplitude and phase of the resultant. See Fig. 4.2.1-2.

The magnitude of the result C can be found by the law of cosines

$$C^2 = A^2 + B^2 - 2AB \cos(\pi - \phi) \qquad [4.2.1\text{-}1]$$

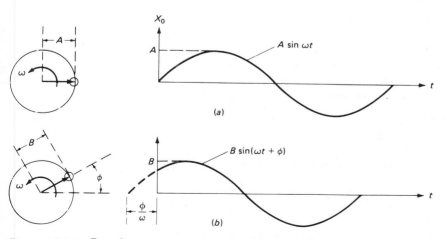

Figure 4.2.1-1   Rotating vector representation for sinusoids. (a) Rotating disk analogy for sine curve. (b) Rotating disk analogy for out of phase sinusoid.

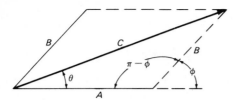

Figure 4.2.1-2 Vector sum of two out-of-phase sinusoids.

The phase angle θ for this can be found by the law of sines

$$\frac{\sin \theta}{B} = \frac{\sin(\pi - \phi)}{C} = \frac{\sin \gamma}{A} \qquad [4.2.1\text{-}2]$$

For the special case of the sum of sin and cos

$$A \sin \omega t + B \cos \omega t = C \sin(\omega t + \theta) \qquad [4.2.1\text{-}3]$$

where

$$C = \sqrt{A^2 + B^2}$$
$$\theta = \sin^{-1}\frac{B}{C} = \tan^{-1}\frac{B}{A} \qquad [4.2.1\text{-}4]$$

4.3 INTRODUCTION TO LAPLACE TRANSFORMS

The Laplace transform technique is an algebraic means for solving differential equations. It includes the initial conditions and the forcing function, and presents fairly straightforward means of including some fairly complicated forcing functions.

4.3.1 Definition of the Laplace Transform

The Laplace Transform $\mathscr{L}$ is defined by the following integral

$$\mathscr{L}f(t) = \int_0^\infty e^{-st}f(t)\, dt = F(s) \qquad [4.3.1\text{-}1]$$

where $f(t)$ = function in time domain
$F(s)$ = function in Laplace domain
s = operator in Laplace domain

Using the above definition, we shall proceed to derive the algebra of the Laplace domain, and also formulate a table of transforms.

(a) Laplace Transform of a Sum

$$\mathscr{L}[f(t) + g(t)] = \int_0^\infty e^{-st}[f(t) + g(t)]\, dt$$
$$= \int_0^\infty e^{-st}f(t)\, dt + \int_0^\infty e^{-st}g(t)\, dt$$
$$= \mathscr{L}f(t) + \mathscr{L}g(t) = F(s) + G(s) \qquad [4.3.1\text{-}2]$$

(b) Laplace Transform of Function with Constant Multiplier

$$\mathscr{L}af(t) = \int_0^\infty e^{-st}af(t)\,dt$$

$$= a\int_0^\infty e^{-st}f(t)\,dt$$

$$= a\mathscr{L}f(t) = aF(s) \qquad [4.3.1\text{-}3]$$

(c) Laplace Transform of First Derivative

$$\mathscr{L}\frac{df(t)}{dt} = \int_0^\infty e^{-st}\frac{df(t)}{dt}\,dt$$

If the function $f(t)$ and its derivative are continuous in the interval, the above may be integrated by parts

$$\int_0^\infty e^{-st}\frac{df(t)}{dt}\,dt = e^{-st}f(t)\Big|_0^\infty + s\int_0^\infty e^{-st}f(t)\,dt$$

$$= sF(s) - f(0) \qquad [4.3.1\text{-}4]$$

(d) Laplace Transform of Second Derivative

$$\mathscr{L}\frac{d^2f(t)}{dt^2} = \int_0^\infty e^{-st}\frac{d^2f(t)}{dt^2}\,dt \quad \text{(again integrate by parts)}$$

$$= e^{-st}\frac{df(t)}{dt}\Big|_0^\infty + s\int_0^\infty e^{-st}\frac{df(t)}{dt}\,dt = \dot{f}(0) + s\mathscr{L}\frac{df(t)}{dt}$$

$$= s^2F(s) - \dot{f}(0) - sf(0) \qquad [4.3.1\text{-}5]$$

(e) Laplace Transform of Exponential

$$\mathscr{L}e^{at} = \int_0^\infty e^{-st}e^{at}\,dt = \int_0^\infty e^{-(s-a)t}\,dt = \frac{1}{s-a}$$

or

$$\mathscr{L}e^{-at} = \frac{1}{s+a} \qquad [4.3.1\text{-}6]$$

(f) Laplace Transform of Sine

$$\mathscr{L}\sin at = \int_0^\infty e^{-st}\sin at\,dt = \frac{a}{s^2+a^2} \qquad [4.3.1\text{-}7]$$

(g) Laplace Transform of Cosine

$$\mathscr{L}\cos at = \int_0^\infty e^{-st}\cos at\,dt = \frac{s}{s^2+a^2} \qquad [4.3.1\text{-}8]$$

Using the results thus found in eqs. 4.3.1-2 through 4.3.1-8, a short table of

TABLE 4.3.1-1   LAPLACE TRANSFORM PAIRS

Laplace domain $\mathscr{L}$	Time domain t	Entry
$F(s) + G(s)$	$f(t) = g(t)$	1
$aF(s)$	$af(t)$	2
$sF(s) - f(0)$	$\dfrac{df(t)}{dt}$	3
$s^2F(s) - sf(0) - \dot{f}(0)$	$\dfrac{d^2f(t)}{dt^2}$	4
$\dfrac{1}{s + a}$	e^{-at}	5
$\dfrac{a}{s^2 + a^2}$	$\sin at$	6
$\dfrac{s}{s^2 + a^2}$	$\cos at$	7

Laplace transform pairs may be compiled. See Table 4.3.1-1. A more complete table appears in Appendix 3.

The power of the Laplace transform technique resides in its definition. It develops, as shown above, that all functions transform to linear combinations of the operator s raised to integer powers. The transformation includes the initial conditions and the forcing function. Thus, the entire problem in differential equations becomes a single problem in algebra.

■ **EXAMPLE 4.3.1-1** Solution to First Order System
Solve the following first order system:

$$\frac{dx}{dt} + ax = 0 \tag{1}$$

$$\text{IC} \quad x(0) = X_0$$

The transformation of eq. (1) will involve a sum. Making use of entry 1 of Table 4.3.1-1, the indicated operations are

$$\mathscr{L}\frac{dx}{dt} + \mathscr{L}ax = \mathscr{L}0 = 0 \tag{2}$$

For the first term, apply entry 3

$$\mathscr{L}\frac{dx}{dt} = sX - X_0 \tag{3}$$

For the second term of eq. (2), make use of entry 2

$$\mathscr{L}ax = aX \tag{4}$$

Thus, the differential equation in the Laplace domain is

$$sX - X_0 + aX = 0 \tag{5}$$

or by using the distributive property,

$$(s + a)X = X_0 \tag{6}$$

Solve algebraically for X [use the notation $X(s)$]

$$X(s) = \frac{X_0}{s + a} \tag{7}$$

The solution in the time domain is found by using entry 5

$$\boxed{x(t) = X_0 e^{-at}} \tag{8}$$

■

■ **EXAMPLE 4.3.1-2** Solution to Second Order System
Solve the following differential equation using Laplace transforms

$$M\ddot{x} + Kx = 0$$
$$\text{IC} \quad x(0) = X_0, \qquad \dot{x}(0) = V_0 \tag{1}$$

For the first term, use entry 4 of Table 4.3.1-1

$$\mathscr{L}M\ddot{x} = M(s^2 X - sX_0 - V_0) \tag{2}$$

For the second term, use entry 2

$$\mathscr{L}Kx = KX \tag{3}$$

Sum eqs. (2) and (3) according to entry 1, and the differential equation in the Laplace domain becomes

$$Ms^2 X - MsX_0 - MV_0 + KX = 0 \tag{4}$$

Using the distributive property, solve for $X(s)$

$$X(s) = \frac{MsX_0 + MV_0}{Ms^2 + K} = \frac{sX_0 + V_0}{s^2 + a^2} \tag{5}$$

where

$$a = \sqrt{\frac{K}{M}} \tag{6}$$

We can anticipate difficulty in proceeding further since the numerator consists of a sum of two independent terms. Thus, break up the two terms (note that it is algebraically correct to break up the numerator, but not the denominator). Then

$$X(s) = X_0 \frac{s}{s^2 + a^2} + V_0 \frac{1}{s^2 + a^2} \tag{7}$$

The first term has the form of entry 7, while the second term lacks the constant a in the numerator.* Thus, the solution is

$$x(t) = X_0 \cos at + \frac{V_0}{a} \sin at \tag{8}$$

Combine cos and sin according to eqs. 4.2.1-3 and 4.2.1-4.   ∎

■ **EXAMPLE 4.3.1-3** Solution to Simultaneous Differential Equations (Coupled System)
Solve the following set of equations.

$$\frac{dx}{dt} + ay = 0 \tag{1}$$

$$\frac{dy}{dt} - bx = 0 \tag{2}$$

$$\text{IC} \quad x(0) = X_0, \qquad y(0) = Y_0 = 0$$

Using Table 4.3.1-1, transform eqs. (1) and (2) into the Laplace domain.

$$sX - X_0 + aY = 0 \tag{3}$$

$$sY - Y_0 - bX = 0 \tag{4}$$

Solve algebraically for $X(s)$ in eq. (4) noting that $Y_0 = 0$.

$$X(s) = \frac{sY}{b} \tag{5}$$

Apply eq. (5) to eq. (3), and transpose X_0

$$s\left(\frac{sY}{b}\right) + aY = X_0 \tag{6}$$

Solve algebraically for $Y(s)$

$$Y(s) = \frac{bX_0}{s^2 + ab} \tag{7}$$

In order to transform back to the time domain, make use of entry 6

$$y(t) = \frac{bX_0}{c} \sin ct \tag{8}$$

where

$$c = \sqrt{ab} \tag{9}$$

---

*As compared to entry 6.

Apply eq. (7) to eq. (5)

$$X(s) = \frac{sX_0}{s^2 + ab} \tag{10}$$

To transform back to the time domain, use entry 7

$$\boxed{x(t) = X_0 \cos ct} \tag{11} \blacksquare$$

4.3.2 Laplace Transform of Various Input Functions

(a) **Unit Impulse Input** A unit impulse is theoretically an infinite force that is applied for an infinitesimal time, where the relationship between the magnitude of the force and the duration of time is shown in Fig. 4.3.2-1.

The unit impulse function $\delta(t)$ is defined as follows

$$\int_0^{\Delta t} \delta(t) \, dt = \lim_{\Delta t \to 0} F\Delta t = \lim_{\Delta t \to 0} \int_0^{\Delta t} F \, dt = 1 \qquad [4.3.2\text{-}1]$$

The area under the curve is unity. Note that as the time is made smaller and smaller, the force becomes larger and larger, but the area remains equal to unity.

Laplace Transform of Unit Impulse

$$\mathscr{L} \, \delta(t) = \int_0^\infty e^{-st} \, \delta(t) = \int_0^{\Delta t} e^{-st} \frac{1}{\Delta t} \, dt = \frac{1 - e^{-s\Delta t}}{s \, \Delta t}$$

Note that as $\Delta t \to 0$ the above expression takes on the form $0/0$, which can be evaluated using L'Hôpital's rule

$$\lim_{\Delta t \to 0} \frac{1 - e^{-s\Delta t}}{s \, \Delta t} = \lim_{\Delta t \to 0} \frac{s e^{-s\Delta t}}{s} = 1$$

Thus,

$$\mathscr{L} \delta = 1 \quad \text{(for unit impulse)} \qquad [4.3.2\text{-}2]$$

$$= P \quad \text{(for impulse of magnitude } P) \qquad [4.3.2\text{-}3]$$

(b) **Step Input** A step input is a suddenly applied input. (See Fig. 4.3.2-2.) It has a value equal to zero at time zero, and then immediately jumps to a value equal to unity and remains at that value forever. The Laplace

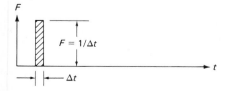

Figure 4.3.2-1   Unit impulse input.

Figure 4.3.2-2   Unit step input.

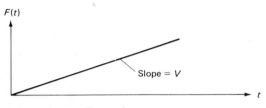

Figure 4.3.2-3   Ramp input.

transform of a unit step $u(t)$ is derived below

$$\mathscr{L}u(t) = \int_0^\infty e^{-st}u(t)\,dt = \int_0^\infty e^{-st}\,dt = \left.\frac{-e^{-st}}{s}\right|_0^\infty \qquad [4.3.2\text{-}4]$$

$$= \frac{1}{s} \quad \text{(for unit step)} \qquad\qquad\qquad [4.3.2\text{-}5]$$

$$= \frac{U}{s} \quad \text{(for step of magnitude } U) \qquad\quad [4.3.2\text{-}6]$$

(c) Ramp Input   A ramp input is one that rises at a constant rate. Its plot appears as a straight line. (See Fig. 4.3.2-3). Such a configuration could be generated by taking the integral of a rectangular area. As an alternate, the same figure could be generated by taking the integral of a step. Referring to the discussion in the previous paragraph, we begin to see a trend. Each of the input functions that was discussed in this section can be generated by successive integrations of the previous ones. This applies in either the time or the Laplace domains. In the time domain, the integral is taken in the conventional manner. In the Laplace domain, the integral is taken by division by the operator S. The process is reversible. Given the response to a ramp input, the responses to a step and to an impulse input is found by the appropriate* multiplication by s in the Laplace domain, and by taking the derivatives in the time domain. Then,

$$\mathscr{L}\{\text{ramp}\} = \frac{1}{s^2} \quad \text{(for ramp of unit slope)} \qquad [4.3.2\text{-}7]$$

$$= \frac{V}{s^2} \quad \text{(for ramp with slope } = V) \qquad [4.3.2\text{-}8]$$

---

*Including the initial conditions.

(d) Summary   The Laplace transforms of three-unit input functions are summarized below:

$$\text{unit impulse input} \qquad 1$$

$$\text{unit step input} \qquad \frac{1}{s} \qquad\qquad [4.3.2\text{-}9]$$

$$\text{unit ramp input} \qquad \frac{1}{s^2}$$

■ **EXAMPLE 4.3.2-1** System With Various Inputs
Solve the following differential equation for each of the listed inputs:

$$\dot{x} + ax = f(t) \qquad \text{IC} = 0$$

where

(a) $f(t)$ = impulse input having magnitude = P
(b) $f(t)$ = step input having magnitude = U
(c) $f(t)$ = ramp input having slope = V

Transform the differential equation to the Laplace domain (including IC)

$$sX - 0 + aX = F(s) \tag{1}$$

or

$$(s + a)X = F(s) \tag{2}$$

Solve for X

$$X(s) = \frac{F(s)}{s + a} \tag{3}$$

The term $F(s)$ is referred to as a general input. It is convenient for cataloging purposes. However, it can *not* be used when taking the inverse Laplace transform. The problem can *not* proceed any further. It is necessary to apply specific functions, which is the reason these were listed in the statement of the problem. Let us apply each one of these, one at a time.

(a) Impulse Input

See eq. 4.3.2-3. For an impulse with magnitude P we have

$$F(s) = P \tag{4}$$

Apply this to eq. (3)

$$X(s) = \frac{P}{s + a} = P\left(\frac{1}{s + a}\right) \tag{5}$$

Use entry 216, Appendix 3,

$$x(t) = Pe^{-at} \tag{6}$$

(b) Step Input

For a step input with magnitude U, see eq. 4.3.2-6

$$F(s) = U/s \tag{7}$$

Apply this to eq. (3)

$$X(s) = \frac{U/s}{s + a} = \frac{U}{s(s + a)} = \frac{U}{a} \frac{a}{s(s + a)} \tag{8}$$

Use entry 215, Appendix 3,

$$x(t) = \frac{U}{a}(1 - e^{-at}) \tag{9}$$

(c) Ramp Input

For a ramp input with slope $= V$, see eq. 4.3.2-8,

$$F(s) = V/s^2 \tag{10}$$

Apply this to eq. (3)

$$X(s) = \frac{V/s^2}{s + a} = \frac{V}{a^2} \frac{a^2}{s^2(s + a)} \tag{11}$$

Use entry 214, Appendix 3,

$$x(t) = \frac{V}{a^2}\left[at - (1 - e^{-at})\right] \tag{12}$$

∎

4.4 TRANSIENT AND STEADY STATE RESPONSES

4.4.1 Meaning of Response

The *response* of a system is the interpretation and physical significance of the solution to its differential equation. The *transient response* involves that portion of the solution during which changes take place The *steady state response* is concerned with the portion of the solution that approaches the condition in which equilibrium prevails. The *frequency response* is a special type of steady state response that is due to a sinusoidal input.

Transient Response   The *transient response* is a fleeting phenomenon, existing for a short time. During the transient, all quantities that can vary are in the process of varying. The transient may be regarded as a state of continual change.

Steady State Response   The *steady state response* represents a condition of dynamic or static equilibrium. The steady state may also be thought of as the limit of the transient state. In most practical systems, the transient lasts but for a short time. The system approaches the steady state rapidly and resides in the near-steady state condition for a long time. Hence, the steady state represents the conditions under which most practical systems operate for most of the time.

4.4.2 Concepts of Transient and Steady State Conditions

The term *steady* implies that a the state of the system is *not* changing. Obviously, something that is standing still is in steady state. However, something that is traveling in a repeating or in a continually regular pattern can also be considered to be in steady state.

4.4.3 Steady State Obtained by the Limit as Time Approaches Infinity

Given the complete solution in the time domain, the steady state response may be found by taking the limit* as time approaches infinity. Several examples will serve to illustrate the method.

   (a) Limit Exists   Refer to eq. 5.1.1-3, and let the term a be positive. Take the limit as time approaches infinity

$$X(t)_{ss} = \lim_{t \to \infty} \left(X_0 e^{-at} \right) = 0$$

   (b) Limit Does NOT Exist   Repeat the above problem, but with the term a negative. We cannot take the limit since the expression e^{at} becomes unbounded as time approaches infinity.

   (c) Mixed Limits   Refer to Ex. 4.3.1-2 (Solution to Second Order System) eq. (8). This is a constant sinusoid whose *displacement* obviously does not approach a limit, even though the *amplitude* does. Since the method discussed in this section operates upon the *displacement*, it cannot be used for this case. The implication is that the limit process shown above will not be successful for this problem. However, there are other techniques that can be applied to problems where the amplitude of oscillation approaches a limit. Such methods are covered in Chapters 7 and 8.

4.4.4 Final Value Theorem

Given the response in the Laplace domain, the steady state response is found by taking the limit,* as s approaches zero, of s times the transformed function. Thus

$$X(t)_{ss} = \lim_{s \to 0} \left[sX(s) \right] \qquad [4.4.4\text{-}1]$$

   (a) Limit Exists   Apply the final value theorem to eq. (7) of Ex. 4.3.1-1.

$$X(t)_{ss} = \lim_{s \to 0} \left[\frac{sX_0}{s + a} \right] = 0$$

which agrees with the result in Section 4.4.3(a).

---

*Provided that the limit exists.

(b) Limit Does NOT Exist   Repeat the previous problem, but where the positive value is used for the term a. If one did *not* know that the limit did *not* exist, the final value theorem might be used in this case. Unfortunately, an answer will be found (-0 would be the erroneous result). The positive term results in an unbounded solution. Thus, it is a good practice to determine beforehand if there is a limit or not.

4.4.5 Initial Value Theorem

If the Laplace transform $Y(s)$ of the variable $y(t)$ is known, then $y(0+)$ can be found from the initial value theorem, which is

$$y(0+) = \lim_{s \to \infty} \left[sY(s) \right] \qquad [4.4.5\text{-}1]$$

Note the similarity in form between the initial value theorem and the final value theorem presented in Section 4.4.4. The final value was found by a limit as $s \to 0$, whereas the initial value is obtained by a limit as $s \to \infty$.

■ EXAMPLE 4.4.5-1

Consider the system shown in Fig. 4.4.5-1, consisting of a mass M restrained by a grounded damper D. The input is an impulse force P applied to the mass; the output is the velocity v of the mass. The differential equation for v is

$$M \frac{dv}{dt} + Dv = P\delta \qquad (1)$$

$$\text{IC} \quad v(0) = 0$$

The Laplace transform of eq. (1) is (see entry 126 in Appendix 3)

$$M\left[sV(s) - v(0-) \right]^0 + DV(s) = P \qquad (2)$$

Solving for $V(s)$, we get

$$V(s) = \frac{P}{Ms + D} \qquad (3)$$

Application of the initial value theorem gives

$$v(0+) = \lim_{s \to \infty} \left[s \frac{P}{Ms + D} \right] = \lim_{s \to \infty} \frac{Ps}{Ms} = \frac{P}{M} \qquad (4)$$

This result can also be obtained from the differential equation by integration

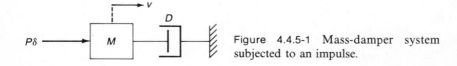

Figure 4.4.5-1 Mass-damper system subjected to an impulse.

between $t = 0-$ and $t = 0+$. We have

$$M \int_{0-}^{0+} \frac{dv}{dt}\, dt + D \int_{0-}^{0+} v\, dt = P \int_{0-}^{0+} \delta\, dt \tag{5}$$

or

$$M\big[v(0+) - v(0-) \big] + D \int_{0-}^{0+} v\, dt = P \tag{6}$$

The term $\int_{0-}^{0+} v\, dt$ is zero, since v is finite. Therefore,

$$Mv(0+) - Mv(0-) = P \tag{7}$$

But $v(0-) = 0$ (given) and, therefore,

$$v(0+) = \frac{P}{M} \tag{8}$$

The above result is recognized as the impulse-momentum principle from dynamics.   ■

4.5 PARTIAL FRACTIONS

4.5.1 Definition

The *partial fraction technique* is an algebraic scheme of transforming a fraction with a high order denominator to the sum of several fractions each with a low order denominator. Let us start with a theorem from the theory of equations. *A polynomial with real coefficients can be factored uniquely into linear and quadratic factors with real coefficients.* It should be pointed out that the theorem merely proves the existence of such factors, but it does not show how to find them. Admittedly, this is often a difficult task. However, once these factors have been found, the expression can be given as a sum of partial fractions. The partial fraction theorem is given as follows.

Given a proper fraction N/D where D is a polynomial in the Laplace variable s, which has been factored into linear and quadratic factors, and where N is a polynomial in the operator s of an order that is less than the order of D, then this can be resolved into the sum of linear and quadratic factors. The numerator of each linear factor is a constant A_i while that for the quadratic factor is a linear term $A_i s + B_i$. If any of these factors is raised to a power (equivalent to repeated factors) then a separate term must be formed consisting each of every power down to unity.

■ **EXAMPLE 4.5.1-1** Partial Fractions
Resolve the following into partial fractions.

$$\frac{N}{D} = \frac{s^6 + es^5 + fs^4 + gs^3 + hs^2 + ks + m}{(s + a)(s + b)^3 (s^2 + cs + d)^2} \tag{1}$$

This can be resolved into partial fractions

$$\frac{N}{D} = \frac{A}{s+a} + \frac{B_1}{(s+b)^3} + \frac{B_2}{(s+b)^2} + \frac{B_3}{s+b} + \frac{C_1 s + D_1}{(s^2 + cs + d)^2}$$

$$+ \frac{C_2 s + D_2}{s^2 + cs + d} \tag{2}$$

The coefficients A, B_1, B_2, B_3, C_1, C_2, D_1, and D_2 are to be determined. Several methods for accomplishing this are presented in the following section.

∎

4.5.2 Methods To Determine Coefficients of Partial Fractions

Method No. 1. Select Numerical Values for the Operator *s*   According to the theorem, the resulting sum of partial fractions is identically* equal to the original expression. An identity is true for all values, hence it is true for any specific value. Then, select several convenient values for *S*. Take these one at a time, substitute into the original expression and into the resulting sum of partial fractions and equate, and form one equation containing all the coefficients. By selecting enough different values, a set of simultaneous equations results. This method is useful when most of the terms drop out upon appropriate selection of the numerical values for *s*.

∎ **EXAMPLE 4.5.2-1** Use Numerical Values for *s*
Given the following expression, resolve it into partial fractions and determine the coefficients by selecting numerical values for the quantity *s*.

$$\frac{N}{D} = \frac{7s^2 - 23s + 10}{(3s - 1)(s - 1)(s + 2)} \tag{1}$$

$$= \frac{A}{3s - 1} + \frac{B}{s - 1} + \frac{C}{s + 2} \tag{2}$$

Cross multiply to clear fractions

$$7s^2 - 23s + 10 = A(s - 1)(s + 2)$$
$$+ B(3s - 1)(s + 2)$$
$$+ C(3s - 1)(s - 1) \tag{3}$$

The opportune numerical values to choose are apparent

$$\begin{array}{lll} \text{Let } s = 1 & \text{then} & B = -1 \\ \text{Let } s = -2 & \text{then} & C = 4 \\ \text{Let } s = \tfrac{1}{3} & \text{then} & A = -2 \end{array} \tag{4}$$

---

*A familiar identity is $\sin^2 \theta + \cos^2 \theta = 1$ for all values of θ.

Thus, the sum of partial fractions becomes

$$\frac{N}{D} = -\frac{2}{3s-1} - \frac{1}{s-1} + \frac{4}{s+2} \tag{5}$$

■

Method No. 2. Equate Coefficients of Like Powers of s   Again making use of the fact that the resulting sum of partial fraction is identically equal to the original expression, and using the expansion obtained for Method No. 1, the identity is satisfied if the coefficients of like powers of the variable are equal.

■ **EXAMPLE 4.5.2-2** Equate Coefficients of Like Powers of s
Given the following expression, resolve it into partial fractions, and determine the coefficients by equating like powers of s.

$$\frac{N}{D} = \frac{5s^2 + 24s + 114}{(s^2 + 36s + 900)(s^2 + 10s + 169)} \tag{1}$$

$$= \frac{A + Bs}{s^2 + 36s + 900} + \frac{C + Es}{s^2 + 10s + 169} \tag{2}$$

Cross multiply to clear fractions

$$(B + E)s^3 + (A + 10B + C + 36E)s^2$$
$$+ (10A + 169B + 36C + 900)s + 169A + 900C$$
$$= 5s^2 + 24s + 114 \tag{3}$$

Equate coefficients of like powers of s

$$
\begin{aligned}
&\text{for } s^3, &&B + E = 0 \\
&\text{for } s^2, &&A + 10B + C + 36E = 5 \\
&\text{for } s, &&10A + 169B + 36C + 900E = 24 \\
&\text{for } s^0, &&169A + 900C = 114
\end{aligned} \tag{4}
$$

Solving these equations simultaneously,

$$A = 6, \quad B = 0, \quad C = -1, \quad E = 0 \tag{5}$$

Thus, the resulting sum of partial fractions becomes

$$\frac{N}{D} = \frac{6}{s^2 + 36s + 900} + \frac{-1}{s^2 + 10s + 169} \tag{6}$$

To check the result, combine the sum of partial fractions into a sum over a common denominator. This will result in obtaining the original expression.   ■

4.6 COMPUTER SIMULATION OF SYSTEM MODELS

For some systems the system model (differential equations and initial conditions) may be too complicated to be solved by Laplace transforms. For example, the differential equation may be nonlinear or of high order. In either case Laplace transforms are impractical. It is possible, however, to obtain a solution in numerical or graphical form by means of computer simulation. One such numerical method approximates the differential equations by finite difference equations. The solution is obtained numerically by one of a number of useful computer algorithms that obtain values of the unknown function at equally spaced time intervals. Another computer simulation method is to use available commercial software specifically designed to handle dynamic system problems. An example of each of these computer methods is described below.

4.6.1 Euler's Method for Integration of Differential Equations

Consider a first order differential equation of the form

$$\frac{dy}{dt} = f(y, t)$$

$$\text{IC} \quad y(0) = y_0$$

[4.6.1-1]

where $f(y, t)$ is determined by the problem under consideration. For example, in Fig. 3.6.1-1(a) the differential equation can be rewritten as

$$\frac{dy}{dt} = \frac{F}{D} - \frac{K}{D}y$$

[4.6.1-2]

If F is time dependent, then the right-hand side of eq. 4.6.1-2 is of the form $f(y, t)$. A finite difference approximation developed by Euler replaces the derivative by a finite difference ratio as follows:

$$\frac{dy}{dt} \doteq \frac{y(t) - y(t - h)}{h}$$

[4.6.1-3]

where h is a small interval of time. Fig. 4.6.1-1 shows a sketch of the above computation.

The differential equation, eq. 4.6.1-1, is approximated as follows:

$$y(t) - y(t - h) = hf(y(t - h), t - h)$$

[4.6.1-4]

The notation in eq. 4.6.1-4 is made simpler if we let n designate one of the

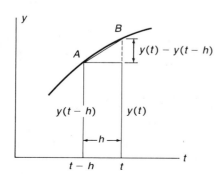

Figure 4.6.1-1 Numerical approximation for dy/dt. The slope dy/dt at $t - h$ is approximated by the slope of the secant line AB. The approximation becomes better as h becomes smaller.

equally spaced time steps. The time t is expressed as $t_n = nh$, and $t - h$ as $t_{n-1} = (n - 1)h$. The value of y at time t is written as y_n, and at $t - h$ as y_{n-1}. Equation 4.6.1-4 then becomes

$$y_n = y_{n-1} + hf(y_{n-1}, t_{n-1}) \qquad [4.6.1\text{-}5]$$

This is a recursion formula that determines y_n at t_n in terms of y_{n-1}, t_{n-1} and h. The computation is started with $n = 1$:

$$y_1 = y_0 + hf(y_0, t_0) \qquad [4.6.1\text{-}6]$$

where $y_0 = $ the given initial value of y
$\qquad t_0 = $ the initial time

Since the function f is given, $f(y_0, t_0)$ can be computed. The next step is to compute y_2 from y_1 as follows:

$$y_2 = y_1 + hf(y_1, t_1) \qquad [4.6.1\text{-}7]$$

where $t_1 = t_0 + h$

The process is continued for as many time steps as required for the given problem.

The computed results will be slightly inaccurate because of the error in approximating the derivative by the finite difference ratio. Discussion of these errors is beyond the scope of this text. To explore this, the student may consult a book on numerical methods (e.g., References [30], [31], and [32]).

Euler's method is easily extended to second order or higher order systems. For example, consider the second order system of Fig. 3.6.2-1(a). The differential equation may be written as

$$\frac{d^2y}{dt^2} = \frac{F}{M} - \frac{D}{M}\frac{dy}{dt} - \frac{K}{M}y \qquad [4.6.1\text{-}8]$$

$$\text{IC} \quad y(0) = y_0 \qquad \dot{y}(0) = v_0$$

Now let

$$\frac{dy}{dt} = v \qquad [4.6.1\text{-}9]$$

$$\text{IC} \quad y(0) = y_0$$

so that eq. 4.6.1-8 can be rewritten thusly:

$$\frac{dv}{dt} = \frac{F}{M} - \frac{D}{M}v - \frac{K}{M}y \qquad [4.6.1\text{-}10]$$

$$\text{IC} \quad v(0) = v_0$$

The pair of equations 4.6.1-9 and 4.6.1-10 are handled in as manner similar to that for the first order system.

■ **EXAMPLE 4.6.1-1** Euler's Method for Second Order System

Solve the following pair of first order differential equations by Euler's method:

$$\frac{dy}{dt} = -3y + 7v + 2t, \qquad y(0) = 1$$

$$\frac{dv}{dt} = -2y + 8v, \qquad v(0) = 0 \tag{1}$$

The corresponding finite difference equations are

$$y_n = y_{n-1} + \left[-3y_{n-1} + 7v_{n-1} + 2t_{n-1}\right]h$$

$$v_n = v_{n-1} + \left[-2y_{n-1} + 8v_{n-1}\right]h \tag{2}$$

Start with $n = 1$, and use $t_0 = 0$, $y_0 = 1$, and $v_0 = 0$. Also, let $h = 0.01$. Then, for $t = 0.01$,

$$y_1 = 1 + \left[-3(1) + 7(0) + 2(0)\right](0.01) = 0.97$$

$$v_1 = 0 + \left[-2(1) + 8(0)\right](0.01) = -0.02 \tag{3}$$

For $t = 0.02$, we have

$$y_2 = 0.97 + \left[-3(0.97) + 7(-0.02) + 2(0.01)\right](0.01) = 0.9203$$

$$v_2 = -0.02 + \left[-2(0.97) + 8(-0.02)\right](0.01) = -0.041 \tag{4}$$

and so forth, for further values of t. Results for $0 < t < 0.10$ are given in Table 4.6.1-1.   ■

4.6.2 Initial Conditions and the Initial Value Theorem

When Euler's numerical method is used to obtain a computer solution for a differential equation, it is necessary to know the initial conditions of the dependent variable. In general, these are given and can be used directly in the

TABLE 4.6.1-1   NUMERICAL RESULTS FOR EQ. (1)

t	y	v
0.00	1.0000	0.0000
0.01	0.9700	−0.0200
0.02	0.9397	−0.0410
0.03	0.9094	−0.0631
0.04	0.8779	−0.0863
0.05	0.8464	−0.1108
0.06	0.8142	−0.1366
0.07	0.7814	−0.1638
0.08	0.7479	−0.1925
0.09	0.7136	−0.2228
0.10	0.6784	−0.2549

solution. There are cases, however, where the initial value changes discontinuously at $t = 0$. The value at $t = 0$ just before the input is applied may be zero and the value just after the application of the input may be nonzero. We say that $y(0-)$ is the value just before, and $y(0+)$ is the value just after, application of the input. Euler's method uses $y(0+)$ as the initial value for the numerical integration to get $y(\Delta t)$, the value at the next time step. $y(0+)$ can be found by using the initial value theorem in this case.

■ EXAMPLE 4.6.2-1

Consider the system whose differential equation and initial conditions are:

$$\ddot{y} + 7\dot{y} + 12y = \dot{y}_i + 6y_i$$

$$\text{IC} \quad y(0) = 0, \qquad \dot{y}(0) = 0 \tag{1}$$

and where

$$y_i(t) = \text{a unit step function} \tag{2}$$

On taking the Laplace transform of eq. (1), we get

$$s^2\left[Y(s) - sy(0-)^0 - y(0-)^0\right] + 7s\left[Y(s) - y(0-)^0\right] + 12Y(s)$$

$$= s\left[\frac{1}{s} - y_i(0-)^0\right] + 6\frac{1}{s} \tag{3}$$

or

$$(s^2 + 7s + 12)Y(s) = \frac{s + 6}{s} \tag{4}$$

The quadratic on the left factors into two linear factors thusly

$$(s + 3)(s + 4)Y(s) = \frac{s + 6}{s} \tag{5}$$

or

$$Y(s) = \frac{s + 6}{s(s + 3)(s + 4)} \tag{6}$$

Using an appropriate combination of entries 225 and 226 in Appendix 3 yields the following for $y(t)$:

$$y(t) = 0.5 - 1.0e^{-3t} + 0.5e^{-4t} \tag{7}$$

The time derivative of $y(t)$ is

$$\dot{y}(t) = 3.0e^{-3t} - 2.0e^{-4t} \tag{8}$$

Initial values obtained from eqs. (7) and (8) are

$$y(0) = 0 \quad \text{and} \quad \dot{y}(0) = 1.0 \tag{9}$$

Note that these values do not agree with the initial conditions specified in eq. (1). The reason for the disagreement is that the values in eq. (9) are for

$t = 0+$, whereas those in eq. (1) are for $t = 0-$. Note that the right-hand side of eq. (1) involves the derivative of y_i, which is a unit step function. The derivative is a unit impulse function $\delta(t)$. A discontinuous change in y at $t = 0$ is to be expected therefore. We would expect $y(0+)$ to be different from $y(0-)$. The results in eq. (9) can be obtained from $Y(s)$ and $V(s)$, the Laplace transform of $\dot{y}(t)$, by means of the initial value theorem. To obtain $y(0+)$, apply the initial value theorem to $Y(s)$ in eq. (6), and get

$$y(0+) = \lim_{s \to \infty} [sY(s)] = \lim_{s \to \infty} \left[\frac{\cancel{s}(s + 6)}{\cancel{s}(s + 3)(s + 4)} \right] = 0 \qquad (10)$$

We need $V(s)$, which is obtained from entry 126 in Appendix 3, as follows:

$$V(s) = sY(s) - y(0^-)^0$$

or

$$V(s) = \frac{\cancel{s}(s + 6)}{\cancel{s}(s + 3)(s + 4)} = \frac{(s + 6)}{(s + 3)(s + 4)} \qquad (11)$$

Now use the initial value theorem to get $y(0+)$.

$$y(0+) = \lim_{s \to \infty} [sV(s)] = \lim_{s \to \infty} \frac{s(s + 6)}{(s + 3)(s + 4)} = 1 \qquad (12)$$

This agrees with the result obtained in eq. (9).

If the differential equation (1) is integrated using Euler's method, the values at $t = 0$, to start the numerical integration, are those at $t = 0+$.

Euler's method is used here to solve eq. (1) with the input given by eq. (2). As in Section 4.6.1 we write the second order equation as two first order equations:

$$\frac{dy}{dts} = v, \qquad\qquad y(0+) = 0$$

$$\frac{dv}{dt} = -12y - 7v + 6, \qquad v(0+) = 1 \qquad (13)$$

Results obtained by Euler's method using $t = h = 0.01$ are given in Table 4.6.2-1. ∎

4.6.3 Computer Simulation by TUTSIM*

A very useful and convenient piece of software TUTSIM enables one to find the time domain solution for system problems of the type just discussed. To operate TUTSIM it is first necessary to draw a block diagram of the system model.

---

*Available in IBM compatible floppy disk form from TUTSIM Products, 200 California Avenue, Suite 212, Palo Alto, California 94306.

TABLE 4.6.2-1   RESULTS OF SECOND ORDER SYSTEM

	Euler		Exact	
t	y	v	y	v
0.00	0.0000	1.0000	0.0000	1.0000
0.02	0.0200	0.9800	0.0198	0.9791
0.04	0.0396	0.9580	0.0392	0.9565
0.06	0.0588	0.9344	0.0560	0.9326
0.08	0.0774	0.9095	0.0764	0.9075
0.10	0.0956	0.8835	0.0943	0.8818

Refer to Section 3.7 in which block diagrams were discussed. Consider the block diagram of the first order system of Fig. 3.7.1-1(b). This is readily converted to a TUTSIM block diagram using special blocks included in the software. The boxes with constant values are gain (GAI) blocks; the box with the integral sign is an integrator (INT) block; and the summing junction (the circle) is represented by a sum (SUM) block. Blocks in the TUTSIM block diagram are numbered 1, 2, 3, and so forth, to aid in construction of what is called in TUTSIM the model structure, which describes the flow of information from block to block. Some blocks, such as the INT and GAI blocks require the specification of parameters. The required parameter for INT is the initial value, and for GAI the parameter is the numerical value of the out/in ratio. Outputs (up to 4) may be taken from the output side of any numbered block in the diagram. Computations are made for equally spaced time values up to a specified time limit. These choices are made when one enters the TUTSIM program. Output may be tabular or graphical.

The TUTSIM program consists of four parts:*

Model Structure

Model Parameters

Plotblocks and Ranges

Timing Data

Model Structure lists blocks by number, specifies the types of blocks, and the inputs by numbers and algebraic signs. *Model Parameters* specifies the required parameters for each numbered block. *Plotblocks and Ranges* specifies the numerical ranges of time and desired output variables. *Timing Data* specifies the time interval and final time for output computations, such as h in the previous section.

■ **EXAMPLE 4.6.3-1** TUTSIM Model for First Order System
Use TUTSIM to model the system of Fig. 3.7.1-1(b) for $D = 4.0$ N · s/m, $K = 20.0$ N/m, and $F = 80.0$ N (step). Let the initial condition be $y(0) = 0$.

---

*A more detailed discussion of programming in TUTSIM is given in Appendix 7.

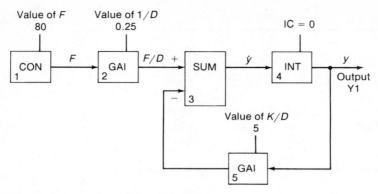

Figure 4.6.3-1   TUTSIM block diagram.

The TUTSIM block diagram is derived from the symbolic block diagram in Fig. 3.7.1-1(b). It is shown above in Fig. 4.6.3-1. Note that the blocks are named and numbered.

The model is entered into the computer from the keyboard by typing the information following each block number as given below.

Model Structure

:1,CON	(Block 1 is a constant.)
:2,GAI,1	(Block 2 is a gain block with input from 1.)
:3,SUM,2, − 5	(Block 3 is a sum block with + input from 2 and − input from 5.)
:4,INT,3	(Block 4 is an integrator with input from 3.)
:5,GAI,4	(Block 5 is a gain block with input from 4.)

Model Parameters

:1,80	(Parameter is the value of the constant.)
:2,0.25	(Parameter is the gain $1/D = 0.25$.)
:3	(No parameter required for a sum block.)
:4,0	(Parameter is the initial value of y.)
:5,5	(Parameter is the gain $K/D = 5$.)

Plotblocks and Ranges

Horz:0,0,1	(Horz means the horizontal axis, time. The first 0 means time, the second 0 is the initial time, and the 1 is the final time.)
Y1:4,0,5	(Output number 1 is taken from block 4; 0 and 5 are the plot limits for the output variable.)
Y2:	
Y3:	(These are not used here.)
Y4:	

Timing Data

.01,1   (Output is computed at intervals of 0.01 up to final
                         time 1.)

After entering the above data, the simulation is run using one of several output choices: SD (simulate to display); SN (simulate to numerical values on the display); and SNP (numerical simulation to the printer.) SNP:10 prints every tenth value.

Results for this example are shown in Fig. 4.6.3-2 and Table 4.6.3-1.   ■

■ **EXAMPLE 4.6.3-2** TUTSIM Model for Second Order System
Use TUTSIM to model the system of Ex. 4.6.2-1. It is more convenient to develop a model for the system using eqs. (14) of Ex. 4.6.2-1. Based on these equations the TUTSIM block diagram is as shown in Fig. 4.6.3-3. The system has just one input, $6y_i = 6$ (step of magnitude 6), represented by the constant

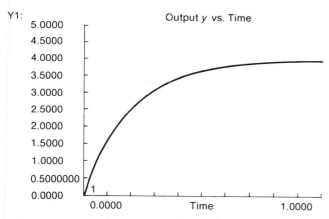

Figure 4.6.3-2   TUTSIM graphical output.

TABLE 4.6.3-1

t	y
0.0000	0.0000
0.1000	1.5757
0.2000	2.5288
0.3000	3.1072
0.4000	3.4582
0.5000	3.6712
0.6000	3.8004
0.7000	3.8789
0.8000	3.9265
0.9000	3.9554
1.0000	3.9729

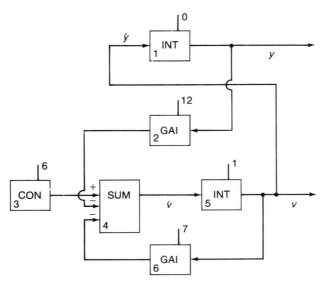

Figure 4.6.3-3   TUTSIM diagram for system of Ex. 4.6.3-2.

block CON, and two outputs y and v. Note that initial conditions for the integrators, INT, are the values at $t = 0+$.

The TUTSIM program is as follows:

Model Structure

:1,INT,5	(Block 1 is an integrator with input from 5.)
:2,GAI,1	(Block 2 is a gain block with input from 1.)
:3,CON	(Block 3 is a constant source.)
:4,SUM, $-$ 2,3, $-$ 6	(Block 4 is a sum block with + input from 3 and $-$ input from 2 and 6.)
:5,INT,4	(Block 5 is an integrator with input from 4.)
:6,GAI,5	(Block 6 is a gain block with input from 5.)

Model Parameters

:1,0	(Initial condition $y(0+)$.)
:2,12	(Gain is 12.)
:3,6	(Value of the input constant equals 6.)
:4	(A sum block has no parameters.)
:5,1	(Initial condition $v(0+)$.)
:6,7	(Gain is 7.)

Plotblocks and Ranges

Horz:0,0,0.5	(Horz means the horizontal axis. The first 0 means time, the second 0 means initial time, and 0.5 is the final time.)

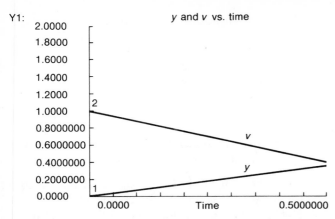

Figure 4.6.3-4   TUTSIM outputs for second order system.

TABLE 4.6.3-2   NUMERICAL RESULTS OF TUTSIM SIMULATION

t	y	v
0.0000	0.0000	1.0000
0.0200	0.0198500	0.9792500
0.0400	0.0392141	0.9565930
0.0600	0.0581108	0.9326160
0.0800	0.0765165	0.9075980
0.1000	0.0944130	0.8817880
0.1200	0.1117870	0.8554040
0.1400	0.1286280	0.8286380
0.1600	0.1449320	0.8016620
0.1800	0.1606950	0.7746220
0.2000	0.1759170	0.7476480
0.2200	0.1906010	0.7208520
0.2400	0.2047520	0.6943310
0.2600	0.2183760	0.6681670
0.2800	0.2314800	0.6424310
0.3000	0.2440740	0.6171820
0.3200	0.2561690	0.5924690
0.3400	0.2677740	0.5683340
0.3600	0.2789040	0.5448070
0.3800	0.2895680	0.5219160
0.4000	0.2997820	0.4996800
0.4200	0.3095570	0.4781120
0.4400	0.3189080	0.4572220
0.4600	0.3278480	0.4370140
0.4800	0.3363900	0.4174900
0.5000	0.3445490	0.3986490

Y1:1,0,2   (The first output, y, is the output of block 1 and
                            has lower and upper plot values 0 and 2.)
Y2:5,0,2   (The second output, v is the output of block 5 and
                            has lower and upper plot values 0 and 2.)
Y3:   (Not used.)
Y4:   (Not used.)

Timing Data

0.01,0.5   (Output is computed at intervals of 0.01 up to the
                            final time 0.5.)

Results of the simulation, using the command SD, are shown in Fig. 4.6.3-4.
Tabulated output for $0 \le t \le 0.5$ is obtained using the command SNP:2 and
is shown in Table 4.6.3-2. Note the near agreement between these results and
exact results in Table 4.6.3-1.   ∎

SUGGESTED REFERENCE READING FOR CHAPTER 4

Solution to differential equations [8], p. 280–291; [22], chap. 2; (37) chap. 1, (46)
    chap. 3.
Linear operators [22], p. 48–50; [37], p. 5, 16.
Introduction to Laplace transforms [17], p. 535–544; [37], p. 51–53.
Laplace Transforms [8], p. 391–400; [17], p. 555–583; [37], chap. 2; [47], chap. 6.
Tables of Laplace transform pairs [8], p. 394; [17], p. 731–755; [47], p. 318–320.
Steady state response [8], p. 398; [17], p. 290–291; [47], p. 351.
Partial fractions [8], p. 386; [17], p. 584–596; [47], p. 330–334.
Computer programming and numerical methods [12]; [30]; [31]; [32]; [34]; [37], chap. 3.
Analog computer methods [8], p. 172–186; [17], p. 299–307; [33].
Nonlinear effects [8], p. 153–155; [12]; [17]; [46], p. 58–62.
Linearization [17], p. 356–360; [46], p. 62–73; [47], p. 195–200.
Solution to nonlinear differential equations [5], chap. 5; [17], p. 360–369.
Phase plane [5], chap. 5; [17], p. 361–369.

PROBLEMS FOR CHAPTER 4

4.1. For the system shown, let $M = 5.0$ kg, $D = 20$ N · s/m, and $f = 100$ N, step.
     The initial velocity of the mass is zero.
     (a) Obtain the solution for $v(t)$ using Laplace transforms.
     (b) Find the time t_1 for the velocity to reach 85% of the steady state value.

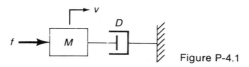

Figure P-4.1

4.2. Solve the following system by Laplace transforms: $\dot{y} + 3y = 4 + 6t$, $y(0) = 1$.

4.3. For the rotational system shown, $R = 0.30$ m, $J = 0.08$ kg $\cdot$ m², $D = 0.45$ N $\cdot$ s/m, and T is an impulse of strength 0.16 N $\cdot$ m $\cdot$ s. The initial conditions are $\theta(0) = 0$ and $\dot{\theta}(0) = 0$.
(a) Derive the differential equation for θ.
(b) Solve for $\theta(t)$ by Laplace transforms.

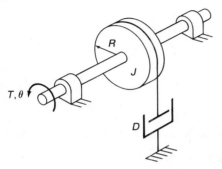

Figure P-4.3

4.4. Find the Laplace transform $X(s)$ for the function $x(t)$, which satisfies $\ddot{x} + 3\dot{x} + 2x = e^{-t}$, with initial conditions $x(0) = 1$ and $\dot{x}(0) = 0$.

4.5. Given

$$F(s) = \frac{1}{s(s + 1)(s + 2)}$$

(a) Expand $F(s)$ into partial fractions.
(b) Find the time domain solution $f(t)$.

4.6. The differential equation for the circuit shown is $R\dot{q} + q/C = V$.
(a) Given $V = 200$ V (step), $C = 10^{-5}$ F, $R = 1000$ Ω, and $q(0) = 0$, obtain $q(t)$ using Laplace transforms. Plot the results.
(b) Obtain $q(t)$ using TUTSIM and compare your results.

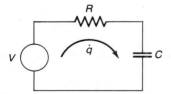

Figure P-4.6

4.7. Obtain the Laplace transforms for the variables indicated in the differential equations for:
(a) The circuit of Fig. 3.6.1-1(e), where $V = kt$, $k =$ constant, and $q(0) = 0$.
(b) The mechanical system of Fig. 3.6.2-1(a), where $F = 0$, $y(0) = 0$, and $\dot{y}(0) = V_0$.

4.8. Find the time domain solutions for:

(a) $Y(s) = \dfrac{1}{s^2 + 16}$;   (b) $X(s) = \dfrac{3s + 1}{s^2}$

4.9. Use partial fractions to obtain the time domain solution for entry 216 in Appendix 3.

4.10. Consider the system of Problem 4.1. Given $M = 2.4$ kg, $D = 4.8$ N · s/m, and $f = 10e^{-2t}$ [f (N), t (s)]:
(a) Assuming $v(0) = 0$, obtain $v(t)$ using Laplace transforms and plot the results.
(b) Use Euler's method to obtain v versus t for $0 < t < 1.0$ s, with a step size of 0.1 s. Compare results with (a).

4.11. Find the time domain solution $f(t)$ if

$$F(s) = \frac{1}{(s + 1)(s + 2)^2}$$

using partial fractions.

4.12. Using Laplace transforms, solve $\ddot{x} + 4x = 3$, with $x(0) = 0$ and $\dot{x}(0) = 1$.

4.13. Given $\ddot{x} + 2\dot{x} + 5x = e^{-t}$, with $x(0) = 0$ and $\dot{x}(0) = 1$:
(a) Find $x(t)$ by Laplace transforms. Plot x versus t.
(b) Obtain x versus t using TUTSIM and compare results.

4.14. Find the time domain solutions for the following:

(a) $X(s) = \dfrac{1}{(s + 2)(s + 5)}$;   (b) $X(s) = \dfrac{1}{s(s^2 + a^2)}$

4.15. Obtain the time domain solution for $y(t)$ if

$$Y(s) = \frac{1}{s(s + 1)^2}$$

Use partial fractions.

4.16. (a) Expand the following Laplace transform into partial fractions:

$$F(s) = \frac{3s + 8}{(s + 2)(s + 3)}$$

(b) Find $f(t)$.

4.17. For the system shown, $x_i(t)$ is a given input and $x(t)$ is the output to be found.
(a) Obtain the differential equation for x.
(b) Given $D = 10$ N · s/m, $K = 20$ N/m, and $x_i = 0.06$ m (step), find $x(t)$ by Laplace transforms.

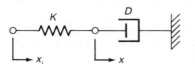

Figure P-4.17

4.18. Obtain the Laplace transforms for each of the $f(t)$ cases shown graphically in Fig. P-4.18.

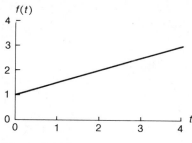

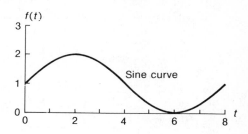

Figure P-4.18

4.19. (a) Derive the differential equation for the system shown.
(b) Given $M = 4$ kg, $D = 8$ N $\cdot$ s/m, and $f =$ a unit impulse δ, use Laplace transforms to get $v(t)$, assuming M to be at rest before f is applied.

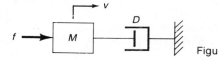

Figure P-4.19

4.20. Obtain the time domain solution for $y(t)$ if

$$Y(s) = \frac{1}{s(s + 1)(s + 2)^2}$$

Use partial fractions.

4.21. The differential equation for p (see Fig. P-4.21) is

$$\frac{R'A}{\gamma}\frac{dp}{dt} + p = p_i$$

The initial condition is $p(0) = p_0$.
(a) Find $p(t)$, assuming p_i is a step.
(b) Use the final value theorem to find the steady state value p_{ss}.

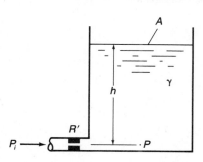

Figure P-4.21

4.22. Consider the system shown in Fig. 3.6.1-1(d). Let $J = 0.08$ kg $\cdot$ m^2, $D = 0.32$ N $\cdot$ s/m, and $T = 160$ N $\cdot$ m (step).
(a) Use Laplace transforms to obtain $\omega(t)$, assuming that $\omega(0) = 0$.
(b) Plot ω versus t.
(c) Obtain the solution for ω versus t using TUTSIM and compare.

4.23. For the system shown, $K = 2500$ N/m, $D = 1000$ N $\cdot$ s/m, $f(t) = 625t$ (N, s), and $x(0) = 0$, find $x(t)$ using Laplace transforms.

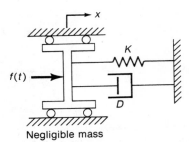

Negligible mass   Figure P-4.23

4.24. The differential equation for x is $D\dot{x} + Kx = Kx_i$. Given $K = 20$ N/m, $D = 10$ N $\cdot$ s/m, $x_i = 0.10e^{-2t}$, and $x(0) = 0$:
  (a) Find $x(t)$ by Laplace transforms and plot the results.
  (b) Find $x(t)$ using TUTSIM and compare your results.

4.25. Consider the nonlinear differential equation $dx/dt = -0.4\sqrt{x}$, $x(0) = 12$.
  (a) Use Euler's method with $h = 0.1$ to obtain values of x for $t = 0$ to $t = 1.0$ in steps of 0.1.
  (b) Use TUTSIM to find corresponding values of x and compare your results.
  (c) Obtain a TUTSIM plot.

4.26. Write each of the following in the Laplace domain
  (a) $M\ddot{X} + D\dot{X} + KX = at^2$ $X(0) = 0$, $\dot{X}(0) = 3$
  (b) $J\ddot{\theta} + D\dot{\theta} + K\theta = b \sin \omega t$ $\theta(0) = 2$, $\dot{\theta}(0) = 5$
  (c) $L\ddot{q} + R\dot{q} + \dfrac{q}{C} = Ee^{-at}$ $q(0) = a$, $\dot{q}(0) = b$
  (d) $L\dot{I} + RI + \dfrac{1}{C}\int I\,dt = $ unit step   IC $= 0$

4.27. Transform each of the following to the time domain.
  (a) $\dfrac{1}{(s + 2)(s + 5)}$
  (b) $\dfrac{2}{s^3(s + 2)}$
  (c) $\dfrac{1}{s(s^2 + a^2)}$
  (d) $\dfrac{1}{s(s + 1)(s + 2)}$
  (e) $\dfrac{1}{s^2(s + 1)}$
  (f) $\dfrac{s^2 - 9}{(s^2 + 9)^2}$
  (g) $\theta(s) = \dfrac{a}{s^2}\dfrac{b}{s^2 + \omega^2}$

*(h) $Y(s) = \dfrac{as}{Ms^2 + Ds + K}$

*(i) $Q(s) = \dfrac{as + b}{Ls^2 + Rs + 1/C}$

*(j) $I(s) = \dfrac{U}{s(Ls + R + 1/Cs)}$

4.28. Derive the Laplace transform for $\sin(at)$ using integration by parts.

4.29. Derive the following entries in Appendix 3.
(a) 217 from entry 216.
(b) 317 from entry 316.
(c) 417 from entry 416.
(d) 448 from entry 447.
(e) 425 from entry 424.

4.30. Find $\dot{x}(t)$, $\ddot{x}(t)$, and $\int x(t)\,dt$ for each of the following:

(a) $x(t) = \dfrac{\omega_n}{\omega_d} e^{-at} \sin[2u(\omega_d t + \phi_1)]$

(b) $x(t) = \dfrac{\omega_n}{\omega_d} e^{-at} \sin(\omega_d t - 2\phi_1)$

(c) $x(t) = 1 + \omega_n t + \dfrac{\omega_n}{\omega_d} e^{-at} \sin(\omega_d t + 3\phi_1)$

4.31. Find the partial fraction expansion for each of the following entries of Appendix 3.
(a) 214   (b) 323   (c) 344
(d) 445   (e) 515   (f) 225
(g) 226   (h) 265   (i) 224
(j) 513   (k) 244   (l) 245

4.32. Using the partial fractions found in Problem 4.31, transform each one to the time domain and compare the result to the corresponding entry in Appendix 3.

4.33. Solve the following equation using TUTSIM.

$$M\ddot{x} + D\dot{x} + Kx = 0, \qquad x(0) = 5 \text{ m}, \quad \dot{x}(0) = 5 \text{ m/s}$$

(a) $M = 0.20$ kg
$K = 5$ N/m
$D = 0.5$ N $\cdot$ s/m
For $0 \le t \le 5$ s at intervals of 0.05 s, plot a graph of x, $\dot{x}$, $\ddot{x}$ using TUTSIM.
(b) $D = (0.5, 1.0, 1.5, 2.0)$ N $\cdot$ s/m
Plot a graph of x for each value of D on one plot by using TUTSIM.

4.34. Solve the differential equation for the following inputs:

$$8\dot{x} + 2x = F \quad (F \text{ in newtons, } x \text{ in meters, } \dot{x} \text{ in meters per second})$$
$$x(0) = 0$$

---

*Hint: normalize and use appropriate ratios listed in Appendix 1.

(a) $F = 5$ (N)
(b) $F = 5$ (N) $0 \le t \le 2$ s
    $= 0$ $t > 2$ s
(c) $F = 2000$ (N) $0 \le t \le 0.005$ s
    $= 0$ $t > 0.005$ s
(d) Impulse $= \int F\,dt = 10$ (N · s)
(e) $F = 5t$ (N)
(f) $F = 2000 e^{-200t}$ (N)
(g) $F = 5 \sin 1.5t$ (N)

4.35. Use whatever method seems appropriate (e.g., Laplace transforms, Euler's method, or TUTSIM) to solve the following differential equation:

$$\dot{X} + 9X = 2e^{-0.3t}, \qquad X(0) = \tfrac{2}{9}$$

for $0 \le t \le 0.5$. Plot the results.

4.36. *Effect of Step Size upon Integration Error.* Solve Problem 4.35 by Euler's method with step sizes $h = 0.04$, 0.02, and 0.01. Comment on the accuracy of the numerical solutions.

4.37. The TUTSIM block PLS (pulse) models a rectangular pulse of magnitude P, turned on at time t_1 and turned off at time t_2. The parameters for this block are t_1, t_2, and P.
(a) Show that if the output of PLS is the input to INT (integrator), with the initial condition equal to 0, the output of INT will be a ramp of slope P, starting at $t = t_1$, ending at $t = t_2$, and followed by a platform of constant magnitude of $P(t_1 - t_2)$.
(b) Set up the PLS, INT model and obtain plots of the pulse and the ramp-platform curves. Let $t_1 = 1$, $t_2 = 3$, and $P = 4$. Scale the time and integrator output to get a suitable plot.

4.38. Each of the following sets of equations constitutes a cascaded system. For each, draw the system and TUTSIM block diagrams and plot the solution. Use the constants as follows: $a = 0.8$, $B = 3$, $\omega_0 = 20$.
(a) $Y = at$, $Z = \int Y\,dt$, $Y(0) = 0$
(b) $X = B \sin \omega_0 t$, $Y = X^2$
(c) $X = B \sin \omega_0 t$, $Y = X^2$, $Z = \int Y\,dt$, $Y(0) = 0$

4.39. Some of the systems shown below contain an algebraic loop, which means that looped back information is instantaneous. These systems cannot be modeled by TUTSIM unless the output of the fed-back signal is delayed by a small time interval. The TUTSIM block DEL (delay) is useful here. Refer to the TUTSIM manual to learn how to use this block. Develop TUTSIM models and plot the results for the following:
(a) $X = B \int_0^t X\,dt$, $X(0) = 2.0$
(b) $X = B \int_0^t - x^2\,dt$, $X(0) = 2.0$
(c) $X = BX^2 + E \int_0^t X^2\,dt$, $X(0) = 2.0$
(d) $X = \int_0^t (1 - 8X + 6Y)\,dt$, $Y = \int_0^t (6X - 17Y)\,dt$
    $X(0) = 0$, $Y(0) = 2.0$
(e) $X = BY e^{-at}$, $Y = CX^2$
    $X(0) = 0$, $Y(0) = 2.0$
(f) $W = W_0(1 - e^{-at})$, $X = B \cos Wt$, $Y = XW$
    $X(0) = 0$, $Y(0) = 2.0$

4.40. Refer to Ex. 3.4.2-3 (Quenching in a Finite Bath). Obtain Laplace transforms of eqs. (1) and (2), and use the final value theorem to obtain the steady state (final) temperature of the rod and bath.

4.41. (a) Use TUTSIM to compare the exact response of a simple pendulum with that of Section 2.2.4(c) (Linearized Model for Small Deflections). Let $M = 0.2$ kg, $L = 0.5$ m, and $\theta = 10, 20, 50, 100,$ and $180°$.

  (b) Write a paragraph discussing the differences observed between the exact nonlinear solution and the approximate linear solution.

4.42. Compare the three functions $Y_1 = X$, $Y_2 = \sin X$, $Y_3 = \tan X$ and tabulate the errors relative to the function Y_1.

4.43. *Transient Response for Turbine System.* (See Fig. P-4.43.) A turbine is driven by a centrifugal pump whose flow rate Q is controlled by a pneumatically operated valve. After a run, the valve is closed and the turbine is allowed to coast down, greatly retarded by internal damping D'. When the turbine speed is $\dot{\theta}_0$, an impulse pressure is accidentally released, which momentarily opens the valve. Neglect the effect of the flow rate Q upon the valve.

  (a) Show that the system is cascaded as a consequence of this.

  (b) Determine the transient response to an impulse input, and the steady state response to an impulse and to a step input. Use the following parameters:

$$M = \text{mass of valve} = 2 \text{ kg}$$

$$D = \text{damping within valve} = 100 \text{ N} \cdot \text{s/m}$$

$$K = \text{control spring in valve} = 2 \times 10^4 \text{ N/m}$$

$$J = \text{moment of inertia of turbine and load} = 200 \text{ kg} \cdot \text{m}^2$$

$$D' = \text{internal damping in turbine} = 10^4 \text{ N} \cdot \text{s} \cdot \text{m}$$

$$A = \text{diaphragm area} = 1.9 \times 10^{-2} \text{ m}^2$$

$$C_1 = \text{valve constant (valve is linear)} = 5 \times 10^3 \text{ m}^4/\text{s}$$

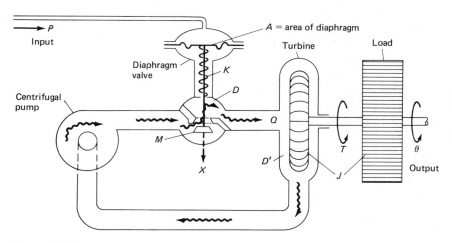

Figure P-4.43

$$C_2 = \text{turbine constant (turbine is linear)} = 4 \times 10^3 \text{ N} \cdot \text{s/m}^2$$

$$\hat{P} = \text{magnitude of pressure impulse} = 10^3 \text{ N} \cdot \text{s/m}^2$$

$$\dot{\theta}_0 = \text{initial velocity of turbine} = 10^5 \text{ rad/s}$$

4.44. In order to reduce the number of unknowns in the partial fraction expansion in Problem 4.43 (Turbine System) factor the term $1/s$ and apply partial fractions to the remaining terms. Upon completion of the operation, multiply each resulting term by $1/s$. Compare to the original result.

4.45. *Transient Response for Cascaded Hydraulic Actuator*. A spool valve (like that in Fig. 3.5.4-2) is positioned by a solenoid actuator. The opening X_1 of the valve determines the fluid flow rate Q. The fluid so controlled is applied to a hydraulic actuator. The system is shown in Fig. P-4.45. Assume that the force F_1 on the spool valve is proportional to the input current i, assume that the fluid flow rate Q is proportional to the valve opening X_1, and assume that the force F_2 against the ram is proportional to the flow rate Q. Assume that the motion of the hydraulic actuator does not affect the valve. Consequently, the system is cascaded. Find the response to a step input current for the following parameters:

$$M_1 = 100 \text{ g } \left(\text{about } \tfrac{1}{4} \text{ lb}\right) \qquad M_2 = 10^4 \text{ g (about 22 lb)}$$

$$D_1 = 400 \text{ dyn} \cdot \text{s/cm} \qquad D_2 = 10^5 \text{ dyn} \cdot \text{s/cm}$$

$$K_1 = 4 \times 10^4 \text{ dyn/cm} \qquad K_2 = 10^6 \text{ dyn/cm}$$

$$C_1 = 10^3 \text{ dyn/A} \qquad C_2 = 10^3 \text{ cm}^3/\text{cm} \quad C_3 = 10^4 \text{ dyn/cm}^3$$

$$U_i = \text{input step} = 10 \text{ A}$$

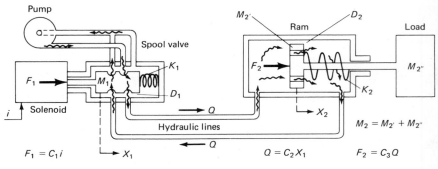

Figure P-4.45

4.46. In a certain fluidic circuit, the flow W due to a unit impulse pressure can be expressed by entry 325 of Appendix 3. Using the format of Ex. 4.3.2-1, make an in-depth study.

4.47. A certain electrical circuit produces a modulated signal. The charge $q(t)$ due to an input voltage function $e(t)$ is given by

$$q(t) = [\cos(\omega + \alpha)t - \cos(\omega - \alpha)t]e(t)$$

Make the following in-depth study:

Find current I, its derivative $\dot{I}$, and its second derivative $\ddot{I}$, due to the following inputs $E(s)$:

(a) Doublet sP';

(b) Impulse P;

(c) Step U/s;

(d) Ramp V/s^2;

(e) Parabola B/s^3.

4.48. Find the steady state response for each part of Ex. 4.3.2-1.

4.49. Find steady state response for each part of Problem 4.46.

4.50. Find steady state response for each part of Problem 4.47.

Chapter 5

Response of First Order Systems

This chapter treats first order systems. The equations are derived and some conclusions are formulated making use of time domain and Laplace domain techniques. Several practical problems are investigated involving linear as well as nonlinear constituents. The first case considers initial conditions and offers as its practical example a flashlight using capacitor discharge for power. The parameters of this system were deliberately chosen poorly so that, in the Problems section of this chapter, the system would of necessity be redesigned.

This chapter considers the response of a first order system to initial conditions, an impulse input, a step, and a ramp. In each case new computer techniques are learned, and at the same time, the computer outputs offer an education in the behavior of the system.

Several nonlinear systems are analyzed using the computer. In each case, the program is written so that the outputs provide considerable insight into the problem of nonlinearity. The text is careful to require only the programming techniques in which the reader has gained sufficient competence at the time.

Three solution techniques are demonstrated in this chapter: solution to the differential equation in the time domain, solution using Laplace transforms, and solutions using computer simulation methods.

5.1 RESPONSE TO INITIAL CONDITIONS AND TO IMPULSE INPUT (GENERALIZED DIFFERENTIAL EQUATION FOR FIRST ORDER SYSTEMS)

A number of first order systems are shown in Fig. 3.6.1-1. The generalized differential equation is given by

$$\frac{dX}{dt} + aX = f(t) \qquad [5.1.1]$$

where X = generalized output or response variable
        a = coefficient (may be constant or variable)
     $f(t)$ = input or forcing function

5.1.1 Response to Initial Conditions

In eq. 5.1.1, let $f(t) = 0$. Then

$$\frac{dX}{dt} + aX = 0 \qquad \text{[5.1.1-1]}$$

$$\text{IC} \quad X(0) = X_0 \qquad \text{[5.1.1-2]}$$

This is the homogeneous equation whose solution for a = constant is given by eq. 4.2.4-2, repeated below

$$X(t) = X_0 e^{-at} \qquad \text{[5.1.1-3]}$$

Normalize eq. 5.1.1-3

$$\frac{X}{X_0} = e^{-at} \qquad \text{[5.1.1-4]}$$

5.1.2 Displacement Versus Time

A plot of eq. 5.1.1-4 is shown in Fig. 5.1.2-1. The curve starts at unity, decays, and approaches zero. The slope is found by differentiating eq. 5.1.1-3

$$\dot{X}(t) = -aX_0 e^{-at} \qquad \text{[5.1.2-1]}$$

To assist in the construction of the curve, without plotting, the initial slope is found by setting $t = 0$ in eq. 5.1.2-1

$$\dot{X}(0) = -aX_0$$

or

$$\frac{\dot{X}}{X_0} = -a \qquad \text{[5.1.2-2]}$$

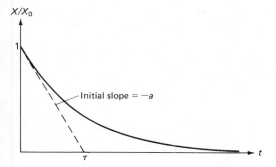

Figure 5.1.2-1   Response of first order system to initial conditions.

If the initial slope line is extended until it strikes the t axis, the abscissa τ can be determined* by the thus formed right triangle;

$$\tau = \frac{1}{a} \qquad [5.1.2\text{-}3]$$

The abscissa τ indicates how quickly the curve approaches the t axis. It is called the *time constant*. An alternate means by which eq. 5.1.2-3 can be derived involves this definition. The time constant τ is defined as the time for which the exponent in eq. 5.1.2-1 becomes equal to unity. By inspection, it is clear[†] that eq. 5.1.2-3 results.

Growth and Decay   Note that a *positive* value of the constant a in eq. 5.1.1-1 results in a *negative* exponential in eq. 5.1.1-4. The resulting curve, shown in Fig. 5.1.2-1, is called a *decay* curve. Thus, a *positive* value of a produces a *decay* curve. A *negative* value of the constant a will produce a *positive* exponential in eq. 5.1.1-4. Such a curve will *rise* without bound and is called a *growth* curve.

5.1.3 Laplace Transform Approach

Write eq. 5.1.1-1 in the Laplace domain, making use of entries 1, 2, and 3 of Table 4.3.1-1.

$$sX(s) - X_0 + aX(s) = 0 \qquad [5.1.3\text{-}1]$$

Solve for $X(s)$

$$X(s) = \frac{X_0}{s + a} \qquad [5.1.3\text{-}2]$$

In order to return to the time domain, take the inverse Laplace transform by using Table 4.3.1-1, entry 5

$$X(t) = X_0 e^{-at} \qquad [5.1.3\text{-}3]$$

which agrees with eq. 5.1.1-3.

Steady State Analysis   In order to determine the steady state value,[‡] apply eq. 4.4.4-1, the final value theorem, to eq. 5.1.3-2

$$X(t)_{ss} = \lim_{s \to 0} \left[\frac{sX_0}{s + a} \right] = 0$$

or

$$\frac{X}{X_0}(t)_{ss} = 0 \qquad [5.1.3\text{-}4]$$

which agrees with Fig. 5.1.2-1.

---

*See Problem 5.28.

[†]See Problem 5.27.

[‡]Since the exponent is negative, we know that steady state exists.

■ **EXAMPLE 5.1.3-1** Capacitor Flashlight

A large capacitor is charged by plugging into a dc source, and it is used to light a 1 W, 100 V lamp. Assuming that when the voltage falls below 50 V, the brightness of the lamp is insufficient, determine how long the flashlight will serve on a single charge using a 0.2 F capacitor. See Fig. 5.1.3-1.

Apply Kirchhoff's laws after the voltage source V_0 is removed and the switch is closed.

$$R\frac{dq}{dt} + \frac{q}{C} = 0 \tag{1}$$

$$\text{IC} \quad q(0) = q_0$$

Separate the variables and integrate

$$\ln q = -\frac{t}{RC} + \ln Q \tag{2}$$

or

$$q = Qe^{-t/RC} \tag{2a}$$

Apply the initial condition

$$q = q_0 e^{-t/RC} \tag{3}$$

Since

$$C = \frac{q}{V} \tag{4}$$

then eq. (3) becomes

$$V(t) = V_0 e^{-t/RC} \tag{5}$$

In order to determine the resistance R, note that power P is given by

$$P = \frac{E^2}{R}$$

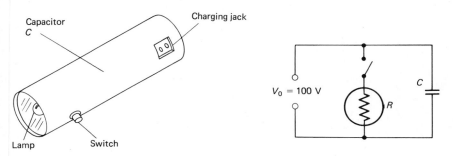

Figure 5.1.3-1   Flashlight using capacitor discharge.

or

$$R = \frac{E^2}{P} = \frac{100^2}{1} = 10,000 \ \Omega \tag{6}$$

Applying numerical values to eq. (5), the answer is

$$t = 1386 \ s \tag{7}$$

A more practical design is found in Problem 5.48.   ■

■ EXAMPLE 5.1.3-2 Water Tower

A water tank is installed on top of a tower for emergency use. Water is pumped into the tank as needed to maintain the level in the tank. A float in the tank senses when the water level falls below the required level. This actuates a switch that turns on the pump. The size of the pump and the source of water are both very limited, which is the reason for the tank. In an emergency, such as a fire, large amounts of water are required for a short time. Neither the pump nor the water supply could meet this high-rate, short-duration demand. But during such a crisis, the tank could provide the necessary water. Given the parameters shown below, determine the time for the tank to empty, and minimum flow rate. See Fig. 5.1.3-2.

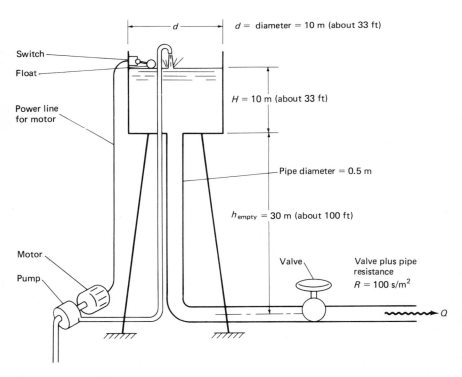

Figure 5.1.3-2   Water tower.

Solution. The minimum flow rate is found by employing eq. 3.5.2-3 at a time when the fluid height is minimum. This occurs when the fluid is at the bottom of the tank, where $h = 30$.

$$Q_{min} = \frac{h_{min}}{R} \tag{1}$$

$$= \frac{30}{100} = 0.3 \text{ m}^3/\text{s (about 4700 gal/min)}$$

The flow in Q_{in} is much smaller than the flow out Q_{out} during the high rate event. Then let

$$Q_{in} = 0 \tag{2}$$

Apply this to eq. 3.5.2-6

$$\frac{dh}{dt} = \frac{-h}{AR} \tag{3}$$

Transform to the Laplace domain

$$sHs(s) - h_0 = -\frac{H(s)}{AR} \tag{4}$$

or

$$H(s) = \frac{h_0}{s + 1/AR} \tag{5}$$

Use entry 5 of Table 4.3.1-1,

$$h(t) = h_0 e^{-t/AR} \tag{6}$$

where

$$h_0 = h_{empty} + H = 30 + 10 = 40 \text{ m}$$
$$h = h_{empty} = 30 \text{ m} \tag{7}$$
$$A = \text{cross sectional area of tank} = 78.5 \text{ m}^2$$

Using numerical values, solve for time t to empty

$$30 = 40e^{-1.27 \times 10^{-4}t} \tag{8}$$

or

$$t = 2.28 \times 10^3 \text{ s (about 38 min)} \tag{9}$$

■

■ **EXAMPLE 5.1.3-3** Computer Simulation of Water Tank with Nonlinearity
A system with assumed linear flow characteristics was considered in Ex. 5.1.3-2. In that problem, it was assumed that,

$$Q = \frac{H}{R} \tag{1}$$

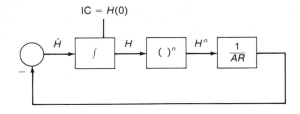

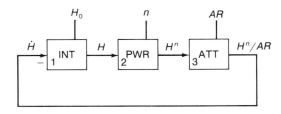

Figure 5.1.3-3   System and TUTSIM block diagrams.

In a real fluid system, this is not exactly the case, particularly if the head H can change by a large proportion. In Ex. 5.1.3-2, the head changed from 40 to 30. Hence, the linear assumption was reasonable. If the bottom of the tank is at the discharge pipe level, the tank may empty, involving a large change in head.

Assume that the flow rate Q is proportional to the head H raised to the nth power. Then

$$Q = \frac{H^n}{R} \qquad (2)$$

The linear case (eq. 1) is a special case of eq. (2) for which $n = 1$.

TABLE 5.1.3-1   TUTSIM MODEL DATA

Model Structure
  :1,INT, − 3
  :2,PWR,1
  :3,ATT,2

Model Parameters
  :1, H_0 (Initial value of H.)
  :2, n (Exponent of H in eq. (3).)
  :3, AR (Denominator of eq. (3).)

Plotblocks and Ranges
  Horz:0,0,20000
  Y1:1,0,50

Timing Data
  300,20000

In order to model the differential equation, use eq. (2) of this example in place of eq. (1) in the original linear example. The resulting differential equation becomes

$$\dot{H} = -\frac{H^n}{AR} \qquad IC = H_0 \qquad (3)$$

This system will be simulated for a value of n equal to 1.25. Note that when a digital computer raises a number to a power that is given by a floating point number, it uses logarithms. This requires that the argument cannot become negative. For this example, the liquid level cannot physically fall below the level in the outlet pipe, and hence cannot become negative. Thus, there is no potential difficulty.

To see the curvature of the decay curve, it is a good practice to let the problem time be equal to twice the time constant. Apply the linear case for this (the time constant is given by eq. 5.1.2-3). Then,

$$\tau = AR = (78.5) \times (100) = 7850 \text{ s} \qquad (4)$$

Let the final time for the simulation be at least twice the time constant, or 20,000 s. For 65 time steps, we require $\Delta t = 300$ s.

The block diagram and model data for the TUTSIM solution* are shown in Fig. 5.1.3-3 and Table 5.1.3-1, respectively. Results are as follows:

Linear case: $n = 1$, $H_0 = 40$, $AR = 7850$.
    Results are shown in Fig. 5.1.3-4(a).

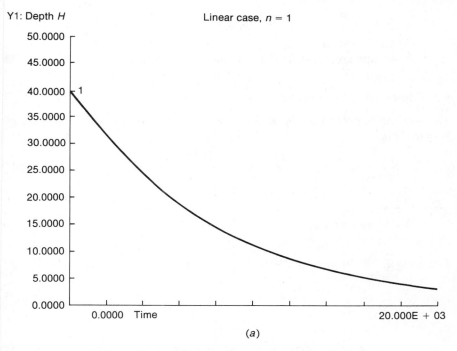

Figure 5.1.3-4   Depth of fluid in tank. (a) Linear case. (b) Nonlinear case.

---

*A more detailed discussion of programming in TUTSIM is given in Appendix 7.

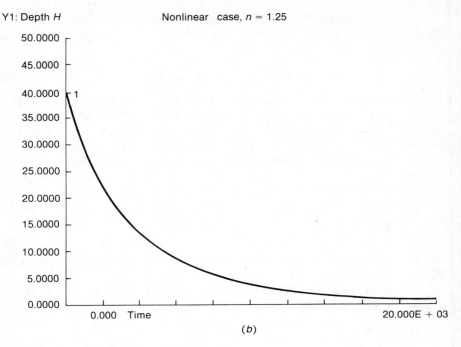

Y1: Depth H Nonlinear case, $n = 1.25$

(b)

Figure 5.1.3-4   Continued

Nonlinear case: $n = 1.25$, $H_0 = 40$, $AR = 7850$.
Results are shown in Fig. 5.1.3-4(b).   ∎

5.1.4 Response to Impulse Input

The relation between impulse and momentum can be found by making use of Newton's law

$$F = M\ddot{x} = \frac{d(Mv)}{dt} \tag{5.1.4-1}$$

where F = force
Mv = momentum
v = velocity

For a small but finite time interval

$$F = \frac{\Delta(Mv)}{\Delta t} \tag{5.1.4-2}$$

or

$$F\Delta t = \Delta(Mv) = \hat{F}, \quad \text{the strength of the impulse} \tag{5.1.4-3}$$

In order to determine the effect of an impulse input, assume that the body is initially at rest with no force applied. Just after the impulse force is applied, the velocity is instantaneously increased by an amount ΔV (assuming that the mass M is constant), where

$$\Delta V = \frac{\hat{F}}{M} \qquad [5.1.4\text{-}4]$$

Now consider the system after the impulse force has been applied. There is now no force applied to the body, and it now has a velocity that has been increased by an amount equal to ΔV. If the body were at rest before the impulse were applied, the resulting velocity would be equal to ΔV. We have thus established the following: A system with an *impulse input and zero initial conditions* is equivalent to one with *zero input and initial velocity*. The magnitude of the equivalent initial velocity V_0 in terms of the actual impulse input δ is given by

$$\text{IC} = V_0 = \Delta V = \frac{\hat{F}}{M} \qquad [5.1.4\text{-}5]$$

The solution will have the form of eq. 5.1.1-3. Then,

$$V = V_0 e^{-at} = \frac{\hat{F}}{M} e^{-at} \qquad [5.1.4\text{-}6]$$

The time-velocity curve will have the same form as the time-displacement curve in Fig. 5.1.2-1.

5.1.5 Laplace Transform Approach

In eq. 5.1.1, change the variable from X to V and let the input $f(t)$ be an impulse of magnitude $\hat{P} = \hat{F}/M$.

$$\frac{dV}{dt} + aV = \hat{P} \text{ (impulse)} \qquad [5.1.5\text{-}1]$$

Write eq. 5.1.5-1 in the Laplace domain, using entries 1, 2, and 3 of Table 4.3.1-1, and eq. 4.3.2-3

$$sV(s) + aV(s) = \frac{\hat{F}}{M} \qquad [5.1.5\text{-}2]$$

Solve for $V(s)$

$$V(s) = \frac{\hat{F}/M}{s + a} \qquad [5.1.5\text{-}3]$$

In order to return to the time domain, apply the inverse Laplace transform by using entry 5 of Table 4.3.1-1

$$V(t) = \frac{\hat{F}}{M} e^{-at} \qquad [5.1.5\text{-}4]$$

which agrees with eq. 5.1.4-6.

One of the advantages of the Laplace transform technique is that all types of inputs (including the IC) may be treated by appropriate functions of the operator s, and subsequently, the problem is handled algebraically. In this particular problem, the Laplace transform technique permitted the handling of the actual problem, an impulse input with zero initial conditions, rather than resorting to the equivalent problem having zero input with an equivalent initial velocity.

Steady State Analysis   In order to determine the steady state value,* apply eq. 4.4.4-1 to eq. 5.1.5-3

$$V(t)_{ss} = \lim_{s \to 0} \left[\frac{s\hat{F}/M}{s + a} \right] = 0$$

or

$$\frac{V}{\hat{F}/M}(t)_{ss} = 0 \qquad\qquad [5.1.5\text{-}5]$$

which agrees with Fig. 5.1.2-1.

5.2 RESPONSE TO A STEP INPUT

5.2.1 Differential Equation

Refer to the general equation, eq. 5.1.1. Assume the forcing function $f(t)$ is a step input of magnitude U, and also assume that a is a constant. The differential equation becomes,

$$\frac{dX}{dt} + aX = U \qquad\qquad [5.2.1\text{-}1]$$

Assume the initial conditions are equal to zero. The homogeneous solution is given by eq. 5.1.1-3. The particular solution is a constant. Then, the complete solution is

$$X = \frac{U}{a} + Ce^{-at} \qquad\qquad [5.2.1\text{-}2]$$

Apply the initial conditions,

$$0 = \frac{U}{a} + C \quad \text{or} \quad C = -\frac{U}{a}$$

Then,

$$X = \frac{U - Ue^{-at}}{a} \qquad\qquad [5.2.1\text{-}3]$$

---

*Solution has the form of a decay curve. Hence, the steady state exists.

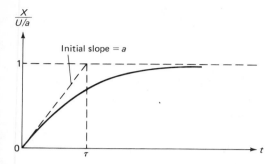

Figure 5.2.2-1   Response of first order systems to step input.

or

$$\frac{X}{U/a} = 1 - e^{-at} \qquad\qquad [5.2.1\text{-}4]$$

5.2.2 Displacement Versus Time

Using the normalized form, eq. 5.2.1-4, the resulting plot is shown on Fig. 5.2.2-1. The curve starts at zero and rises rapidly at first, and then leveling off, it approaches unity. For reasons similar to those in Section 5.1.2, the initial slope is equal to a and the time constant is equal to $\tau = 1/a$.

5.2.3 Laplace Transform Approach

Write eq. 5.2.1-1 in the Laplace domain, making use of entries 1, 2, and 3 of Table 4.3.1-1, and eq. 4.3.2-6

$$sX(s) + aX(s) = \frac{U}{s} \qquad\qquad [5.2.3\text{-}1]$$

Solve for $X(s)$

$$X(s) = \frac{U}{s(s + a)} \qquad\qquad [5.2.3\text{-}2]$$

In order to return to the time domain, apply the inverse Laplace transform by using entry 215 of Appendix 3.

$$X(t) = \frac{U(1 - e^{-at})}{a} \qquad\qquad [5.2.3\text{-}3]$$

which agrees with eq. 5.2.1-4.

Steady State Analysis   In order to determine the steady state value,* apply the final value theorem, eq. 4.4.4-1 to eq. 5.2.3-2

$$X(t)_{ss} = \lim_{s \to 0} \left[\frac{sU}{s(s + a)} \right] = \frac{U}{a}$$

or

$$\frac{X}{U/a}(t)_{ss} = 1 \qquad\qquad [5.2.3\text{-}4]$$

which agrees with Fig. 5.2.2-1.

■ **EXAMPLE 5.2.3-1** Exponential Rise in Motor Speed
Assume that a motor rotor produces a constant torque T_a and that the motor shaft is resisted by magnetic damping D_1, which is assumed to be linear. Also assume that the load on the motor can be lumped as an equivalent damper D_2. See Fig. 5.2.3-1.

Note that the two damping terms oppose the motion. Sum the torques according to Newton's law

$$\sum T = J\ddot{\theta} = T_a - D_1\dot{\theta} - D_2\dot{\theta} \qquad\qquad (1)$$

Let

$$D = D_1 + D_2 \qquad\qquad (2)$$

Apply eq. (2) to eq. (1) and separate the variables

$$J\ddot{\theta} + D\dot{\theta} = T_a \qquad\qquad (3)$$

Since the variable θ does not appear in this equation, but instead only its derivatives appear, then the order of the equation can be reduced by a change of variables. Then let

$$\dot{\theta} = \omega \qquad\qquad (4)$$

and eq. (3) becomes

$$J\dot{\omega} + D\omega = T_a \qquad\qquad (5)$$

or normalizing,

$$\dot{\omega} + \frac{D}{J}\omega = \frac{T_a}{J} \qquad\qquad (6)$$

Equation (6) is now a first order differential equation in the variable ω. This has the form of eq. 5.2.1-1, and its solution will have the form of eq. 5.2.1-4.

$$\frac{\omega/T_a}{J/a} = \frac{\omega}{T_a/D} = 1 - e^{-at} \qquad\qquad (7)$$

---

*Models of systems of this type have been built and tested. They all are known to reach a limit. Hence, steady state exists.

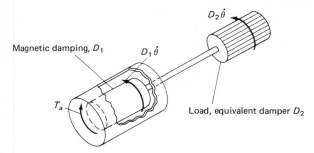

Figure 5.2.3-1   Motor driving load at constant speed.

where

$$a = \frac{D}{J} = \frac{1}{\tau} \tag{8}$$

The steady state speed ω_{ss} may be found by letting time t approach infinity in eq. (7). Then

$$\omega_{ss} = \frac{T_a}{D} \tag{9}$$

Motor-driven pumps and generators are practical examples of this system. It is known that such machines reach a limiting speed $\dot{\theta}$. Let the derivative $\ddot{\theta}$ approach zero, and note that there are no derivatives lower than the first $\dot{\theta}$. Then, the steady state equation becomes

$$D\dot{\theta}_{ss} = T_a$$

or

$$\dot{\theta}_{ss} = \frac{T_a}{D} \tag{10}$$

Making use of eq. (4), eq. (10) becomes

$$\omega_{ss} = \frac{T_a}{D} \tag{11}$$

which agrees with eq. (9).   ■

5.3 RESPONSE TO A RAMP INPUT

5.3.1 Differential Equation

In the general differential equation given by eq. 5.1.1, if the forcing function $f(t)$ varies linearly with time (ramp input), and if a is constant, the differential equation becomes

$$\frac{dX}{dt} + aX = bt \tag{5.3.1-1}$$

5.3.2 Laplace Transform Approach

Transforming eq. 5.3.1-1 to the Laplace domain yields

$$sX(s) + aX(s) = \frac{b}{s^2} \qquad [5.3.2\text{-}1]$$

Solving for $X(s)$,

$$X(s) = \frac{b}{s^2(s + a)} \qquad [5.3.2\text{-}2]$$

The time domain response is found by using entry 214 of Appendix 3. We get

$$X(t) = \frac{b}{a^2}\left[at - (1 - e^{-at})\right] \qquad [5.3.2\text{-}3]$$

Multiplication by a gives

$$aX = bt - \frac{b}{a}(1 - e^{-at}) \qquad [5.3.2\text{-}4]$$

5.3.3 Displacement versus Time

In order to construct the time-displacement curve, note that the first term bt is a straight line with slope equal to b, which is identical to that of the input. Also note that the second term resembles the exponential rise described in Section 5.2.2. The steady state value of this rise is equal to $-b/a$. Since this term is subtracted from the straight line (which is identical to the input), we can conclude that the output will lag the input. This is referred to as an error ϵ. Refer to Fig. 5.3.3-1.

Equation 5.3.2-4 may be rewritten

$$aX = \text{input ramp} - \epsilon \qquad [5.3.3\text{-}1]$$

where

$$\epsilon = \text{error} = (1 - e^{-at})\frac{b}{a} \qquad [5.3.3\text{-}2]$$

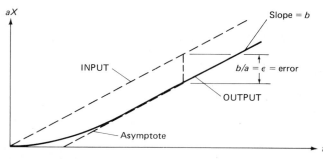

Figure 5.3.3-1   Response of first order systems to ramp input.

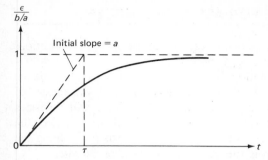

Figure 5.3.4-1   Error due to ramp input to first order system.

5.3.4 Error due to Ramp Input

The error ϵ of the output with respect to the input (ramp bt) is given by the second term of eq. 5.3.2-4. Then,

$$\epsilon = \frac{b}{a}(1 - e^{-at}) \qquad [5.3.4\text{-}1]$$

Normalizing,

$$\frac{\epsilon}{b/a} = 1 - e^{-at} \qquad [5.3.4\text{-}2]$$

Interestingly enough, it develops that the error due to a ramp input resembles the exponential rise due to a step input. See Fig. 5.3.4-1.

5.3.5 Steady State Analysis

In order to determine the steady state value, one might be tempted to apply the final value theorem. However, we have a priori knowledge of this system indicating that the displacement X is unbounded. On the other hand, we also know that the velocity $\dot{X}$ approaches a limit. Then, modify eq. 5.3.2-2 to describe the system output velocity $\dot{X}$ (instead of displacement X) by multiplying by the operator s (as discussed in Section 4.3.1, eq. 4.3.1-4).

$$\dot{X}(s) = sX - x(0) = \frac{sb}{s^2(s + a)} + 0 \qquad [5.3.5\text{-}1]$$

Since we know that the velocity approaches a limit, we may apply the final value theorem, eq. 4.4.4-1

$$\dot{X}(t)_{ss} = \lim_{s \to 0}\left[\frac{ssb}{s^2(s + a)}\right] = \lim_{s \to 0}\left[\frac{b}{s + a}\right] = \frac{b}{a} \qquad [5.3.5\text{-}1]$$

which agrees with Fig. 5.3.3-1.

5.4 TRANSIENT RESPONSE FOR ASSORTED INPUTS

The previous sections treated systems with only one input or initial condition. Often, real systems have combinations of these.

It is possible to seek the solution by comparison to solved problems. However, this approach should not be overdone since it may rob the analyst of the opportunity to perform the task without assistance. There will be times when the problem does not conform to a solved one. Without practice, the analyst may spend more time in seeking a similar problem than in solving the problem from the beginning. Such is the case for nonlinear problems.

■ **EXAMPLE 5.4.1-1** Computer Simulation for Nonlinear Annealing Oven
Glass items are formed when the material is in a semimolten state. If the manufactured item were allowed to cool naturally, the glass would crack. To avoid this, glass items are annealed. They are reheated uniformly and slowly in a special oven. Of practical importance is the rate of heating and the degree of forced circulation of the heated air. In order to study this problem, the following assumptions will be made:

(a) The surrounding air is an infinite source. That is, its temperature T_a remains constant.
(b) The temperature T of the glass body is uniform throughout.
(c) The surface heat transfer coefficient h is uniform over the surface, and it is constant with respect to time, but is a function of temperature.

The time rate of change of temperature is given by eq. 3.4.1-1, repeated below.

$$mc\frac{dT}{dt} = hA(T_a - T) \tag{1}$$

where A = surface area for the transfer of heat from the glass body
        m = mass of the glass body
        c = specific heat of the glass body

Normalizing and rearranging terms, eq. (1) becomes

$$\frac{dT}{dt} + \frac{hA}{mc}T = \frac{hA}{mc}T_a \tag{2}$$

If the coefficient of heat transfer h is constant for all temperatures, then eq. (2) has the form of eq. 5.2.1-1, and its solution would have the form of eq. 5.2.1-4. However, h is *not* constant. It is known that the coefficient h will vary with temperature, and that this variation depends upon the degree of forced circulation. This condition constitutes a nonlinearity which may be expressed in the form

$$h = K|T_a - T|^n \tag{3}$$

where K is a constant and the exponent n is a function of the circulation. Note that eq. (2) reduces to the linear form when $n = 0$. Typical values for

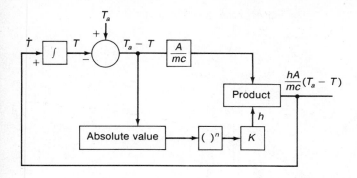

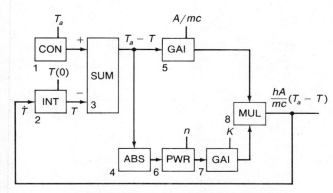

Figure 5.4.1-1   Block diagram and TUTSIM diagram.

TABLE 5.4.1-1   TUTSIM MODEL DATA

Model Structure	Model Parameters
:1,CON	:1,400 (T_a)
:2,INT,8	:2,100 (T_0)
:3,SUM,1, − 2	:3
:4,ABS,3	:4
:5,GAI,3	:5,28.57 (A/mc)
:6,PWR,4	:6,0 or 0.2 for nonlinear
:7,GAI,6	:7,0.2 (K)
:8,MUL,5,7	:8

Plotblocks and Ranges
 Horz:0,0,0.5
 Y1:2,0,500 (convenient temperature scale)

Timing Data
 0.01,0.5 (Δt, final time)

Y1: Temperature   Linear case, $N = 0$

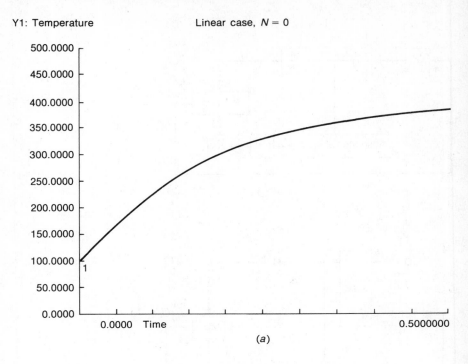

(a)

Y1: Temperature   Nonlinear case, $N = 0.2$

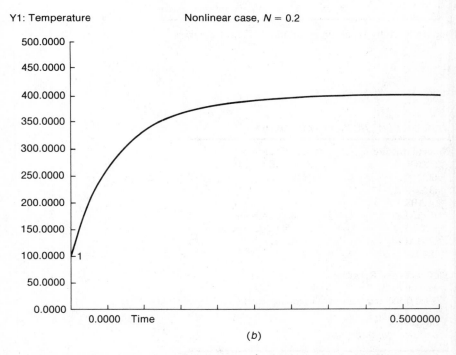

(b)

Figure 5.4.1-2   Specimen temperature in oven. (a) Linear case. (b) Nonlinear case.

this problem are:

$$T_a = 400°F \qquad\qquad K = 2 \times 10^{-1} \qquad\qquad A = 0.1 \text{ ft}^2$$

$$c = 0.007 \text{ Btu/slug} \cdot °F \qquad m = 0.5 \text{ slug} \qquad T(0) = 100°F$$

The nonlinear system will be simulated on the computer using TUTSIM. The block diagram and the corresponding TUTSIM diagram are shown in Fig. 5.4.1-1, and the TUTSIM model formulation is shown in Table 5.4.1-1. Results for the linear and nonlinear cases are shown in Fig. 5.4.1-2(a) and (b). ∎

5.4.2 Laplace Transform Approach

∎ **EXAMPLE 5.4.2-1** In-Depth Study of First Order Systems
Given a first order system, determine its transient and steady state response of $x(t)$ and $\dot{x}(t)$ for each of the following:

(a) Zero input with $x(0) = x_0$.
(b) Impulse input with IC = 0.
(c) Step input with IC = 0.
(d) Ramp input with IC = 0.

Solution. The differential equation with its IC is given by eq. 5.1.1-1, repeated below:

$$\dot{x} + ax = f(t)$$
$$x(0) = x_0 \tag{1}$$

Transform to the Laplace domain

$$sX - x_0 + aX = F(s) \tag{2}$$

Solve for X

$$X(s) = \frac{F(s) + x_0}{s + a} \tag{3}$$

(a) Initial Conditions

$$F(s) = 0, \qquad x(0) = x_0 \tag{4}$$

Apply to eq. (3)

$$X(s) = \frac{x_0}{s + a} \tag{5}$$

For transient response, use entry 216, Appendix 3,

$$x(t) = x_0 e^{-at} \tag{6}$$

which agrees with eq. 5.1.1-3 and is plotted in Fig. 5.1.2-1. The slope of the curve is given by the derivative by applying entry 126, Appendix 3, to eq. (5)

$$\mathscr{L}\dot{x}(t) = \frac{sx_0}{s + a} - x_0 \tag{7}$$

For transient response, use entries 217 and 146, Appendix 3,

$$\dot{x}(t) = x_0(\delta - ae^{-at} - \delta) = -x_0 ae^{-at} \tag{8}$$

which agrees with eq. 5.1.2-1.

(b) Impulse Input

$$F(s) = P, \qquad x(0) = 0 \tag{9}$$

Apply to eq. (3)

$$X(s) = \frac{P}{s + a} \tag{10}$$

Use entry 216, Appendix 3, for transient response

$$x(t) = Pe^{-at} \tag{11}$$

From eq. (11) it is clear that the response to an impulse resembles the response for IC, eq. (5), where the magnitude P of the impulse is used in place of the magnitude x_0 of the initial condition.

For steady state response, apply eq. 4.4.4-1 to eq. (10)

$$x(t)_{ss} = \lim_{s \to 0} \frac{sP}{s + a} = 0 \tag{12}$$

The slope of the curve is given by the derivative, which in turn makes use of entry 126, Appendix 3,

$$\mathscr{L}\dot{x}(t) = sX(s) - x(0) \tag{13}$$

Apply eqs. (10) and (9)

$$\mathscr{L}\dot{x}(t) = \frac{sP}{s + a} - 0 \tag{14}$$

For transient response, use entry 217, Appendix 3,

$$\dot{x}(t) = P(\delta - ae^{-at}) \tag{15}$$

For steady state response, apply eq. 4.4.4-1 to eq. (14)

$$\dot{x}(t)_{ss} = \lim_{s \to 0} \frac{s^2 P}{s + a} = 0 \tag{16}$$

(c) Step Input

$$F(s) = \frac{U}{s}, \qquad x(0) = 0 \tag{17}$$

Apply to eq. (3)

$$X(s) = \frac{U}{s(s + a)} \tag{18}$$

For transient response, use entry 215, Appendix 3,

$$x(t) = \frac{U}{a}(1 - e^{-at}) \tag{19}$$

which agrees with eq. 5.2.1-3.

(d) Ramp Input

$$F(s) = \frac{V}{s^2}, \qquad x(0) = 0 \tag{20}$$

Apply to eq. (3) and multiply by a

$$aX(s) = \frac{aV}{s^2(s + a)} \tag{21}$$

For transient response, use entry 214, Appendix 3,

$$ax(t) = \frac{VA}{a}(at - 1 + e^{-at}) \tag{22}$$

which agrees with eq. 5.3.2-4. For steady state response, apply eq. 4.4.4-1 to eq. (21)

$$ax(t)_{ss} = \lim_{s \to 0} \frac{sVa}{s^2(s + a)} \to \infty \tag{23}$$

The fact that the output $x(t)$ approaches infinity does not indicate any difficulty. It is the natural outcome of a system that is trying to follow an input that approaches infinity. It will prove more valuable in such a case to study the slope of the curve.

For the derivative, apply entry 126, Appendix 3, to eq. (21)

$$\mathcal{L}a\dot{x}(t) = \frac{sVa}{s^2(s + a)} - 0 = \frac{Va}{s(s + a)} \tag{24}$$

For transient response, use entry 215, Appendix 3,

$$a\dot{x}(t) = V(1 - e^{-at}) \tag{25}$$

For steady state response, apply eq. 4.4.4-1 to eq. (24)

$$a\dot{x}(t)_{ss} = \lim_{s \to 0} \frac{s^2V}{s^2(s + a)} = V \tag{26}$$

which agrees with Fig. 5.3.3-1.

The difference between the input $F(s)$ and the output $X(s)$ is defined as the error $\epsilon(s)$. Apply eqs. (20) and (21)

$$\epsilon(s) = \frac{V}{s^2} - \frac{aV}{s^2(s + a)} \tag{27}$$

For transient response, use entries 144 and 214, Appendix 3,

$$\epsilon(t) = Vt - \frac{V}{a}(at - 1 + e^{-at}) \qquad (28)$$

$$= \frac{V}{a}(1 - e^{-at}) \qquad (29)$$

which agrees with eq. 5.3.4-1. The in-depth study is complete. Note that it summarized every concept in Chapter 5.   ■

5.5 SUMMARY OF TRANSIENT RESPONSES

5.5.1 Common Characteristics

The four first order system treated in Sections 5.1 through 5.3 are but a sampling of the many first order systems. However, these are encountered fairly often, making the limited survey reasonably extensive. Upon examination of the response curves, Figs. 5.1.2-1 through 5.3.4-1, there is an apparent commonality. The curves are either the same, or are mirror images of one another. The time constant τ for all four can be found by extending the line of initial slope, and the slope* of this line is equal to a.

It is desirable to expand upon this commonality. Consider the departure of each output curve from its steady state value (then this notion will allow us to include Fig. 5.3.3-1 in this general grouping). We find there is a common expression that defines the departure of each of the four curves from its respective steady state value. This departure is given by the exponential

$$e^{-at} = \text{departure} \qquad [5.5.1\text{-}1]$$

Geometrically, the time constant τ was found to be

$$\tau = \frac{1}{a} \qquad [5.5.1\text{-}2]$$

Summarizing, the common characteristics are shown in Fig. 5.5.1-1.

5.5.2 Time Constant and Settling Time

The *time constant* τ is an index to the rate of decay, and is defined as the *time for the system to reach a value equal to e^{-1}*. Because of the similarity of the curves, and if this is applied to the deviation (or the departure) from steady state, this applies to all the systems in the previous section. Thus, referring to the exponents of each term of Fig. 5.5.1-1, where the time t is set equal to the

---

*Absolute value.

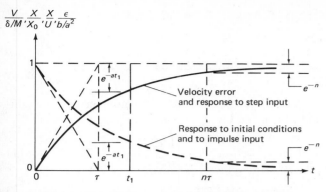

Figure 5.5.1-1   Transient response for first order systems.

time constant τ, we have

$$e^{-a\tau} = e^{-1} \qquad [5.5.2\text{-}1]$$

or

$$\tau = \frac{1}{a} \qquad [5.5.2\text{-}2]$$

In addition to indicating the rate of decay, the time constant also indicates the rate at which the system approaches steady state. Mathematically speaking, the system never reaches steady state, but practically speaking, the *settling time* T_s is defined as the *time* for an actual system to reach a practical value of steady state. Note that on Fig. 5.5.1-1, the value of X decays to e^{-1} after one time constant τ. Then, after n time constants, applying eq. 5.5.2-2 to eq. 5.5.1-1, the value of X will decay to e^{-n}. Thus, for a particular application, the steady state may be defined by a certain percentage* P of the initial value. Consequently, the time to reach this practical value of steady state may be determined by the number of time constants. This becomes a problem in exponential algebra for which the table shown below will provide some assistance.

$$P = e^{-n} \times 100\% \qquad [5.5.2\text{-}3]$$

$$t = n\tau \qquad [5.5.2\text{-}4]$$

■ **EXAMPLE 5.5.2-1** Radioactive Decay
Radioactive materials decay at a rate that is proportional to the number of atoms that are still radioactive. If the "half-life" (time for the material to have decayed to one-half its initial value) is one year, determine the decay coefficient, the time constant, and the time to decay to 10% of the initial value.

---

*This will be referred to as *transient residue*.

TABLE 5.5.2-1   DECAY IN TERMS OF NUMBER OF TIME CONSTANTS

$P\%$	50	36.8	25	13.5	10	5	1.8	0.67
n	0.7	1	1.4	2	2.3	3	4	5

Solution. Using Table 5.5.2-1, note that for 50%, the number of time constants n is 0.7. The actual time given for this is one year.

$$0.7\tau = 1 \text{ yr} \quad \text{or} \quad \tau = \frac{1 \text{ yr}}{0.7} \tag{1}$$

$$a = \frac{1}{\tau} = 0.7/\text{yr} \tag{2}$$

Using Table 5.5.2-1, the time to decay to 10% requires 2.3 time constants. Then the actual time is

$$t = n\tau = \frac{2.3}{0.7} = 3.29 \text{ yr} \tag{3}$$

■

■ **EXAMPLE 5.5.2-2** Motor Rise Time
Refer to Ex. 5.2.3-1 (Exponential Rise in Motor Speed). Given the following parameters:

$$J = \text{moment of inertia of rotor and load} = 0.1 \text{ kg} \cdot \text{m}^2$$
$$D_2 = \text{equivalent damping of the load} = 0.03 \text{ N} \cdot \text{m} \cdot \text{s}$$

Determine D_1 so that the motor attains 95% of its steady state speed in 4.2 s.

Solution. Refer to Fig. 5.5.1-1. 95% of steady state speed is equivalent to

$$P = 100 - 95 = 5\% \tag{1}$$

Using Table 5.5.2-1, this requires $n = 3$ time constants. Apply this to eq. 5.5.2-2 and the given time

$$4.2 \text{ s} = 3\tau = \frac{3}{a} \quad \text{or} \quad a = 0.714 \tag{2}$$

Apply eq. (2) to eq. (8) of the original example

$$0.714 = a = \frac{D}{J} \quad \text{or} \quad D = 0.0714 \tag{3}$$

or

$$D + 0.714J = 0.0714 \text{ N} \cdot \text{m} \cdot \text{s} \tag{4}$$

Use eq. (2) of the original example

$$D_1 = D - D_2 = 0.0714 - 0.03 = 0.0414 \text{ N} \cdot \text{m} \cdot \text{s} \tag{5}$$

■

SUGGESTED REFERENCE READING FOR CHAPTER 5

First order systems [8], p. 124–128; [12]; [17], p. 182–197, 210–214; [47], p. 352–357.
Computer simulation for first order systems [12].

PROBLEMS FOR CHAPTER 5

5.1. (a) By means of a free body diagram, show that the differential equation for v in
Fig. P-5.1 is $M\dot{v} + Dv = f$.
(b) Given $M = 2.0$ kg, $D = 4.0$ N $\cdot$ s/m, $f = 0.1$ N (step), and $v(0) = 0$, find
$v(t)$.
(c) Draw a graph to scale of v versus t for four time constants.

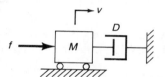

Figure P-5.1

5.2. For the fluid system shown, $R_1' = 4000$ N $\cdot$ s/m^5, $R_2' = 6000$ N $\cdot$ s/m^5, $h(0) =$
0.6 m, and $p_i = 30,000$ N/m^2 (step).
(a) Obtain the differential equation for h.
(b) Draw a curve to scale of h versus t for three time constants.

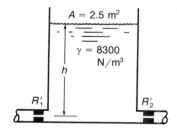

Figure P-5.2

5.3. The differential equation describing the velocity of a bomb of mass M falling
vertically in air with velocity-square resistance is $M\dot{v} + cv^2 = Mg$, $v(0) = 0$.
(a) Let $M = 10$ kg and $c = 0.0392$ N $\cdot$ s^2/m^2. Use Euler's method of numerical
integration to find the velocity for $0 < t < 5.0$ s in steps of 0.5 s.
(b) Obtain a TUTSIM solution and compare results.

5.4. (a) Derive the differential equation for $x(t)$ for the system in Fig. P-5.4.
(b) Given $K = 80$ N/m, $D = 40$ N $\cdot$ s/m, and $f = 2.0$ N (step), find the time
constant and the steady state value of x.
(c) Draw a graph to scale of x versus t.

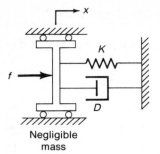

Negligible
mass

Figure P-5.4

5.5. A hollow aluminum sphere 4.0 in. in diameter, weighing 2.0 lb, is surrounded by a layer of insulation 0.25 in. thick for which $k = 0.05$ Btu/h · ft · °F. An electric heater inside the sphere supplies heat at a constant rate of 40.0 Btu/h. Assuming only conduction heat transfer, find the temperature of the sphere after 20 min. The initial temperature of the sphere, which is also the (constant) temperature of the surrounding air, is 80°F.

5.6. The differential equation for the fluid system shown is $A\dot{h} + h/R = 0$.
  (a) Use Laplace transforms to obtain the solution for h, assuming that the initial value of h is h_0.
  (b) Given $A_0 = 3.0$ m², $h_0 = 2.5$ m, and $R = 4.0$ s/m², find the time (s) for h to decrease to 1.0 m.

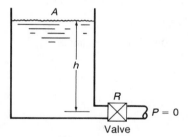

Figure P-5.6

5.7. For the rotational system shown, $K = 200$ N · m/rad, $D = 100$ N · m · s/rad, and $\theta_a = 0.6$ rad (step).
  (a) Obtain the differential equation for θ.
  (b) Find $\theta(t)$ by Laplace transforms, assuming $\theta(0) = 0$.

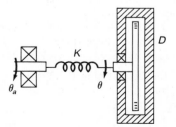

Figure P-5.7

5.8. For the fluid system shown, $A = 0.6$ m^2 and $\gamma = 8000$ N/m^3. The flow through the orifice is governed by the equation $Q = 0.008\sqrt{(P_i - P)}$ [Q (m^3/s), P (N/m^2)].

(a) Obtain the (nonlinear) differential equation for P.

(b) If $P_i = 4.0 \times 10^5$ N/m^2, and the operating point is at $P = 2.0 \times 10^5$, linearize the differential equation.

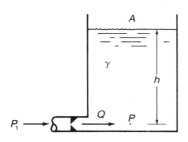

Figure P-5.8

5.9. For the mechanical system shown, $M = 2.0$ kg, $D = 4.0$ N $\cdot$ s/m, $v(0) = 0$, and $f = 20t$ [f (N), t (s)]. The differential equation for v is $M\dot{v} + Dv = f$.

(a) Solve the differential equation by Laplace transforms.

(b) Draw to scale a graph of v versus t for $0 < t < 2.0$ s.

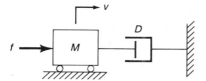

Figure P-5.9

5.10. For the system shown in Fig. 3.6.1-1(c), let $M = 2.0$ kg, $D = 8.0$ N $\cdot$ s/m, and $f = 160$ N (step).

(a) Find the time constant and the final value of v.

(b) Sketch a graph of v versus t making use of the results of part (a).

5.11. For the system in Fig. 3.6.1-1(a), let $K = 40$ N/m, $D = 20$ N $\cdot$ s/m, and $f = 5t$ (N, s). Assume $y(0) = 0$. Draw a sketch of v versus t for $0 < t < 2$ s.

5.12. Consider the water tower shown in Fig. 5.1.3-2. The valve has the property $Q = K\sqrt{h}$.

(a) Show that the differential equation for h is

$$\frac{dh}{dt} = -\frac{Q}{A} = -\frac{K\sqrt{h}}{A}$$

where h is in meters and Q in m^3/s.

(b) Let $h(0) = 15$ m, $K = 0.8$ (in units compatible with the above equation), and $A =$ tank area $= 1.4$ m^2. Use Euler's method to obtain h for $0 < t < 2.0$ s in steps of 0.2 s.

5.13. Consider the nonlinear differential equation $dx/dt = -0.5\sqrt{x + 1}$, $x(0) = 10.0$.

(a) Use Euler's method to obtain x for $0 < t < 1.0$ in steps of 0.1.

(b) Obtain a TUTSIM solution and compare your results.

5.14. For the tank shown, $A = 70$ m², $\gamma = 9806$ N/m³, and $R = 80$ s/m². The tank is drained by opening the valve at $t = 0$, at which time the water depth in the tank $h_T = 10$ m.
(a) Find the time constant.
(b) Find the depth h_T at $t = 20$ min.

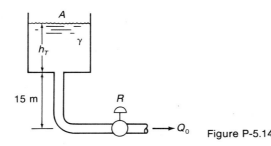

15 m

R

Figure P-5.14

5.15. For the system of Fig. 5.2.3-1, $J = 0.028$ kg · m², $D = 0.014$ N · m · s. Assuming T_a to be a step torque of magnitude 12.6 N · m:
(a) Obtain the time constant.
(b) Obtain the steady state value of ω (rad/s).
(c) Find ω at $t = 4$ s, assuming $\omega(0) = 0$.

5.16. The differential equation for the system shown is $D\dot{x} + f_s = f$, where $f_s = 100 - 5x^3$ (N, m) and $D = 50$ N · s/m (constant).
(a) Assume $x(0) = 0$, and use Euler's method to obtain x for $0 < t < 1.0$ s in steps of 0.1 s. Assume $f = 200$ N (step).
(b) Obtain the solution by TUTSIM and compare your results.

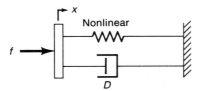

Figure P-5.16

5.17. (a) For the system shown, show that the differential equation for x is $D\dot{x} + Kx = Kx_1$.
(b) Let $D = 60$ N · s/m, $K = 240$ N/m, and $x_1 = 0.05$ m (step). Find the solution for $x(t)$, and sketch the graph for x versus t for four time constants.

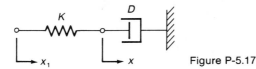

Figure P-5.17

5.18. For the system shown in Fig. 5.2.3-1, assume that the motor torque T_a is given by $T_a = bt$. Let $J = 0.01$ kg · m², $D = 0.02$ N · m · s, and $b = 0.40$ N · m/s.
(a) Find $\omega(t)$ by Laplace transforms, assuming $\omega(0) = 0$.
(b) Sketch a graph of ω versus t to scale for $0 < t < 2.0$ s.
(c) Obtain the solution by TUTSIM and compare results.

5.19. For the circuit shown:
   (a) Obtain the differential equation for $e_0(t)$, given L, R, and $e_i(t)$.
   (b) Obtain the time constant.
   (c) Given $L = 5.0$ H, $R = 100$ Ω, $e_i = 20$ V (step), find and plot e_0 versus t.

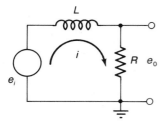

Figure P-5.19

5.20. The differential equation for h in the system shown is $R'Ah + h = P_i$. The valve is opened gradually so that $R' = 4000 - 1000t$. Let $A = 2.0$ m² and $\gamma = 8000$ N/m³.
   (a) Use Euler's method to find h for $0 < t < 2.0$ s, using an interval of 0.2 s. Assume $h(0) = 0$.
   (b) Obtain a TUTSIM solution and compare your results.

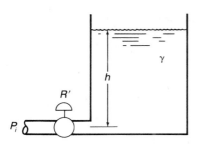

Figure P-5.20

5.21. (a) Derive the differential equation for h in the fluid system shown. The left reservoir is large enough to neglect its depth change.
   (b) If $h(0) = 0$, find the solution for $h(t)$.

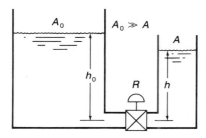

Figure P-5.21

5.22. The differential equation for the temperature T of a glass object placed at $t = 0$ in an oven at temperature T_a is given by $\dot{T} + (hA/mc)T = (hA/mc)T_a$. Let $m = 1.2$ lb$_m$, $c = 0.2$ Btu/lb$_m$ · °F, $h = 0.3$ Btu/h · ft² · °F, $A = 1.6$ ft², $T(0) = 100$°F, and $T_a = 400$°F. Obtain the solution for T versus t and draw a graph.

5.23. For the fluid system shown, $A = 2.0 \text{ m}^2$ and $\gamma = 9806 \text{ N/m}^3$.

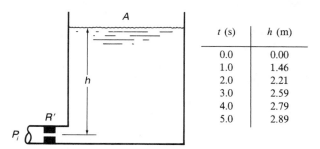

t (s)	h (m)
0.0	0.00
1.0	1.46
2.0	2.21
3.0	2.59
4.0	2.79
5.0	2.89

Figure P-5.23

(a) Given the data in the table, plot a graph of h versus t.
(b) From the graph estimate the time constant and the steady state value of h.
(c) Estimate the value of R' $(\text{N} \cdot \text{s/m}^5)$.

5.24. A hot rock with a mass of 40 lb_m, a specific heat of 0.25 Btu/$\text{lb}_m \cdot$ °F, a temperature of 180°F is placed in a container of water (infinite sink) at 60°F. Heat is transferred from the rock to the water by convection according to q (Btu/h) $= 200(T_{rock} - T_{water})$.
(a) Derive the differential equation for the temperature of the rock.
(b) Use Laplace transforms to obtain the solution for the temperature of the rock as a function of time.

5.25. For the thermal system shown, $R_{T1} = 2.0$°F $\cdot$ h/Btu, $T_1 = 200$°F, $R_{T2} = 4.0$°F $\cdot$ h/Btu, and $T_2 = 60$°F.
(a) Obtain the differential equation for T, the temperature of the specimen.
(b) Find the time constant. Assume that $M = 3.0 \text{ lb}_m$ and $c = 0.2$ Btu/$\text{lb}_m \cdot$ °F.

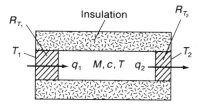

Figure P-5.25

5.26. Assume that the term a in eq. 5.1.1-1 is not constant (you may select any function that is convenient). How does this alter the approach to solving the equation?

5.27. Prove that the initial slope of the curve in Fig. 5.1.2-1 is equal to $-a$.

5.28. Prove geometrically that in Fig. 5.1.2-1 $\tau = 1/a$.

5.29. Is the capacitor flashlight in Ex. 5.1.3-1 a practical device? What changes would improve the design? How large is a 0.2-F capacitor?

5.30. Why was it satisfactory to use a linear relationship in Ex. 5.1.3-2, but not in Ex. 5.1.3-3? Indicate how, if at all, to solve the resulting nonlinear differential equation. Was the simulation necessary in Ex. 5.1.3-3?

5.31. A bullet having velocity V and mass M_1 is fired into a block whose mass is M_2 and is initially at rest. Assume the collision approximates an impulse. Assuming nonlinear damping, write a TUTSIM program describing the motion of the block.

5.32. Refer to Ex. 5.2.3-1 (Exponential Rise in Motor Speed). Using appropriate gearing, this motor is used to drive a slider crank mechanism. Show how this can be used to provide a sinusoidal input with a frequency sweep. Determine the speed-time relationship.

5.3.3. Rework Ex. 5.4.1-1 (Annealing Oven), using $n = 0.2$, initial body temperature $T_0 = 400°F$, and air temperature over a range of temperatures that would involve heating and cooling. It is suggested that you use $T_a = 0°$, $200°$, $400°$, $600°$, and $800°F$.

5.34. *Dial Movement with Ramp Input.* Many instruments make use of a dial output display. To minimize jerky movements, the input (which may have sudden jumps and discontinuities) is not applied directly to the dial. Instead, the input must pass through a weak spring, which tends to "filter" such disturbances and thus "smooth" out the otherwise unreadable information. A typical device is shown in Fig. 7.2.3-1. Using values of D from one-half to eight times that given in the figure, plot the input and output for a ramp input equal to KBt where $B = 0.4$ rad/s.

5.35. *Ballistic Testing Device.* A ballistic testing device makes use of a piston type damper, like that in Fig. 2.3.1-2. A bullet having a mass M_B and a velocity V strikes a special pad on the end of the piston rod. Assume that the collision can be approximated as an impulse input. In order to examine the effect of the pulse width, assume five different widths, using a rectangular pulse as derived in Problem 4.37(a), given the following system parameters

$$M = \text{mass of piston, piston rod, and pad} = 100 \text{ kg}$$

$$M_B = \text{mass of bullet} = 0.04 \text{ kg}$$

$$D = \text{equivalent damping coefficient} = 200 \text{ N} \cdot \text{s/m}$$

$$V = \text{velocity of bullet before impact} = 2000 \text{ m/s}$$

Develop a TUTSIM model and obtain outputs for each pulse width.

5.36. *Effect of Step Size in Solving a First Order Differential Equation.* Using a procedure that is similar to that in Problem 4.36, solve eq. 5.1.1-1 on the computer. Let $a = 2.0$ and $X_0 = 1.0$. Compare the results to the exact solution given by eq. 5.1.1-3.

5.37. The system of Problem 5.32 represents the start-up of any reciprocating machine. Let the steady state speed (frequency) be 105.83 rad/s, and let the final time be 2.4 s, with an interval size $\Delta t = 0.005$ s. Use TUTSIM to obtain plots of ω versus time and of displacement versus time if ω increases according to:
(a) ω (rad/s) $= 105.3(1 - e^{-0.625t})$
(b) ω (rad/s) $= 105.83(1 - t/1.6)$, $0 < t < 1.6$ s
   $= 105.83$, $t > 1.6$ s

5.38. Simulate using TUTSIM the problem of finite quenching as modeled in Ex. 3.4.2-3.

5.39. In an induction motor, torque is proportional to slip. Slip is defined as the difference in speed between the motor and the excitation. An induction motor runs slower than the excitation, and this represents a *velocity* error. Note that this is not the same as the *displacement* error due to a ramp input. The torque T_a is proportional to the difference in velocities

$$T_a = C(\omega_s - \omega)$$

where C = constant of proportionality = 0.2 N · m · s
ω_s = excitation frequency = 1000 rad/s
ω = actual motor speed
J = moment of inertia of rotor = 0.1 N · m ·^{2}
Plot the motor speed and determine the time to reach 95% of the excitation speed.

5.40. For Problem 3.25(a), let $M = 1000$ kg, $D = 50$ N · s/m, and $V(0) = 20$ m/s.
(a) What is the time constant of the system?
(b) How long does it take for the velocity to decrease to 2 m/s?
(c) What is the velocity at the end of 40 s?
(d) What is the steady state velocity?

5.41. For Problem 3.25(b), $M = 200$ kg and $D = 0.07$ N · s/m. Determine
(a) The steady state velocity.
(b) The time required to reach 0.95 the steady state velocity.
(c) Approximate this nonlinear equation by a linear first order equation which most closely approximates the nonlinear response from $t = 0$ to time found in part (b). Give the time constant for the linear system.

5.42. For the system described in Problem 3.25(c), $M = 2$ kg and $D = 0.005$ N · s/m. Find the velocity after 10 s.

5.43. Refer to Problem 3.25(f), $J_1 = 10$ N · m · s^2, $J_2 = 20$ N · m · s^2, $N_1 = 50$, $N_2 = 20$, and $D = 2$ N · m · s. For a step input torque of 20 N · m, find
(a) The time to reach 95% of steady state speed.
(b) The system time constant.

5.44. A piston at rest receives a force pulse of 2000 N over a time of 0.04 s. The piston mass is 100 kg and when in motion the piston experiences a drag which has an equivalent viscous damping coefficient of 200 N · s/m. Derive the equation of motion where the response is the velocity of the piston.

5.45. *Transient Flow in Accumulator Tank (Line Surge Filter).* An accumulator tank is used in fluid systems where a head is needed to be maintained. It is also used to help filter the line surges and the pulsations of the supply pump. See Fig. P-5.45. Given an accumulator tank with an orifice restriction whose fluid resistance is R (assuming laminar flow) determine the following:
(a) Tank is initially full, zero flow-in, find the time for the head to drop to 10 cm.
(b) Tank is initially empty, a volume of fluid is injected instantaneously (impulse input). Determine the percentage of this volume that still remains after 2 s.
(c) Tank is initially empty, determine the rate of constant flow Q_{in} that will have a steady state volume equal to one-half the capacity of the tank. Find the time to reach 95% of this value.
(d) Represent a line surge as an impulse pressure input. Show how effectively this tank reduces the surge transmitted to the rest of the fluid system.

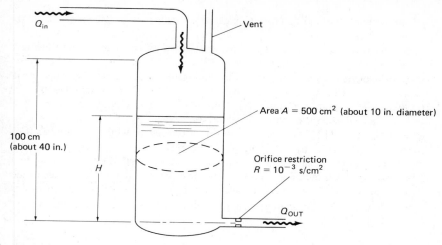

Figure P-5.45

5.46. *Design of Linear Quenching System.* In order to affect the quality of steel, it is heated until it is cherry red, and is then dropped into a bath. The sudden cooling affects the physical properties such as strength, hardness, and modulus of elasticity. Refer to Fig. 3.4.2-1. In order to formulate a rough approximation, the system is modeled using the following assumptions. Defend each one.
1. Bath is an infinite sink.
2. Body temperature of the steel part is uniform throughout its volume (but changes with time).
3. Surface heat transfer coefficient is uniform over the surface of the body and is constant with respect to time and temperature.

Design the system and proportion the body having a 1 lb mass so that it will cool from 1300°F to 500°F in 30 s.

5.47. *Design of Nonlinear Quenching System.* Redesign the system in Problem 5.48 using the following:

$$h = K|T_a - T|^n$$

$$n = 0, 0.1, 0.2, \text{ and } 0.3$$

5.48. *Design of Capacitor Flashlight.* The parameters used in Ex. 5.1.3-1 were deliberately chosen poorly so as to necessitate a redesign. It will be seen that a 0.2-F capacitor is very large. Also, a 1-W lamp for a flashlight is very large. A better design is sought in this problem.
(a) Design a parallel plate capacitor having an air gap for $C = 0.2$ F, $V = 100$ V. (Ans. $A = 2.3 \times 10^6 \text{ m}^2$ or about 1 mile^2.)
(b) Redesign the capacitor in part (a) using mica. (Ans. $A = 3.9 \times 10^2 \text{ m}^2$ or about 1 acre.)
(c) Redesign the capacitor in part (b) by stacking the plates. Use plate thickness $= 2.5 \mu m$ (10^{-6} m × 2.5, or about 0.0001 in.). Check a number of designs using from 1000 to 128,000 plates. Select what you consider the best design

and defend your choice. (Ans. Capacitor is a cube where each side is 0.2 m or about 8 in.)

(d) Redesign the capacitor in part (c) using titanium dioxide and a plate thickness = 1.0 μm. (Ans. Capacitor is a cube 0.1 m per side or 4 in.)

(e) Redesign the capacitor in part (d) by rolling the plates. (Ans The capacitor is a cylinder whose diameter and length are equal to 0.22 m or about 9 in.)

(f) Redesign the capacitor in part (e) so that the shape is compatible either with a hand-held flashlight or with a lantern type device.

(g) Examine several light sources. Note that a typical flashlight lamp draws 0.25 A at 3.0 V, or it uses 0.75 W. Cold light sources are more efficient and produce as much light, but by consuming only 0.1 W. Redesign the system using this power requirement and allow for constant use for 2 h, at which time the power level drops to one-half the starting value. Note that since power is equal to V^2/R, this implies that the voltage drops to a value equal to $\sqrt{\frac{1}{2}}$ or $V_{final} = 70.7$ V.

Chapter 6

Free Vibration and
Transient Response
of Second Order Systems

This chapter is concerned with the dynamic response of second order systems that have had an initial excitation, and then are left free to vibrate or respond. One cannot discuss vibration without also discussing second order systems and vice versa. Hence, it appears profitable to treat these two topics simultaneously rather than separately.

This chapter begins with the concepts of vibration and periodic motion. Undamped (conservative) systems are first treated with energy methods. Following the discussion of damping ratio, the undamped system is revisited, this time, using free body diagrams and Newton's law. The results are compared to those that were obtained with energy methods.

The next portion of the chapter considers response to initial conditions, a step, impulse, and ramp. In some cases, the problem is repeated using Laplace transforms. The illustrative examples include: a landing system for rocket vehicles, a common spring scale, and a fluid flow device. In some cases, nonlinearities are introduced and are solved on the computer. These nonlinearities are due to nonlinear springs, velocity squared damping, and Coulomb damping.

Four solution techniques are demonstrated in this chapter: solution to the differential equation in the time domain, solution in the Laplace domain, use of the digital computer, and comparison to a solved problem.

6.1 FREE UNDAMPED VIBRATION

6.1.1 Energy Method for Vibrating Systems

Kinetic Energy Due to Moving Mass   The energy E_k associated with a mass M moving at velocity $\dot{X}$ is given by

$$E_k = \tfrac{1}{2}M\dot{X}_{\max}^2 \qquad [6.1.1\text{-}1]$$

The analogous case for a rotating flywheel whose moment of inertia is J is given by

$$E_k = \tfrac{1}{2} J \dot{\theta}^2 \qquad [6.1.1\text{-}2]$$

Potential Energy Stored in Spring   The energy E_p stored in a coiled spring is determined by the work required to compress the spring. Then,

$$E_p = \int_0^{X_{max}} F\, dX = \int_0^{X_{max}} KX\, dX$$

$$= \tfrac{1}{2} KX_{max}^2 \qquad [6.1.1\text{-}3]$$

Similarly, the energy stored in a torsion spring is

$$E_p = \tfrac{1}{2} K \theta_{max}^2 \qquad [6.1.1\text{-}4]$$

Interchange of Energy Between Spring and Mass   Considering the properties of sinusoids (see Section 4.2.1), the displacement and velocity periodically go through maximum and zero values. It is important to note that when displacement is maximum, velocity is zero, and therefore all the system energy is potential E_p. At another time, the velocity is maximum while the displacement is zero, in which case, all the system energy is kinetic E_k. (It should also be pointed out that this property is true for sinusoids, and not necessarily true for any other type of motion.) Since the system goes from one maximum to the other, periodically, there is the periodic exchange of energy. Thus, there is an *interchange* of *kinetic energy* of the *moving mass* with the *potential energy* of *deformation of a spring*. The process is periodic, it is completely reversible, and no energy is dissipated.

Energy Dissipated by a Damper   In Section 2.3.4, it was established that a damper dissipates energy. In such a case, energy so dissipated is irretrievable and is permanently lost. Note that problems utilizing the energy method are simple only when there is no damping.

6.1.2 Vibration Requires Source of Energy

From the previous discussions, it is apparent that once a system is vibrating, it has energy in either kinetic form $E_k = \tfrac{1}{2} MV^2$ or potential form $E_p = \tfrac{1}{2} KX^2$. Prior to vibrating, the system was at rest, in which case its energy was zero. Thus, the immediate conclusion is that *to initiate a vibration, energy must be put into the system.*

If the energy input is applied in a single burst, the system is said to undergo *free vibration*. This implies that the introduction and application of the burst does not interfere with the vibration. The vibration proceeds free or unimpeded. If the energy is applied either continuously or in periodic bursts, the system is said to undergo *forced vibration*. In this case, energy is applied during the vibration, and thus will affect the behavior of the system.

6.1.3 Frequency of Vibration

The interchange of energy in a conservative system was discussed in Section 6.1.1. The undamped system is a conservative system. In this case, there is an interchange of energy between the spring and the mass. Previously, it was shown that in a vibrating system at different points, the system experiences either a maximum velocity (and with it a maximum kinetic energy) or a maximum displacement (and with it a maximum potential energy). In the undamped system, there is no energy dissipated in damping, making it convenient to equate the maximum potential and maximum kinetic energies. Using eqs. 6.1.1-1 and 6.1.1-3,

$$E_k = \tfrac{1}{2}M\dot{X}_{max}^2 = E_p = \tfrac{1}{2}KX_{max}^2 \qquad [6.1.3\text{-}1]$$

Assume the motion is sinusoidal,

$$X = A \sin \omega t \quad \text{then} \quad X_{max} = A \qquad [6.1.3\text{-}2]$$

$$\dot{X} = A\omega \cos \omega t \quad \text{then} \quad \dot{X}_{max} = A\omega \qquad [6.1.3\text{-}3]$$

Substitute eqs. 6.1.3-2 and 6.1.3-3 into eq. 6.1.3-1

$$\tfrac{1}{2}M(A\omega)^2 = \tfrac{1}{2}K(A)^2$$

or

$$\omega = \sqrt{\frac{K}{M}} \qquad [6.1.3\text{-}4]$$

For the torsional or rotary system, follow a similar procedure, using eqs. 6.1.1-2 and 6.1.1-4, and get

$$\omega = \sqrt{\frac{K}{J}} \qquad [6.1.3\text{-}5]$$

In the translatory as well as the torsional systems, the *frequency* of vibration is *in*dependent of the *amplitude*.

6.1.4 Amplitude of Vibration

If a system has initial displacement X_0 and initial velocity V_0 the total energy input is

$$E_{tot} = \tfrac{1}{2}KX_0^2 + \tfrac{1}{2}MV_0^2 \qquad [6.1.4\text{-}1]$$

The amplitude of vibration is defined as the maximum displacement. Assuming a sinusoidal vibration, where the maximum displacement occurs, velocity will be zero. Then at the point where the maximum displacement occurs, the total energy will be potential, or

$$E_{tot} = \tfrac{1}{2}KX_{max}^2 \qquad [6.1.4\text{-}2]$$

Equate eq. 6.1.4-1 to eq. 6.1.4-2

$$\tfrac{1}{2}KX_0^2 + \tfrac{1}{2}MV_0^2 = \tfrac{1}{2}KX_{max}^2$$

or

$$X_{max}^2 = X_0^2 + \frac{M}{K}V_0^2 \qquad [6.1.4\text{-}3]$$

Substitute eq. 6.1.3-4 and extract the square root

$$X_{max} = \sqrt{X_0^2 + \left(\frac{V_0}{\omega}\right)^2} \qquad [6.1.4\text{-}4]$$

Since the amplitude is a function of energy, one may conclude that the amplitude of vibration is an indicator of the amount of energy in a system.

6.2 DAMPED VIBRATION

6.2.1 Some Second Order Systems

There are many second order systems. Some simple examples were shown in Fig. 3.6.2-1. It should be pointed out that for each type of system there are many variations. For example, in a mechanical system, there could be more than one spring and more than one damper. In an electrical circuit, there could be additional capacitors and resistances. In a fluid system, there could be a closed tank that is pressurized. In addition to the above variations, there could be combinations of inputs, inputs could be applied at different points in the system, there could be assorted initial conditions, there could be simultaneous applications of combinations of all of these, and there could be mixed disciplines.

6.2.2 Generalized Second Order Differential Equation

For the first half of this chapter, only the simplest systems will be considered. Those systems shown in Fig. 3.6.2-1 utilize only one of each type of element of a single discipline. The differential equation for the simple generalized second order system may be expressed as

$$J\ddot{\theta} + D\dot{\theta} + K\theta = T(t)$$
$$\theta(0) = \theta_0, \qquad \dot{\theta}(0) = \dot{\theta}_0 \qquad [6.2.2\text{-}1]$$

where

$$\theta = \text{generalized output or response variable}$$
$$T(t) = \text{generalized input or forcing function} \qquad [6.2.2\text{-}2]$$
$$J, D, K = \text{constant coefficients}$$

The solution to the differential equation will be achieved in two parts, the homogeneous solution and the particular solution. The homogeneous solution will depend upon initial conditions, while the particular solution will depend upon the forcing function.

The homogeneous solution for the generalized system is of the form

$$\theta_h = C_1 e^{p_1 t} + C_2 e^{p_2 t} \qquad [6.2.2\text{-}3]$$

where p_1 and p_2 are roots of the characteristic equation and C_1 and C_2 are constants of integration, to be determined by initial conditions.

6.2.3 Characteristic Equation and Its Roots

The *characteristic equation* is an *algebraic* equation that is formulated to assist in solving the homogeneous differential equation. It is formulated by using an algebraic operator p which is raised to a power equal to the order of the corresponding derivative in the differential equation. Thus, the characteristic equation becomes

$$Jp^2 + Dp + K = 0 \qquad [6.2.3\text{-}1]$$

Solving algebraically, the roots (p) become

$$p = -\frac{D}{2J} \pm \sqrt{\left(\frac{D}{2J}\right)^2 - \frac{K}{J}} \qquad [6.2.3\text{-}2]$$

6.2.4 Damping Ratio

The homogeneous solution will take on very different forms, depending upon the terms under the radical in eq. 6.2.3-2. This depends upon the relative values of the systems parameters J, D, and K. The full range of possibilities will be sampled by selecting four discrete cases

$$\left.\begin{array}{ll} \text{case 1} & \left(\dfrac{D}{2J}\right)^2 = 0 \\[2em] \text{case 2} & \left(\dfrac{D}{2J}\right)^2 < \dfrac{K}{J} \\[2em] \text{case 3} & \left(\dfrac{D}{2J}\right)^2 = \dfrac{K}{J} \\[2em] \text{case 4} & \left(\dfrac{D}{2J}\right)^2 > \dfrac{K}{J} \end{array}\right\} \qquad [6.2.4\text{-}1]$$

As a reference or datum for damping to which all cases may be compared, the critical damping D_c is used. *Critical damping* is the value of damping which satisfies case 3. Thus,

$$\left(\frac{D_c}{2J}\right)^2 = \frac{K}{J}$$

or

$$D_c = 2\sqrt{KJ} \qquad\qquad [6.2.4\text{-}2]$$

Using the critical damping as the datum, define the ratio of any particular case to this case as the *damping ratio* ζ,

$$\zeta = \frac{D}{D_c} = \frac{D}{2\sqrt{KJ}} = \frac{D}{2J\omega_n} \qquad [6.2.4\text{-}3]$$

The four cases may be defined in terms of the damping ratio ζ. This also provides a more suitable form for the exponent p. Using these notations, the four cases become:

case 1 undamped $\qquad \zeta = 0 \qquad p = \pm j\omega_n$, pure imaginary roots
case 2 underdamped $\qquad \zeta < 1 \qquad p = -a \pm j\omega_d$, complex roots
case 3 critically damped $\quad \zeta = 1 \qquad p = -a$, repeated real roots
case 4 overdamped $\qquad \zeta > 1 \qquad p = -a \pm \omega_0$, two real roots

where*

$$\left.\begin{array}{l} a = \text{decay coefficient} = \dfrac{D}{2J} = \zeta\omega_n \\[2em] \omega_n = \text{undamped natural frequency} = \sqrt{\dfrac{K}{J}} \\[2em] \omega_d = \text{damped natural frequency} \\[1em] \quad = \sqrt{\dfrac{K}{J} - \left(\dfrac{D}{2J}\right)^2} = \sqrt{\omega_n^2 - a^2} = \omega_n\sqrt{1 - \zeta^2} \\[2em] \omega_v = \text{argument for hyperbolic functions} \\[1em] \quad = \sqrt{\left(\dfrac{D}{2J}\right)^2 - \dfrac{K}{J}} = \sqrt{a^2 - \omega_n^2} \end{array}\right\} \quad [6.2.4\text{-}4]$$

6.3 RESPONSES TO INITIAL CONDITIONS AND TO STEP INPUT

6.3.1 Differential Equation

The differential equation in terms of the generalized coordinate θ is given by eq. 6.2.2-1, repeated below

$$J\ddot{\theta} + D\dot{\theta} + K\theta = T(t)$$

$$\text{IC} \quad \theta(0) = \theta_0, \qquad \dot{\theta}(0) = \dot{\theta}_0 \qquad [6.3.1\text{-}1]$$

---

*These are listed in Appendix 1.

6.3.2 Homogeneous Solution

The homogeneous solution is given by eq. 6.2.2-3, repeated below

$$\theta_h = C_1 e^{p_1 t} + C_2 e^{p_2 t} \qquad [6.3.2\text{-}1]$$

The roots p_1 and p_2 depend upon the damping ratio ζ while the constants C_1 and C_2 depend upon the initial conditions. Equation 6.3.2-1 will be studied for each of the four cases of damping.

6.3.3 Response to Initial Conditions (Homogeneous Solution)

Case 1. Undamped, $\zeta = 0$

$$p = \pm j\omega_n$$

then

$$\theta(t) = B_1 e^{j\omega_n t} + B_2 e^{-j\omega_n t} \qquad [6.3.3\text{-}1]$$

Use Euler's identities,* collect terms, and apply IC

$$\theta(t) = \theta_0 \cos \omega_n t + \frac{\dot{\theta}_0}{\omega_n} \sin \omega_n t \qquad [6.3.3\text{-}2]$$

or by combining the sine and cosine terms

$$\theta(t) = A_0 \sin(\omega_n t + \phi_0) \qquad [6.3.3\text{-}2a]$$

where

$$A_0 = \sqrt{(\theta_0)^2 + \left(\frac{\dot{\theta}_0}{\omega_n}\right)^2} \qquad [6.3.3\text{-}2b]$$

$$\phi_0 = \tan^{-1} \frac{\theta_0 \omega_n}{\dot{\theta}_0} \qquad [6.3.3\text{-}2c]$$

Note that eq. 6.3.3-2b is identical to that found by the energy method, eq. 6.1.4-4. The results are plotted in Fig. 6.3.3-1. For initial displacement θ_0, making use of eq. 6.3.3-2, the result is a *cosine* curve of constant amplitude. For initial velocity $\dot{\theta}_0$, the result is a *sine* curve. Thus, we see that the responses to these two initial conditions are 90° out-of-phase. The period λ_n is given by $2\pi/\omega_n$.

Case 2. Underdamped, $\zeta < 1$

$$p = -a \pm j\omega_d$$

---

*See Appendix 1.

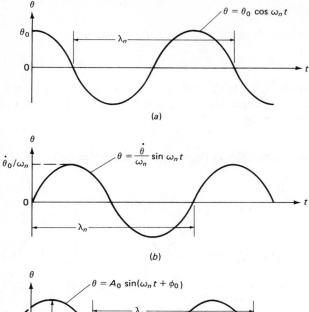

Figure 6.3.3-1   Response of undamped system to initial conditions. (a) Initial displacement. (b) Initial velocity. (c) Initial displacement and initial velocity combined.

then

$$\theta(t) = B_1 e^{(-a+j\omega_d)t} + B_2 e^{(-a-j\omega_d)t} \qquad [6.3.3\text{-}3]$$

Use Euler's identities,* collect terms and apply initial conditions

$$\theta(t) = e^{-at}\left[\theta_0 \cos \omega_d t + \frac{a\theta_0 + \dot{\theta}_0}{\omega_d}\sin \omega_d t\right] \qquad [6.3.3\text{-}4]$$

This can be simplified if we take one initial condition at a time and solve for the corresponding dimensionless ratio, as follows.

(a) Initial Displacement

$$\frac{\theta(t)}{\theta_0} = e^{-at}\left(\cos \omega_d t + \frac{a}{\omega_d}\sin \omega_d t\right) \qquad [6.3.3\text{-}4a]$$

---

*See Appendix 1.

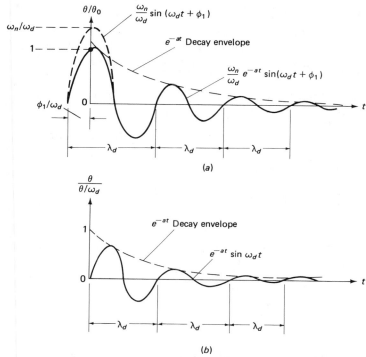

Figure 6.3.3-2 Response of underdamped system to initial conditions. (a) Initial displacement. (b) Initial velocity.

or by combining sine and cosine terms (using eq. 4.2.1-3)

$$\frac{\theta(t)}{\theta_0} = \frac{\omega_n}{\omega_d} e^{-at} \sin(\omega_d t + \phi_1) \qquad [6.3.3\text{-}4b]$$

where*

$$\phi_1 = \tan^{-1} \frac{\omega_d}{a} = \cos^{-1} \zeta \qquad [6.3.3\text{-}4c]$$

(b) Initial Velocity   (Note that $\dot{\theta}_0/\omega_d$ is dimensionless.)

$$\frac{\theta(t)}{\dot{\theta}_0/\omega_d} = e^{-at} \sin \omega_d t \qquad [6.3.3\text{-}4d]$$

Note that when $\zeta = 0$, either eq. 6.3.3-4b or eq. 6.3.3-4d reduces to the undamped case, eq. 6.3.3-2.

When the damping ratio is less than unity, the system will vibrate or oscillate. The amplitude of this vibration decreases with time. The manner in which it decreases is exactly like the decay curve for a first order system. Thus, the sinusoid can be drawn within the confines of a decay curve. This decay

---

*See Appendix 1.

curve is referred to as the "envelope" of the sinusoid. The period λ_d of this vibration is given by $2\pi/\omega_d$.

Refer to Fig. 6.3.3-2. The decaying vibration due to initial displacement is shown in part (a), while that due to initial velocity appears in part (b).

Laplace Transform Approach   Transform eq. 6.3.1-1 to the Laplace domain, and let $T = 0$.

$$J\left(s^2\theta(s) - s\theta_0 - \dot{\theta}_0\right) + D\left(s\theta(s) - \theta_0\right) + K\theta(s) = 0 \quad [6.3.3\text{-}4e]$$

Solve for $\theta(s)$, normalize, and separate IC.

$$\theta(s) = \frac{(s + 2a)\theta_0}{s^2 + 2as + \omega_n^2} + \frac{\dot{\theta}_0}{s^2 + 2as + \omega_n^2} \quad [6.3.3\text{-}4f]$$

In order to return to the time domain, apply the inverse Laplace transform, making use of entries 424 and 425 of Appendix 3.

$$\left.\begin{aligned}
\frac{\theta}{\theta_0}(t) &= \frac{\omega_n}{\omega_d}e^{-at}\sin(\omega_d t + \phi_1) \\[2em]
\frac{\theta}{\dot{\theta}_0}(t) &= \frac{e^{-at}}{\omega_d}\sin\omega_d t
\end{aligned}\right\} \quad [6.3.3\text{-}4g]$$

where

$$a = \zeta\omega_n = \frac{D}{2J} \qquad \phi_1 = \cos^{-1}\zeta$$

$$\omega_d = \omega_n\beta \qquad \beta = \sqrt{1 - \zeta^2}$$

which agree with eqs. 6.3.3-4 a through d.

Steady State Analysis   Since it is known that the solution to the underdamped case does approach a limit, apply the final value theorem, eq. 4.4.4-1, to eq. 6.3.3-4f (collect the IC into one expression)

$$\theta(t)_{ss} = \lim_{s \to 0} s\left[\frac{(s + 2a)\theta_0 + \dot{\theta}_0}{s^2 + 2as + \omega_n^2}\right] = 0 \quad [6.3.3\text{-}4h]$$

which agrees with Fig. 6.3.3-2.

Case 3. Critically Damped, $\zeta = 1$

$$p = -a \quad [6.3.3\text{-}5]$$

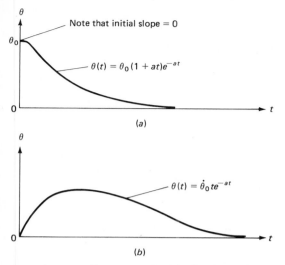

Figure 6.3.3-3   Response of critically damped system to initial conditions. (a) Initial displacement. (b) Initial velocity.

Note that the two roots are equal. For repeated roots, multiply the second term of the particular solution by time t. The solution becomes

$$\theta(t) = C_1 e^{-at} + C_2 t e^{-at} \qquad [6.3.3\text{-}5a]$$

Apply initial conditions

$$\theta(t) = e^{-at}\left[\theta_0(1 + at) + \dot{\theta}_0 t\right] \qquad [6.3.3\text{-}5b]$$

The result for each initial condition is plotted in Fig. 6.3.3-3. Note that there is no oscillation. The curve for initial displacement θ_0 resembles a decay curve.* The response to initial velocity $\dot{\theta}_0$ is a curve that rises to a maximum and then follows a decay-like approach to zero. See Fig. 6.3.3-3.

Laplace Transform Approach   The solution in the Laplace domain is given by eq. 6.3.3-4f. Before attempting to transform back to the time domain, note that the damping ratio $\zeta = 1$. Then the denominators are perfect squares, or

$$\frac{\theta}{\theta_0}(s) = \frac{s}{(s + a)^2} + \frac{2a}{(s + a)^2}$$

and $\qquad\qquad\qquad\qquad\qquad\qquad\qquad\qquad\qquad\qquad\qquad [6.3.3\text{-}5c]$

$$\frac{\theta}{\dot{\theta}_0}(s) = \frac{1}{(s + a)^2}$$

---

*However, the initial slope is zero.

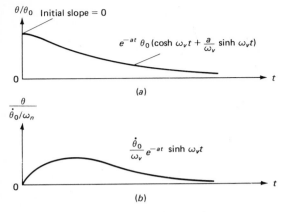

Figure 6.3.3-4 Response of overdamped system to initial conditions. (a) Initial displacement. (b) Initial velocity.

For the inverse Laplace transform, refer to Appendix 3, entries 285 and 286. Then

$$\frac{\theta}{\theta_0}(t) = (1 + at)e^{-at}$$

and

$$\frac{\theta}{\dot{\theta}_0}(t) = te^{-at} \qquad [6.3.3\text{-}5d]$$

The above agrees with eq. 6.3.3-5b.

Case 4. Overdamped, $\zeta > 1$

$$p = -a \pm \omega_v \qquad [6.3.3\text{-}6]$$

then

$$\theta(t) = B_1 e^{(-a+\omega_v)t} + B_2 e^{-(a+\omega_v)T}$$

$$= e^{-at}\left(B_1 e^{\omega_v t} + B_2 e^{-\omega_v t}\right) \qquad [6.3.3\text{-}6a]$$

Use Euler's identities,* collect terms, and apply IC

$$\theta(t) = \theta_0 e^{-at}\left(\cosh \omega_v t + \frac{a}{\omega_v}\sinh \omega_v t\right)$$

$$+ \frac{\dot{\theta}_0}{\omega_v}e^{-at}\sinh \omega_v t \qquad [6.3.3\text{-}6b]$$

---

*See Appendix 1.

A curve for each initial condition is shown in Fig. 6.3.3-4. Note that these curves resemble those for critical damping, and that the obvious difference is that the curves for the overdamped case are much flatter, demonstrating the sluggish nature of this case.

■ **EXAMPLE 6.3.3-1** Capacitor Discharge
The circuit shown in Fig. 6.3.3-5 has a capacitor C that was charged by using a $1\frac{1}{2}$-V dc source. When the switch is closed, determine the frequency of oscillation, the initial amplitude, and explain what is happening.

Solution. The differential equation for this system is given by eq. 3.3.1-3, which conforms to the form of the general equation eq. 6.3.1-1. Consequently, we may use eq. 6.2.4-4.

$$\omega_n = \sqrt{\frac{1}{LC}} = \sqrt{\frac{1}{0.1 \times 1.6 \times 10^{-6}}} \tag{1}$$

$$= 2500 \text{ rad/s} \tag{2}$$

Use eq. 6.2.4-3 to determine the damping ratio.

$$\zeta = \frac{R}{2L\omega_n} = \frac{250}{2 \times 0.1 \times 2500} = 0.5 \tag{3}$$

The frequency of oscillation is the damped natural frequency, or

$$\omega_d = \omega_n\sqrt{1 - \zeta^2} = 2500\sqrt{1 - 0.5^2} \tag{4}$$

$$= 2175 \text{ rad/s} \tag{5}$$

In order to determine the initial conditions, note that the switch was open before the problem began. Hence, initial current is zero.

$$i(0) = \dot{q}(0) = 0 \tag{6}$$

The initial charge was not given, but it can be computed with the use of eq. 2.4.3-2.

$$q(0) = CV(0) = 1.6 \times 10^{-6} \times 1.5 \tag{7}$$

or

$$q_0 = 2.40 \times 10^{-6} \text{ C} \tag{8}$$

Thus, there is an initial charge, but no initial current. The damping ratio is

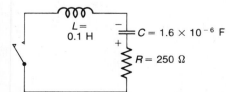

Figure 6.3.3-5 Capacitor discharge.

equal to 0.5. This system is an underdamped circuit with initial displacement. The solution is given by eq. 6.3.3-4b, the initial amplitude of which is given by

$$q_0 \omega_n / \omega_d = 2.40 \times 10^{-6} \times 2500/2175 \tag{9}$$

$$= 2.76 \times 10^{-6} \text{ C} \tag{10}$$

Description: Before the switch was closed, potential energy was stored as a negative charge $-q_0$ on the top plate of the capacitor. When the switch is closed, current flows through the inductor (reluctantly) and through the resistor to the other plate of the capacitor. At the instant when both plates have the same charge, current is still flowing; consequently (by the action of the inductor), current *continues* to flow, now building up a charge differential again. The current oscillates back and forth, but each time some energy is consumed in the resistor, producing the characteristic decaying sinusoid. ■

■ **EXAMPLE 6.3.3-2** Recoil Landing System

Rocket transport vehicles, which usually land on terrestrial bodies, use a three-legged landing system. Each leg consists of a spring and damper within a tube. Refer to Fig. 6.3.3-6(a). This system was discussed and modeled in Ex. 3.2.1-5. Using the parameters, free body diagram, and differential equation from that example, sketch the resulting motion at landing and discuss the design and how effective it is.

Free Body Diagram

In Ex. 3.2.1-5, the cluster of three springs has been replaced by a single vertical spring K which connects the mass M to the ground. The cluster of three dampers has been replaced by a single vertical damper D which connects the mass M to the ground. Neglect the mass of the foot, rod, and piston. Consequently, we may assume that the spring is at its free length before

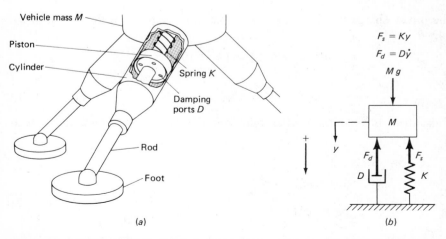

Figure 6.3.3-6   Recoil landing system. (a) Actual system. (b) Free body diagram.

landing. Whether the feet have touched the ground or not, there is a constant gravity force Mg acting downward. Then, the free body diagram is shown in Fig. 6.3.3-6(b).

Assuming the convention to be positive in the downward direction, sum the forces according to Newton's law

$$\sum F = M\ddot{y} = -F_s - F_d + Mg$$

or

$$M\ddot{y} + D\dot{y} + Ky = Mg \tag{1}$$

$$\text{IC} \quad y(0) = 0, \qquad \dot{y}(0) = V_0 \tag{2}$$

[Equations (1) and (2) are identical to eq. (6) in Ex. 3.2.1-4.] The differential equation is given by eq. (1) with its initial conditions in eq. (2). Note that eq. (1) is not homogeneous and that the force Mg is constant. This implies that it would be present even in steady state. We are certain that this landing system would come to rest. Hence, the limit exists, and then all derivatives would vanish. Hence, in steady state, eq. (1) becomes

$$Ky_{ss} = Mg$$

or

$$y_{ss} = \frac{Mg}{X} \tag{3}$$

This implies that in steady state, the spring would compress by an amount equal to y_{ss} and it would produce an equal and opposite force to oppose gravity. With this in mind, if we write the differential equation with respect to the steady state condition, the gravity force Mg would be cancelled by the spring force Ky_{ss}, and we would have a homogeneous system. To accomplish the above, we will change the variable from y (with spring compression initially zero) to Y (which is displacement with respect to the steady state position). In the new variable Y, the steady state position has been taken as zero; then this condition is equivalent to an *initial expansion* equal to $-Y_0$ of the spring. Hence, the initial conditions are $V_0, -Y_0$, where

$$Y_0 = \frac{Mg}{K} \tag{4}$$

The differential equation in the new variable becomes

$$M\ddot{Y} + D\dot{Y} + KY = 0 \tag{5}$$

where

$$M = \text{mass of vehicle} = 2000 \text{ kg (about 2.2 tons)}$$

$$D = \text{equivalent damping for three legs} = 24{,}000 \text{ N} \cdot \text{s/m} \tag{6}$$

$$K = \text{equivalent spring for three legs} = 72{,}000 \text{ N/m}$$

Equation (5) describes the system response as measured from the equilibrium

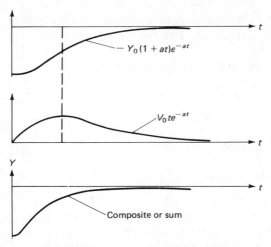

Figure 6.3.3-7   Response of recoil landing system.

or steady state position. Let us solve this equation by comparison to a solved problem. In order to accomplish this, it is necessary to determine the natural frequency and damping ratio as follows:

$$\omega_n = \sqrt{\frac{K}{M}} = \sqrt{\frac{72{,}000}{2000}} = 6 \text{ rad/s}$$

and   (7)

$$\zeta = \frac{D}{2M\omega_n} = \frac{24{,}000}{2 \times 2000 \times 6} = 1$$

Thus, the system is critically damped. The solution is given by eq. 6.3.3-5b, and the resulting motion is shown in Fig. 6.3.3-3.

$$Y = e^{-at}\left[-Y_0(1 + at) + V_0 t\right]$$ (8)

The numerical values of $-Y_0$ and V_0 were propitiously chosen so that the two effects are properly balanced. Note that the peak of the response to initial velocity V_0 is offset by the instantaneous value of the response to initial displacement $-Y_0$. If V_0 were larger, there would be an overshoot or bounce. See Fig. 6.3.3-7. The parameters chosen for this system have made it an effective design.   ■

■ **EXAMPLE 6.3.3-3** Overfilled Accumulator
An accumulator near the ocean is filled to a level that is 10 m above sea level when the valve is closed. When the valve is suddenly opened, determine the following: undamped natural frequency ω_n, damping ratio ζ, decay coefficient a, and period λ. See Fig. 6.3.3-8.

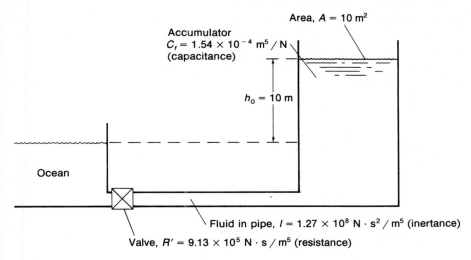

Figure 6.3.3-8   Overfilled accumulator.

Solution. The system configuration is similar to that in Fig. 3.6.2-1, but without a pump. The differential equation conforms to the form of the general system in eq. 6.3.1-1, permitting us to use eqs. 6.2.4-4 and 6.2.4-3. Then

$$\omega_n = \sqrt{\frac{1}{IC_f}} = \sqrt{\frac{1}{1.27 \times 10^8 \text{ N} \cdot \text{s}^2/\text{m}^5 \times 1.54 \times 10^{-4} \text{ m}^5/\text{N}}}$$

$$= 7.17 \times 10^{-3} \text{ rad/s} \tag{1}$$

$$\zeta = \frac{R'_f}{2I\omega_n} = \frac{9.13 \times 10^5 \text{ N} \cdot \text{s/m}^5}{2 \times 1.27 \times 10^8 \text{ N} \cdot \text{s}^2/\text{m}^5 \times 7.17 \times 10^{-3} \text{ rad/s}}$$

$$= 0.501 \tag{2}$$

$$a = \zeta\omega_n = 0.501 \times 7.17 \times 10^{-3} \text{ rad/s}$$

$$= 3.59 \times 10^{-3} \text{ rad/s} \tag{3}$$

$$\lambda = \frac{2\pi}{\omega_n} = \frac{2 \times 3.14}{7.17 \times 10^{-3} \text{ rad/s}} = 870 \text{ s} \tag{4}$$

Although the differential equation is given in terms of flow W, by virtue of the area A of the accumulator, the relationship may be thought of in terms of height h. Initially, the tank is overfilled to a height of 10 m. When the valve is opened, the water starts to flow, slowly overcoming its inertance I. The flow picks up speed and, in time, the level in the tank is equal to sea level. However, the flow doesn't stop; due to inertance, it continues to flow, rapidly at this point, and starts to empty the accumulator. The level in the accumulator begins to fall below sea level. This continues until the level is almost as far below sea level as it was above at the beginning. Finally, the flow stops and then reverses as water flows from the ocean into the accumulator. Thus, the water flows back and forth, where a certain amount of energy, initially stored as potential energy, is dissipated during each cycle. Thus, the amplitude of oscillation decays (as it does for all forms of this type of system) until it reaches the steady state condition—the level in the accumulator is equal to sea level.   ■

6.3.4 Response to Step Input

Let the forcing function in eq. 6.3.1-1 be a step. Then

$$J\ddot{\theta} + D\dot{\theta} + K\theta = \text{step input} = T \qquad [6.3.4\text{-}1]$$

Equation 6.3.4-1 is not homogeneous. Hence, its solution will be found in two parts, the homogeneous and the particular solutions. The homogeneous solution is given by eq. 6.2.2-3. The particular solution is a constant and is given by

$$\theta_p = A \qquad [6.3.4\text{-}2]$$

Apply the particular solution to the inhomogeneous equation, eq. 6.3.4-1, and we have

$$\theta_p = \frac{T}{K}$$

The complete solution is given by the sum of the particular solution, eq. 6.3.4-2 and the homogeneous solution, eq. 6.2.2-3

$$\theta = \frac{T}{K} + C_1 e^{p_1 t} + C_2 e^{p_2 t} \qquad [6.3.4\text{-}3]$$

where the constants C_1 and C_2 will be determined by applying the initial conditions IC = 0.

Case 1. Undamped, $\zeta = 0$

$$p = \pm\sqrt{-\frac{K}{J}} = \pm j\omega_n \qquad [6.3.4\text{-}4]$$

Then,

$$\theta = \frac{T}{K} + A_1 e^{j\omega_n t} + A_2 e^{-j\omega_n t} \qquad [6.3.4\text{-}5]$$

Substitute zero initial conditions and apply Euler's identities*

$$\theta = \frac{T}{K}(1 - \cos \omega_n t) \qquad [6.3.4\text{-}6]$$

---

*See Appendix 1.

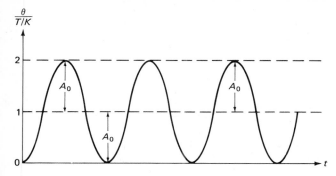

Figure 6.3.4-1   Response of undamped system to step input.

When the damping ratio is zero, no energy is removed from the system. Hence, the oscillation is maintained at constant amplitude A_0. Refer to Fig. 6.3.4-1.

Laplace Transform Approach   In eq. 6.3.1-1, set $D = 0$ and let $T(t)$ be a step input. Write the differential equation in the Laplace domain. Note that the initial conditions are equal to zero.

$$Js^2\theta(s) + K\theta(s) = \frac{T}{s} \qquad [6.3.4\text{-}7]$$

Solve for $\theta(s)$

$$\theta(s) = \frac{T}{s(Js^2 + K)} \qquad [6.3.4\text{-}7a]$$

Normalize

$$\frac{\theta}{T/K}(s) = \frac{\omega_n^2}{s(s^2 + \omega_n^2)} \qquad [6.3.4\text{-}7b]$$

In order to return to the time domain, apply the inverse Laplace transform, making use of entry 314 of Appendix 3

$$\frac{\theta}{T/K}(t) = 1 - \cos \omega_n t \qquad [6.3.4\text{-}7c]$$

which agrees with eq. 6.3.4-6.

Steady State Analysis   Since the solution does not approach a limit (it oscillates at constant amplitude), the steady state value will not be sought.

Case 2. Underdamped, $\zeta < 1$

$$p = -a \pm j\omega_d \qquad [6.3.4\text{-}8]$$

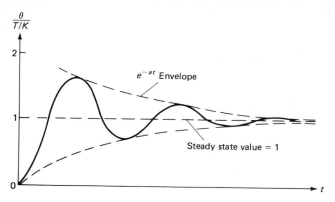

Figure 6.3.4-2   Response of underdamped system to step input.

Then

$$\theta = \frac{T}{K} + C_1 e^{(-a+j\omega_d)t} + C_2 e^{(-a-j\omega_d)t} \qquad [6.3.4\text{-}8a]$$

Use Euler's identities* and apply zero initial conditions

$$\theta = \frac{T}{K}\left[1 - e^{-at}\left(\cos \omega_d t + \frac{a}{\omega_d}\sin \omega_d t\right)\right] \qquad [6.3.4\text{-}8b]$$

Combine sine and cosine terms (refer to Section 4.2.1 and Appendix 1).

$$\theta = \frac{T}{K}\left[1 - \frac{\omega_n}{\omega_d}e^{-at}\sin(\omega_d t + \phi_1)\right] \qquad [6.3.4\text{-}8c]$$

where

$$\phi_1 = \cos^{-1}\zeta \qquad [6.3.4\text{-}8d]$$

When the system is underdamped, it will oscillate, but the amplitude of oscillation diminishes within a decay envelope. See Fig. 6.3.4-2.

Laplace Transform Approach   In eq. 6.3.1-1 let $T(t)$ be a step input. Write the equation in the Laplace domain

$$Js^2\theta(s) + DS\theta(s) + K\theta(s) = \frac{T}{s} \qquad [6.3.4\text{-}8e]$$

Solve for $\theta(s)$

$$\theta(s) = \frac{T}{s(Js^2 + Ds + K)} \qquad [6.3.4\text{-}8f]$$

---

*See Appendix 1.

Normalize

$$\frac{\theta}{T/K}(s) = \frac{\omega_n^2}{s(s^2 + 2as + \omega_n^2)} \qquad [6.3.4\text{-}8g]$$

where*

$$a = \zeta\omega_n = \frac{D}{2J} \qquad [6.3.4\text{-}8h]$$

In order to return to the time domain, apply the inverse Laplace transform, making use of entry 414 of Appendix 3

$$\frac{\theta}{T/K}(t) = 1 - \frac{\omega_n}{\omega_d}e^{-at}\sin(\omega_d t + \phi_1) \qquad [6.3.4\text{-}8i]$$

which agrees with eq. 6.3.4-8c.

Steady State Analysis   Since the solution approaches a limit, apply the final value theorem, eq. 4.4.4-1, to eq. 6.3.4-8g

$$\frac{\theta}{T/K}(t)_{ss} = \lim_{s\to 0} s\left[\frac{\omega_n^2}{s(s^2 + 2as + \omega_n^2)}\right] = 1 \qquad [6.3.4\text{-}8j]$$

which agrees with Fig. 6.3.4-2.

Case 3. Critically Damped, $\zeta = 1$

$$p = -a \quad \text{(note there will be repeated roots)} \qquad [6.3.4\text{-}9]$$

Then

$$\theta = \frac{T}{K} + C_1 e^{-at} + C_2 t e^{-at} \qquad [6.3.4\text{-}9a]$$

Apply zero initial conditions

$$\theta = \frac{T}{K}\left[1 - (1 + \cdot at)e^{-at}\right] \qquad [6.3.4\text{-}9b]$$

Case 4. Overdamped, $\zeta > 1$

$$p = -a \pm \omega_v \qquad [6.3.4\text{-}9c]$$

Then

$$\theta = \frac{T}{K} + C_1 e^{(-a+\omega_v)t} + C_2 e^{(-a-\omega_v)t} \qquad [6.3.4\text{-}9d]$$

Using Euler's identities, introducing new constants, applying zero initial

---

*See Appendix 1.

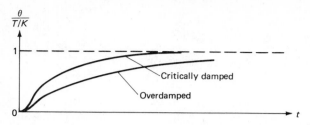

Figure 6.3.4-3   Response of critically damped and overdamped systems to step input.

conditions, and referring to Appendix 1, we have

$$\theta = \frac{T}{K}\left[1 - e^{-at}\left(\cosh \omega_v t - \frac{a}{\omega_v}\sinh \omega_v t\right)\right] \qquad [6.3.4\text{-}9e]$$

The critically damped and the overdamped cases have no oscillation. While the critically damped case represents the minimum amount of damping that will result in no oscillation, the question arises concerning the practical need for the overdamped case. The latter is sometimes designed to be sure that even where manufacturing tolerances may accumulate adversely, there will never be less than critical damping. If the damping ratio is much larger than unity, the system will appear very sluggish. See Fig. 6.3.4-3.

■ **EXAMPLE 6.3.4-1** Spring Scale
Consider a spring scale, in which the weighing platform is supported on a spring and lubricated piston (which provides damping). The translation of the platform is magnified by appropriate gearing to a calibrated dial. See Fig. 3.2.1-3. Various items are placed suddenly on the platform. Describe the motion. The system was modeled in Ex. 3.2.1-2.
    Given

M = equivalent lumped mass of all moving parts = 500 g

K = spring rate = 3.2×10^6 dyn/cm (about 18 lb/in.)

$D = 6.4 \times 10^4$ dyn · s/cm

m = mass of item to be weighed = 1500, 7500, 31,500 g (items are about 3, 17, and 70 lb)

Solution.   This system consists of a mass $M + m$, a spring K, and a damper D. The item of mass m that is dropped or suddenly placed on the platform represents a step input. The free body diagram is shown in Fig. 3.2.1-3(b), where $F_a = mg$. Note that the spring and damper are held constant, while the mass $M + m$ and the magnitude of the step input vary. The differential equation is given by eq. (8) of Ex. 3.2.1-2, repeated below,

$$(M + m)\ddot{y} + D\dot{y} + Ky = mg \qquad (1)$$

where y is measured from the initial position when the spring force was in equilibrium with the gravity force. Using this convention, the initial displacement is zero by definition. Since the scale was not moving initially, the initial velocity is also equal to zero. Then,

$$y(0) = \dot{y}(0) = 0 \tag{1a}$$

Transform eq. (1) to the Laplace domain

$$[(M + m)s^2 + Ds + K]Y(s) = \frac{mg}{s} \tag{2}$$

Solve for $Y(s)$ and normalize

$$\frac{Y}{mg/K}(s) = \frac{\omega_n^2}{s(s^2 + 2as + \omega_n^2)} \tag{2a}$$

where

$$\omega_n = \sqrt{\frac{K}{M + m}}$$

$$\zeta = \frac{D}{2(M + m)\omega_n} \tag{3}$$

$$a = \zeta\omega_n$$

Damped natural frequency (frequency of free vibration) is

$$\omega_d = \omega_n\sqrt{1 - \zeta^2} \tag{4}$$

The results are tabulated in Table 6.3.4-1. Note that the damping ratio ζ lies in the range between zero and unity. Hence, the system is under damped. The solution is given by entry 414 of Appendix 3. Then

$$\frac{y}{mg/K}(t) = 1 - \frac{\omega_n}{\omega_d}e^{-at}\sin(\omega_d t + \phi) \tag{5}$$

■

TABLE 6.3.4-1   RESPONSE OF SPRING SCALE

Item	$M + m$	ω_n	ζ	ω_d	Sketch of motion
1500	2000	40	0.4	36.8	
7500	8000	20	0.2	19.6	
31,500	32,000	10	0.1	9.95	

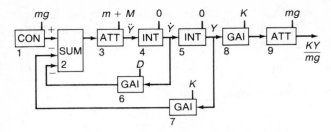

Figure 6.3.4-4   TUTSIM block diagram for spring scale.

■ **EXAMPLE 6.3.4-2** TUTSIM Simulation for Spring Scale

Determine the response of the spring scale in Ex. 6.3.4-1, using a TUTSIM simulation. In preparation for the TUTSIM solution, algebraically solve for the highest derivative in eq. (1):

$$\ddot{y} = \frac{mg - D\dot{y} - Ky}{m + M} \qquad (6)$$

Select the final time so that at least two full cycles appear in the output. Thus let final time = 2.0 s. Select the time interval $\Delta t = 0.002$ s. In order that the three curves will have compatible scales, normalize the output with respect to their static values. Let

Normalized output $R = y/y_0$,   where $y_0 = mg/K$ = static value

The TUTSIM block diagram for eq. (6) is shown in Fig. 6.3.4-4. The TUTSIM model data are given in Table 6.3.4-2. Results for the three cases are shown in Fig. 6.3.4-5.   ■

TABLE 6.3.4-2   TUTSIM MODEL FORMULATION

Model Structure	Model Parameters
:1,CON	:1,mg (three values)
:2,SUM,1, − 6, − 7	:2
:3,ATT,2	:3,$(m + M)$ (three values)
:4,INT,3	:4,0
:5,INT,4	:5,0
:6,GAI,4	:6,6.4 × 10⁴
:7,GAI,5	:7,3.2 × 10⁶
:8,GAI,5	:8,3.2 × 10⁶
:9,ATT,8	:9,mg (three cases)

Plotblocks and Ranges
  Horz:0,0,2
  Y1:9,0,2

Timing
  0.002,2

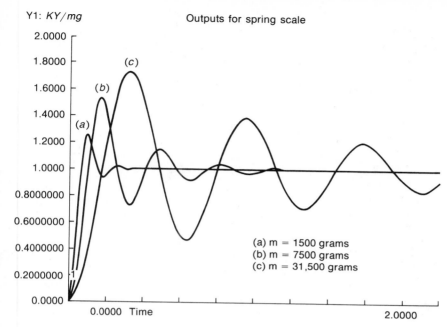

Figure 6.3.4-5   Spring scale response—linear damping.

■ **EXAMPLE 6.3.4-3** Computer Simulation of Spring Scale with Velocity Squared Damping

Note that damping in the previous example was assumed to be linear. Now consider damping that is proportional to the square of the velocity. Note that the *magnitude* would be determined by squaring the velocity, but the *sign* would be lost by the process. Since the damping force is a vector quantity, having magnitude and direction, and since the direction is opposite to that of the velocity vector, the damping force must be found by multiplying the velocity by its absolute value. This will provide both the magnitude and direction. The damping force vector is given by

$$F = DV|V| \tag{1}$$

Thus the differential equation becomes

$$\ddot{Y} = (mg - KY - DV|V|)/(m + M) \tag{2}$$

In order to be able to compare this case, velocity squared damping, to the previous case of viscous (linear) damping, it will be necessary to use an appropriate value for the coefficient D. Let D be equal to one-eighth of that in the viscous case. Since the velocity is squared, this value will approximately compensate for the resulting increase.

The TUTSIM block diagram for this example is shown in Fig. 6.3.4-6. The TUTSIM model data are listed in Table 6.3.4-3, and the output for three different inputs is shown in Fig. 6.3.4-7. Comparing these to the linear case,

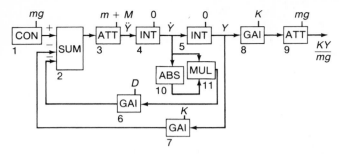

Figure 6.3.4-6   TUTSIM diagram for velocity squared damping.

TABLE 6.3.4-3   TUTSIM MODEL FORMULATION

Model Structure	Model Parameters
:1,CON	:1,1.47E5
:2,SUM,1, − 6, − 7	:2
:3,ATT,2	:3,2000
:4,INT,3	:4,0
:5,INT,4	:5,0
:6,GAI,11	:6,8000
:7,GAI,5	:7,3.2E6
:8,GAI,5	:8,3.2E6
:9,ATT,8	:9,1.47E5
:10,ABS,4	:10
:11,MUL,4,10	:11

Plotblocks and Ranges
    Horz:0,0,2
    Y1:9,0,2

Timing Data
    0.002,2

we note that they are similar during the first quarter cycle. After that, in the case of velocity squared damping, there is little further damping because the velocity is small. In velocity squared damping most of the decay occurs early (while the velocity is high). ∎

■ **EXAMPLE 6.3.4-4** Computer Simulation of Spring Scale with Coulomb Damping

As another nonlinear case, consider the spring scale with Coulomb damping. Coulomb damping is characterized by a constant damping force vector whose direction is opposite to that of the velocity vector. In order to simulate this, use will be made of the TUTSIM block REL (relay). For this example, the inputs to REL are set to provide an output of $+1$ when the velocity is positive, and -1 when the velocity is negative. The differential equation is

$$\ddot{Y} = (\text{input} - KY - DU)/(m + M) \qquad (1)$$

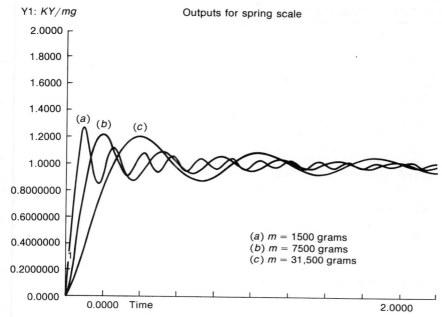

Figure 6.3.4-7   Velocity squared damping.

where

$$U = \text{Output of REL } (+1 \text{ or } -1) \tag{2}$$

In this example, only one value of m is used (namely, 7500 g), together with two different values of the damping coefficient D (which are 3.84×10^5 and 7.68×10^5). These values were selected to produce curves whose decay resembles that of the linear (viscous) case of Ex. 6.3.4-2. Since the differential equation contains a discontinuous function, integration requires a very small

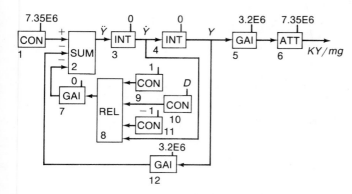

Figure 6.3.4-8   TUTSIM block diagram for Coulomb friction damping.

TABLE 6.3.4-4   TUTSIM MODEL DATA

Model Structure	Model Parameters
:1,CON	:1,7.35E6
:2,SUM,1, − 7, − 12	:2,8000
:3,INT,2	:3,0
:4,INT,3	:4,0
:5,GAI,4	:5,3.2E6
:6,ATT,5	:6,7.35E6
:7,GAI,8	:7,3.84E5 (or 7.68E5)
:8,REL,9,10,11,3	:8,0
:9,CON	:9,1
:10,CON	:10,0
:11,CON	:11, − 1
:12,GAI,4	:12,3.2E6

Plotblocks and Ranges
  Horz:0,0,2
  Y1:6,0,2

Timing
  0.002,2

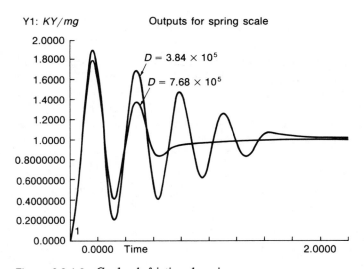

Y1: KY/mg Outputs for spring scale

$D = 3.84 \times 10^5$

$D = 7.68 \times 10^5$

Figure 6.3.4-9   Coulomb friction damping.

time step Δt. The TUTSIM block diagram for this example is shown in Fig. 6.3.4-8, and the model data are shown in Table 6.3.4-4.

The computed outputs for $D = 3.84 \times 10^5$ and 7.68×10^5 are shown in Fig. 6.3.4-9. Note that the decay envelope for Coulomb damping is a straight line rather than an exponential function.   ∎

6.4 RESPONSE TO RAMP AND IMPULSE INPUTS

6.4.1 Response Using Time Domain Methods

Both the ramp and the impulse inputs produce an inhomogeneous differential equation. The homogeneous portion of the solution is given by eq. 6.2.2-3, and the particular solution is formed by the linear combination of the input and its derivatives.

6.4.2 Response Using Laplace Transform Methods

In the Laplace domain, particularly with zero initial conditions, these two types of inputs become fairly simple problems.

6.4.3 Response to Ramp Input for Underdamped System

In eq. 6.2.2-1, let $F(t)$ be a ramp input Bt, and let the initial conditions be equal to zero. Write the equation in the Laplace domain

$$Ms^2X(s) + DsX(s) + KX(s) = \frac{B}{s^2} \qquad [6.4.3\text{-}1]$$

where

$$B = KV \qquad [6.4.3\text{-}2]$$

Solve for $X(s)$

$$X(s) = \frac{KV}{s^2(M^2 + Ds + K)} \qquad [6.4.3\text{-}3]$$

Normalize

$$\frac{X}{V/\omega_n}(s) = \frac{\omega_n^3}{s^2(s^2 + 2as + \omega_n^2)} \qquad [6.4.3\text{-}4]$$

In order to return to the time domain, apply the inverse Laplace transform, making use of entry 413 of Appendix 3

$$\frac{X}{V/\omega_n}(t) = \omega_n t - \frac{2a}{\omega_n} + \frac{\omega_n}{\omega_d}e^{-at}\sin(\omega_d t + 2\phi_1) \qquad [6.4.3\text{-}5]$$

This result is asymptotic to a straight line whose slope is equal to $\omega_n t$ that does not pass through zero, but is instead displaced down by $2a/\omega_n$. Superimposed upon this asymptote is a decaying sinusoid. See Fig. 6.4.3-1.

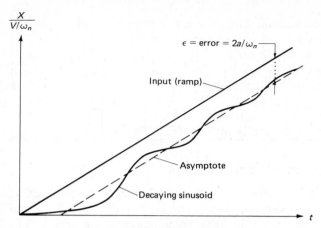

Figure 6.4.3-1   Response to ramp input.

In steady state, eq. 6.4.3-5 becomes

$$\left.\frac{X}{V/\omega_n}\right|_{ss} = \omega_n t - \frac{2a}{\omega_n} \qquad [6.4.3\text{-}6]$$

Equation 6.4.3-6 is the equation of a straight line having a slope equal to ω_n and an intercept equal to $-2a/\omega_n$. This line is therefore parallel to the input and is $2a/\omega_n$ below the input. This vertical displacement of the straight line represents an error ϵ, similar to that in Section 5.3.4 but for the second order system this error is equal to $2a/\omega_n$ for all cases. Then

$$\epsilon = -\frac{2a}{\omega_n} \qquad [6.4.3\text{-}7]$$

Steady State Analysis   The displacement obviously does not approach a limit, but the velocity does. Then, take the derivative of eq. 6.4.3-3

$$\dot{X}(s) = sX(s) - x(0) = \frac{KV}{s(Ms^2 + Ds + K)} - 0 \qquad [6.4.3\text{-}8]$$

Apply* the final value theorem, eq. 4.4.4-1

$$\dot{X}(t)\big|_{ss} = \lim_{s \to 0} s\left[\frac{KV}{s(Ms^2 + Ds + K)} \right] = V \qquad [6.4.3\text{-}9]$$

which agrees with Fig. 6.4.3-1.

---

*Since we know that the velocity approaches a limit.

6.4.4 Response to an Impulse Input

In eq. 6.2.2-1, let the input $F(t)$ be an impulse δ. Write the equation in the Laplace domain.

$$Ms^2X + DsX + KX = \hat{F} = F\delta \qquad [6.4.4\text{-}1]$$

or

$$(Ms^2 + Ds + K)X = F \qquad [6.4.4\text{-}2]$$

or

$$X(s) = \frac{F}{Ms^2 + Ds + K} \qquad [6.4.4\text{-}3]$$

Normalize (make use of Appendix 1)

$$\frac{X}{F\omega_n/K}(s) = \frac{\omega_n}{s^2 + 2as + \omega_n^2} \qquad [6.4.4\text{-}4]$$

To transform back to the time domain, use Appendix 3, entry 415

$$\frac{X}{\delta\omega_n/K}(t) = \frac{\omega_n}{\omega_d}e^{-at}\sin\omega_d t \qquad [6.4.4\text{-}5]$$

Note that the response to an impulse input is the same as that for an initial velocity. Compare eqs. 6.4.4-5 and 6.3.3-4d.

Steady Sate Analysis   Since the solution approaches a limit, apply the Final Value Theorem, eq. 4.4.4-1, to eq. 6.4.4-3.

$$X(t)_{ss} = \lim_{s\to 0} s\left[\frac{F}{Ms^2 + Ds + K}\right] = 0 \qquad [6.4.4\text{-}6]$$

which agrees with Fig. 6.3.3-2(b).

6.5 TRANSIENT RESPONSE FOR ASSORTED INPUTS

The previous sections treated systems with only one input or initial condition. Often, real systems have combinations of these. It is possible to seek the solution by comparison to solved problems. However, this approach should not be overdone since it may rob the analyst of the opportunity to perform the task without assistance. There will be times when the problem does not conform to a solved one. In a futile search, the analyst may spend more time in seeking a similar problem than in solving the problem from the beginning. In this section, the analyst is encouraged to set up the problem and seek an independent solution using the general concepts, but not to imitate those presented in this chapter.

■ **EXAMPLE 6.5.1-1** Coupled Fluid Flow System
Two reservoirs are connected as shown in Fig. 6.5.1-1. Neglecting inertance, show that this system is of second order.

Solution.   Apply eqs. 3.5.2-3 and 3.5.2-4 to reservoir no. 1

$$Q_1 = \frac{h_1 - h_2}{R_1} \tag{1}$$

$$Q_{in} - Q_1 = A_1 \dot{h}_1 \tag{2}$$

Substitute eq. (1) into eq. (2) and simplify

$$A_1 \dot{h}_1 + \frac{h_1}{R_1} = Q_{in} + \frac{h_2}{R_1} \tag{3}$$

Apply a similar procedure to reservoir no. 2

$$Q_2 = \frac{h_2}{R_2} \tag{4}$$

$$Q_1 - Q_2 = A_2 \dot{h}_2 \tag{5}$$

and

$$A_2 \dot{h}_2 + h_2 \left(\frac{1}{R_1} + \frac{1}{R_2} \right) = \frac{h_1}{R_1} \tag{6}$$

Take the derivative of eq. (6)

$$\frac{\dot{h}_1}{R_1} = A_2 \ddot{h}_2 + \dot{h}_2 \left(\frac{1}{R_1} + \frac{1}{R_2} \right) \tag{7}$$

Equations (3) and (6) are simultaneous equations in the variables h_1 and h_2. Eliminate one of them, say h_1, by applying eqs. (6) and (7) to eq. (3). Then divide both sides by R_2 and simplify.

$$F\ddot{h}_2 + G\dot{h}_2 + h_2 = R_2 Q_{in} \tag{8}$$

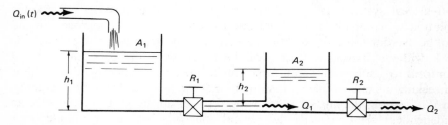

Figure 6.5.1-1   Coupled fluid flow system.

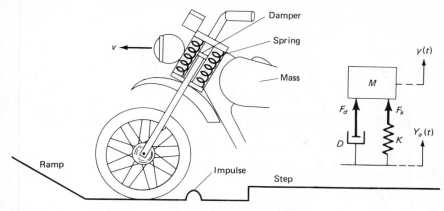

Figure 6.5.1-2   Suspension system for front wheel of motorcycle.

where

$$F = R_1 R_2 A_1 A_2$$

$$G = R_1 A_1 + R_2 A_2 + R_2 A_1 \qquad (9)$$

Thus, we see that the two coupled first order systems produce a single second order system. ■

■ **EXAMPLE 6.5.1-2** Suspension System for Motorcycle

The objective of a suspension system is to minimize the transmission of the road bumps to the rider. Since road conditions may vary, the design is often a compromise. In order to analyze the system, the road and the vehicle must be modeled. Making use of the results of Problem 3.29, the suspension system can be represented by an equivalent vertical spring-damper parallel network which lies between the road and the vehicle mass M. The actual system and its free body diagram are shown in Fig. 6.5.1-2. Determine the response of the system to each of the following road conditions.

(a) A single bump (approximately an impulse input).
(b) Vehicle is driven off the curb (step input).
(c) Vehicle climbs hill (ramp input).

Solution.   Sum the forces according to Newton's law

$$\sum F = F_k + F_d = M\ddot{y} \qquad (1)$$

where $F_k = K(Y_a - y)$
        $F_d = D(\dot{Y}_a - \dot{y})$

then

$$M\ddot{y} = K(Y_a - y) + D(\dot{Y}_a - \dot{y})$$

or, collecting terms,

$$M\ddot{y} + D\dot{y} + Ky = D\dot{Y}_a + KY_a \tag{2}$$

Transform to Laplace domain and let initial conditions be equal to zero

$$Ms^2Y + DsY + KY = DsY_a + KY_a$$

Collect terms and solve for Y

$$Y(s) = \frac{(Ds + K)Y_a(s)}{Ms^2 + Ds + K} \tag{3}$$

Equation (3) represents the system with the generalized input $Y_a(s)$.

(a) Response to Impulse Input

Assume that there is a bump, but no gully. The input is approximately an impulse displacement. Then, $Y_a(s) = A\delta$ and

$$Y(s) = \frac{(Ds + K)A}{Ms^2 + Ds + K} \tag{4}$$

Normalize

$$\frac{Y}{A\omega_n}(s) = \frac{2as + \omega_n^2}{\omega_n(s^2 + 2as + \omega_n^2)} \tag{5}$$

The solution in the time domain is found by using entry 447, Appendix 3.

$$\frac{y}{A\omega_n}(t) = -\frac{e^{-at}}{\beta}\sin(\omega_d t - 2\phi) \tag{6}*$$

(b) Response to Step Input

Assume the motorcycle is driven off a curb. This represents a downward step input. Then, $Y_a(s) = -U/S$ and eq. (3) becomes

$$Y(s) = \frac{(Ds + K)(-U/s)}{Ms^2 + Ds + K} \tag{7}$$

Normalize

$$\frac{Y}{U}(s) = \frac{-(2as + \omega_n^2)}{s(s^2 + 2as + \omega_n^2)} \tag{8}$$

---

*For definition of terms, see Appendix 1.

The transient response in the time domain is found by using entry 446, Appendix 3.

$$\frac{y}{U}(t) = -\left[1 + \frac{e^{-at}}{\beta}\sin(\omega_d t - \phi)\right] \qquad (9)$$

(c) Response to Ramp Input

Assume that the motorcycle has been riding along level ground which suddenly becomes an upgrade. Refer to Fig. 6.5.1-2. Let time equal zero just as the motorcycle makes contact with the upgrade. Then the input is a ramp that starts at time = 0. Then, $Y(s) = V/S^2$ and

$$Y(s) = \frac{(Ds + K)V/s^2}{Ms^2 + Ds + K} \qquad (10)$$

Normalize

$$\frac{Y(s)}{V/\omega_n} = \frac{\omega_n(2as + \omega_n^2)}{s^2(s^2 + 2as + \omega_n^2)} \qquad (11)$$

The transient response in the time domain is found by using entry 445 in Appendix 3. Then,

$$\frac{y(t)}{V/\omega_n} = \omega_n t - \frac{\omega_n}{\omega_d}e^{-at}\sin(\omega_d t) \qquad (12)$$

∎

6.6 PROPERTIES OF DECAYING VIBRATION

6.6.1 Time Constant and Settling Time

The *time constant* is an index to the rate of decay where *time* is the reference. The *settling time* relates the rate of decay in terms of the *number of time constants*. The *logarithmic decrement* describes the rate of decay in terms of the *number of cycles*.

The amplitude of damped vibration is given approximately* by the decay envelope. See Fig. 5.1.2-1. The decay curve was discussed in Section 5.5. Since all second order systems have the same decay envelope, then the time constant

---

*The decay curve does not pass through the *peak* points exactly.

τ and the settling time T_s are common to all second order systems. Thus,

$$\tau = \frac{1}{a} = \frac{1}{\zeta\omega_n} \qquad [6.6.1\text{-}1]$$

$$T_s = n\tau \qquad [6.6.1\text{-}2]$$

For numerical values of n see Table 5.5.2-1.

6.6.2 Log Decrement

The *logarithmic decrement* Λ is defined as the natural log of the ratio of any two successive amplitudes of oscillation.

$$\Lambda = \ln\frac{A_0}{A_1} = \ln\frac{A_1}{A_2} = \ln\frac{A_{n-1}}{A_n} \qquad [6.6.2\text{-}1]$$

Forming the appropriate triangle on log-log paper, the ratio is given by the slope a multiplied by the side, period λ_d

$$\Lambda = \ln\frac{A_0}{A_1} = a\lambda_d = \zeta\omega_n\lambda_d \qquad [6.6.2\text{-}2]$$

Apply Appendix 1 to eq. 6.6.2-2.

$$\Lambda = \frac{2\pi\zeta}{\sqrt{1 - \zeta^2}} \qquad [6.6.2\text{-}3]$$

The above expression may be written for r cycles, where A_0 is the amplitude of the first, and A_r of the last,

$$\ln\frac{A_0}{A_r} = \ln\left(\frac{A_0}{A_1}\right)^r = r\Lambda \qquad [6.6.2\text{-}4]$$

or

$$\Lambda = \frac{1}{r}\ln\frac{A_0}{A_r} \qquad [6.6.2\text{-}5]$$

Similarly, for half a cycle

$$\Lambda = 2\ln\frac{A_0}{A_{1/2}} \qquad [6.6.2\text{-}6]$$

The above functions are listed in Table 6.6.2-1.

TABLE 6.6.2-1 DECAY PROPERTIES FOR ALL SECOND ORDER SYSTEMS

ζ	β	Λ	A_1/A_0	$A_{1/2}/A_0$	Curve
0	1	0	1	1	
0.01	1.000	0.0628	0.943	0.971	
0.02	1.000	0.125	0.893	0.943	
0.05	0.999	0.315	0.730	0.854	
0.07	0.997	0.440	0.641	0.800	
0.10	0.995	0.631	0.527	0.729	
0.15	0.989	0.955	0.385	0.621	
0.2	0.980	1.28	0.278	0.527	
0.25	0.968	1.62	0.196	0.444	
0.3	0.954	1.98	0.137	0.372	
0.4	0.917	2.74	0.0645	0.254	
0.5	0.866	3.63	0.0256	0.163	
0.6	0.800	4.71	0.0091	0.095	
0.7	0.714	6.16	0.0021	0.0461	
0.8	0.600	8.38	0.0006	0.0150	
0.9	0.436	13.0	0.0001	0.0015	
1.0	0	∞	0	0	

SUGGESTED REFERENCE READING FOR CHAPTER 6

Introduction to vibration [1], p. 1–9; [4], p. 1–4; [5], p. 1–9; [17], p. 225–229; [46], chap. 5.
Energy method [1], p. 24–25; [4], p. 15–19; [5], p. 9–13; [8], p. 112; [17], p. 229–235, 482–488.
Damping ratio [1], p. 75–79; [4], p. 25; [5], p. 37; [8], p. 305–308; [17], p. 251.
Free vibration [1], chap. 3; [4], chap. 1; [5], chaps. 1, 2; [17], chaps. 7, 8; [46], chap. 7.
Computer simulation [4], chaps. 6, 7; [5], p. 118–123; [12]; [8], p. 117–121; [17], p. 299–308; [46], p. 410–432.
Properties of vibration [1], p. 85; [4], p. 28; [5], p. 43.
Distributed parameters [4], chap. 5; [5], chap. 8; [8], p. 158; [15]; [17], p. 462–488; [47], chap. 10.
Nonlinear systems [4], p. 29–38; [5], chap. 5; [17], chap. 10.

PROBLEMS FOR CHAPTER 6

6.1. For the system shown in Fig. P-6.1, $M = 8.0$ kg, $K = 200$ N/m, and $D = 16.0$ N · s/m.
(a) Find the differential equation for x, using a free body diagram of M.
(b) Find ω_n, ζ, and ω_d.
(c) Given $x(0) = 0$ and $\dot{x}(0) = 10$ m/s, find $x(t)$.

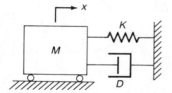

Figure P-6.1

6.2. The system shown consists of a block of mass $M = 5.0$ kg, to which is attached a spring of modulus $K = 2500$ N/m. The initial velocity is $V_0 = 20$ m/s.
(a) Find the maximum spring compression.
(b) Find the maximum deceleration of the mass.

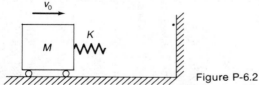

Figure P-6.2

6.3. Consider the spring scale of Ex. 6.3.4-1, with $m = 7500$ g placed on the scale pan and released at $t = 0$.
(a) Find the value of D which must be used to make $\zeta = 0.5$.
(b) Find y (cm) at $t = (\pi/\omega_d)$ s.

6.4. The differential equation for x, and the initial conditions, for the system shown are:

$$\left(M + \frac{J}{R^2} \right) \ddot{x} + D\dot{x} + Kx = f, \qquad x(0) = 0, \quad \dot{x}(0) = 0$$

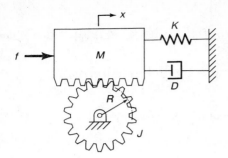

Figure P-6.4

Given $M = 2.0$ kg, $J = 0.01$ kg-m^2, $D = 7.2$ N $\cdot$ s/m, $K = 27.0$ N/m, $f = 5.4$ N (step), and $R = 0.1$ m:
(a) Find ω_n, ζ, and ω_d.
(b) Find the steady state value of x.

6.5. The mass M with the spring K attached is dropped from rest from a height h. Given $M = 20$ kg, $K = 50,000$ N/m, and $h = 2.0$ m:
(a) Find the velocity of M when the spring first touches the floor.
(b) Find the maximum spring compression.

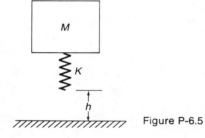

Figure P-6.5

6.6. For the system shown, $J = 3.75 \times 10^{-4}$ kg $\cdot$ m^2, $K = 12.0$ N/m, $D = 2.0$ N $\cdot$ s/m, and $R = 0.05$ m; $\theta = 0$ when the spring is unstressed.
(a) Derive the differential equation for θ.

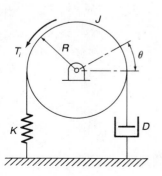

Figure P-6.6

(b) Find ω_n and ζ.

(c) Find the steady state value of θ if $T_i = 0.015$ N $\cdot$ m (step), $\theta(0) = 0$, and $\dot{\theta}(0) = 0$.

6.7. For the spring-mass system shown, $M = 2.5$ kg, $K = 1000$ N/m, $x(0) = 0$, and $\dot{x}(0) = 1$ m/s:

(a) Find $x(t)$.

(b) Find the maximum compression of the spring.

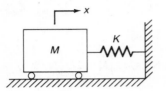

Figure P-6.7

6.8. Consider the rotational system shown. For steel, $G = 8.27 \times 10^{10}$ N/m^2 and $\rho = 7800$ kg/m^3, and for the fluid, $\mu = 0.309$ N $\cdot$ s/m^2. Given $d_1 = 0.01$ m, $d_2 = 0.40$ m, $L_1 = 0.50$ m, $L_2 = 0.30$ m, and $h = 0.2$ mm:

(a) Find the torsional stiffness K of the shaft;

(b) Find the moment of inertia J of the steel rotor;

(c) Find the torsional damping constant D, ignoring the viscous effects of the oil on the left and right ends of the rotor;

(d) Find the natural frequency ω_n and the damping ratio ζ.

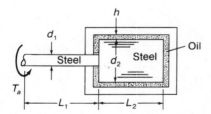

Figure P-6.8

6.9. For the system shown, $M = 12$ kg, $K = 240$ N/m, $D = 48$ N $\cdot$ s/m, and f is an impulse of magnitude 24 N $\cdot$ s.

(a) If $x(0) = \dot{x}(0) = 0$ prior to the impulse application, find $x(t)$ using Laplace transforms.

(b) Draw a reasonably accurate sketch of x versus t.

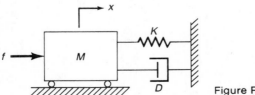

Figure P-6.9

6.10. For the system shown, $M = 2.0$ kg, $K = 5000$ N/m, and $V_0 =$ the initial velocity of the mass $= 5.0$ m/s.

(a) Use the energy method to find the maximum compression of the spring.

(b) Find the time between initial contact of the spring and maximum compression.

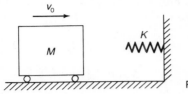

Figure P-6.10

6.11. The graph shown is for the motion of a second order damped linear system. Using data from the graph, estimate ζ, ω_n, and ω_d.

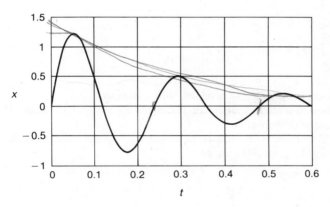

Figure P-6.11

6.12. The circuit in Fig. P-6.12 provides a means to listen to a decaying sinusoid. The average person can hear a sound when it is one-millionth of the loudest sound

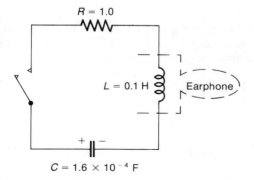

Figure P-6.12

that can be heard comfortably. Avoiding the extremes, how long will the sound in the circuit be heard if a level of 4.5×10^{-5} is arbitrarily selected as the cutoff point? The capacitor C is initially charged to a level that will make an initial sound that is loud but comfortable.

6.13. For the rotational system shown, $J = 0.70$ kg $\cdot$ m^2, $D = 2.1$ N $\cdot$ m $\cdot$ s, $R = 0.5$ m, $K = 70.0$ N/m, $\theta(0) = 0.1$ rad, and $\dot{\theta}(0) = 0$. The spring is unstressed when $\theta = 0$. Find $\theta(t)$ using Laplace transforms.

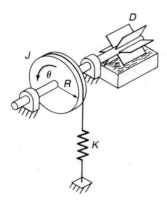

Figure P-6.13

6.14. The system shown is a model of a vehicle with a bumper.
   (a) Obtain the differential equation for the displacement x of the mass M just after impact with the wall.
   (b) If $x(0) = 0$ and $\dot{x}(0) = V$, find $x(t)$ by means of Laplace transforms. Assume the system to be critically damped.

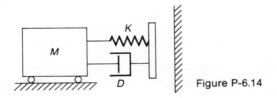

Figure P-6.14

6.15. A steel shaft 2.5 m long and 1.0 cm in diameter drives a flywheel of $J = 8.25$ kg $\cdot$ m^2 at 120 rev/min. The drive pulley suddenly stops. Find the angle of twist of the shaft when the flywheel first comes to rest. Ignore damping.

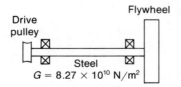

Figure P-6.15

6.16. For the bullet-block system shown, $M_1 = 0.01$ kg, $v_1 = 600$ m/s, $M_2 = 1.49$ kg, $K = 150$ N/m, $D = 12$ N $\cdot$ s/m, and the initial velocity v_2 of the block is zero.

(a) Find the velocity of the bullet-block combination just after the bullet is stopped.

(b) Consider the velocity found in (a) to be the initial velocity, with initial displacement approximately zero. Find $x_2(t)$ after impact.

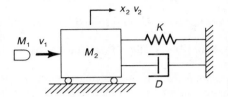

Figure P-6.16

6.17. For the system shown, $J = 0.000375$ N $\cdot$ m $\cdot$ s^2, $R = 0.05$ m, $M = 0.25$ kg, $K = 10.0$ N/m, and $D = 1.6$ N $\cdot$ s/m.

(a) Derive the differential equation for x.

(b) Find ω_n and ζ.

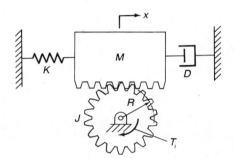

Figure P-6.17

6.18. Equations (1) and (2) in Ex. 6.3.3-2 pertain to the recoil system of Fig. 6.3.3-6. Given $M = 1000$ kg, $g = 1.6$ m/s^2 (moon's surface), and $V_0 = 2.0$ m/s:

(a) Find K and D if $\omega_n = 5.0$ rad/s and $\zeta = 0.6$.

(b) Find $y(t)$ by Laplace transforms and plot y versus t.

(c) Obtain $y(t)$ using TUTSIM and compare your results.

6.19. Consider the two tank system shown. $A_1 = 4.0$ m^2, $A_2 = 3.0$ m^2, $h_1(0) = 5.2$ m, and $h_2(0) = 2.0$ m.

(a) Obtain two first order differential equations for h_1 and h_2.

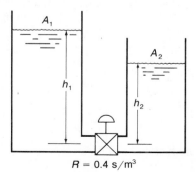

$R = 0.4$ s/m^3 Figure P-6.19

(b) Eliminate h_2 and obtain a second order differential equation for h_1.

(c) Find $h_1(t)$ using TUTSIM.

6.20. Refer to Fig. 6.5.1-1 (Coupled Fluid Flow System). A short pulse of fluid flow Q_{in} causes the level in tank no. 1 to rise to h_{10} at a time when the level in tank no. 2 is h_{20}.

(a) Determine the conditions that would cause the fluid to oscillate between the tanks indefinitely. (Can $I = 0$?)

(b) Determine the conditions that would never allow flow rate Q_1 to become negative. What is the equivalent ζ?

(c) Determine the conditions that would result in a decaying oscillation where each succeeding cycle has an amplitude that is one-half that of the previous cycle.

6.21. (a) Given K, M, D, and x_a for the system of Fig. P-6.21, derive the differential equation for x.

(b) If $x(0) = \dot{x}(0) = 0$, and x_a is a step of magnitude A, find the initial acceleration of M.

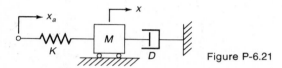

Figure P-6.21

6.22. For the system of Fig. P-6.22, draw the free body diagram of the mass and derive the differential equation for x, the displacement parallel to the inclined plane after the bumper strikes the wall.

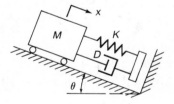

Figure P-6.22

6.23. Draw the necessary free body diagrams for the system of Fig. P-6.23, and derive the differential equation for θ. T, K, J, and D are given, and the spring is unstressed when $\theta = 0$.

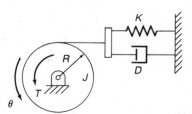

Figure P-6.23

6.24. A beam of length L, mass M, and flexural rigidity EI vibrates in a pattern similar to that shown in Fig. P-6.24.

    (a) Obtain the approximate natural frequency ω_n of the beam using the equivalent mass and equivalent spring constant at the center.

    (b) The exact value of ω_n^2 is $(\pi^4 EI)/(ML^3)$. Find the percent error for the approximate value.

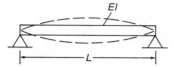

Figure P-6.24

6.25. For the system shown in Fig. 6.5.1-1, let $A_1 = 3.5$ m², $A_2 = 2.4$ m², $R_1 = 5.1$ s/m², and $R_2 = 2.6$ s/m².

    (a) Find ω_n and ζ.

    (b) Find the steady state value of h_2 if $Q_{in} = 1.3$ m³/s (constant).

6.26. A square block of mass $M = 4.0$ kg has attached to it, at its corners, four identical springs of modulus $K = 1250$ N/m. The free length of each spring is $L = 0.3$ m. Find the height through which the assembly must fall so that all springs are compressed an amount $L/2$ when the block comes to rest. (See Fig. P-6.26.)

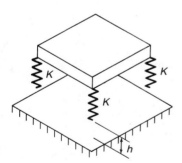

Figure P-6.26

6.27. For the system shown, D, J, and K are given.

    (a) Derive the differential equation for θ.

    (b) If $\dot{\theta}(0) = 0$ and $\theta(0) = \theta_0$, find the initial angular acceleration of the disk. The spring is untwisted when $\theta = 0$.

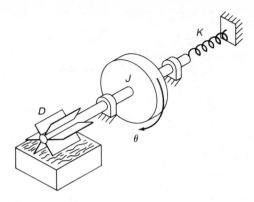

Figure P-6.27

6.28. (a) Obtain the differential equation for the depth h for the fluid system shown.
(b) Given $I = 0.08$ N $\cdot$ s^2/m^5, $R' = 160$ N $\cdot$ s/m^5, $A = 0.4$ m^2, and $P = 8000$ N/m^3, find ω_n and ζ.

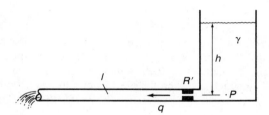

Figure P-6.28

6.29. Derive a rule for obtaining the nth derivative of
(a) $\sin \omega t$ (b) $\cos \omega t$

6.30. Resolve each of the following into a single sinusoid.
(a) $A \sin \omega t - B \cos \omega t$
(b) $-A \sin \omega t + B \cos \omega t$
(c) $A \sin \omega t + B \sin(\omega t + \phi)$
(d) $A \sin(\omega t + \phi_1) + B \sin(\omega t + \phi_2)$

6.31. Refer to Problem 3.25(g). Let

$$J = 2 \times 10^{-3} \text{ N} \cdot \text{m} \cdot \text{s}^2, \qquad R = 0.1 \text{ m}, \qquad M = 0.3 \text{ kg}, \qquad K = 50 \text{ N/m}$$

(a) What value of D will give a damping ratio equal to one?
(b) What is the undamped natural frequency?
(c) What is the damped natural frequency if the damping ratio is 0.2.

6.32. Refer to Prob. 3.25(j).
(a) If the "U" tube were filled with water at 20°C, and if $L = 0.3$ m and $R = 0.05$ m, determine the natural frequency (neglect viscous effects).
(b) Rework the problem in terms of the algebraic terms shown in the figure.

6.33. Refer to Problem 3.25(k).
   (a) If $J = 20$ N · m · s^2, determine K and D so that the undamped natural frequency is equal to 15 rad/s and the damping ratio is equal to 0.5.
   (b) Determine the angular displacement input step that would result in a response of 0.1 rad when the time is equal to 0.01 s.
   (c) Explain how the device acts as a fluid drive.

6.34. Refer to Problem 3.25(k). Given $J = 20$ N · m · s^2, $D = 200$ N · m · s, and $K = 2000$ N · m. For each of the following, determine the transient response, and sketch the output displacement curve as a function of time. Label all pertinent points.
   (a) Input displ. = 0
      IC $\theta_0 = 0.25$ rad, $\dot{\theta}_0 = 0$
   (b) Input displ. = 0
      IC $\theta_0 = 0$, $\dot{\theta}_0 = 8$ rad/s

6.35. In each of the following electrical systems, determine the transient response to an impulse, a step, and a ramp input.
   (a) Fig. 3.6.2-1(c) (Series Circuit).
   (b) Fig. 3.6.2-1(d) (Parallel Circuit).
   (c) Ex. 5.1.3-1 (Capacitor Flashlight).
   (d) Fig. 7.6.3-1 (RC Output of Series Circuit).
   (e) Fig. 7.7.2-1 (Transformer Output of Series Circuit).

6.36. A parallel electrical circuit with a current source, like that shown in Fig. 3.6.3-1(d) is driven by long pulses. The pulse width is equal to three time constants. Sketch the response for three pulses.

6.37. An elevator in an office building is descending at a velocity $V_0 = 5$ m/s. When the car is 100 m below the cable drum, the drum is suddenly stopped. Determine

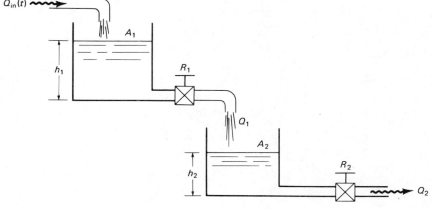

Figure P-6.38

the maximum stress in the cable for the following parameters:

$$M = \text{mass of car and passengers} = 1000 \text{ kg}$$

$$A = \text{cross sectional area of cable} = 10^{-4} \text{ m}^2$$

$$E = 2.07 \times 10^{11} \text{ N/m}^2 \text{ (steel)}$$

$$\rho = 7800 \text{ kg/m}^3 \text{ (steel)}$$

Assume the cable acts like a single strand (and include its mass).

6.38. *Cascaded Reservoirs.* (See Fig. P-6.38.) Two reservoirs have valves at their exits. The valves can be adjusted in order to alter the fluid resistances R_1 and R_2.
(a) Model the system for $Q_{in}(t) = Q_0$ (a constant).
(b) Determine the valve settings so that the heights h_1 and h_2 will remain constant.
(c) Model the system for $Q_{in}(t) = $ general function of time.

Chapter **7**

Forced Vibration and Frequency Response of Second Order Systems

This chapter is concerned with the dynamic response of second order systems that are continuously excited by a sinusoidal input. The subject of frequency response makes allowance for inputs at all frequencies. Hence, the subject is really treating an infinite number of inputs, which is quite different from that in the previous chapter where there was only one input at a time. This chapter shows several means by which the frequency response can be determined.

The chapter opens the subject with the significance of frequency response. This is reinforced by presenting several techniques for obtaining the frequency response. Among these are: method of undetermined coefficients, Euler's identities (complex frequency method), transformation from Laplace domain to complex domain, and computer simulation.

The subject is introduced on a simple level using first order systems. A dial movement, approximated as a first order system, is studied using the method of undetermined coefficients. The problem is then solved on the computer in an effort to show the reader how vibration might appear when the input frequency varies.

The method of undetermined coefficients is applied to a simple second order system. It is pointed out that there are many different second order systems and that each one has its individual frequency response. The desirability of short cuts to finding frequency response becomes evident when one attempts to apply the methods to practical systems. Some of the well-known shortcut methods are introduced and are applied.

Some of the illustrative examples include: a phono pickup, series electrical circuits, a motorcycle on a bumpy road, and a seismic instrument. Some of these are solved on the computer in addition to being solved analytically. Here we see that the computer can have shortcomings. For sinusoidally forced systems, the resulting response becomes unwieldy on the computer. However, if the problem were first formulated in the operator notation (either complex or Laplace domain) and if this were reduced to mere tables of computational needs, the combination of analytical and computer methods becomes an ideal approach. Once again we see that no one tool is the overall best, and that there is need for the coexistence and the collaboration of several of them.

7.1 FREQUENCY RESPONSE

7.1.1 Meaning of Frequency Response

Using several methods, it will be shown that when a system is excited by a sinusoidal input, the output in steady state is also sinusoidal, and the frequency of the output becomes equal to the frequency of the input. It will be shown that the amplitude and the phase of the output *vary* with input frequency. This variation is referred to as "frequency response."

If a typical second order system is so excited using three frequencies, the typical response is as shown in Fig. 7.1.1-1. In Fig. 7.1.1-1, we find that in each of the three cases, in steady state, the output frequency becomes equal to the input frequency. Also note that while the *displacement* does not approach a limit, the *amplitude* does, but this limit is different for each frequency. Another phenomenon, which is not apparent from Fig. 7.1.1-1, is the variation

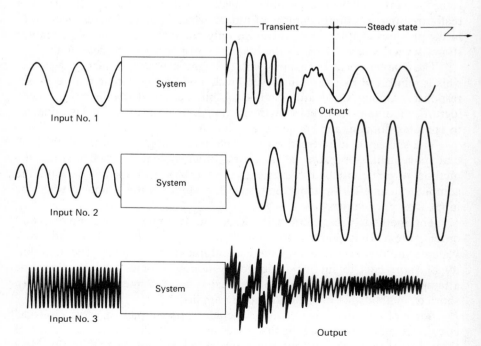

Figure 7.1.1-1   Frequency response for typical second order system.

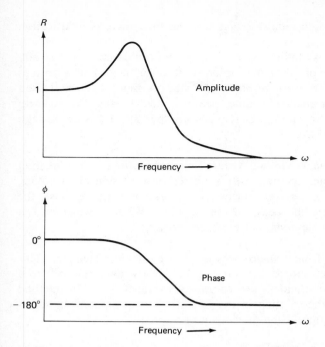

Figure 7.1.1-2   Frequency response for typical system.

in the steady state value of the *phase* ϕ, which also varies with frequency. Defining the out/in amplitude ratio as R, a plot of R and ϕ each versus frequency is referred to as the *frequency response*. See Fig. 7.1.1-2.

7.1.2 Methods for Determining Frequency Response

A number of techniques are available with which frequency response may be determined. These are listed below.

(a) **Method of Undetermined Coefficients** The method of undetermined coefficients is a classical method of solving a linear inhomogeneous differential equation. It provides the complete solution, requiring some effort to extract the steady state response. This is accomplished by taking the limit as time approaches infinity. This method is best suited for determining the transient response.

(b) **Computer Simulation** A computer simulation provides the complete solution to a differential equation. With a certain amount of effort, one may extract the transient and steady state responses. The run must be sufficiently long to permit the extrapolation of the steady state response. The major drawback to the use of a computer simulation for frequency response is the large amount of data that must be taken and analyzed. However, if the transient response to a sinusoidal input is desired, then there is justification for the effort.

The transient response indicates how the system builds up to the steady state vibration. This is of particular interest at resonance. The examples appearing in this chapter are: the response of a vibrator pump to several fixed input frequencies taken one at a time (see Ex. 7.3.3-1) and the start-up transient in a reciprocating pump as it picks up speed and passes through resonance (see Ex. 7.9.1-1).

(c) Complex Frequency Method Using Euler's Identities This method provides a direct approach to the steady state response to sinusoidal inputs. This method may be considered a shortcut for the steady state response, the price being the loss of any knowledge of the transients. While this method is a shortcut, there are other methods that are even shorter.

(d) Transformation from Laplace to Complex Frequency Domain This method provides a direct approach to finding the steady state response to a sinusoidal input with the attendant loss of transient information. This method operates directly upon the solution in the Laplace domain, making it somewhat easier to use than that in the previous paragraph.

(e) Log-Log Plots (Bode Diagrams) The log-log plot is a graphical means of performing the numerical work for the method described in paragraph (d). To further assist in this effort, tables and curves are provided in a later chapter. The method replaces the use of complex numbers with the use of log-log graphs. As such, the multiplication of two functions is accomplished by adding their curves. For design, Chapter 8 is devoted to this method. The chapter provides means to simplify the construction of the curves and to add them. Applications include analysis, design, and synthesis.

7.2 FREQUENCY RESPONSE OF FIRST ORDER SYSTEMS

7.2.1 Method of Undetermined Coefficients

The differential equation for a first order system with a sinusoidal input is given by

$$\frac{dX}{dt} + aX = B \sin \omega t \qquad [7.2.1\text{-}1]$$

The homogeneous solution is given by

$$X_h = Ce^{-at} \qquad [7.2.1\text{-}2]$$

The particular solution is of the form

$$X_p = G \sin \omega t + H \cos \omega t \qquad [7.2.1\text{-}3]$$

It will be more convenient to apply the IC before we determine the coefficients G and H. The complete solution becomes

$$X = Ce^{-at} + G \sin \omega t + H \cos \omega t \qquad [7.2.1\text{-}4]$$

Apply the IC, $X(0) = 0$, then,

$$C = -H \qquad [7.2.1\text{-}5]$$

Now evaluate the undetermined coefficients by applying the particular solution X_p, eq. 7.2.1-3, to the differential equation, eq. 7.2.1-1. We have

$$G\omega \cos \omega t - H\omega \sin \omega t$$
$$+ aG \sin \omega t + aH \cos \omega t = B \sin \omega t \qquad [7.2.1\text{-}6]$$

Consider the above equation as an identity, which must be true for all values of ω. Then, equate coefficients of similar terms ($\sin \omega t$ and $\cos \omega t$)

$$-H\omega + aG = B$$
$$G\omega + aH = 0$$

Solve the above simultaneously for G and H

$$G = \frac{Ba}{E^2}$$

$$H = -\frac{B\omega}{E^2} \qquad [7.2.1\text{-}7]$$

where

$$E^2 = \omega^2 + a^2$$

Apply the above to eq. 7.2.1-4, and the solution becomes

$$X = \frac{B\omega}{E^2} e^{-at} + \frac{Ba}{E^2} \sin \omega t - \frac{B\omega}{E^2} \cos \omega t$$

or

$$X = \frac{B\omega}{E^2}\left(e^{-at} + \frac{a}{\omega} \sin \omega t - \cos \omega t\right) \qquad [7.2.1\text{-}8]$$

Sum the two sinusoids, making use of eqs. 4.2.1-3 and 4.2.1-4. Then, after some algebra

$$X = \frac{B\omega}{E^2} e^{-at} + A \sin(\omega t + \phi) \qquad [7.2.1\text{-}9]$$

where

$$A = \frac{B}{\sqrt{\omega^2 + a^2}}$$

$$\phi = -\arctan\left(\frac{\omega}{a}\right) \qquad [7.2.1\text{-}10]$$

7.2.2 Steady State Response

The steady state response can be obtained by allowing time t to approach infinity. In that case, the exponential term in eq. 7.2.1-9 vanishes. The steady state response is a sinusoid having a frequency ω equal to that of the input, and an amplitude A and a phase angle* ϕ, both of which vary with frequency ω (as shown by eqs. 7.2.1-10).

7.2.3 Laplace Transform Approach

Apply Laplace Transforms to eq. 7.2.1-1, making use of entries 1, 2, 3, and 6 of Table 4.3.1-1;

$$sX(s) + aX(s) = \frac{B\omega}{s^2 + \omega^2}$$ [7.2.3-1]

Solve for $X(s)$

$$X(s) = \frac{B\omega}{(s + a)(s^2 + \omega^2)}$$ [7.2.3-2]

In order to transform back to the time domain, the inverse Laplace transform can be obtained by using entry 515 (with $\zeta = 0$) of Appendix 3:

$$X(t) = \frac{B\omega}{E^2}e^{-at} + A\sin(\omega t + \phi)$$ [7.2.3-3]

which agrees with eq. 7.2.1-9.

■ **EXAMPLE 7.2.3-1** Dial Movement with Sinusoidal Inputs
Consider a dial that is not driven by a rigid shaft, but instead is actuated by a weak spring. Assume that the needle of the dial has very small mass, and also assume that the system has linear damping. The system is shown on Fig. 7.2.3-1. Determine the response to a set of sinusoidal inputs, each having unity amplitude at frequencies from 4 to 64 rad/s.

Solution. Summing torques on the mass, we have

$$\sum T = J\ddot{\theta} = K(\theta_a - \theta) - D\dot{\theta}$$ (1)

Assume that the acceleration $\ddot{\theta}$ is small, and thus the product of this with the

---

*The output lags (has a negative phase angle) the input.

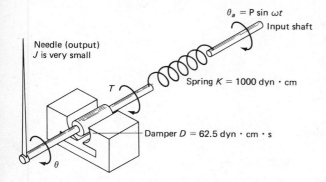

Figure 7.2.3-1   Dial movement with sinusoidal inputs.

moment of inertia J can be neglected. Then,

$$K(\theta_a - \theta) = D\dot{\theta} \tag{1a}$$

where

$$\theta_a = P \sin \omega t \tag{1b}$$
$$P = 1.0$$

Using this in eq. (1a), the differential equation becomes:

$$K(P \sin \omega t - \theta) = D\dot{\theta} \tag{2}$$

or by rearranging and normalizing

$$\dot{\theta} + \frac{K}{D}\theta = \frac{K}{D}P \sin \omega t \tag{3}$$

Compare the terms of the above equation to corresponding terms of eq. 7.2.1-1. We have

$$a = \frac{K}{D} = \frac{1000}{62.5} = 16$$
$$B = \frac{KP}{D} = aP = 16 \times 1.0 = 16 \tag{4}$$

The time constant τ is given by

$$\tau = \frac{1}{a} = \frac{1}{16} = 0.0625 \text{ s} \tag{5}$$

From eqs. (1a)–(5) we have

$$\dot{\theta} = 16(\theta_a - \theta), \qquad \theta(0) = 0 \tag{6}$$

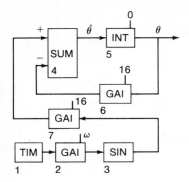

Figure 7.2.3-2

where

$$\theta_a = P \sin \omega t, \qquad P = 1 \text{ rad} \tag{7}$$

By computer simulation determine the steady state output $\theta(t)$ for $\omega = 4, 8, 16, 32, 64$ rad/s. Select the final time for the simulation to include at least one complete cycle at the lowest frequency. The period at the lowest frequency is $2\pi/4 = 1.57$ s. Select the final time to be 2.0 s for all runs. To get a feeling for how the output amplitude and phase angle vary with frequency, we shall plot the output and the input sine wave on the same graph. Five graphs are produced using TUTSIM, one for each frequency. The TUTSIM block diagram for the dial system is obtained from eqs. (6) and (7). It is shown in Fig. 7.2.3-2. The model data is given in Table 7.2.3-1. Values of output amplitude and phase obtained from the computer output [shown in Fig. 7.2.3-3(a)–(e)] are shown in Table 7.2.3-2, and are compared with analytical results obtained from eq. 7.2.1-10. Note that the results agree well. ∎

TABLE 7.2.3-1   MODEL DATA

Model Structure	Model Parameters
:1,TIM	:1
:2,GAI,1	:2,4 (also 8, 16, 32, and 64)
:3,SIN,2	:3
:4,SUM,7, − 6	:4
:5,INT,4	:5,0
:6,GAI,5	:6,16

Plotblocks and Ranges
  Horz:0,0,2
  Y1:3, − 1,1
  Y2:5, − 1,1

Timing Data
  0.005,2

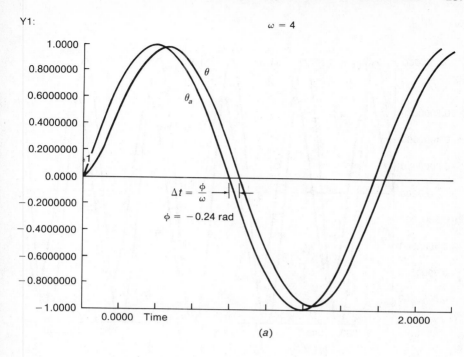

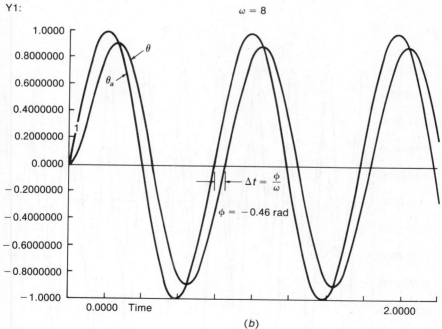

Figure 7.2.3-3   Output response for dial movement: (a) $\Omega = 4$; (b) $\Omega = 8$; (c) $\Omega = 16$; (d) $\Omega = 32$; (e) $\Omega = 64$.

Y1: $\omega = 16$

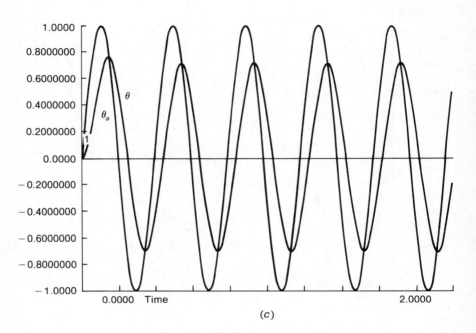

(c)

Y1: $\omega = 32$

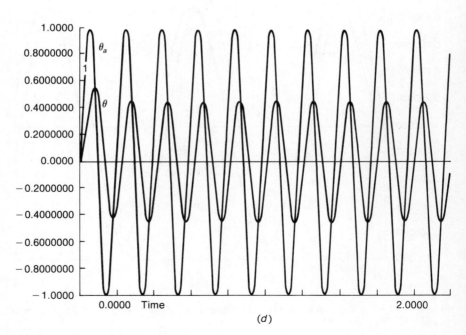

(d)

Figure 7.2.3-3 Continued

Y1: $\omega = 64$

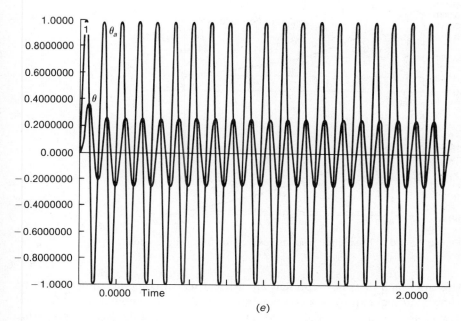

(e)

Figure 7.2.3-3   Continued

TABLE 7.2.3-2

ω	A_{anal}	A_{comp}	ϕ_{anal}	Δt	ϕ_{comp}
4	0.987	0.969	−0.244	0.060	−0.24
8	0.894	0.894	−0.463	0.055	−0.44
16	0.707	0.707	−0.785	0.050	−0.80
32	0.448	0.444	−1.107	0.035	−1.1
64	0.246	0.243	−1.326	0.020	−1.3

■ **EXAMPLE 7.2.3-2** Frequency Response for First Order System
Using the data from the previous example, plot the frequency response.

Solution. The frequency response is obtained by extrapolating the steady state amplitude and phase for each run. For the five runs in the previous example, there will be five points for each curve. These are plotted in Fig. 7.2.3-4.

It becomes apparent that this technique may suffer in accuracy due to the lack of sufficient time for the solution to reach steady state. It is also apparent that for any meaningful data, a great amount of computer time is required both for each run and for the many runs required. Hence, in its present form,

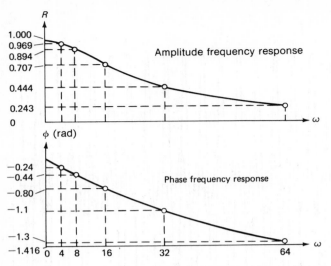

Figure 7.2.3-4   Frequency response curves for first order system.

this approach does not represent an efficient use of computer time. It will be shown subsequently that a combination of analytical, computer, and graphical techniques is a successful blend that accomplishes the task of preparing frequency response curves.   ■

7.3 FREQUENCY RESPONSE OF SECOND ORDER SYSTEMS

7.3.1 Family of Second Order Systems

There are a number of arrangements of springs, dampers, and inputs. Consequently, there are a number of different second order systems. It will be seen that certain of these will be grouped together when they have similar frequency response curves. However, there will be different frequency response curves, depending upon system configuration. These will be discussed in Sections 7.5 through 7.8. We will consider only the very simplest second order system at first. Such systems are shown in Fig. 3.6.2-1.

7.3.2 Method of Undetermined Coefficients

A series electrical circuit is shown in Fig. 7.3.2-1. The input is a sinusoidal voltage of magnitude E_{a0}; the differential equation is

$$L\ddot{q} + R\dot{q} + \frac{q}{C} = E_{a0} \cos \omega t \qquad [7.3.2\text{-}1]$$

Upon normalizing, this becomes

$$\ddot{q} + 2\zeta\omega_n\dot{q} + \omega_n^2 q = E_{a0}C\omega_n^2 \cos \omega t \qquad [7.3.2\text{-}2]$$

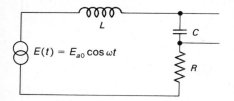

Figure 7.3.2-1   Series circuit.

Since the system is linear, we may determine the homogeneous and particular solutions individually, and then superpose the results. The homogeneous solution is given by eq. 6.2.2-3

$$q_h = C_1 e^{P_1 t} + C_2 e^{P_2 t} \qquad [7.3.2-3]$$

The particular solution is formulated by a linear combination of the forcing function and its derivatives. For input frequency ω

$$q_p = A \cos \omega t + B \sin \omega t \qquad [7.3.2-4]$$

Substitute the particular solution into the differential equation, eq. 7.3.2-2.

$$-\omega^2 A \cos \omega t - \omega^2 B \sin \omega t + 2\zeta\omega_n(\omega B \cos \omega t - \omega A \sin \omega t)$$

$$+ \omega_n^2(A \cos \omega t + B \sin \omega t) = E_{a0}\omega_n^2 C \cos \omega t \qquad [7.3.2-5]$$

Since eq. 7.3.2-5 is true for all ω then we may equate coefficients of $\cos \omega t$ and $\sin \omega t$. Then,

$$-\omega^2 A + 2\zeta\omega_n\omega B + \omega_n^2 A = E_{a0}\omega_n^2 C$$

$$-\omega^2 B - 2\zeta\omega_n\omega A + \omega_n^2 B = 0 \qquad [7.3.2-6]$$

Divide through by ω_n^2 (which is not equal to zero), use the dimensionless frequency ratio $\gamma = \omega/\omega_n$, and solve simultaneously for A and B.

$$A = \frac{1 - \gamma^2}{\Delta} E_{a0} C$$

$$\qquad [7.3.2-7]$$

$$B = \frac{2\zeta\gamma}{\Delta} E_{a0} C$$

where

$$\Delta = (1 - \gamma^2)^2 + (2\zeta\gamma)^2 \qquad [7.3.2-8]$$

Substitute eqs. 7.3.2-7 and 7.3.2-8 in the particular solution eq. 7.3.2-4 and combine sine and cosine terms. The particular solution becomes

$$q_p = E_{a0} CP \cos(\omega t - \phi) \qquad [7.3.2-9]$$

where

$$P = (A^2 + B^2)^{1/2} = \left[(1 - \gamma^2)^2 + (2\zeta\gamma)^2\right]^{1/2}/\Delta$$

$$= \frac{\Delta^{1/2}}{\Delta} = \Delta^{-1/2}$$

$$= \sqrt{\frac{1}{(1 - \gamma^2)^2 + (2\zeta\gamma)^2}} \qquad [7.3.2\text{-}10]$$

and

$$\phi = \tan^{-1} \frac{B}{A} = \tan^{-1} \left[\frac{2\zeta\gamma/\Delta}{(1 - \gamma^2)/\Delta}\right]$$

$$= \tan^{-1} \frac{2\zeta\gamma}{1 - \gamma^2} \qquad [7.3.2\text{-}11]$$

The homogeneous solution is given by eq. 7.3.2-3. The complete solution is formed by adding the particular solution, eq. 7.3.2-9, to the homogeneous solution. The complete solution becomes

$$q = q_h + q_p$$

$$= e^{-at}\left[q_0 \cos \omega_d t + \frac{aq_0 + \dot{q}_0}{\omega_d} \sin \omega_d t\right]$$

$$+ E_{a0}CR(\cos \omega t - \phi) \qquad [7.3.2\text{-}12]$$

The complete solution expressed by eq. 7.3.2-12 depends upon the initial conditions q_0 and $\dot{q}_0$ as well as the input frequency ω. Sketches of the complete solution for three different frequencies are given in Fig. 7.1.1-1.

The steady state solution is found by letting time t approach infinity. In that case, the exponential term vanishes and only the particular solution, eq. 7.3.2-9, remains.

7.3.3 Use of the Computer for Second Order Systems

If the system is simulated on the computer, the transient response to any input may be found. For a sinusoidal input, the differential equation for a dynamically similar mechanical system is

$$M\ddot{x} + D\dot{x} + Kx = F\cos \omega t \qquad [7.3.3\text{-}1]$$

Solve for the highest derivative

$$\ddot{x} = \frac{(F\cos \omega t - D\dot{x} - Kx)}{M} \qquad [7.3.3\text{-}2]$$

where

$$\dot{x} = \int \ddot{x}\,dt \qquad \text{initial condition } \dot{x}(0) = \dot{x}_0 \qquad [7.3.3\text{-}3]$$

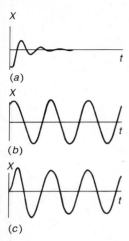

X

t

(a)

X

t

(b)

X

t

(c)

Figure 7.3.3-1   Transient response to sinusoidal input: (a) homogeneous solution; (b) particular solution; (c) total solution.

and

$$x = \int \dot{x}\, dt \quad \text{initial condition } x(0) = x_0 \qquad [7.3.3\text{-}4]$$

Equations 7.3.3-2, 7.3.3-3, and 7.3.3-4 will be used in the simulation. However, before proceeding, let us consider the anticipated result. In order to do this, consider the form of the solution as it would be found by means of the method of undetermined coefficients.

The homogeneous solution will be of the form of an exponentially decaying sinusoid having a frequency ω_d, as sketched in Fig. 7.3.3-1(a). The particular solution will have the form of a sinusoid having frequency ω with a phase shift ϕ, as shown in Fig. 7.3.3-1(b). The final result will be obtained by summing the two curves, point by point, as sketched in Fig. 7.3.3-1(c).

■ **EXAMPLE 7.3.3-1** TUTSIM Simulation for Forced Vibration in Second Order System

Given the system modeled by eq. 7.3.3-1, determine the response to a range of frequencies from one-quarter to twice the natural frequency,

where M = mass = 5 kg
D = damper = 1500 N · s/m
K = spring = 2.8×10^6 N/m
F_a = sinusoidal input force = $F \cos \omega t$
F = force amplitude = 1.12×10^5 N

From the figure we can obtain the differential equation for X as follows:

$$M\ddot{X} + D\dot{X} + KX = F_a, \qquad X(0) = \dot{X}(0) = 0 \qquad (1)$$

To model this system by means of TUTSIM, we rewrite the differential

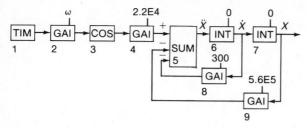

Figure 7.3.3-2   TUTSIM block diagram.

equation as

$$\ddot{X} = \frac{F_a}{M} - \frac{D}{M}\dot{X} - \frac{K}{M}X \tag{2}$$

Using the given values of M, D, K, and F, we get

$$\ddot{X} = 22{,}400 \cos \omega t - 300 \dot{X} - 560{,}000 X \tag{3}$$

The undamped natural frequency $\omega_n = \sqrt{K/M} = 750$ rad/s and the damping ratio $\zeta = D/(2M\omega_n) = 0.2$. We shall run simulations for $\omega = \frac{1}{4}$, $\frac{1}{2}$, 1, and 2 times ω_n. That is, $\omega = 187.4$, 375, 750 and 1500 rad/s. In order to obtain at least three cycles of output at $\omega = 375$ rad/s, the final time should be $t_f = 3(2\pi)/375 = 0.05$ s. The period for the highest frequency is $2\pi/1500 = 0.004$ s. To make good plots it is necessary to have at least 10 points per quarter period, which requires that $\Delta t = 0.0001$ s.

The TUTSIM block diagram and the model data are given in Fig. 7.3.3-2 and Table 7.3.3-1, respectively. The results of the simulation are given in Fig. 7.3.3-3(a) through (d). ∎

TABLE 7.3.3-1   TUTSIM MODEL DATA

Model Structure	Model Parameters
:1,TIM	:1
:2,GAI,1	:2,187.5 (also 375, 750, 1500)
:3,COS,2	:3
:4,GAI,3	:4,2.24E4
:5,SUM,4, −8, −9	:5
:6,INT,5	:6,0
:7,INT,6	:7,0
:8,GAI,6	:8,300
:9,GAI,7	:9,5.6E5
Plotblocks and Ranges	Timing Data
Horz:0,0,.05	0.0001,0.05
Y1:4, −5E4,5E4	
Y2:7, −0.1,0.1	

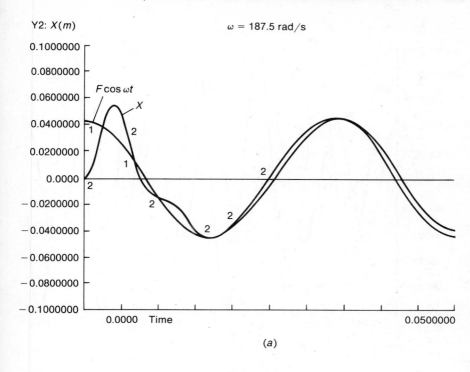

(a)

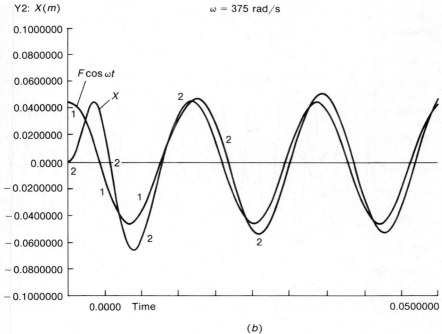

(b)

Figure 7.3.3-3   Second order system response: (a) $\omega = 187.5$ rad/s; (b) $\omega = 375$ rad/s; (c) $\omega = 750$ rad/s; (d) $\omega = 1500$ rad/s.

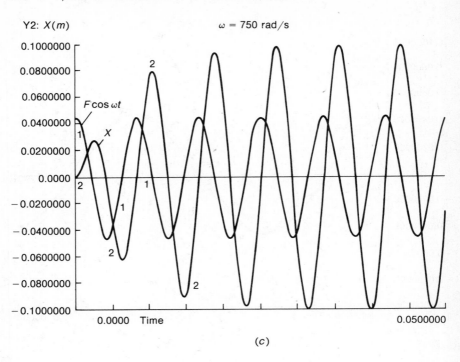

(c)

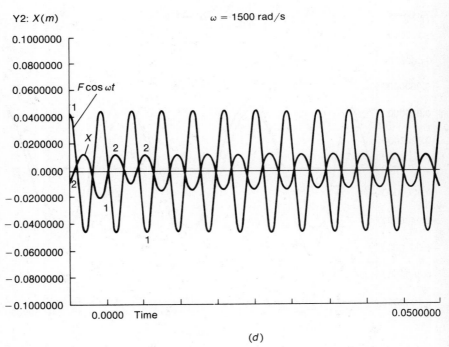

(d)

Figure 7.3.3-3   Continued

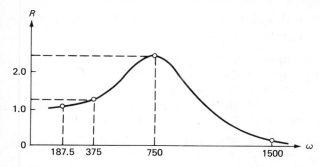

Figure 7.3.3-4   Frequency response curve for second order system.

■ **EXAMPLE 7.3.3-2** Frequency Response for Second Order System
Using the data from the previous example, plot the frequency response.

Solution.   Each run in the previous example provides one point on the
frequency response curve. These are summarized below.

u	X (output amplitude)	$R\left(\dfrac{\text{out}}{\text{in}}\right) = \dfrac{X}{F/K}$
187.5	0.044	1.1
375	0.05	1.25
750	0.0995	2.49
1500	0.013	0.322

The frequency response curve is plotted in Fig. 7.3.3-4. Needless to say, four
points are insufficient to draw an accurate curve. On the other hand, each
point requires a computer run. While the computer runs have shown the
meaning of frequency response, this approach does not appear practical for
quantitative results. Hence, other approaches will be sought in the subsequent
sections.   ■

7.4 FREQUENCY RESPONSE USING STEADY STATE METHODS

In the previous sections, two methods were employed, undetermined coefficients and the computer. These methods provided the transient and (extrapo-
lated) steady state responses. If one were interested in only the steady state
response, then it will prove profitable to employ other methods that bypass the
transient response and proceed immediately to the steady state response.

In order to accomplish this goal, employing the results of the previous methods, several conclusions may be drawn:

1. In steady state, the output is sinusoidal.
2. In steady state, the output amplitude A is constant.
3. In steady state, the output phase ϕ is constant.
4. In steady state, the output frequency ω is equal to that of the input.
5. In steady state, the output is of the form $X = A \sin(\omega t + \phi)$.

These conclusions suggest several shortcut methods which will be investigated in the subsequent sections.

7.4.1 Frequency Response Using Sinusoidal Steady State Solution

It has been concluded that the steady state solution will be sinusoidal of the form

$$X = A \sin(\omega t + \phi) \qquad [7.4.1\text{-}1]$$

where

$$A = \text{amplitude (to be found)}$$

$$\phi = \text{phase angle (to be found)} \qquad [7.4.1\text{-}2]$$

$$\omega = \text{input frequency (known)}$$

In order to determine the unknowns A and ϕ, apply the solution, eq. 7.4.1-1, to the differential equation, eq. 7.3.2-1. While the concept behind this method promised a more direct approach, in practice it appears that it will be similar to that for undetermined coefficients. In addition, the argument is equal to the sum of two terms, which will require careful attention to the trigonometric manipulations.

It should be pointed out that the method has promise. It is only the computational operation that is objectionable. There are other means to represent sinusoidal functions, such as Euler's identities, which make use of exponentials. Such an approach will make use of complex numbers. To facilitate these methods, a brief review appears in Appendix 5.

7.4.2 Frequency Response Using Euler's Identities

Refer to Appendix 1 where the relationships between complex and sinusoidal quantities are indicated. We can say

$$F_0 e^{j\omega t} = F_0(\cos \omega t + j \sin \omega t) \qquad [7.4.2\text{-}1]$$

or

$$F_0 \cos \omega t = \text{Re } F_0 e^{j\omega t} \qquad [7.4.2\text{-}1a]$$

$$F_0 \sin \omega t = \text{Im } F_0 e^{j\omega t} \qquad [7.4.2\text{-}1b]$$

Corresponding to any of the above three forms of the input, the output (solution) will have a similar form, but it will include a phase angle ϕ. For example, using eq. 7.4.2-1(b) for the input, the corresponding output would be

$$X = A \sin(\omega t - \phi) = \text{Im } A e^{j(\omega t - \phi)} \qquad [7.4.2-2]$$

Apply the general input, eq. 7.4.2-1, and specialize later.

$$X = A e^{j(\omega t - \phi)} = A e^{j\omega t} e^{-j\phi} = \overline{A} e^{j\omega t} \qquad [7.4.2-2a]$$

where

$$\overline{A} = A e^{-j\phi} \quad \text{(a complex number)} \qquad [7.4.2-3]$$

Upon differentiating eq. 7.4.2-2a, we have

$$\dot{X} = j\omega \overline{A} e^{j\omega t} \qquad [7.4.2-4]$$

Apply eqs. 7.4.2-1, 7.4.2-2a, and 7.4.2-4 to eq. 7.2.1-1

$$j\omega \overline{A} e^{j\omega t} + a\overline{A} e^{j\omega t} = B e^{j\omega t} \qquad [7.4.2-5]$$

Divide by $e^{j\omega t} \neq 0$

$$j\omega \overline{A} + a\overline{A} = B \qquad [7.4.2-6]$$

Substitute eq. 7.4.2-3, and regroup terms

$$\overline{A} = A e^{-j\phi} = \frac{B}{a + j\omega} \qquad [7.4.2-7]$$

where

$$A = \frac{B}{\sqrt{a^2 + \omega^2}} \qquad [7.4.2-8]$$

$$\phi = -\arctan\frac{\omega}{a} \qquad [7.4.2-9]$$

which agrees with eqs. 7.2.1-10.

7.4.3 Transformation from Laplace to Complex Frequency Domain

Up to now, we have been transforming from the time to the Laplace domain. In this section, we are going to transfer from the Laplace domain to another domain. To facilitate this transformation, it will be useful to define a special function called the Laplace transfer function.

The *Laplace transfer function* is defined as the Laplace transform of the ratio of the output to the input with all initial conditions equal to zero. Also note that in forming the Laplace transfer function the forcing function (input), $f(t)$, is transformed symbolically only. That is, the amplitude F is employed, and the sin ωt is *not* transformed to $\omega/(S^2 + \omega^2)$.

■ **EXAMPLE 7.4.3-1** Laplace Transfer Function

Given the differential equation below with zero initial conditions, write the Laplace transfer function.

$$M\ddot{x} + D\dot{x} + Kx = F \sin \omega t = f(t) \qquad (1)$$

Solution.   Transform the differential equation, eq. (1), to the Laplace domain, noting that the initial conditions are zero.

$$(Ms^2 + Ds + K)X = F(s) \qquad (2)$$

Note that the input has been expressed in terms of the general function $F(s)$ instead of the actual sinusoid. This has been done to avoid the temptation to transform the sinusoid. Solve for the out/in ratio

$$\frac{X}{F}(s) = \frac{1}{Ms^2 + Ds + K} \qquad (3)$$

■

Equation (3) is the required Laplace transfer function. An interesting notion can be unfolded by examining eq. 7.4.2-4. The following conclusion is apparent; a derivative in the time domain is equivalent to multiplication by $j\omega$ in the complex frequency domain. Thus,

$$\frac{d}{dt}x(t) = j\omega X(j\omega) \qquad [7.4.3\text{-}1]$$

If IC = 0, multiplication by s in the Laplace domain is equivalent to the derivative in the time domain. See eq. 4.3.1-4. Then,

$$\frac{d}{dt}x(t) = sX(s) \quad \text{when } x(0) = 0 \qquad [7.4.3\text{-}2]$$

Considering the above relationships, one has a choice. In order to determine the frequency response, either substitute $j\omega$ for each *derivative* of the differential equation in the *time domain*, or substitute $j\omega$ for the operator s in the Laplace transfer function.

In symbols, the frequency response is found by letting

$$s = j\omega \qquad [7.4.3\text{-}3]$$

■ **EXAMPLE 7.4.3-2** Frequency Response Using Laplace Transforms

Refer to eq. 7.2.1-1, which is repeated below

$$\frac{dx}{dt} + ax = B \sin \omega t = f(t) \qquad (1)$$

Transform to the Laplace domain, assuming zero initial conditions.

$$(s + a)X = F(s) \qquad (2)$$

Solve for X/F, where F is amplitude of input

$$\frac{X}{F}(s) = \frac{1}{s + a} \qquad (3)$$

The input, $F(s)$, has an amplitude equal to B. Then, upon taking the limit for the steady state response to a sinusoid,

$$X(j\omega) = \frac{B}{j\omega + a} = \frac{B}{a + j\omega} \tag{4}$$

Equation (4) represents a complex number whose magnitude A and whose phase ϕ are given below.

$$A = |x(j\omega)| = \left|\frac{B}{a + j\omega}\right| = \frac{B}{\sqrt{a^2 + \omega^2}} \tag{5}$$

$\frac{B^2(a^2+\omega^2)}{(a^2+\omega^2)^2}$

$$\phi = -\tan^{-1}\frac{\omega}{a} \tag{6}$$

$\frac{B\cdot(a-jw)}{(a+jw)(a-jw)} = \frac{Ba - Bjw}{a^2+w^2}$

which agree with eqs. 7.4.2-8 and 7.4.2-9, respectively. ∎

7.5 FREQUENCY RESPONSE OF SYSTEMS WITH DIRECT EXCITATION

We have observed that each system has its own individual frequency response curve. (See Figs. 7.2.3-4 and 7.3.3-4.) On the one hand, each system has its own curve, but on the other hand, certain differing systems will have the same curve. With this in mind, Sections 7.5 through 7.8 have organized groups of systems with common frequency response curves.

7.5.1 Sinusoidal Force Applied Directly to Mass

Consider a system in which a force F is applied directly to a mass M that is restrained by a damper D and spring K. The free body diagram is shown in Fig. 7.5.1-1.

Sum the forces according to Newton's law and obtain

$$M\ddot{X} + D\dot{X} + KX = F_a(t) = B \sin \omega t \tag{7.5.1-1}$$

We will be interested in the steady state response to a sinusoidal input. Hence, we may employ any of the methods described in Section 7.4. Let us employ Euler's identities, as described in Section 7.4.2. However, as an interesting variation, apply a phase lead ϕ to the input (instead of a phase lag ϕ applied

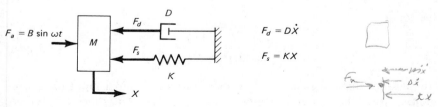

$F_a = B \sin \omega t$ $F_d = D\dot{X}$

$F_s = KX$

Figure 7.5.1-1   Sinusoidal force.

to the output as was done in Section 7.4.2). Since the input is given by the sine, we will understand that the imaginary part of all terms is implied. Then let

$$X = Ae^{j(\omega t - \phi)} \qquad F_a(t) = Be^{j(\omega t)} \qquad [7.5.1-2]$$

Then

$$-M\omega^2 Ae^{j(\omega t - \phi)} + j\omega DAe^{j(\omega t - \phi)} + KAe^{j(\omega t - \phi)} = Be^{j(\omega t)} \quad [7.5.1-2a]$$

Factor out $e^{j\omega t}$, collect terms.

$$(K - M\omega^2 + j\omega D)Ae^{-j\phi} = B$$

or

$$\frac{A}{B}e^{-j\phi} = \frac{1}{K - M\omega^2 + j\omega D} \qquad [7.5.1-3]$$

Normalize by dividing numerator and denominator of the right-hand side by M. Then, multiply both sides of the equation of K. Referring to Appendix 1, note that $K/M = \omega_n^2$ and that $D/M = 2\zeta\omega_n$. Introduce the dimensionless frequency ratio γ and divide numerator and denominator by ω_n^2

$$\gamma = \frac{\omega}{\omega_n} \qquad [7.5.1-3a]$$

$$\frac{A}{B/K}e^{-j\phi} = \frac{1}{1 - \gamma^2 + j2\zeta\gamma} = Re^{-j\phi} \qquad [7.5.1-4]$$

This is a complex number whose magnitude R is given by

$$R = \sqrt{\frac{1}{(1 - \gamma^2)^2 + (2\zeta\gamma)^2}} = \frac{A}{B/K} \qquad [7.5.1-5]$$

which agrees with eq. 7.3.2-10 (which was derived by the method of undetermined coefficients). Before the frequency response curves are plotted, the same result will be obtained using the transformation from the Laplace to the complex domain. Write eq. 7.5.1-1 in the Laplace domain, noting that the initial conditions are equal to zero.

$$Ms^2 X + DsX + KX = F_a(s) \qquad [7.5.1-6]$$

Solve for X

$$X(s) = \frac{F_a(s)}{Ms^2 + Ds + K} \qquad [7.5.1-6a]$$

In preparation for the steady state response to a sinusoidal input (frequency response), form the output over input ratio X/B, where B is the amplitude of the sinusodial input. Normalize by dividing the numerator and denominator by M and then multiply both sides by K. To simplify the result, make use of Appendix 1.

$$\frac{X}{B/K}(s) = \frac{\omega_n^2}{s^2 + 2\zeta\omega_n s + \omega_n^2} \cdot \qquad [7.5.1\text{-}7]$$

The right-hand side of eq. 7.5.1-7 will become very familiar. It is perhaps the most common expression in the subject. It has the form of a constant divided by a quadratic. It is sometimes referred to as a "reciprocal quadratic."

The steady state response in the complex frequency domain is found by setting $s = j\omega$. Then eq. 7.5.1-7 becomes

$$\frac{X}{B/K}(j\omega) = \frac{\omega_n^2}{-\omega^2 + j\omega 2\zeta\omega_n + \omega_n^2} \qquad [7.5.1\text{-}7a]$$

Divide the numerator and denominator by ω_n^2 and apply the dimensionless frequency ratio γ as given by eq. 7.5.1-3a

$$\frac{X}{B/K}(j\omega) = \frac{1}{1 - \gamma^2 + j2\zeta\gamma} \qquad [7.5.1\text{-}7b]$$

which agrees with eq. 7.5.1-4. Equation 7.5.1-5 will be plotted for a range of γ from 0 to 2.5. To assist in making the plot, use the damping ratio ζ as parameter, where the following values of γ will be employed:

$$\text{when } \gamma = 0 \qquad \text{then } R = 1.0 \qquad\qquad [7.5.1\text{-}8]$$

$$\text{when } \gamma = 1 \qquad \text{then } R = \frac{1}{2\zeta} \qquad\qquad [7.5.1\text{-}9]$$

$$\text{when } \gamma \gg 1 \qquad \text{then } R \rightarrow \frac{1}{\gamma^2} \qquad\qquad [7.5.1\text{-}10]$$

Note that when ζ is less than $\sqrt{\frac{1}{2}}$, then R reaches a maximum value referred to as the resonant peak. This peak occurs at a frequency that is *less* than the undamped natural frequency ($\gamma = 1$). As a rough approximation, the peak may be found by using eq. 7.5.1-9. For a more exact value, take the derivative of eq. 7.5.1-5. Using either the approximate or the more exact value, it is clear that the *lower* the damping ratio ζ the *higher* is the resonant peak. The frequency response curves are plotted in Fig. 7.5.1-2.

Like the reciprocal quadratic (eq. 7.5.1-7), the frequency response curves (Fig. 7.5.1-2) will become very familiar as the reader delves into the subject of system dynamics. It will be disclosed later that two other systems will have the same frequency response as that shown above. To assist in recognizing such systems, the out/in function in the Laplace domain, and the amplitude function R are shown in the figure.

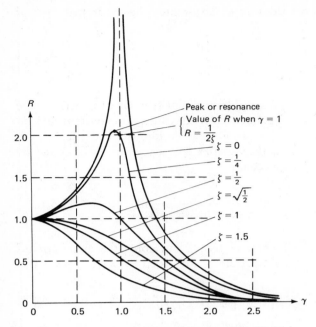

Figure 7.5.1-2   Frequency response for sinusoidal force.

$$\frac{\text{out}}{\text{in}}(s) = \frac{\omega_n^2}{s^2 + 2as + \omega_n^2}$$

$$R = \sqrt{\frac{1}{\left(1 - \gamma^2\right)^2 + \left(2\zeta\gamma\right)^2}}$$

7.5.2 Sinusoidal Base Motion Applied Through a Spring

Consider a system in which the base motion X_a is applied through a spring K_1 to mass M which is restrained by a spring K_2 and a damper D. The free body diagram is shown in Fig. 7.5.2-1, where

$$\begin{aligned} X_a(t) &= B \sin \omega t \\ F_1 &= K_1(X_a - X) \\ F_2 &= K_2 X \\ F_d &= D\dot{X} \end{aligned} \qquad [7.5.2\text{-}1]$$

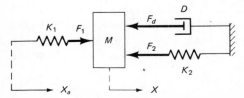

Figure 7.5.2-1   Transmission through spring.

Sum the forces according to Newton's law and separate variables

$$M\ddot{X} + D\dot{X} + (K_1 + K_2)X = K_1 X_a(t) \qquad [7.5.2\text{-}2]$$

Assuming zero initial conditions, transform to the Laplace domain

$$(Ms^2 + Ds + K)X(s) = K_1 X_a(s) \qquad [7.5.2\text{-}3]$$

where

$$K = K_1 + K_2 \qquad [7.5.2\text{-}4]$$

Solve eq. 7.5.2-3 for $X(s)$

$$X(s) = \frac{K_1 X_a(s)}{Ms^2 + Ds + K} \qquad [7.5.2\text{-}5]$$

Normalize by dividing numerator and denominator by M,* and formulate the dimensionless ratio $X/(BK_1/K)$

$$\frac{X}{BK_1/K}(s) = \frac{\omega_n^2}{s^2 + 2\zeta\omega_n s + \omega_n^2} \qquad [7.5.2\text{-}6]$$

Note that eq. 7.5.2-6 is identical to eq. 7.5.1-7. Hence, the frequency response is given by eq. 7.5.1-5 and Fig. 7.5.1-2.

■ **EXAMPLE 7.5.2-1** Response of Phono Pickup

A phono pickup is a mechanical device which follows a groove cut in a record and actuates an electrical sensor to produce signals in accordance with vibrations induced by the groove. A sharp needle follows the groove, which wavers from side to side, appearing like a displacement input θ_a. See Fig. 3.2.1-4(a). The needle is connected through a torsional elastic rod K to the electrical sensor. Determine the frequency response.

Solution. For the free body diagram, note that the input is θ_a which is connected through a torsional spring K to the mass J. Motion θ of J is resisted by a damper D connected to ground, as shown in Fig. 3.2.1-4(b).

Sum the forces

$$\sum F = J\ddot{\theta} = K(\theta_a - \theta) - D\dot{\theta} \qquad (1)$$

or

$$J\ddot{\theta} + D\dot{\theta} + K\theta = K\theta_a(t) \qquad (2)$$

Transform to the Laplace domain, noting that the initial conditions are equal to zero.

$$(Js^2 + Ds + K)\theta(s) = K\theta_a(s) \qquad (3)$$

Solve for the output $\theta(s)$

$$\theta(s) = \frac{K\theta_a(s)}{Js^2 + Ds + K} \qquad (4)$$

---

*Note $K/M = \omega_n^2$.

Normalize by dividing numerator and denominator by J

$$\theta(s) = \frac{\omega_n^2 \theta_a(s)}{s^2 + 2as + \omega_n^2} \tag{5}$$

where

$$\theta_a(s) = \text{generalized input}$$

For frequency response (steady state response to sinusoidal input) let $s = j\omega$, divide by ω_n^2, and let $\gamma = \omega/\omega_n$

$$\frac{\theta}{\theta_a}(j\omega) = \frac{1}{1 - \gamma^2 + j2\zeta\gamma} \tag{6}$$

Equation (6) is identical to eq. 7.5.1-4. Hence, the frequency response is given by eq. 7.5.1-5 and Fig. 7.5.1-2.

The nature of frequency response can be demonstrated with the help of Fig. 7.5.1-2. The groove, which wanders laterally, represents a sinusoidal input. The output will also be sinusoidal. At very low frequencies relative to the natural frequency ω_n the frequency ratio γ is very small. Then, the output amplitude θ will be equal to the input amplitude θ_a. Since the phono pickup is attempting to reproduce the sound signal that has been impressed in the record, it is desirable that the out/in ratio be equal to unity. Hence, at low input frequencies, the fidelity is good. This implies that two low frequency musical instruments (such as a bass and drum) would be reproduced faithfully. At the resonant frequency (γ is approximately unity, or the input frequency ω is close to the natural frequency ω_n) the out/in ratio is greater than unity, or there is magnification of the sound. Such musical instruments would sound much louder than they actually were during the recording, which is not desirable. When the input frequency ω is much greater than the natural frequency ω_n, the frequency ratio γ is very large. According to Fig. 7.5.1-2, the out/in ratio becomes negligible. This implies that high pitched instruments (such as a violin or fife) would hardly be audible. Consequently, for faithful reproduction, the natural frequency of a phono pickup should be very high.

The frequency response curves show that a low damping ratio ζ results in high magnification at resonance, but that a large damping ratio causes the output to drop off early. The optimum value is approximately equal to one-half. ∎

7.5.3 Sinusoidal Voltage Applied to a Series Electric Circuit

Consider a series electrical circuit with a sinusoidal input voltage $E(t)$. The circuit is shown in Fig. 7.5.3-1. This is a single closed loop, for which the loop analysis of Section 3.3.1 applies. Sum the voltages around the loop according to Kirchhoff's law

$$E_a \sin \omega t - L\ddot{q} - \frac{q}{C} - \dot{q}R = 0 \tag{7.5.3-1}$$

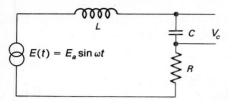

Figure 7.5.3-1   Series circuit.

Transform to the Laplace domain and use the term $E(s)$ for the input.

$$E(s) - \left(Ls^2 + \frac{1}{C} + Rs \right) Q(s) = 0 \qquad [7.5.3\text{-}2]$$

or

$$\frac{Q}{E}(s) = \frac{1}{Ls^2 + Rs + \frac{1}{C}} \qquad [7.5.3\text{-}3]$$

The voltage drop across the capacitor is

$$V_c = \frac{Q}{C}$$

Substitute eq. 7.5.3-3 and normalize

$$\frac{V_c}{E}(s) = \frac{\omega_n^2}{s^2 + 2\zeta\omega_n s + \omega_n^2} \qquad [7.5.3\text{-}4]$$

where

$$\omega_n^2 = \frac{1}{LC} \qquad [7.5.3\text{-}5]$$

$$2\zeta\omega_n = \frac{R}{L} \qquad [7.5.3\text{-}6]$$

Equation 7.5.3-4 is identical to eq. 7.5.2-6 and to eq. 7.5.1-7. Hence, the frequency response is given by eq. 7.5.1-5 and Fig. 7.5.1-2.

■ **EXAMPLE 7.5.3-1** Voice Actuated Circuit
Design a voice actuated circuit that responds to male and female voices but not to stray sound like a dog barking, a baby crying, or a siren.

Solution.   A microphone picks up the person's voice, which is amplified and appears as the input signal $E(t)$ in Fig. 7.5.3-1. However, the microphone will also pick up stray sounds. Fortunately, there is a large frequency difference between the voices and the stray sounds. The male voice has a frequency of 1250 rad/s (about 200 Hz), the female voice is 2500 rad/s (about 400 Hz), and the stray sounds are all above 10,000 rad/s (about 1600 Hz). The frequency response curve is shown in Fig. 7.5.1-2, where we will let

$$\gamma = 1 \quad \text{(corresponds to frequency 2500 rad/s)} \tag{1}$$

$$\zeta = 0.5 \quad \text{(an appropriate damping ratio, since curve}$$
$$\text{neither rises too high nor drops too low)} \tag{2}$$

From the curve it is clear that

$$V_c/E \simeq 1.0 \quad \text{for } \gamma = 0.5 \quad \text{(frequency = 1250 rad/s)} \tag{3}$$

$$V_c/E \simeq 1.0 \quad \text{for } \gamma = 1.0 \quad \text{(frequency = 2500 rad/s)} \tag{4}$$

$$V_c/E \ll 1.0 \quad \text{for } \gamma > 4 \quad \text{(frequencies > 10,000 rad/s)} \tag{5}$$

It is clear that the choices made in eqs. (1) and (2) produced the results requested in the statement of the problem, as shown in eqs. (3) through (5). Then proceed with the design.

Equation (1) implies that

$$\omega_n = 2500 \text{ rad/s} \tag{6}$$

From practical experience, a good choice of inductance for the frequency range is

$$L = 0.1 \text{ H} \tag{7}$$

Solve for C in eq. 7.5.3-5

$$C = \frac{1}{L\omega_n^2} = \frac{1}{0.1 \times 2500^2} = 1.60 \times 10^{-6} \text{ F} \tag{8}$$

$$= 1.60 \ \mu\text{F} \tag{9}$$

Solve for R in eq. 7.5.3-6

$$R = 2\zeta\omega_n L = 2 \times 0.5 \times 2500 \times 0.1 = 250 \ \Omega \tag{10}$$

Check on the attenuation* of the stray sounds at frequencies above 10,000 rad/s

$$\gamma > \frac{\omega}{\omega_n} = \frac{10,000}{2500} = 4.0 \tag{11}$$

Apply eqs. (2) and (7) to eq. 7.5.1-5, where the output/input ratio is V_c/E,

$$\frac{V_c}{E} < \sqrt{\frac{1}{\left(1 - 4^2\right)^2 + 4^2}} = 0.0644 \tag{12}$$

which is much smaller than the value of 1.0 for the voice signals. Hence, the system will operate properly.   ■

---

*reduction in amplitude

7.6 TRANSMISSIBILITY

7.6.1 Force Transmitted to Base (or Foundation)

Given the system used in Fig. 7.5.1-1, determine the force that is transmitted to the base (or foundation). The free body diagram for this investigation is shown in Fig. 7.6.1-1.

The force F_t that is transmitted to the foundation is given by

$$F_t = F_d + F_s = D\dot{X} + KX \qquad [7.6.1-1]$$

or in the Laplace domain

$$F_t(s) = DsX + KX = (Ds + K)X(s) \qquad [7.6.1-2]$$

In order to formulate the quantity $X(s)$, write eq. 7.5.1-1 in the Laplace domain [writing the input as $F(s)$] and solve for $X(s)$

$$X(s) = \frac{F(s)}{Ms^2 + Ds + K} \qquad [7.6.1-3]$$

Substituting for X, the transmitted force F_t becomes

$$F_t(s) = \frac{(Ds + K)F(s)}{Ms^2 + Ds + K} \qquad [7.6.1-4]$$

Normalize by dividing numerator and denominator by M

$$\frac{F_t}{F}(s) = \frac{2\zeta\omega_n s + \omega_n^2}{s^2 + 2\zeta\omega_n s + \omega_n^2} \qquad [7.6.1-5]$$

The ratio of the force transmitted F_t to the input force F for various input frequencies ω is referred to as the "transmissibility." Using damping ratio ζ as parameter, plots of transmissibility are frequency response curves.

Let $s = j\omega$, divide by ω_n^2, and let $\gamma = \omega/\omega_n$

$$\frac{F_t}{F}(j\omega) = \frac{1 + j2\zeta\gamma}{1 - \gamma^2 + j2\zeta\gamma} \qquad [7.6.1-6]$$

This is a complex number whose magnitude R is given by

$$R = \sqrt{\frac{1 + (2\zeta\gamma)^2}{(1 - \gamma^2)^2 + (2\zeta\gamma)^2}} \qquad [7.6.1-7]$$

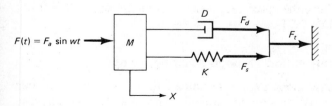

Figure 7.6.1-1   Transmission to base.

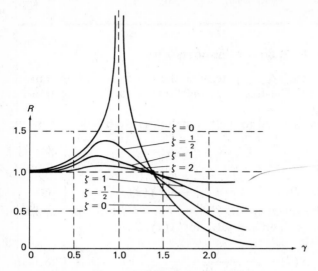

Figure 7.6.1-2   Transmissibility.

$$\frac{\text{out}}{\text{in}}(s) = \frac{2\zeta\omega_n s + \omega_n^2}{S^2 + 2\zeta\omega_n s + \omega_n^2}$$

$$R = \sqrt{\frac{1 + (2\zeta\gamma)^2}{(1 - \gamma^2)^2 + (2\zeta\gamma)^2}}$$

To assist in sketching the curves, note that

$$\text{when } \omega \ll \omega_n, \quad \gamma \ll 1 \qquad \text{then } R = 1$$

$$\text{when } \omega = \omega_n, \quad \gamma = 1 \qquad \text{then } R = \frac{1}{2\zeta}\sqrt{1 + (2\zeta)^2} \qquad \text{[7.6.1-8]}$$

$$\text{when } \omega \gg \omega_n, \quad \gamma \gg 1 \qquad \text{then } R \to \frac{2\zeta}{\gamma}$$

See Fig. 7.6.1-2.

At low frequencies, the out/in ratio R is equal to unity. At a frequency somewhat below the natural frequency, the curves peak, but not as high as those in Fig. 7.5.1-2. Above the natural frequency, the curves drop off, but not as rapidly as those in Fig. 7.5.1-2.

7.6.2 Base Motion Transmitted Through a Spring-Damper

In Section 7.5.2, the base motion was transmitted through a spring. In this section, the base motion is transmitted through a spring and a damper. The free body diagram for a typical system is shown in Fig. 7.6.2-1.

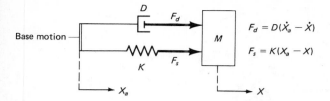

Figure 7.6.2-1   Base motion transmitted through spring-damper.

Sum the forces according to Newton's law

$$\sum F = M\ddot{X} = F_d + F_s = D(\dot{X}_a - \dot{X}) + K(X_a - X) \qquad [7.6.2\text{-}1]$$

Separate the variables

$$M\ddot{X} + D\dot{X} + KX = D\dot{X}_a + KX_a \qquad [7.6.2\text{-}2]$$

Transform to the Laplace domain. Note that there are two variables, X and X_a. Assume zero initial conditions.

$$(Ms^2 + Ds + K)X(s) = (Ds + K)X_a(s) \qquad [7.6.2\text{-}3]$$

Solve for $X(s)$

$$X(s) = \frac{(Ds + K)X_a(s)}{Ms^2 + Ds + K} \qquad [7.6.2\text{-}4]$$

Note that eq. 7.6.2-4 is formally identical to eq. 7.6.1-4. Hence, the frequency response is given by eq. 7.6.1-7 and Fig. 7.6.1-2.

Compare this case to that in Section 7.5.2, where the base motion was transmitted through a pure spring. There, at the high frequencies, a large damping ratio was effective in reducing the transmission. Here, the opposite condition prevails.

■ **EXAMPLE 7.6.2-1** Motorcycle on Bumpy Road
In Ex. 6.5.1-2 (Motorcycle) determine the frequency response when the road may be approximated by a sequence of equally spaced bumps (approximately a sinusoidal input). Assume that the spacing between bumps is equal to L and that the forward speed of the vehicle is equal to V. (See Fig. 7.6.2-2.)

Solution.   The response to a generalized input $Y_a(s)$ is given by eq. (3) of Ex. 6.5.1-2, which is repeated below

$$Y(s) = \frac{(Ds + K)Y_a(s)}{Ms^2 + Ds + K} \qquad (1)$$

For the steady state response (frequency response) to a sinusoidal input, formulate the transfer function Y/Y_a.

$$\frac{Y}{Y_a}(s) = \frac{Ds + K}{Ms^2 + Ds + K} \qquad (2)$$

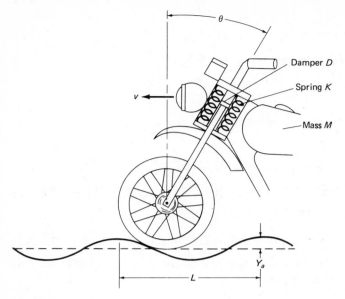

Figure 7.6.2-2

Let $s = j\omega$, and separate real and imaginary parts

$$\frac{Y}{Y_a}(j\omega) = \frac{K + jD\omega}{K - M\omega^2 + jD\omega} \tag{3}$$

Normalize, divide by ω_n^2 and let $\gamma = \omega/\omega_n$

$$\frac{Y}{Y_a}(j\omega) = \frac{1 + j2\zeta\gamma}{1 - \gamma^2 + j2\zeta\gamma} \tag{4}$$

Note that eq. (2) of this example, which involves the response to a sinusoidal base motion, has exactly the same form as eq. 7.6.1-6, which was concerned with the force transmitted through the spring and damper to the foundation. In that case we were dealing with *transmissibility of force*. In this example we are dealing with *transmissibility of motion.*

The input frequency is given by

$$\omega = \frac{2\pi V}{L} \tag{5}$$

From eq. (5), we learn that for a fixed spacing L between bumps, the input frequency ω is proportional to the forward velocity V. Hence, as the motorcycle picks up speed, it will proceed through a frequency range. At each frequency, (assuming there is time to approach the steady state condition) we can appreciate the meaning of Fig. 7.6.1-2. At low speed V, the input frequency ω is low, and the curve indicates that the out/in ratio is approximately unity. This implies that the motorcycle frame M will be displaced sinusoidally with an amplitude equal to the input Y_a. As the speed increases to a value as which, by means of eq. (5), the input frequency ω is close to the resonant frequency, the out/in ratio is greater than unity. This implies that the motorcycle frame will be displaced sinusoidally with an amplitude that is greater than that of the bump (a rather uncomfortable response). At very high speed V, the input frequency ω is high, and according to Fig. 7.6.1-2, the output displacement Y is negligible.

From the above, we conclude that the most comfortable ride occurs at high speed. Consequently as a design goal, let the natural frequency ω_n be low, so that normal speeds would correspond (relative to ω_n) to high input frequencies. In order that the natural frequency ω_n be low, the spring K should be small (very soft spring) while the mass M of the frame should be large. Practical limitations conflict with these design goals. Soft springs result in large static deflection due to the weight of the vehicle. A large frame mass results in a heavy machine.

Now consider the selection of the optimum damping ratio ζ. In Fig. 7.6.1-2, note that the damping ratio affects the height of the resonant peak. Since the out/in ratio is greater than unity, this is also referred to as a magnification. We see that the *more* damping, the *less* the magnification. At high speeds, the out/in ratio is smaller than unity, but here, the *less* damping the better. There is a conflict in the choice of damping. A large damping ratio is desirable to minimize the magnification at speeds corresponding to the resonant frequency. On the other hand, a small damping ratio is desirable to reduce the vibration at high speeds. Before making the choice, one must compare the percentage of the time at speeds near the resonant frequency to that at high speeds. This example indicates the need to examine all properties in order to make the appropriate design decision. See Problems 7.22 through 7.24. ∎

7.6.3 Sinusoidal Voltage Transmitted Across Series RC Circuit

Consider the electric circuit that was given in Section 7.5.3. In that case, the voltage drop across C was found. In this section, the objective is to find the voltage drop across the series RC circuit. The circuit is shown in Fig. 7.6.3-1.

The output voltage E_t is given by

$$E_t(t) = R\dot{q} + \frac{q}{C} \qquad [7.6.3\text{-}1]$$

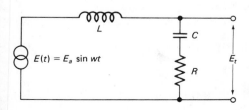

Figure 7.6.3-1   Output of RC of series circuit.

Transform eq. 7.6.3-1 to the Laplace domain

$$E_t(s) = \left(Rs + \frac{1}{C} \right) Q(s) \qquad [7.6.3\text{-}2]$$

Substitute the expression for charge $Q(s)$ in the Laplace domain as given by eq. 7.5.3-3 into eq. 7.6.3-2, and divide by E

$$\frac{E_t}{E}(s) = \frac{Rs + 1/C}{Ls^2 + Rs + 1/C} \qquad [7.6.3\text{-}3]$$

Note that eq. 7.6.3-3 has the same form as that in eq. 7.6.1-4. Hence, the frequency response is given by eq. 7.6.1-7 and by Fig. 7.6.1-2.

This circuit is often used as coupling between stages of an amplifier. For an amplifier of many frequencies, like one used for reproduction of music, it is desirable that all frequencies in the specified range be transmitted with approximately the same amplitude ratio. Hence, referring to Fig. 7.6.1-2, the most desirable design is one that has out/in = 1 for the greatest range. A fair approximation is afforded by a damping ratio ζ equal to unity. The frequency range is from zero to about $1.5\omega_n$ (note that $\gamma = \omega/\omega_n$).

■ **EXAMPLE 7.6.3-1** Inductive Pickup for Telephone

An inductive pickup for a telephone consists of a coil of wire in a snap-on cap that fits over the ear piece. The cap is perforated so that it doesn't interfere with the normal use of the telephone. When the ear piece receives its normal signal, there is a small magnetic signal that is transmitted for a distance of only 0.002 m (about 0.08 in.). The coil is capable of picking up this signal, which permits it to listen to the conversation. The pickup's signal is amplified for various purposes: for the hard-of-hearing; for office use; to permit several people to participate in the conversation; and so on.

The circuit is shown in Fig. 7.6.3-1, where the input signal $E(t)$ is generated by the inductive pickup, and where L is its inductance. Given $L = 0.05$ H, design the circuit so that it will handle frequencies up to 31,400 rad/s (5000 Hz, the typical frequency range for a telephone).

Solution. The frequency response curve is shown in Fig. 7.6.1-2, where the following applies:

$$\text{frequency range corresponds to} \quad \gamma = 1.5 \qquad (1)$$

$$\text{ideal damping ratio} \qquad \zeta = 1.0 \qquad (2)$$

From eq. (1) we may state

$$31{,}400 = 1.5\omega_n \qquad (3)$$

or

$$\omega_n = 20{,}930 \text{ rad/s} \qquad (4)$$

Solve for C in eq. 7.5.3-5

$$C = \frac{1}{L\omega_n^2} = \frac{1}{0.05 \times 20{,}930^2} = 0.0457 \times 10^{-6} \qquad (5)$$

$$= 0.0457\,\mu\text{F} \qquad (6)$$

Solve for R in eq. 7.5.3-6

$$R = 2\zeta\omega_n L = 2 \times 1.0 \times 20{,}930 \times 0.05 \qquad (7)$$

$$= 2{,}093\,\Omega \qquad (8)$$

■

7.7 SELF-EXCITED VIBRATION

7.7.1 Rotating Unbalanced Machine

Consider a rotating rigid body that is rotating in trunnions that are mounted on a deformable foundation. No matter how carefully the rotor is balanced, there will always be some eccentricity. The equivalent unbalance may be represented by a lumped mass M_1 that has an eccentricity r. See Fig. 7.7.1-1. Assume the trunnion structure, mass M_2, is constrained to vertical motion Y. The displacement in the vertical direction of mass M_1 is

$$Y_1 = Y + r \sin \omega t \qquad [7.7.1\text{-}1]$$

The acceleration of mass M_1 becomes, upon differentiation,

$$\ddot{Y}_1 = \ddot{Y} - r\omega^2 \sin \omega t \qquad [7.7.1\text{-}1a]$$

Apply Newton's law to Fig. 7.7.1-1 and note that there are two masses M_1 and

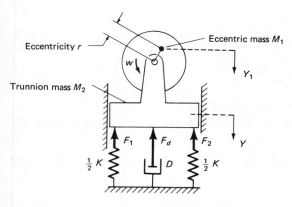

Figure 7.7.1-1   Unbalanced rotating mass.

M_2 with different accelerations

$$M_1\ddot{Y}_1 + M_2\ddot{Y} = -D\dot{Y} - KY \qquad [7.7.1\text{-}2]$$

Substitute eq. 7.7.1-1a into eq. 7.7.1-2

$$M_1(\ddot{Y} - r\omega^2 \sin \omega t) + M_2\ddot{Y} = -D\dot{Y} - KY \qquad [7.7.1\text{-}2a]$$

Perform the indicated operations, rearrange terms, and let

$$M = M_1 + M_2, \qquad \mu = \frac{M_1}{M} \qquad [7.7.1\text{-}3]$$

then

$$M\ddot{Y} + D\dot{Y} + KY = \mu M r\omega^2 \sin \omega t \qquad [7.7.1\text{-}4]$$

This system will be analyzed using Euler's identities as was done in Section 7.5.1. Let

$$Y = \text{Im}(Ae^{j(\omega t - \phi)}) \quad \text{and} \quad \sin \omega t = \text{Im}[e^{j(\omega t)}] \qquad [7.7.1\text{-}5]$$

It will be understood that the imaginary part of all complex quantities will be implied. Apply eqs. 7.7.1-5 to eq. 7.7.1-4

$$-M\omega^2 Ae^{j(\omega t - \phi)} + j\omega D Ae^{j(\omega t - \phi)} + KAe^{j(\omega t - \phi)} = \mu M r\omega^2 e^{j\omega t} \qquad [7.7.1\text{-}5a]$$

Divide through by $e^{j\omega t} \neq 0$, collect terms.

$$(K - M\omega^2 + j\omega D)Ae^{-j\phi} = \mu M r\omega^2 \qquad [7.7.1\text{-}5b]$$

Normalize and let

$$\gamma = \frac{\omega}{\omega_n} \qquad [7.7.1\text{-}5c]$$

$$\frac{Ae^{-j\phi}}{\mu r} = Re^{-j\phi} = \frac{\gamma^2}{1 - \gamma^2 + j2\zeta\gamma} \qquad [7.7.1\text{-}5d]$$

where

$$R = \frac{\gamma}{\sqrt{(1 - \gamma^2)^2 + (2\zeta\gamma)^2}}$$

$$\qquad [7.7.1\text{-}6]$$

$$-\phi = \arctan\frac{2\zeta\gamma}{1 - \gamma^2}$$

Curves for eq. 7.7.1-6 are plotted in Fig. 7.7.1-2.

The same result will be obtained using the transformation from the Laplace to complex domain. Let

$$Y_a(t) = r \sin \omega t \qquad [7.7.1\text{-}7]$$

Then, eq. 7.7.1-1 becomes

$$Y_1(t) = Y(t) + Y_a(t) \qquad [7.7.1\text{-}7a]$$

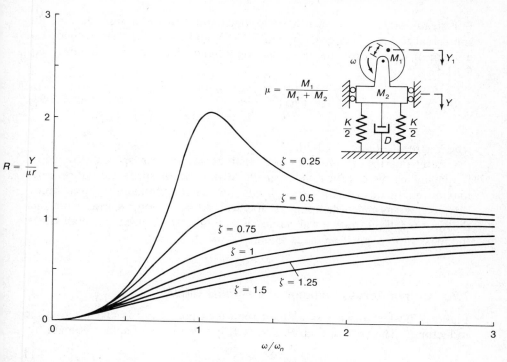

Figure 7.7.1-2   Frequency response for self-excited vibration.

Transform eqs. 7.7.1-7a and 7.7.1-2 to the Laplace domain, noting that IC = 0.

$$Y_1(s) = Y(s) + Y_a(s) \qquad [7.7.1\text{-}7b]$$

$$M_1 s^2 Y_1(s) + M_2 s^2 Y(s) = -(Ds + K)Y(s) \qquad [7.7.1\text{-}7c]$$

Eliminating $Y_1(s)$

$$M_1 s^2 [Y(s) + Y_a(s)] + M_2 s^2 Y(s) = -(Ds + K)Y(s) \quad [7.7.1\text{-}8]$$

or

$$Y(s) = \frac{-M_1 s^2 Y_a(s)}{(M_1 + M_2)s^2 + Ds + K} \qquad [7.7.1\text{-}8a]$$

Normalize and apply eqs. 7.7.1-3 and 7.7.1-7, and use only the amplitude r of $Y_a(s)$

$$\frac{Y}{\mu r}(s) = \frac{-s^2}{s^2 + 2\zeta\omega_n s + \omega_n^2} \qquad [7.7.1\text{-}9]$$

For steady state response to a sinusoidal input, use the amplitude r of the sinusoidal input Y_a (as given by eq. 7.7.1-7), and let $s = j\omega$. Then divide the numerator and denominator of the right-hand side by ω_n^2 and apply eq. 7.7.1-5c

$$\frac{Y}{\mu r}(j\omega) = \frac{\gamma^2}{1 - \gamma^2 + j2\zeta\gamma} \qquad [7.7.1\text{-}9a]$$

which agrees with eq. 7.7.1-5d.

Referring to Fig. 7.7.1-2, note that at low motor speeds (low input frequency ω) the out/in ratio is small. At the *critical* speed the out/in ratio maximizes. In this case, damping limits the peak amplitude, making it necessary to have as much damping as practical. At high motor speeds, note that the out/in ratio is unity for all values of damping. To reduce the overall effect, the mass ratio μ must be made small.

7.7.2 Output Across Inductor of Electrical Circuit

Figure 7.7.2-1 shows a series circuit for which the output is taken across the inductor L. The voltage drop V_l across L is given in the Laplace domain by:

$$V_l = Ls^2Q \qquad [7.7.2\text{-}1]$$

Substitute eq. 7.5.3-3 and normalize

$$\frac{V_l}{E}(s) = \frac{s^2\omega_n^2}{s^2 + 2\zeta\omega_n s + \omega_n^2} \qquad [7.7.2\text{-}2]$$

where ω_n and ζ are given by eqs. 7.5.3-5 and 7.5.3-6. Note that eq. 7.7.2-2 has the form of eq. 7.7.1-9. Hence, the frequency response curve is shown in Fig. 7.7.1-2.

There are a number of applications of this circuit that make effective use of its unity response to high frequencies with attenuation of low frequencies. Some of the applications are: remote control system; proximity switch; metal detector; door opener; displacement sensor; intruder alarm; security system; and so on.

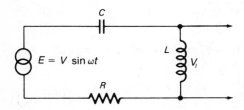

Figure 7.7.2-1   Series circuit with output across inductance.

■ **EXAMPLE 7.7.2-1** Proximity Switch

A proximity switch is an electrical switch, with no moving or exposed parts, that is actuated by touch. See Fig. 7.7.2-2(a). The presence of a human finger acts as the mating capacitor plate for each of two capacitors. When the finger is not present, there is an open circuit. [See Fig. 7.7.2-2(b).] With finger in place, it forms two capacitors C_b, in series with the electrical resistance R_b (body resistance) of the person's finger. Design the system so that the intrusion of the finger produces the maximum voltage across the inductor L. Given:

capacitor plates $= 0.001$-m squares (about 0.04 in.)

mylar thickness $= 2.5 \times 10^{-4}$ m (about 0.01 in.)

body resistance $R_b = 5 \times 10^4$ Ω

excitation frequency $\omega_x = 2 \times 10^7$ rad/s

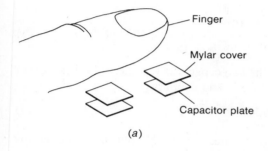

(a)

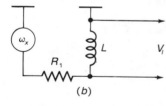

(b)

Finger in place

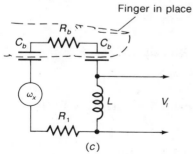

(c)

Figure 7.7.2-2   Proximity Switch: (a) actual system; (b) circuit is open; (c) circuit is closed by finger.

Solution.   Assume the finger lies flat on the mylar covering for at least the area of the capacitor plate. This forms two parallel plate capacitors, each C_b, whose capacitance is given in entry 1, Table 2.4.2-1, Appendix 2.

$$C_b = \frac{\epsilon A}{X_0} \tag{1}$$

where

$$\epsilon = 4.16 \times 10^{-11} \text{ F/m} \quad \text{(Appendix 6, Dielectric} \atop \text{properties, mylar)} \tag{2}$$

$$A = (0.001)^2 = 10^{-6} \text{ m}^2 \tag{3}$$

$$X_0 = 2.5 \times 10^{-4} \text{ m} \tag{4}$$

Apply numerical values to eq. (1)

$$C_b = \frac{4.16 \times 10^{-11} \times 10^{-6}}{2.5 \times 10^{-4}} = 1.664 \times 10^{-13} \text{ F} \tag{5}$$

Since this is a series circuit, place the two capacitors together, so it is apparent that they are in series. Apply eq. 2.4.3-7, and simplify

$$C = 0.832 \times 10^{-13} \text{ F} \tag{6}$$

In order to be sure that the system operates in the unity out/in region, let the natural frequency be one-half the excitation frequency.

$$\omega_n = \omega_x/2 = 10^7 \text{ rad/s} \tag{7}$$

Apply numerical values to eq. 7.5.3-5 and solve for L

$$L = \frac{1}{\omega_n^2 C} \tag{8}$$

$$= \frac{1}{(10^7)^2 \times 0.832 \times 10^{-13}} = 0.1202 \text{ H} \tag{9}$$

The resistances are in series. Solve for R in eq. 7.5.3-6.

$$R_b + R_1 = R = 2\zeta\omega_n L \tag{10}$$

$$= 2 \times 0.5 \times 0.1202 \times 10^7 = 1.202 \times 10^7 \ \Omega \tag{11}$$

Then

$$R_1 = R - R_b = 1.15 \times 10^6 \ \Omega \tag{12}$$

The design is complete.   ∎

7.8 SEISMIC INSTRUMENTS

7.8.1 General

A seismic instrument measures the motion of its case with respect to inertial space. Four seismic instruments will be discussed in this section: the seismograph, a vibrometer, a torsiograph, and an accelerometer. An important commonality among seismic instruments is that the output is obtained as the difference between the motion of the case (the input) and that of the mass (also referred to as the "proof mass"). Typical seismic instruments are illustrated in Fig. 7.8.1-1.

The motion of the mass is found by using the analysis of a system with base motion, as was done in Section 7.6.2. The result is given by eq. 7.6.2-4, which is repeated below

$$\frac{Y}{Y_a}(s) = \frac{(Ds + K)}{Ms^2 + Ds + K} \qquad [7.8.1\text{-}1]$$

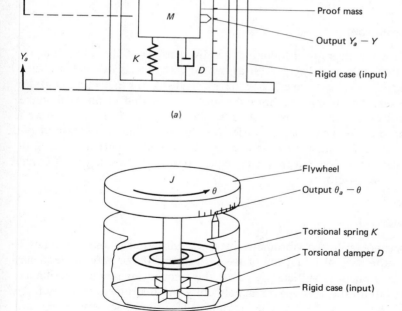

(a)

(b)

Figure 7.8.1-1   Typical seismic instruments. (a) Translatory seismic device. (b) Rotary or torsional seismic device.

The output of a seismic instrument is not Y (the motion of the proof mass) but it is $Y_a - Y$ (the relative motion between the input or case and the proof mass). Form a dimensionless ratio of this quantity

$$\frac{Y_a - Y}{Y_a}(s) = 1 - \frac{Y}{Y_a}(s) = 1 - \frac{(Ds + K)}{Ms^2 + Ds + K} \qquad [7.8.1\text{-}2]$$

Collect both terms over a common denominator

$$\frac{Y_a - Y}{Y_a}(s) = \frac{Ms^2 + Ds + K - (Ds + K)}{Ms^2 + Ds + K} = \frac{Ms^2}{Ms^2 + Ds + K} \qquad [7.8.1\text{-}3]$$

Normalize by dividing by M

$$\frac{Y_a - Y}{Y_a}(s) = \frac{s^2}{s^2 + 2\zeta\omega_n s + \omega_n^2} \qquad [7.8.1\text{-}4]$$

Note that eq. 7.8.1-4 is identical* to eq. 7.7.1-9. Strange as it seems, the frequency response for a seismic device is identical to that for a self-excited device. Equation 7.8.1-4 applies to all seismic devices. However, there are some small but significant differences which will be discussed subsequently.

7.8.2 Seismograph

The complete seismograph installation usually consists of three instruments, similar to that shown in Fig. 7.8.1-1(a), where each one is oriented to sense each of three coordinate axes. The device is used for detecting and for measuring *very low* frequency vibration such as that due to an earthquake, underground explosion, and mine or tunnel cave-in. The device must have a *very low* natural frequency ω_n, in order for the operation to be carried out where the output/input ratio R is approximately unity. Since there is little danger of the input having a frequency equal to ω_n, the damping ratio ζ may be chosen close to zero.

7.8.3 Vibrometer

Like the seismograph, the vibrometer measures vibration whose frequency is higher than the natural frequency of the instrument. If the natural frequency ω_n of the instrument is sufficiently *below* the frequencies under surveillance, the output/input ratio R is approximately unity. An instrument with electrical output is shown in Fig. 7.8.3-1. The instrument is designed for a wide range of frequencies, and since the input frequency may be close to ω_n, the damping ratio is chosen high (about one-half to unity).

---

*Except for the sign. However, the magnitude curves, Fig. 7.7.1-2, are unchanged.

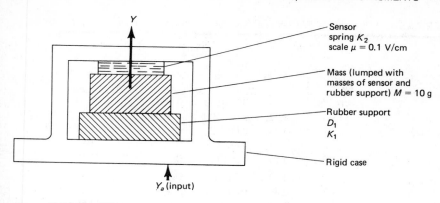

Figure 7.8.3-1   Vibrometer.

7.8.4 Torsiograph

The torsiograph may be considered the torsional counterpart of the vibrometer. The instrument is shown in general in Figs. 7.8.1-1(b) and 8.7.2-3.

7.8.5 Accelerometer

The vibrometer, discussed in Section 7.8.3, measures the *displacement* Y_a of the input vibration, where

$$Y_a = A \sin \omega t \qquad [7.8.5\text{-}1]$$

The *acceleration* $\ddot{Y}_a$ of the sinusoidal input vibration is given by

$$\ddot{Y}_a = -A\omega^2 \sin \omega t \qquad [7.8.5\text{-}2]$$

Usually the sign is not important when measuring vibration. Then, the only difference between eq. 7.8.5-1 (the displacement) and eq. 7.8.5-2 (the acceleration) is the multiplier ω^2. In that case, several possibilities exist for measuring acceleration of vibration. For the first, use a vibrometer, measure the displacement Y_a and the frequency ω and compute the acceleration by means of eq. 7.8.5-2. For the second, use a special vibrometer whose natural frequency ω_n is much *higher* than the input frequency. Then, $\gamma \ll 1$ and eq. 7.7.1-6 becomes

$$R \approx \gamma^2 = \left(\frac{\omega}{\omega_n}\right)^2 \qquad [7.8.5\text{-}3]$$

Thus, the frequency response curve automatically multiplies the result by the necessary quantity ω^2, where the remaining terms can be lumped into the instrument scale factor.

■ **EXAMPLE 7.8.5-1** Frequency Range for Seismic Device
Given a seismic instrument as shown in Fig. 7.8.1-1, show that the instrument behaves as a vibrometer for high frequencies and as an accelerometer for low

TABLE 7.8.5-1   FREQUENCY RANGE FOR SEISMIC DEVICES

| | Input | | Output / input[a] | | Error (%) |
	ω	γ	R_a actual	R_i ideal	$\dfrac{R_a - R_i}{R_i}$
Range for accelerometer	0.529	0.01	0.000100	0.0001	0
	1.06	0.02	0.000400	0.0004	0
	2.12	0.04	0.00162	0.0016 $\}\gamma^2$	0.13
	4.23	0.08	0.00644	0.0064	0.6
	8.47	0.16	0.02617	0.0256	2.2
	16.9	0.32	0.1123	0.1024	8.8
	33.8	0.64	0.6156	0.4096	33.4
	67.7	1.28	1.804	1.0	44.6
	106	2	1.264	1.0	20.8
Range for vibrometer	212	4	1.058	1.0	5.5
	423	8	1.014	1.0	1.4
	847	16	1.003	1.0 $\}1$	0.3
	1690	32	1.001	1.0	0.1
	3380	64	1.000	1.0	0

[a]R_a is given by eq. 7.7.1-6. R_i for an accelerometer is given by eq. 7.8.5-3. R_i for a vibrometer is equal to unity as explained in Section 7.8.3.

frequencies. Define each range given the following parameters.

$$M_1 = \text{proof masss} = 9 \text{ g}$$

$$M_2 = \text{mass of support (spring-damper)} = 3 \text{ g}$$

$$D = \text{damper} = 267 \text{ dyn} \cdot \text{s/cm}$$ (1)

$$K = \text{spring} = 2.8 \times 10^4 \text{ dyn/cm}$$

Solution.

$$M = M_1 + \frac{M_2}{3} = 9 + \frac{3}{3} = 10 \text{ g}$$ (2)

$$\omega_n = \sqrt{\frac{K}{M}} = \sqrt{\frac{2.8 \times 10^4}{10}} = 52.92 \text{ rad/s}$$ (3)

$$\zeta = \frac{D}{2\omega_n M} = \frac{267}{2 \times 52.92 \times 10} = 0.25$$ (4)

Using eq. 7.7.1-6, tabulate the results for an input with constant amplitude over a range of frequencies from 0.529 to 3380 rad/s. (See Table 7.8.5-1.)   ■

7.9 FREQUENCY RESPONSE INVOLVING TRANSIENTS

7.9.1 Transients for Sinusoidal Input

■ **EXAMPLE 7.9.1-1** TUTSIM Simulation for Start-up Transient in Reciprocating Pump

The pump in a refrigerator or in an air conditioner usually makes use of a reciprocating device. The pump mechanism is usually mounted on vibration isolators to minimize the transmission of the vibration to the surroundings. Whenever the pump starts up, it must pass through a range of speeds. This provides a range of input frequencies ω. Determine the transient response. See Fig. 7.9.1-1, where

$$D = \text{equivalent lumped damping}$$

$$K = \text{equivalent lumped spring} = 2.8 \times 10^4 \text{ N/m}$$

$$M = M_1 + M_p$$

$$\mu = \frac{M_p}{M}$$

Assume that the L/r ratio is very large. Consequently, the displacement of the piston is approximately sinusoidal.

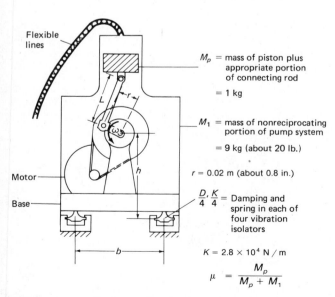

M_p = mass of piston plus appropriate portion of connecting rod

= 1 kg

M_1 = mass of nonreciprocating portion of pump system

= 9 kg (about 20 lb.)

r = 0.02 m (about 0.8 in.)

$\frac{D}{4}, \frac{K}{4}$ = Damping and spring in each of four vibration isolators

$K = 2.8 \times 10^4 \text{ N/m}$

$$\mu = \frac{M_p}{M_p + M_1}$$

Figure 7.9.1-1   Reciprocating pump in air conditioner.

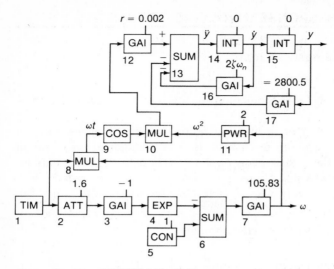

Figure 7.9.1-2   TUTSIM block diagram.

TABLE 7.9.1-1   TUTSIM MODEL DATA

Model Structure	Model Parameters
:1,TIM	:1
:2,ATT,1	:2,1.6
:3,GAI,2	:3,-1
:4,EXP,3	:4
:5,CON	:5,1
:6,SUM,-4,5	:6
:7,GAI,6	:7,105.83 (ω_s)
:8,MUL,1,7	:8
:9,COS,8	:9
:10,MUL,9,11	:10
:11,PWR,7	:11,2
:12,GAI,10	:12,0.002 (μr)
:13,SUM,12,-17,-16	:13
:14,INT,13	:14,0
:15,INT,14	:15,0
:16,GAI,14	:16,1.058 (also 2.117, 5.292)
:17,GAI,15	:17,2800.5 (ω_n^2)

Plotblocks and Ranges	Timing Data
Horz:0,0,2	0.001,2
Y1:15,-0.01,0.01	

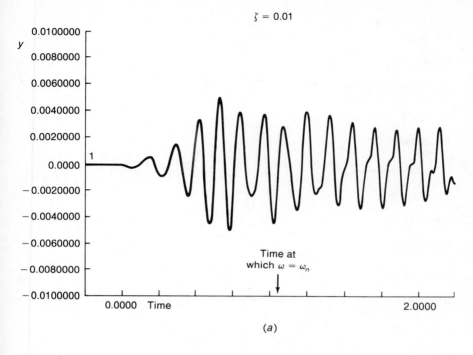

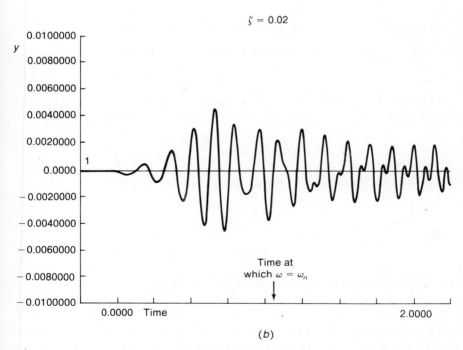

Figure 7.9.1-3   Start-up transient: (a) $\zeta = 0.01$; (b) $\zeta = 0.02$; (c) $\zeta = 0.05$.

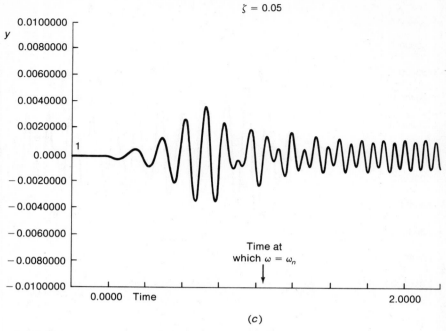

$\zeta = 0.05$

Time at which $\omega = \omega_n$

(c)

Figure 7.9.1-3   Continued

The differential equation is given by eq. 7.7.1-4, repeated below

$$M\ddot{y} + D\dot{y} + Ky = \mu M r\omega^2 \cos \omega t$$

$$= \text{input} \tag{1}$$

The cosine was chosen instead of the sine to improve the appearance of the resulting computer output. Solve eq. (1) for the highest derivative and get

$$\ddot{y} = \mu r\omega^2 \cos \omega t - \frac{D}{M}\dot{y} - \frac{K}{M}y \tag{2}$$

The natural frequency ω_n and the damping constant D satisfy

$$\omega_n = \sqrt{\frac{K}{M}} \tag{3}$$

$$D = 2\zeta\omega_n M \tag{4}$$

Assume that the system has little damping and that

$$0.01 \le \zeta \le 0.05 \tag{5}$$

Since the damping ratio is very low, the resonant frequency will be very close to the natural frequency. In order to simulate the system, the machine will be assumed to increase speed in an exponential manner, as described in Ex.

5.2.3-1 and Problem 5.32. If the steady state speed is ω_s, then the instantaneous speed ω is given by

$$\omega = \omega_s(1 - e^{-t/\tau}) \tag{6}$$

where $\tau = 1.6$ s and $\omega_s = 105.83$ rad/s.

In terms of ω_n and ζ, eq. (2) becomes

$$\ddot{y} = \mu r \omega^2 \cos \omega t - 2\zeta \omega_n \dot{y} - \omega_n^2 y \tag{7}$$

with ω given by eq. (6). Let $\zeta = 0.01$, 0.02, and 0.05 (three different simulations). The TUTSIM block diagram is shown in Fig. 7.9.1-2. The TUTSIM model data are stated in Table 7.9.1-1. Results of the simulation are shown in Figures 7.9.1-3(a), (b), and (c), for three different values of damping. Notice that in spite of very low damping, the amplitude near the resonant frequency does not become very large, since the system spends very little time near the resonant condition. ∎

SUGGESTED REFERENCE READING FOR CHAPTER 7

Forced vibration [1], chap. 4; [4], chap. 2; [5], chap. 3; [17], chap. 9; [46], chap. 7.
Frequency response [8], p. 360–375; [17], chap. 9; [46], p. 125–137; [47], p 383–394.
Complex plane [4], p. 54–56; [5], p. 57; [8], p. 361–364; [17], p. 236–243, 342; [47], p. 388–390.
Root locus [8], p. 332–335; [17], p. 384, 401; chap. 21.
Computer methods [12]; [47], p. 422–426.
Transmissibility [1], p. 141–143; [4], p. 58–60; [5], p. 62–66; [46], p. 146; [47], p. 398–403.
Self-excited vibration and unbalance [1], p. 124–130; [4], p. 64–67; [5], p. 58–61, 79–91; [46], p. 137–141; [47], p. 396–398.
Seismic instruments [4], p. 60–63; [5], p. 75–78; [17], p. 345; [47], p. 403–406.

PROBLEMS FOR CHAPTER 7

7.1. For the circuit in Fig. 3.6.1-1(f), let $L = 2.0$ H and $R = 50.0$ Ω. The voltage V is sinusoidal with an amplitude of 100 V. Obtain and plot the amplitude of the current versus angular frequency ω for $0 < \omega < 50$ rad/s.

7.2. For the system shown:
(a) Show that the output x satisfies the differential equation $D\dot{x} + Kx = Kx_a$.
(b) Given $D = 6.25$ N · s/m, $K = 100$ N/m, and $x_a = 0.04 \sin 5t$ (m, s), find the steady state output x in the form $x = X \sin(5t + \phi)$.

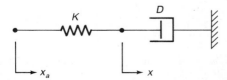

Figure P-7.2

7.3. For the system of Fig. 7.5.1-1, $M = 5.0$ kg, $K = 180$ N/m, and $D = 10.0$ N · s/m.
   (a) Find ω_n and ζ.
   (b) If the steady state amplitude for F_a is 5.4 N at a frequency of 3.6 rad/s, find the steady state amplitude of x.

7.4. For the system in Fig. 3.6.2-1(a), do the following:
   (a) Find the amplitude response $|KY/F(j\omega)|$ and the phase angle versus ω.
   (b) If $K = 20$ N/m and $D = 3$ N · s/m, find ω for which $|KY/F| = 0.25$.

7.5. The amplitude response for a motorcycle driven over a sinusoidal test road is given by eq. (3) or (4) of Ex. 7.6.2-1. Show that all amplitude response curves cross at $\gamma = 2$ and that they have the value $|Y/Y_a| = 1$.

7.6. Refer to Fig. 7.2.3-1. The differential equation relating the output to the input is given by eq. (1a). Derive the amplitude response $|\theta/\theta_a|$ versus ω and the phase response ϕ versus ω. Use the values for K and D shown in the figure.

7.7. Refer to Fig. 7.5.2-1. Given $K_1 = 68$ N/m, $K_2 = 220$ N/m, $D = 12$ N · s/m and $M = 2$ kg, find the steady state amplitude X of the mass, and the phase angle if $x_a = 0.08 \sin 9t$ (m, s).

7.8. For the system shown, $D = 640$ N · s/m, $M = 80$ kg, and $K = 5120$ N/m.
   (a) Obtain the differential equation for x.
   (b) Given $x_a = 0.02 \sin(12t)$ (m, s), find the amplitude of the force transmitted to the wall.

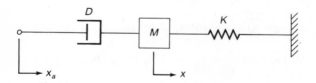

Figure P-7.8

7.9. Consider the phonograph pickup in Fig. 3.2.1-4. Let $J = 2.4 \times 10^{-4}$ g-cm^2, $K = 3.0 \times 10^5$ dyn-cm, and $D = 10.18$ dyn-cm-s.
   (a) Draw a freehand graph to scale of $|\theta/\theta_a|$ versus ω.
   (b) Estimate the bandwidth in Hz.

7.10. In Fig. 3.6.1-1(a), let $K = 240$ N/m and $D = 48$ N · s/m. A sinusoidal force $f_a = 50 \sin \omega t$ is applied. Find the amplitude and phase for y when $\omega = 2$ rad/s and $\omega = 5$ rad/s. Use Euler's identities.

7.11. The complex frequency response for a motorcycle traveling over the sinusoidal road profile shown is

$$\frac{Y}{Y_a}(j\gamma) = \frac{1 + j(2\zeta\gamma)}{(1 - \gamma^2) + j(2\zeta\gamma)}$$

Let $M = 250$ kg, $K = 100,000$ N/m, $L = 5.0$ m, $Y_a = 0.05$ m and $V = 30$ m/s. Find D so that $Y = 0.03$ m. (See Fig. P-7.11.)

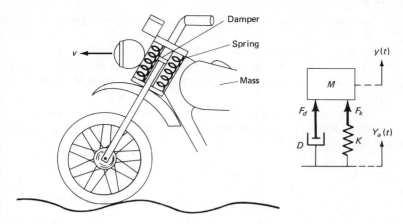

Figure P-7.11

7.12. For the system shown in Fig. 7.5.1-1, $M = 0.07$ kg, $K = 2800$ N/m, and $D = 11.2$ N · s/m. The applied force is $F_a = 120 \sin \omega t$ (N, s).
 (a) Find the amplitude of the steady state displacement for $\omega/\omega_n = 0.5, 1.0, 1.5,$ and 2.0.
 (b) Sketch a graph of X versus ω to scale.

7.13. A rotating unbalanced machine like that in Fig. 7.7.1-1 has $M_1 = 4.0$ kg, $M_2 = 16.0$ kg, $K = 3.2 \times 10^6$ N/m, $D = 3200$ N · s/m, and $r = 0.001$ m. The machine rotates at 2400 rpm.
 (a) Find the steady state displacement amplitude of the supporting mass.
 (b) Find the amplitude of the steady state sinusoidal force transmitted to the floor.

7.14. For the system shown, derive the differential equation for the output x, and obtain the expression for the amplitude response.

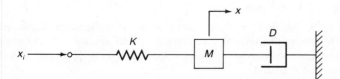

Figure P-7.14

7.15. A turbine rotor has a mass of 170 kg, an unbalance eccentricity $r = 1.25$ mm, and runs at 7200 rpm. The turbine is attached to a trunnion of mass 850 kg. The trunnion is supported on springs with combined stiffness of $K = 14.5 \times 10^7$ N/m, and a damper of value $D = 2.31 \times 10^5$ N · s/m.
 (a) Find ω_n and ζ.
 (b) Find the amplitude Y of the trunnion vibration.

7.16. For the system of Fig. 7.5.1-1, let $M = 4$ kg, $K = 10$ N/m, $D = 16$ N · s/m, and $F_a = 20 \sin \omega t$ (N, s). Obtain an expression for x in the form $x = X_0 \sin(\omega t + \phi)$ for $\omega = 6$ rad/s.

7.17. The curves in Fig. 7.5.1-2 may be used for the output/input response $|\theta/\theta_a|$ for a phonograph pickup. To keep the response relatively flat over the desired frequency range of 20 to 10,000 Hz, the damping ratio ζ is chosen to be $\sqrt{\frac{1}{2}}$. The damping constant D for the pickup is 5.0 dyn-cm · s/rad. If $|\theta/\theta_a|$ is to be greater than 0.707 for the desired frequency range, what must the values of K and J be?

7.18. For the system shown, $M = 2.0$ kg, $D = 100$ N · s/m, $K = 5000$ N/m, and $x_a = 0.04 \sin(100t)$ (m, s). Find the amplitude and phase angle for steady state sinusoidal motion of the mass.

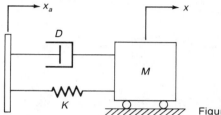

Figure P-7.18

7.19. For the system in Fig. 7.6.1-1, let $M = 300$ kg, $K = 14{,}700$ N/m, and $D = 1680$ N · s/m. For $F = 50 \sin(6t)$, find the amplitude of the force transmitted to the wall.

7.20. For the system of Fig. 7.5.2-1, $M = 2.0$ kg, $K_1 = 20$ N/m, $K_2 = 50$ N/m, and $D = 5.92$ N · s/m. The curves in Fig. 7.5.1-2 apply. Estimate the amplitude of x if the amplitude of x_a at $\omega = 7.5$ rad/s is 0.15 m.

7.21. For the system in Fig. 3.6.1-1(a), let $K = 1200$ N/m, and $D = 240$ N · s/m. A sinusoidal force $F = 270 \sin 5t$ is applied. Find the steady state amplitude Y and the phase angle ϕ.

7.22. Sketch the frequency response for Ex. 7.6.2-1 (Suspension System for Motorcycle) for the following damping ratios: 0, $\frac{1}{4}$, $\frac{1}{2}$, and 1. Select the optimum value (in your opinion) and defend your choice.

7.23. The discomfort to the rider is related to the vertical acceleration due to road conditions. Determine the vertical acceleration due to an impulse and a step. Sketch the transient response for damping ratios equal to 0, $\frac{1}{4}$, $\frac{1}{2}$, and 1. Select optimum value and defend it.

7.24. Considering Problems 7.22 and 7.23, select an overall optimum and defend your choice.

7.25. In Fig. 7.6.1-2, prove that all the curves cross at a single point. Find the coordinates of this point.

7.26. The circuit in Fig. 7.5.3-1 is used in a radio control system. As such, it must respond to only one frequency, the resonant frequency ω_r. This would be a simple matter if it were not for stray or noise disturbances at many different frequencies, some of which occur uncomfortably close to the resonant frequency. Since the radio control signal and the noise both appear as inputs, but only the signal is desired, it will be necessary to separate the two. The damping ratio $\zeta = 0.05$. The signal has an amplitude = 1.0 V at the resonant frequency ω_r. The noise has an amplitude = 0.2 V at a frequency = 90% ω_r. Determine the signal/noise ratio at the resonant frequency ω_r.

7.27. In Ex. 7.5.2-1 (Phono Pickup) use a damping ratio $= \frac{1}{2}$ and sketch the transient response to (a) impulse, (b) step, and (c) frequency response.

7.28. In Ex. 7.5.2-1, assuming that the natural frequency is 30,000 rad/s, sketch the frequency response for the following damping ratios: 0, $\frac{1}{4}$, $\frac{1}{2}$, and 1. Select and defend the optimum value of ζ.

7.29. Repeat Problem 7.28 for transient response to an impulse and to a step. Considering both the frequency and transient responses, select the optimum damping and defend your choice.

7.30. An amplifier uses the circuit in Fig. 7.6.3-1 as the coupling between stages. If the output/input ratio is allowed to vary from 0.9 to 1.1 ($\pm 10\%$ from unity), determine the allowable frequency range (bandwidth) if $\omega_n = 60,000$ rad/s.

7.31. The minimum volume level change that the human ear can detect is approximately $\pm 40\%(\sqrt{\frac{1}{2}}$ to $\sqrt{2})$. Rework Problem 7.30 using this tolerance.

7.32. A generator is slightly unbalanced and its support structure (including the bearings) is elastic. The following parameters apply (refer to Section 7.7.1):

$$M = \text{mass of rotor} = 10 \text{ kg}$$

$$r = \text{eccentricity of rotor} = 2 \times 10^{-4} \text{ m}$$

$$D = \text{equivalent damping in support structure} = 106 \text{ N} \cdot \text{s/m}$$

$$K = \text{equivalent elasticity of support structure } 2.8 \times 10^4 \text{ N/m}$$

Determine the amplitude ratio $A/\mu r$ of steady state vibration at the following speeds: 26.46, 52.92, 105.83 rad/s.

7.33. Repeat Problem 7.32 where $D = 26.5$ N $\cdot$ s/m.

7.34. A vibrometer has a damping ratio $= \frac{1}{2}$. How far above the natural frequency must the input be so that the output/input ratio is within $\pm 5\%$ of unity?

7.35. An accelerometer has $\zeta = 0.01$. How far below the natural frequency must the input be so that the error is within $\pm 5\%$ of the factor $(\omega/\omega_n)^2$?

7.36. In Problem 3.25(d), let $M = 150$ kg and $F(t) = 5 \sin 30t$ (N). The following springs and dampers are available: $K = 1.35(10)^5$, $5.4(10)^5$ N/m, and $D = 2250$, 12000 N $\cdot$ s/m. What combinations of one spring and one damper will give a steady state amplitude of $15(10)^{-6}$ m or less?

7.37. In Problem 3.25(d), let $M = 150$ kg, $F(t) = 5 \sin \omega t$ (N) where $0 \le \omega \le 30$ rad/s and the machines are run at full speed most of the time. Select values for K and D so that the motion of the machine y is less than 1 mm amplitude when running at full speed. Justify the values you select.

7.38. An RLC series circuit has an input $e_1 = E_1 \cos \omega t$, e_2 is the voltage across R, $L = 1$ H, $C = 1 \mu$F.
(a) Find R for critical damping.
(b) Find the natural frequency (in radians per second) of the circuit.
(c) Sketch the frequency response e_2/e_1 for $R = 1000 \ \Omega$.
(d) What is the maximum value for $|e_2/e_1|$ when $R = 1000 \ \Omega$? At what frequency and phase angle does this occur?

7.39. In Problem 4.43 (Air Controlled Turbine) assume that the air supply has a sinusoidal variation (due to the pump). As the pump comes up to speed, this introduces a range of input frequencies. Sketch the frequency response for the system and explain the significance of the result.

7.40. Given the frequency response curves shown in Fig. P-7.40, synthesize the corresponding system (sketch the free body diagram and write the differential equation) for each one.

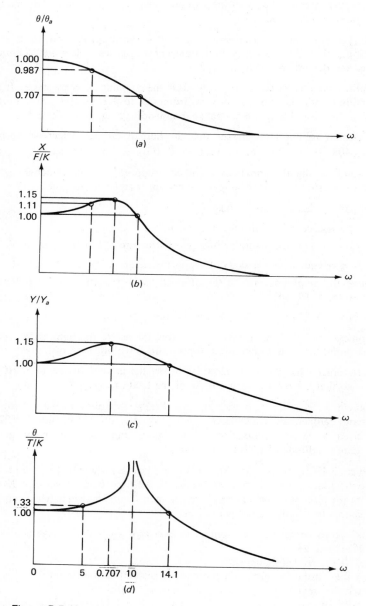

Figure P-7.40

7.41. *Output Across Resistor of Series Circuit.* A series circuit is shown in Fig. 7.5.3-1. If the voltage drop across the resistor R is V_r, determine the ratio V_r/E and sketch the frequency response curve.

7.42. Is the function or the curve of Problem 7.41 of the form of any of the three plotted in this chapter (Figs. 7.5.1-2, 7.6.1-2, and 7.7.1-2)?

7.43. Will the function derived in Problem 7.41 serve the needs of Problem 7.26? Explain. Determine the signal/noise ratio at resonance, and at $\pm95\%$ of resonance.

7.44. *Determination of R and L.* A household appliance uses a large electric motor whose R and L are unknown, but it is known that they are in series. Problem 7.1 suggests a means to determine R and L by making the following tests:

$$I_{ac} = 10 \text{ A}, \quad \text{when } V = 100 \text{ V and } \omega = 377 \text{ rad/s (about 60 Hz)}$$

$$I_{dc} = 10 \text{ A}, \quad \text{when } V = 5 \text{ V dc } (\omega = 0)$$

7.45. *Circuit Loading at Various Frequencies.* Using the results of Problem 7.1, determine the current flow for each of the following:
  (a) the motor in Problem 7.44, where $V = 1.0$ V and $\omega = 6.28 \times 10^5$ rad/s (100 kHz);
  (b) the motor in Problem 7.44, where $V = 110$ V and $\omega = 377$ rad/s (about 60 Hz);
  (c) a toaster oven, pure resistance, $R = 10$ Ω, where $V = 1.0$ V and $\omega = 6.28 \times 10^5$ rad/s;
  (d) a toaster oven, pure resistance, $R = 10$ Ω, where $V = 110$ V and $\omega = 377$ rad/s.

7.46. *Magnetic Phono Pickup.* A phono pickup may be modeled as a series RL device. The movement of the needle generates a voltage $E(t)$. The phono pickup is made part of a series circuit as shown in Fig. 7.6.3-1. Design the system given the following:
  frequency range 0 to 6.28×10^5 rad/s (10 kHz)
  $L = 0.01$ H
  $\zeta = 0.6$ (very small rise in frequency response curve).

7.47. Refer to Ex. 7.7.2-1 (Proximity Switch). Determine the out/in ratio when the person's finger or hand is the following distance from the pads:
  (a) 10^{-3} m (about 0.04 in.).
  (b) 10^{-2} m.
  (c) 10^{-1} m.

7.48. *Metal Detector.* A metal detector consists of a large flat coil with an air core. The presence of even a small amount of metal changes the inductance of the coil so that the circuit resonates, announcing the presence of metal. The coil was tested and found to induce resonance when a coin is placed 0.1 m (about 4 in.) from the coil. When the coin is removed, the inductance decreases by 14%. (Note that this test has accounted for the actual magnetic circuit, the permeability of the combined air-metal medium, and the geometry.) Design the system, given the following:

$$d = \text{coil diam} = 0.2 \text{ m (about 8 in.)}$$

$$N = \text{number of turns} = 1000$$

$$l = \text{coil length} = 0.02 \text{ m}$$

$$\zeta = \text{system damping ratio} = 0.005$$

$$\omega_r = \text{resonant and excitation frequencies} = 1.000 \times 10^4 \text{ rad/s}$$

7.49. *Intruder Alarm.* The presence of an intruder—without touching or damaging the homeowner's property—triggers this system. It operates on a principle similar to that in Problems 7.47 and 7.48. The system consists of two large capacitor plates that face the direction of the expected intruder. The presence of either an animal or human changes the medium, thus changing the capacitance. The output is taken across an inductor in the circuit. Design the system for the following:

plate size 0.1 m (square) (about 4 in.)

proximity of intruder 0.3 m (about 12 in.)

use several damping ratios and excitation frequencies.

7.50. *Magnetic Displacement Sensor.* A displacement sensor (sometimes called a Shavitz transformer) operates by the change of inductance due to the very small movement of a specially shaped armature. See Fig. P-7.50. When the shaft rotates through an angle θ the armature attached to the shaft produces better magnetic coupling for the coil on the right, thus increasing voltage V_2. At the same time, voltage V_1 is decreased. Design the device for the following:

$$h = \text{gap between rotor and stator} = 10^{-4} \text{ m (about 0.004 in.)}$$

$$\theta = \text{angular sensitivity} = 10^{-3} \text{ rad (about 3 arcmin)}$$

$$(V_2 - V_1)V_1 = \text{relative voltage change} = 0.1\%$$

$$r = 0.01 \text{ m}, \qquad b = 0.001 \text{ m}$$

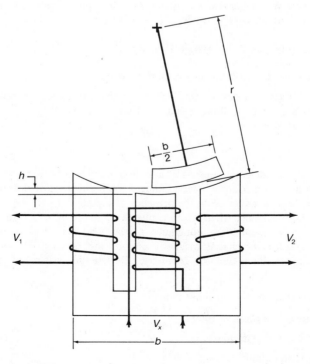

Figure P-7.50

7.51. *Capacitive Displacement Sensor.* Similar to Problem 7.50, but using capacitor plates, curved as the poles in Problem 7.50. Note that this device permits large travel of the rotor. Use square plates, 0.002 on a side.

7.52. A more sensitive device with very small travel is possible if the capacitor plates in Problem 7.51 are set parallel to the radius vector r. Note that the travel is limited to $\pm h/2$. Using the value of h in Problem 7.51, determine the angular sensitivity for the same $(V_2 - V_1)/V_1$.

7.53. *Carrier Current Transmission.* Lights and household appliances may be turned on or off from a central panel by using the existing household wiring. A 100-kHz signal is placed across the 110-V, 60-Hz line. Any number of receivers may be used to sense the 100-kHz signal. (See Fig. P-7.53.) Design the receiver for the following:

$$L = 0.001 \text{ H}, \qquad \zeta = 0.5$$

$$E = 1.0 \text{ V} = V_x, \qquad \omega_x = 6.28 \times 10^5 \text{ rad/s} \ (100 \text{ kHz})$$

Note that two systems are sharing the same line:
100 V at 60 Hz
1.0 V at 100 kHz
At any time during the operation of the carrier current transmission system, any member of the household may plug in a toaster, iron, washing machine, or vacuum cleaner across the 110-V line. The transmitter is plugged in across the line (note that no wiring is permanent) and any number of receivers are plugged in across the line. Each receiver actuates an ac relay, where the relay is the inductance in a series circuit. Then this is a system where the output is across the

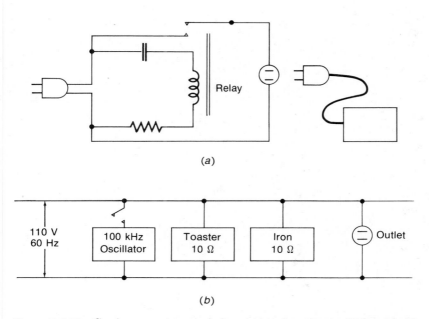

(a)

(b)

Figure P-7.53   Carrier current transmission: (a) receiver circuit; (b) household loading.

inductance, since that is what actuates the relay. Is this simple scheme sufficient to keep all the signals straight without interference?

(a) Prove that the 110-V, 60-Hz power across the line will not affect either the transmitter (not to be designed for this problem) or the receiver.

(b) Prove that if both a toaster and an electric iron (each with pure resistance $R = 10\ \Omega$) are placed across the line, it will not affect either the transmitter (assume it has a capacity of 0.5 A) or the receiver. Make use of the results of Problem 7.45.

(c) Prove that if the typical household appliance motor discussed in Problem 7.44 is placed across the line, it will not affect either the transmitter or receiver.

(d) Prove that the concept of output across an inductance of a series circuit accomplishes the task.

Chapter **8**

Design Oriented Approach to Frequency Response

This chapter considers means to determine the frequency response of systems whose Laplace transfer functions can be expressed in terms of products. The methods employed can be used for analysis, and design. Design tables and graphs are presented in order to shorten the trip from cause to effect. In this way, propitious choices of constituents can be made.

The technique in this chapter is basically graphical, where the asymptotes of the various curves can be determined with sufficient accuracy to satisfy most engineering requirements. The graphs are drawn using log-log scales. Consequently, many curves are approximately straight lines. Needless to say, such curves can be drawn instantly and modified by inspection. In addition to this, some methods are shown that greatly assist in extending the ideas to high order systems.

Some of the illustrative examples are: high and low pass filters, resonators, broadband devices, design of a stepper motor, design of the suspension system for a motorcycle, design of a capacitor microphone, and design of a vibrating reed tachometer.

8.1 FREQUENCY RESPONSE USING PRODUCTS

8.1.1 Laplace Transfer Function

The Laplace transfer function was defined in Section 7.4.3, and is briefly repeated here. This function is defined as the Laplace transform of the out/in ratio where only the amplitude of the input is used and where all initial conditions are equal to zero.

289

8.1.2 Frequency Response Obtained from Transfer Function

In general, the Laplace transfer function $G(s)$ will appear in the form of one polynomial $G_n(s)$ divided by another $G_m(s)$. An expression of this form is given by

$$G(s) = \frac{G_n(s)}{G_m(s)} \qquad [8.1.2\text{-}1]$$

$$= \frac{s^4 + as^3 + \cdots + n}{s^6 + gs^5 + hs^4 + \cdots + m} \qquad [8.1.2\text{-}2]$$

Given the Laplace transfer function in the form of eq. 8.1.2-2, the frequency response could be determined by letting $s \rightarrow j\omega$, as was done in Section 7.4.3. This would produce a complex number $A_n + jB_n$ in the numerator, and another one $A_m + jB_m$ in the denominator. Using the appropriate complex algebra, as discussed in Appendix 5, this would result in a single complex number as shown in eq. 8.1.2-3.

$$G(j\omega) = \frac{A_n + jB_n}{A_m + jB_m} = Re^{j\phi} \qquad [8.1.2\text{-}3]$$

8.1.3 Transfer Function Given as Product of Factors

Suppose that each of the polynomials $G_n(s)$ and $G_m(s)$ could be factored. It is proposed that we plot the frequency response curve for each factor, and then multiply (or divide) these as required. The Laplace transfer function would then have the form

$$G(s) = \frac{(s + a)(s + b) \cdots (s + n)}{(s + p)(s + q) \cdots (s + m)} \qquad [8.1.3\text{-}1]$$

for first order factors. For quadratic factors, the Laplace transfer function would have the form

$$G(s) = \frac{(s^2 + as + b) \cdots}{(s^2 + ps + q)(s^2 + rs + t) \cdots} \qquad [8.1.3\text{-}2]$$

Examples of Laplace transfer functions in the form of one polynomial divided by another are given by eqs. 7.5.1-7, 7.6.1-5, and 7.7.1-9, respectively, repeated below.

$$\frac{X}{B/K}(s) = \frac{\omega_n^2}{s^2 + 2\zeta\omega_n s + \omega_n^2} \qquad [8.1.3\text{-}3]$$

$$\frac{F}{F_t}(s) = \frac{2\zeta\omega_n s + \omega_n^2}{s^2 + 2\zeta\omega_n s + \omega_n^2} \qquad [8.1.3\text{-}4]$$

$$\frac{Y}{\mu Y_a}(s) = \frac{s^2}{s^2 + 2\zeta\omega_n s + \omega_n^2} \qquad [8.1.3\text{-}5]$$

Note that all three functions listed in eqs. 8.1.3-3–8.1.3-5 have the same denominator. The numerator of eq. 8.1.3-3 is a constant, the numerator of eq. 8.1.3-4 is a first order factor, and that of eq. 8.1.3-5 is an exponential. In Chapter 7, each of these functions was evaluated individually producing the magnitude functions given by eqs. 7.5.1-5, 7.6.1-7, and 7.7.1-6, respectively. Each of these in turn was plotted separately in Figs. 7.5.1-2, 7.6.1-2, and 7.7.1-2.

Reconsidering these three functions, a suggestion is to start with a curve for the denominator that is common to all three, and multiply this curve by each of the corresponding numerators of eqs. 8.1.3-3–8.1.3-5. It appears reasonable to construct a catalog of all possible curves and multiply these as needed. The question is, "How many curves are needed?"

8.1.4 Number of Curves Required

From the theory of equations, there is a theorem that proves that a polynomial having real coefficients can always be factored into first order and quadratic factors that have real coefficients. If we also include the exponential factor, then five curves should be listed: the exponential, first order factor in numerator, first order factor in denominator, quadratic in numerator, and quadratic in denominator. Note that all possibilities are covered by multiplication only (no division).

8.2 FREQUENCY RESPONSE USING GRAPHICAL MEANS

8.2.1 Concept of Multiplying Curves Graphically

In the previous section, it was shown that the Laplace transfer function could be expressed as the product of a small number of factors. If one collected the frequency response curves for each of the factors, then the frequency response curve for the system would be obtained by the product of these curves.

Assume that we have started compiling a table of such frequency response curves, and that two have been selected for illustrative purposes, say, the

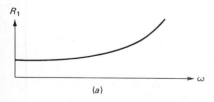 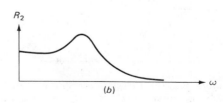

Figure 8.2.1-1   Some sample frequency response curves. (a) Frequency response curve for $2\zeta\omega_n + \omega_n^2$. (b) Frequency response curve for

$$\frac{1}{s^2 + 2\zeta\omega_n s + \omega_n^2}$$

curves for the functions shown in Fig. 8.2.1-1. These curves will be multiplied in the subsequent sections.

8.2.2 Product of Curves on Linear Scales

Consider plotting the frequency response curve obtained in Ex. 7.6.2-1 (Suspension System for Motorcycle) by the proposed multiplication. Normalize eq. (1) of that example and write the transfer function as a product

$$\frac{y}{Y_a}(s) = \left[2\zeta\omega_n s + \omega_n^2\right]\left[\frac{1}{s^2 + 2\zeta\omega_n s + \omega_n^2}\right]$$

The curves are given by Figs. 8.2.1-1(a) and (b). Multiplying these curves point by point, the product or composite thus obtained is identical to that in Ex. 7.6.2-1. See Fig. 8.2.2-1.

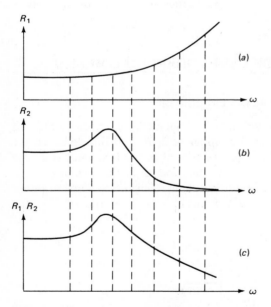

Figure 8.2.2-1 Frequency response by multiplying point by point. (a) Frequency response curve for $2\zeta\omega_n s + \omega_n^2$. (b) Frequency response curve for

$$\frac{1}{s^2 + 2\zeta\omega_n s + \omega_n^2}$$

(c) Frequency response curve for system. This composite is obtained by multiplying the above curves.

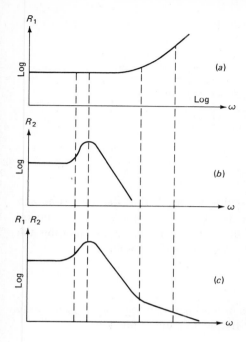

Figure 8.2.3-1 Frequency response by summing on log-log scales. (a) Frequency response curve for $2\zeta\omega_n s + \omega_n^2$. (b) Frequency response curve for

$$\frac{1}{s^2 + 2\zeta\omega_n s + \omega_n^2}$$

(c) Frequency response curve for system. This composite was obtained by adding the above curves.

8.2.3 Product of Curves on Log-Log Scales

Recall that multiplication in the arithmetic domain becomes equivalent to a sum in the log domain. Suppose the two curves employed in Fig. 8.2.2-1 were plotted on log-log paper. Then, instead of *multiplying* these curves point by point, they could be *added* point by point. This is shown in Fig. 8.2.3-1.

To prepare for the application of log-log methods of multiplying curves, let us briefly review the algebra and geometry of log-log plots, as covered in the next section.

8.3 BRIEF REVIEW OF THE ALGEBRA AND GEOMETRY OF LOG-LOG PLOTS

8.3.1 Log Scales

A log scale is shown below. First note that it is nonuniform. Multiplication (in the arithmetic domain) is obtained by addition (in the log domain). This being the case, then division is obtained by subtraction. Consequently, the distance from one number N_1 to another number N_2 represents the division of one by the other. Or one may regard a *distance* on the log scale as the *ratio* of the two numbers at its ends. Note in the figure below that the distance from 0.5 to 2 is exactly equal to the distance from 1 to 4 or the distance from 2 to 8. Note that there cannot be a number zero on a log scale.

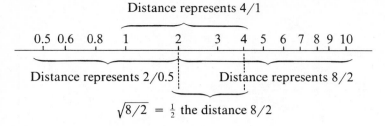

Log scales.

Referring to another property of log scales, we recall that the log of the nth root of a number is given by $1/n$ times the log of the number. Thus, on the log scale, the square root (for instance) of 4 would be found by dividing the distance representing 4 (such as 2/0.5 or 4/1 or 8/2) in half.

8.3.2 The Decibel Scale

The decibel scale is an alternate means of working with logs. It is applied only to the magnitude (not to the phase or frequency) and is given by

$$dB = 20 \log_{10} \frac{A_1}{A_0} \qquad [8.3.2\text{-}1]$$

where A_1 = amplitude under investigation
        A_0 = reference amplitude

Two convenient ratios to know are

$$\text{ratio of } 2 \approx 6 \text{ dB} \quad \text{and} \quad \text{ratio of } \sqrt{2} \approx 3 \text{ dB}$$

8.3.3 Exponential Curve on Log-Log Scales

Given the exponential

$$R = a\omega^n \qquad [8.3.3\text{-}1]$$

where a = constant real positive number
        n = constant real number

plot the curve on log-log scales.
    Take the log of both sides

$$\log R = \log a + n \log \omega \qquad [8.3.3\text{-}2]$$

Using log-log scales, eq. 8.3.3-2 is the point-slope form of a straight line, whose slope is n and whose intercept is a. Since there is no zero on log scales, the intercept must be defined. The intercept may be chosen anywhere that is convenient, and the choice must be identified. A convenient point is $\omega = 1$,

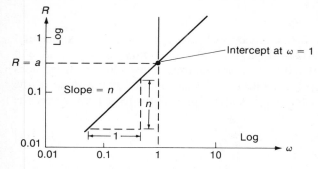

Figure 8.3.3-1   Exponential curve on log-log scales.

since unity raised to any power (even a decimal) is still unity. Then the intercept at $\omega = 1$ is equal to the constant a. The exponential is plotted on log-log scales in Fig. 8.3.3-1.

8.3.4 Multiplication of a Curve on Log-Log Scales

Given a curve in linear scales, to multiply this by a constant, one would proceed point by point, multiply, and then draw the new curve by connecting the points. On log-log scales, one could proceed point by point and add instead of multiply. Of special interest is the multiplication of a curve by a constant. On log-log scales, this is achieved by translating the entire curve like a rigid body. Thus, to multiply a curve by the constant b, raise the curve a distance equal to b. To divide by a constant b, lower the curve by a distance equal to b. See Fig. 8.3.4-1. (Actually log b.)

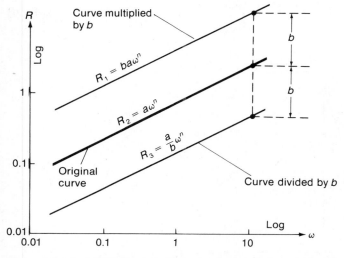

Figure 8.3.4-1   Multiplication of a curve by a constant on log-log scales.

8.3.5 Log-Log Triangle

A log-log triangle will be defined as one whose hypotenuse is an exponential (therefore a straight line whose slope is n and whose unity intercept is a) and whose sides are horizontal and vertical, respectively. The lengths of these sides of the log-log triangle represent ratios of the end points. If the slope of the exponential is unity, then the two sides are equal. For any general slope, the ratio of the sides is

$$D_v = D_h^n \qquad [8.3.5\text{-}1]$$

where

$$D_v = \text{length of vertical side of triangle}$$
$$\text{Do} \rightarrow D_h = \text{length of horizontal side of triangle} \qquad [8.3.5\text{-}2]$$
$$n = \text{slope of exponential curve on log-log scales}$$

Consider three such triangles whose slopes are respectively 1, 2, and 3 and let all three triangles have a common base as shown on Fig. 8.3.5-1. Determine the ratio of the vertical side to the horizontal side for each triangle.

$$D_0 = \frac{6}{2} = 3 \qquad [8.3.5\text{-}3]$$

$$D_1 = D_0 = 3 = \frac{6}{2} \qquad [8.3.5\text{-}4]$$

$$= \frac{1.5}{0.5} \qquad [8.3.5\text{-}5]$$

$$D_2 = D_0^2 = 3^2 = 9 \qquad [8.3.5\text{-}6]$$

$$= \frac{4.5}{0.5} \qquad [8.3.5\text{-}7]$$

$$D_3 = D_0^3 = 3^3 = 27 \qquad [8.3.5\text{-}8]$$

$$= \frac{13.5}{0.5} \qquad [8.3.5\text{-}9]$$

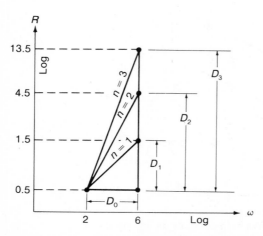

Figure 8.3.5-1 Sides of a triangle on log-log scales.

8.3.6 Curves Approximated by Asymptotes

The multiplication of curves, even when accomplished by adding on log-log scales can be simplified greatly if the curves are straight lines. While the curves are not exactly straight lines, they approach asymptotes* on log-log scales. Hence, the curves could be approximated by a set of intersecting straight line asymptotes. These could be multiplied (by adding on log-log scales) and then once the composite has been constructed, the actual departures from straight lines could be sketched in. These are constructed in the subsequent sections.

8.4 CATALOG OF FREQUENCY RESPONSE CURVES

8.4.1 Frequency Response for Any Function

Given any Laplace function $G(s)$, determine its frequency response. The general approach is listed as follows:

1. Write the function $G(s)$ in the form of one polynomial $G_n(s)$ divided by another $G_m(s)$.
2. Let $s = j\omega$.
3. Treat each polynomial separately. For each, collect real and imaginary terms.
4. Write each polynomial in polar form. Now form the quotient of the two quantities and obtain the final polar form $G(j\omega) = Re^{j\phi}$, where R = the magnitude (as a function of ω) and ϕ = the phase angle (as a function of ω).
5. Plot the magnitude R on log-log scales and plot the phase angle ϕ on semilog scales. Where possible, make use of the straight line asymptotes.

8.4.2 Frequency Response for a Constant

If the Laplace transfer function is a constant, then upon using the rules set in Section 8.4.1, the magnitude plot will be a horizontal line, while the phase angle will be zero. It will seldom be required to plot the constant since the constant can readily be included in any one of the other factors.

8.4.3 Frequency Response for Exponential Factor

Given

$$G(S) = as^n \qquad [8.4.3\text{-}1]$$

---

*Asymptotic plots, often credited to H. W. Bode, have been in use by controls engineers since the early 1940s. Such straight line approximations using only the asymptotes are also referred to as Bode plots.

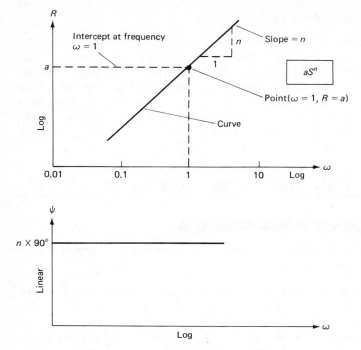

Figure 8.4.3-1   Frequency response for exponential factor.

Let

$$s = j\omega \qquad G(j\omega) = aj^n\omega^n \qquad [8.4.3\text{-}2]$$

The magnitude R is

$$R = a\omega^n \qquad [8.4.3\text{-}3]$$

Plotted on log-log paper, eq. 8.4.3-3 becomes a straight line whose *slope* is equal to n and whose *intercept** is equal to a. The phase angle ψ for eq. 8.4.3-2 is given by

$$\psi = n90° \qquad [8.4.3\text{-}4]$$

The magnitude (eq. 8.4.3-3) is plotted on log-log scales, while the phase (eq. 8.4.3-4) is plotted on semilog[†] scales, as shown in Fig. 8.4.3-1.

---

*There is no zero on log-log scales. Hence, any convenient abscissa (frequency) may be used. Refer to eq. 8.4.3-3, for $\omega = 1$ then $a\omega^n = a$ for all n. Hence, unity is a convenient abscissa.

[†]The phase angle ψ can be zero. Since zero cannot be shown on a log scale, the coordinate for phase angle will be linear, while frequency will still use the log coordinate.

8.4.4 Frequency Response of First Order Factor in Numerator

Given

$$G(s) = \frac{s+a}{a} = \frac{s}{a} + 1 \qquad [8.4.4\text{-}1]$$

The frequency response is found by letting $s = j\omega$. Then,

$$G(j\omega) = \frac{j\omega}{a} + 1 = 1 + j\frac{\omega}{a} \qquad [8.4.4\text{-}2]$$

Transform to polar form

$$R = \sqrt{1 + \left(\frac{\omega}{a}\right)^2}$$
$$\qquad\qquad\qquad\qquad [8.4.4\text{-}3]$$
$$\psi = \tan^{-1}\frac{\omega}{a}$$

Before making the plot, note the behavior of the curve at extreme values of frequency. When ω is very small, the asymptote becomes equal to unity. When $\omega = a$, then the radical becomes equal to the square root of 2. When ω is very large, unity under the radical may be neglected, and the asymptote becomes $R = \omega/a$. See Fig. 8.4.4-1.

Using log-log geometry, it can be shown that the two asymptotes intersect at the point $(a, 1)$. This point is referred to as the *break point* or the *corner frequency*. Now the curve can be constructed knowing the slopes of the two

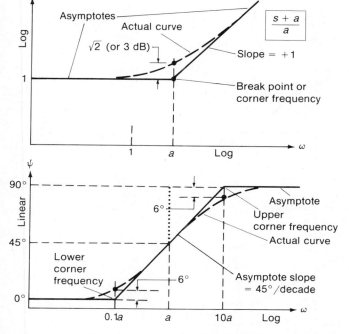

Figure 8.4.4-1   Frequency response for first order factor in numerator.

asymptotes and the point at which they intersect. Summarizing, for the magnitude plot, note that

$$
\begin{aligned}
\text{when } \omega \ll a \quad &\text{then } R \to 1 \\
\text{when } \omega = a \quad &\text{then } R = \sqrt{2} \\
\text{when } \omega \gg a \quad &\text{then } R \to \frac{\omega}{a}
\end{aligned}
\qquad [8.4.4\text{-}4]
$$

Following similar steps, the phase angle curve can be constructed. However, this curve will have three asymptotes. The phase angle ψ is positive, and is referred to as *phase lead*. In making the phase plot, note the following;

$$
\begin{aligned}
\text{when } \omega \ll 0.1a \quad &\text{then } \psi \to 0 \\
\text{when } \omega = 0.1a \quad &\text{then } \psi = 6° \\
\text{when } \omega = a \quad &\text{then } \psi = 45° \\
\text{when } \omega = 10a \quad &\text{then } \psi = 84° = 90° - 6° \\
\text{when } \omega \gg 10a \quad &\text{then } \psi \to 90°
\end{aligned}
\qquad [8.4.4\text{-}5]
$$

In order to include a constant other than a, note the following.

$$
\frac{s + a}{b} = \frac{a}{b}\frac{s + a}{a} = \frac{a}{b}\left[\frac{s}{a} + 1\right]
\qquad [8.4.4\text{-}6]
$$

8.4.5 Frequency Response of Reciprocal First Order Factor

Given

$$
G(s) = \frac{b}{s + a} = \frac{b/a}{(s/a) + 1}
\qquad [8.4.5\text{-}1]
$$

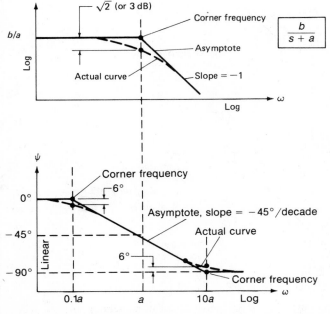

Figure 8.4.5-1   Frequency response for reciprocal first order factor.

Let $s = j\omega$

$$G(j\omega) = \frac{b/a}{1 + j\omega/a} \qquad [8.4.5\text{-}2]$$

Magnitude

$$R = \sqrt{\frac{1}{1 + (\omega/a)^2}} \left(\frac{b}{a}\right) \qquad [8.4.5\text{-}3]$$

Phase

$$\psi = -\tan^{-1}\frac{\omega}{a} \qquad [8.4.5\text{-}4]$$

This is similar to the procedure in the previous section, but here the magnitude drops and the phase curve *lags* (has phase lag). See Fig. 8.4.5-1.

8.4.6 Frequency Response of Quadratic Factor in Numerator

Given

$$G(s) = \frac{s^2 + 2\zeta\omega_n s + \omega_n^2}{\omega_n^2} = \left(\frac{s}{\omega_n}\right)^2 + 2\zeta\left(\frac{s}{\omega_n}\right) + 1 \qquad [8.4.6\text{-}1]$$

Let

$$s = j\omega \quad \text{and} \quad \gamma = \frac{\omega}{\omega_n} \qquad [8.4.6\text{-}2]$$

Then

$$G(j\omega) = 1 - \gamma^2 + j2\zeta\gamma \qquad [8.4.6\text{-}3]$$

Magnitude

$$R = \sqrt{\left(1 - \gamma^2\right)^2 + (2\zeta\gamma)^2} \qquad [8.4.6\text{-}4]$$

Phase

$$\psi = \tan^{-1}\frac{2\zeta\gamma}{1 - \gamma^2} \qquad [8.4.6\text{-}5]$$

Be careful to use the correct quadrant for ψ. Similar to the treatment in Section 8.4.4, determine the asymptotes for extreme values of frequencies. When the frequency is very low, the asymptote becomes $R = 1$. When $\omega = \omega_n$, then $\gamma = 1$, and $R = 2\zeta$. When the frequency is high $\gamma \gg 1$, and unity in the first term may be neglected. Consequently, γ^2 becomes greater than the term $2\zeta\gamma$. Hence, for high frequencies, the asymptote becomes $R = \gamma^2 = (\omega/\omega_n)^2$. These are summarized below.

$$\text{when } \omega \ll \omega_n \quad \text{or} \quad \gamma \ll 1 \qquad \text{then } R \to 1$$

$$\text{when } \omega = \omega_n \quad \text{or} \quad \gamma = 1 \qquad \text{then } R = 2\zeta$$

$$\text{when } \omega \gg \omega_n \quad \text{or} \quad \gamma \gg 1 \qquad \text{then } R \to \left(\frac{\omega}{\omega_n}\right)^2$$

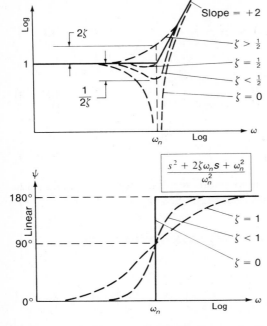

Figure 8.4.6-1   Frequency response for quadratic factor.

At low frequencies, the asymptote is a horizontal line whose intercept is unity. At high frequencies, the asymptote is a straight line whose slope is $+2$ and whose intercept is $(1/\omega_n)^2$. Using log-log geometry, it can be shown that the two asymptotes intersect at the point $\omega = \omega_n$, $R = 1$. This point is referred to as the corner frequency or break point. See Fig. 8.4.6-1. Note that if $\zeta \geq 1$, then the quadratic may be factored into two first order factors. Also note that the phase is positive. This implies that the output *leads* the input.

8.4.7 Frequency Response for Reciprocal Quadratic Factor

Given

$$G(s) = \frac{\omega_n^2}{s^2 + 2\zeta\omega_n s + \omega_n^2} \qquad [8.4.7\text{-}1]$$

Following a procedure similar to that in Section 8.4.6, but where the magnitude R is inverted, and where the phase ψ is taken as the negative of eq. 8.4.6-5,

$$R = \sqrt{\frac{1}{\left(1 - \gamma^2\right)^2 + \left(2\zeta\gamma\right)^2}} \qquad [8.4.7\text{-}2]$$

$$\psi = -\tan^{-1}\frac{2\zeta\gamma}{1 - \gamma^2} \qquad [8.4.7\text{-}3]$$

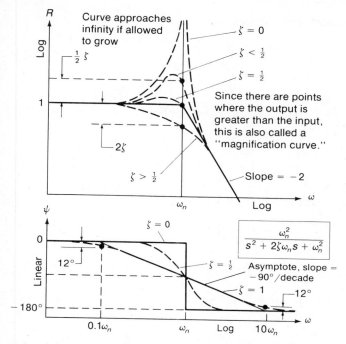

Figure 8.4.7-1   Frequency response for reciprocal quadratic.

where

$$\gamma = \frac{\omega}{\omega_n}$$ [8.4.7-4]

The frequency response curves are sketched in Fig. 8.4.7-1.

If the damping ratio is greater than 1, then the quadratic can be factored into two first order factors. Note that the phase lags.

8.5 RESONANCE

8.5.1 Resonance Is the Maximum Point

The point at which the frequency response curve for the reciprocal quadratic (see Fig. 8.4.7-1) reaches a maximum for a given damping ratio ζ is called the resonance. The equation of this curve is given by eq. 8.4.7-2, repeated below. Note $\gamma = \omega/\omega_n$.

$$R = \sqrt{\frac{1}{\left(1 - \gamma^2\right)^2 + \left(2\zeta\gamma\right)^2}}$$ [8.5.1-1]

Since resonance is a maximum point, this can be found by maximizing the magnitude R with respect to the ratio γ. Then

$$0 = \frac{dR}{d\gamma} = -\frac{1}{2}U^{-3/2}\frac{dU}{d\gamma} \qquad [8.5.1\text{-}2]$$

where

$$U = (1 - \gamma^2)^2 + (2\zeta\gamma)^2 \qquad [8.5.1\text{-}3]$$

and

$$\frac{dU}{d\gamma} = 4\gamma(\gamma^2 + 2\zeta^2 - 1) \qquad [8.5.1\text{-}4]$$

There are three cases which satisfy eq. 8.5.1-2:

$$\gamma_r = 0, \qquad \zeta = \sqrt{\tfrac{1}{2}} \qquad [8.5.1\text{-}5]$$

for which

$$R_r = 1 \qquad [8.5.1\text{-}5a]$$

$$\gamma_r = 1, \qquad \zeta = 0 \qquad [8.5.1\text{-}6]$$

for which

$$R_r \to \infty \qquad [8.5.1\text{-}6a]$$

$$0 \le \gamma_r \le \sqrt{\tfrac{1}{2}} \qquad [8.5.1\text{-}7]$$

then

$$\gamma_r = \sqrt{1 - 2\zeta^2} \qquad [8.5.1\text{-}8]^*$$

for which

$$R_r = \frac{1}{2\zeta}\sqrt{\frac{1}{1 - \zeta^2}} \qquad [8.5.1\text{-}9]^*$$

8.5.2 Additional Points to Aid in Sketching Frequency Response Curve

(a) **Magnitude $R_{1/2}$ Where Frequency Is $\frac{1}{2}\omega_n$** Substitute $\gamma = \frac{1}{2}$ in eq. 8.5.1-1;

$$R_{1/2} = \sqrt{\frac{1}{9/16 + \zeta^2}} \qquad [8.5.2\text{-}1]^*$$

---

*Numerical values are tabulated in Table 8.5.2-1 and the meaning of all terms is shown graphically in Fig. 8.5.2-1(a).

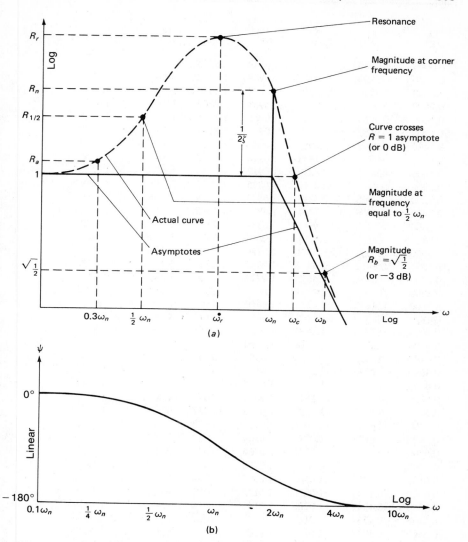

Figure 8.5.2-1 Additional points for sketching frequency response. (a) Magnitude curve. (b) Phase curve.

(b) Frequency ω_b Where Curve Crosses $R = \sqrt{\frac{1}{2}}$ Substitute $R = \sqrt{\frac{1}{2}}$ into eq. 8.5.1-1, solve for γ^2

$$\gamma_b^2 = 1 - 2\zeta^2 \pm \sqrt{2}\sqrt{1 - \zeta^2} = \gamma_r^2 \pm \sqrt{2}\,\beta$$

or

$$\gamma_b = \sqrt{\gamma_r^2 \pm \sqrt{2}\,\beta} \qquad\qquad [8.5.2\text{-}2]*$$

---

*Numerical values are tabulated in Table 8.5.2-1 and the meaning of all terms is shown graphically in Fig. 8.5.2-1(a).

TABLE 8.5.2-1 ADDITIONAL POINTS FOR SKETCHING THE RECIPROCAL QUADRATIC

ζ ZETA	Resonance		Ratios		Gammas	
	γ_r GAMMA	R_r RATIO	R_a .30W_N	$R_{1/2}$ .50W_N	γ_c R = 1	γ_b R = SQRT(.5)
0.0000	1.0000	—	1.0989	1.3333	1.4142	1.5538
0.0100	0.9999	50.0024	1.0989	1.3332	1.4141	1.5537
0.0200	0.9996	25.0050	1.0988	1.3329	1.4136	1.5533
0.0500	0.9975	10.0125	1.0983	1.3304	1.4107	1.5510
0.1000	0.9899	5.0252	1.0965	1.3216	1.4000	1.5428
0.1200	0.9855	4.1970	1.0955	1.3166	1.3937	1.5379
0.1400	0.9802	3.6069	1.0942	1.3107	1.3862	1.5322
0.1600	0.9741	3.1658	1.0928	1.3040	1.3775	1.5255
0.1800	0.9671	2.8239	1.0912	1.2965	1.3676	1.5180
0.2000	0.9592	2.5516	1.0895	1.2883	1.3565	1.5096
0.2200	0.9605	2.3298	1.0875	1.2794	1.3440	1.5002
0.2400	0.9406	2.1461	1.0854	1.2699	1.3303	1.4900
0.2600	0.9299	1.9916	1.0831	1.2598	1.3151	1.4788
0.2800	0.9183	1.8601	1.0806	1.2491	1.2986	1.4667
0.3000	0.9053	1.7471	1.0780	1.2380	1.2806	1.4537
0.3200	0.8917	1.6492	1.0752	1.2264	1.2611	1.4397
0.3400	0.8768	1.5637	1.0723	1.2144	1.2400	1.4248
0.3600	0.8607	1.4887	1.0692	1.2020	1.2172	1.4090
0.3800	0.8433	1.4225	1.0660	1.1894	1.1926	1.3922
0.3827	0.8409	1.4142	1.0655	1.1877	1.1892	1.3899
0.4000	0.8246	1.3639	1.0626	1.1765	1.1662	1.3745
0.4200	0.8045	1.3118	1.0590	1.1633	1.1377	1.3559
0.4400	0.7828	1.2654	1.0554	1.1500	1.1071	1.3363
0.4600	0.7595	1.2242	1.0516	1.1366	1.0741	1.3158
0.4800	0.7343	1.1974	1.0477	1.1230	1.0385	1.2943
0.5000	0.7071	1.1547	1.0436	1.1094	1.0000	1.2720
0.5200	0.6776	1.1257	1.0395	1.0957	0.9583	1.2486
0.5400	0.6456	1.1001	1.0352	1.0820	0.9130	1.2248
0.5600	0.6106	1.0777	1.0309	1.0684	0.8635	1.2000
0.5800	0.5720	1.0583	1.0264	1.0547	0.8090	1.1745
0.6000	0.5292	1.0417	1.0218	1.0412	0.7483	1.1482
0.6100	0.5058	1.0344	1.0195	1.0344	0.7153	1.1349
0.6200	0.4808	1.0278	1.0172	1.0277	0.6800	1.1214
0.6300	0.4541	1.0220	1.0145	1.0209	0.6422	1.1078
0.6400	0.4252	1.0168	1.0125	1.0142	0.6013	1.0941
0.6500	0.3937	1.0122	1.0100	1.0076	0.5568	1.0802
0.6600	0.3589	1.0084	1.0076	1.0010	0.5075	1.0663
0.6700	0.3197	1.0053	1.0052	0.9943	0.4521	1.0523
0.6800	0.2742	1.0028	1.0027	0.9878	0.3878	1.0383
0.6900	0.2166	1.0011	1.0003	0.9812	0.3092	1.0242
0.7000	0.1414	1.0002	0.9978	0.9747	0.2000	1.0100
0.7071	0.0044	1.0000	0.9960	0.9701	0.0266	1.0000
0.7500			0.9850	0.9428		0.9396
0.8000			0.9720	0.9119		0.8709
0.8500			0.9586	0.8822		0.8059
0.9000			0.9450	0.8536		0.7461
0.9500			0.9313	0.8262		0.6919
1.0000			0.9174	0.8000		0.6436

$$R = \sqrt{\frac{1}{\left(1 - \gamma^2\right)^2 + (2\zeta\gamma)^2}}$$

TABLE 8.5.2-2   POINTS FOR RECIPROCAL QUADRATIC PHASE CURVES

ZETA	GAMMA values						
	0.1	0.25	0.5	1.0	2	4	10
0.00	0.000	0.000	0.000	0.0	−180.000	−180.000	−180.000
0.02	−0.232	−0.611	−1.528	−90.0	−178.472	−179.389	−179.768
0.04	−0.463	−1.223	−3.054	−90.0	−176.946	−178.777	−179.537
0.06	−0.695	−1.634	−4.576	−90.0	−175.424	−178.166	−179.305
0.08	−0.926	−2.444	−6.092	−90.0	−173.908	−177.556	−179.074
0.10	−1.158	−3.054	−7.598	−90.0	−172.402	−176.946	−178.842
0.12	−1.389	−3.664	−9.095	−90.0	−170.905	−176.336	−178.611
0.14	−1.621	−4.272	−10.579	−90.0	−169.421	−175.728	−178.379
0.16	−1.852	−4.880	−12.049	−90.0	−167.951	−175.120	−178.148
0.18	−2.084	−5.486	−13.503	−90.0	−166.497	−174.514	−177.916
0.20	−2.315	−6.092	−14.939	−90.0	−165.061	−173.908	−177.685
0.22	−2.546	−6.695	−16.356	−90.0	−163.644	−173.305	−177.454
0.24	−2.777	−7.298	−17.754	−90.0	−162.246	−172.702	−177.233
0.26	−3.008	−7.899	−19.129	−90.0	−160.871	−172.101	−176.992
0.28	−3.239	−8.498	−20.483	−90.0	−159.517	−171.502	−176.761
0.30	−3.470	−9.095	−21.812	−90.0	−158.188	−170.905	−176.530
0.32	−3.701	−9.690	−23.118	−90.0	−156.882	−170.310	−176.299
0.34	−3.931	−10.283	−24.389	−90.0	−155.601	−169.717	−176.069
0.36	−4.162	−10.874	−25.654	−90.0	−154.346	−169.126	−175.838
0.38	−4.392	−11.463	−26.883	−90.0	−153.117	−168.537	−175.698
0.40	−4.622	−12.049	−28.087	−90.0	−151.913	−167.951	−175.378
0.42	−4.852	−12.632	−29.264	−90.0	−150.736	−167.368	−175.148
0.44	−5.082	−13.213	−30.414	−90.0	−149.586	−166.787	−174.918
0.46	−5.312	−13.791	−31.538	−90.0	−148.462	−166.209	−174.688
0.48	−5.541	−14.367	−32.636	−90.0	−147.364	−165.633	−174.459
0.50	−5.771	−14.939	−33.707	−90.0	−146.293	−165.061	−174.229
0.52	−6.000	−15.508	−34.752	−90.0	−145.248	−164.492	−174.000
0.54	−6.229	−16.075	−35.772	−90.0	−144.228	−163.925	−173.771
0.56	−6.458	−16.638	−36.766	−90.0	−143.234	−163.362	−173.542
0.58	−6.686	−17.197	−37.735	−90.0	−142.265	−162.803	−173.314
0.60	−6.915	−17.754	−38.679	−90.0	−141.321	−162.246	−173.085
0.62	−7.143	−18.307	−39.599	−90.0	−140.401	−161.693	−172.857
0.64	−7.371	−18.856	−40.496	−90.0	−139.504	−161.144	−172.629
0.66	−7.598	−19.402	−41.369	−90.0	−138.631	−160.598	−172.402
0.68	−7.826	−19.944	−42.219	−90.0	−137.781	−160.056	−172.174
0.70	−8.053	−20.483	−43.047	−90.0	−136.953	−159.517	−171.947
0.72	−8.280	−21.017	−43.853	−90.0	−136.147	−158.983	−171.720
0.74	−8.507	−21.548	−44.638	−90.0	−135.362	−158.452	−171.493
0.76	−8.733	−22.076	−45.402	−90.0	−134.598	−157.924	−171.267
0.78	−8.959	−22.599	−46.147	−90.0	−133.853	−157.401	−171.041
0.80	−9.185	−23.118	−46.871	−90.0	−133.129	−156.882	−170.815
0.82	−9.411	−23.633	−47.577	−90.0	−132.423	−156.367	−170.589
0.84	−9.636	−24.145	−48.264	−90.0	−131.736	−155.855	−170.364
0.86	−9.861	−24.652	−48.933	−90.0	−131.067	−155.348	−170.139
0.88	−10.086	−25.155	−49.585	−90.0	−130.415	−154.845	−169.914
0.90	−10.310	−25.654	−50.220	−90.0	−129.780	−154.346	−169.690
0.92	−10.534	−26.149	−50.838	−90.0	−129.162	−153.851	−169.466
0.94	−10.758	−26.640	−51.441	−90.0	−128.559	−153.360	−169.242
0.96	−10.981	−27.126	−52.028	−90.0	−127.972	−152.874	−169.019
0.98	−11.204	−27.609	−52.600	−90.0	−127.400	−152.391	−168.796
1.00	−11.427	−28.087	−53.157	−90.0	−126.843	−151.913	−168.573

$$\psi = -\tan^{-1}\frac{2\zeta\gamma}{1-\gamma^2}$$

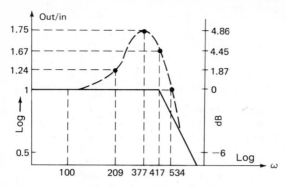

Figure 8.5.2-2 Frequency response using additional points.

(c) Frequency ω_c Where Curve Crosses $R = 1$ Asymptote Substitute $R = 1$ in eq. 8.5.1-1, solve for γ

$$\gamma_c = \sqrt{2(1 - 2\zeta^2)} = \gamma_r\sqrt{2} \qquad [8.5.2-3]*$$

(d) Phase Curve. (See eq. 8.4.7-4.)

$$\psi = -\tan^{-1}\frac{2\zeta\gamma}{1 - \gamma^2} \qquad [8.5.2-4]^\dagger$$

■ **EXAMPLE 8.5.2-1** Use of Additional Points

Sketch the actual frequency response curve for the following transfer function

$$\frac{\text{out}}{\text{in}}(s) = \frac{\omega_n^2}{s^2 + 2\zeta\omega_n s + \omega_n^2} \qquad (1)$$

where

$$\omega_n = 417 \text{ rad/s}$$

$$\zeta = 0.3 \qquad (2)$$

Using Table 8.5.2-1 and Fig. 8.5.2-1, the following is obtained:

$$\gamma_r = 0.905$$

---

*Numerical values are tabulated in Table 8.5.2-1 and the meaning of all terms is shown graphically in Fig. 8.5.2-1(a).

†Numerical values are tabulated in Table 8.5.2-2 and the meaning of all terms is shown in Fig. 8.5.2-1(b).

then

$$\omega_r = 0.905 \times 417 = 377 \text{ rad/s}$$

$$R_r = 1.75 \text{ (resonant magnitude)}$$

$$R_{1/2} = 1.24 \qquad \tfrac{1}{2}\omega_n = \tfrac{1}{2}417 = 209$$

$$\gamma_c = 1.28 \qquad \text{then } \omega_c = 1.28 \times 417 = 534$$

$$\frac{1}{2\zeta} = \frac{1}{2 \times 0.3} = 1.67$$

(3)

The result is shown in Fig. 8.5.2-2.   ∎

■ **EXAMPLE 8.5.2-2** Design of Stepper Motor

A powerful miniature variable speed electric motor can be made with a permanent magnet vibrator. Alternating current is applied to a coil that attracts and repels the permanent magnet. The magnet is permitted to slide back and forth (Fig. 8.5.2-3) and two ratchet pawls engage (alternately) a ratchet wheel. Thus, the vibration is converted into rotation. The spring K_s (designed for tension and compression) keeps the magnet centered in the solenoid coil. Note that there is a magnetic spring K_m and a magnetic damper D_m. Design the system to meet the following specifications:

(a) Output speed $\dot{\theta}$ is proportional to input frequency.
(b) Speed range $0 \le \dot{\theta} \le 89$ rad/s (about 870 rpm).
(c) Maximum frequency corresponding to maximum speed $\omega_{max} = 534$ rad/s.
(d) Equivalent mass of all translating parts $M = 2 \times 10^{-3}$ kg.
(e) Minimum power dissipation.
(f) Overall package size = 1 cm^3 (about the size of a single green pea).
(g) Nominal current $I_0 = 5$ mA $= 5 \times 10^{-3}$ A.

Solution.

Free Body Diagram

In order to construct the free body diagram, it will be necessary to study the operation of the device. An alternating current is applied to the coil that alternately attracts and repels the magnet. This applies a sinusoidal force to the mass of the device. If the moment of inertia of the ratchet is neglected, the equivalent lumped mass is given by the sum of the masses of all the translating parts. As the mass M moves, it alternately compresses and extends a spring K_s. The motion of the magnet in the coil produces an equivalent magnetic spring K_m and an equivalent magnetic damper D_m. The two springs K_s and K_m are in parallel, while the damper D_m is alone. Then, let

$$K = K_s + K_m \quad \text{(recall that } K_m < 0; \text{ thus } K_s > |K_m|)$$

$$D = D_m$$

(1)

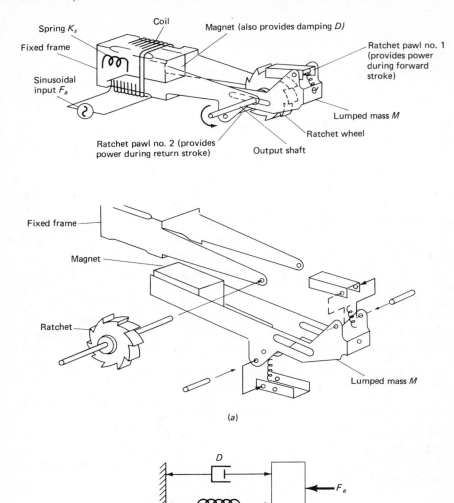

Figure 8.5.2-3   Miniature stepper motor. (a) Actual system. (b) Free body diagram.

The translation of the mass M of the device is converted into rotation by means of a pair of pawls, which make this a double stroke engine. As the mass travels in either direction, one or the other of the pawls engages the ratchet, providing two power strokes per cycle. If the stroke is adjusted in length, note the following. If the stroke is less than the arc length of one tooth of the ratchet, there will be no engagement, and thus no output rotation even though there is a translation. If the stroke is greater than the arc length of one tooth, but less than two teeth of the ratchet, the ratchet will be advanced one tooth.

The free body diagram will be drawn for the mass M. Note that the spring K and the damper D connect the mass M to the ground. The input is a sinusoidal force. The free body diagram is shown in Fig. 8.5.2-3(b).

Sum the forces, according to Newton's law,

$$\sum F = M\ddot{X} = F_a - D\dot{X} - KX \tag{2}$$

Assume zero initial conditions, and transform to the Laplace domain. The differential equation becomes

$$Ms^2X = F_a - DsX - KX$$

or

$$X = \frac{F_a}{Ms^2 + Ds + K} \tag{3}$$

Normalize, and formulate the dimensionless ratio of out/in

$$\frac{X}{F/K}(s) = \frac{\omega_n^2}{s^2 + 2\zeta\omega_n s + \omega_n^2} \tag{4}$$

The transfer function is a reciprocal quadratic. The general shape of the frequency response curve is given by Fig. 8.4.7-1. A more exact, but still general curve is shown in Fig. 8.5.2-1, which will be employed in the design. At this point in the problem, we have no points on the curve and we do not know the damping ratio or natural frequency.

Selection of Damping Ratio and Natural Frequency

At low frequencies, let the out/in ratio be unity. If the device is proportioned so that this corresponds to the correct stroke to advance the ratchet wheel one tooth, then for this range of speeds, there will be a correct engagement of the ratchet wheel. For each cycle, there will be two power strokes. Then, the output rate of rotation of the ratchet wheel will be proportional to the input frequency, as required by specification (a).

Now consider the resonant condition. At the resonant frequency ω_r, the out/in ratio will be greater than the correct stroke for correct engagement. If we let this ratio be less than two (the distance for an engagement of two teeth), there will be some lost motion, but the ratchet wheel will still advance only *one* tooth. Thus, for satisfactory operation of the device, the resonant peak $R_r < 2$.

Specification (e) requires minimum dissipation of energy. This in turn requires minimum damping. The smaller the damping ratio, the higher the resonant peak. As a compromise, let $R_r = 1.75$. Then, using Table 8.5.2-1,

$$\boxed{\zeta = 0.3} \tag{5}$$

At frequencies above the resonant frequency, the out/in ratio R will decrease and eventually fall below unity. When this occurs, the stroke will be too short for any engagement and the ratchet wheel will no longer turn. Up to this point, the ratchet wheel had been turning at a rate that was proportional to the input frequency. Then, the frequency (shown as ω_c in Fig. 8.5.2-1) at which the curve crosses unity magnitude is the highest frequency for operation of the stepper motor. Specification (c) lists this as $\omega_{max} = 534$ rad/s. For $\zeta = 0.3$ in Table 8.5.2-1, $\gamma_c = 1.28$. Since

$$\gamma_c = \frac{\omega_c}{\omega_n}$$

then

$$\omega_n = \frac{\omega_c}{\gamma_c} = \frac{534}{1.28} = \boxed{417 \text{ rad/s}} \tag{5a}$$

The frequency response curve is shown in Fig. 8.5.2-2.

Selection of K and D

Since

$$\omega_n = \sqrt{\frac{K}{M}}$$

then

$$K = M\omega_n^2 = 2 \times 10^{-3}(417)^2 = \boxed{348 \text{ N/m}} \tag{6}$$

Since

$$2\zeta\omega_n = \frac{D}{M}$$

then

$$D = 2\zeta\omega_n M = 2(0.3)(417)(2 \times 10^{-3})$$

$$= \boxed{0.5 \text{ N} \cdot \text{s/m}} \tag{6a}$$

Design of Ratchet

The number n of teeth on the ratchet will be selected to meet specifications (b) and (c). Noting that there are two power strokes per cycle,

$$n = \frac{2\omega_{max}}{\theta_{max}} = \frac{(2)(534)}{89} = \boxed{12 \text{ teeth}} \tag{7}$$

Considering specification (f), the overall length is equal to 10^{-2} m. Let the diameter d of the ratchet be equal to one fifth of this.

$$d = \frac{10^{-2}}{5} = \boxed{2 \times 10^{-3} \text{ m}} \tag{8}$$

The distance between teeth is

$$P = \pi \frac{d}{n} = \frac{\pi 2 \times 10^{-3}}{12} = 5.23 \times 10^{-4} \text{ m} \tag{9}$$

Let the stroke be about 10% greater than this. The stroke is twice the amplitude of vibration. Then,

$$2X = (100\% + 10\%)5.23 \times 10^{-4}$$

or

$$X = \boxed{2.88 \times 10^{-4} \text{ m}} \tag{10}$$

Design of Coil

Assume that the magnetic actuator behaves as one with a large air gap. Then, referring to Appendix 2, Table 2.4.1-1, item (2), we have

$$F = \frac{AN^2 I_0^2}{2 X_0^2} \tag{11}$$

$$K_m = -\frac{AN^2 I_0^2}{X_0^3}, \qquad K = K_s + K_m \tag{12}$$

$$D_m = \frac{A^2 I_0^2 N^4}{R_c X_0^4} \tag{13}$$

For most of its range, the out/in ratio = 1. Then

$$\frac{\text{out}}{\text{in}} = \frac{X}{F/K} = 1$$

or

$$F = KX = 348(2.88 \times 10^{-4}) = 0.1 \text{ N} \tag{14}$$

Let the dimensions of the magnetic material be 10% of the overall dimensions. Then the cross sectional area is

$$A = (10^{-3})(10^{-3}) = 10^{-6} \text{ m}^2 \tag{15}$$

It is good practice to let the nominal air gap X_0 be at least five times the amplitude of vibration. Then

$$X_0 = 5X = 5(2.88 \times 10^{-4}) = \boxed{1.44 \times 10^{-3} \text{ m}} \tag{16}$$

Apply eqs. (14), (15), and (16) as well as specification (g) to eq. (11) and solve for the number of turns N.

$$N = \sqrt{\frac{2 X_0^2 F}{A I_0^2}} = \boxed{81 \text{ turns}} \tag{17}$$

Using eqs. (11) and (12), form the ratio F/K_m

$$\left|\frac{F}{K_m}\right| = \frac{X_0}{2} \qquad (18)$$

Apply eqs. (14) and (16) to eq. (18) and solve for K_m. This design is completed in Problem 8.36. ∎

■ **EXAMPLE 8.5.2-3** Actual Phase Curve Using Additional Points
Sketch the actual phase curve for the following function:

$$G(s) = \frac{1}{s^2 + 10s + 100} \qquad (1)$$

Compute the natural frequency and damping ratio.

$$\omega_n = \sqrt{100} = 10 \text{ rad/s} \qquad (2)$$

$$\zeta = 10/(2\omega_n) = 10/(2 \times 10) = 0.5 \qquad (3)$$

Refer to Table 8.5.2-2 (Points for Reciprocal Quadratic Phase Curves). For ZETA = 0.5, read the following (rounded off to the nearest three significant figures):

$$\gamma = 0.1 \quad 0.25 \quad 0.5 \quad 1 \quad 2 \quad 4 \quad 10 \qquad (4)$$

$$\psi = -5.77° \quad -14.9° \quad -33.7° \quad -90.0° \quad -146° \quad -165° \quad -174° \qquad (5)$$

Apply the definition of γ in eq. 8.4.7-4 to determine the corresponding frequencies.

$$\omega = 1 \quad 2.5 \quad 5 \quad 10 \quad 20 \quad 40 \quad 100 \qquad (6)$$

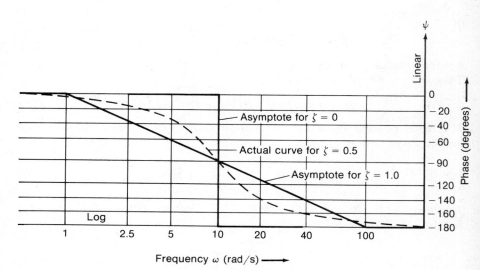

Figure 8.5.2-4   Actual phase curve using additional points.

The curve is plotted in Fig. 8.5.2-4. For comparison purposes, the asymptotic curves for $\zeta = 0$ and $\zeta = 1$ are included. ∎

8.6 COMPOSITE CURVES

Given the system transfer function $G(s)$, where this may be expressed as the product of two functions $G_1(s)$ and $G_2(s)$, then the frequency response curve may be constructed by first drawing the individual curves for the functions $G_1(s)$ and $G_2(s)$ and then multiplying them. The resulting curve, formed by the product of the two, is called a "composite" curve. If the two functions $G_1(s)$ and $G_2(s)$ are one of the curves listed in Sections 8.4.3 through 8.4.7, then the construction of the two individual curves is relatively simple. The objective in this section is to show equally simple means by which to multiply the curves.

8.6.1 Magnitude Curve Involving the Product of Exponentials

If the individual curves for $G_1(s)$ and $G_2(s)$ are exponentials, find their product. The individual functions are expressed as follows:

$$G_1(s) = as^n \qquad [8.6.1\text{-}1]$$

and

$$G_2(s) = bs^m \qquad [8.6.1\text{-}2]$$

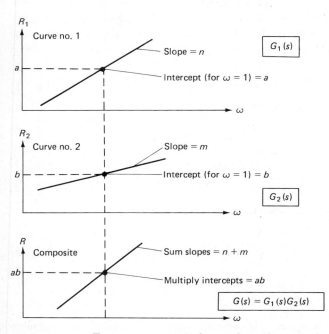

Figure 8.6.1-1   Frequency response for product of two exponentials.

Let us first multiply these expressions algebraically in order to formulate a useful rule that subsequently can be performed graphically. The product is given by

$$G = G_1 G_2 = (as^n)(bs^m) \qquad [8.6.1\text{-}3]$$
$$= abs^{n+m} \qquad [8.6.1\text{-}4]$$

From eq. 8.5.1-4, we can state the rule. Given two frequency response curves in the form of exponentials, their product is given by a single straight line whose slope is equal to the *sum* of the slopes and whose intercept is equal to the *product** of the intercepts. See Fig. 8.6.1-1.

8.6.2 Magnitude Curve Involving the Product of Broken Lines

If the individual curves $G_1(s)$ and $G_2(s)$ are formed by broken lines, then their product is accomplished piecewise. In this case, the rule derived in the previous section must be applied piecewise. In order to accomplish this, divide the frequency coordinate into a finite regions, each one terminating at a frequency where any one curve changes slope. In this way, each region will contain a pair of straight lines, conforming to the model in the previous section.

Applying the rule, *summing slopes* and *multiplying intercepts*, to each region, we find that the multiplication of intercepts need be done at only *one* point. If it is repeated in every region, consistent results will develop. Since any one point will provide the necessary condition, then select the most convenient one. It should be pointed out that the multiplication of intercepts must be accomplished at only *one* frequency, but this frequency is arbitrary.

When there are a number of factors, and particularly when there is an exponential factor, the computation of the composite intercept I_c can become unwieldy. Consequently, an alternative is sought. Consider a system function of the form

$$G(s) = s^n G_1(s) \qquad [8.6.2\text{-}1]$$

Then the composite intercept I_c becomes

$$I_c = I_0 I_e \qquad [8.6.2\text{-}2]$$

where

$$I_0 = \lim_{s \to 0} G_1(s) \qquad [8.6.2\text{-}3]$$
$$I_e = \omega^n \qquad [8.6.2\text{-}4]$$

Note that if there is no exponential factor, then

$$\left. \begin{array}{l} n = 0 \\ I_e = 1 \\ I_c = I_0 \end{array} \right\} \quad \text{(no exponential factor)} \qquad [8.6.2\text{-}5]$$

---

*Using decibel scales, *sum* the intercepts. Here is one advantage to the decibel scale. The rule is simple: sum everything.

Both terms I_0 and I_e must be computed at the *same* frequency ω, but it may be any frequency that is convenient. It should be pointed out that the limit in eq. 8.6.2-3 applies at frequency equal to zero, but is a fair approximation at a frequency sufficiently below the lowest frequency of any factor in either the numerator or denominator.

■ **EXAMPLE 8.6.2-1** Magnitude Curve for Broken Lines
Given

$$G(s) = \frac{1}{s^2 + 210s + 2000} \tag{1}$$

Construct the magnitude frequency response curve. The given transfer function appears, at first, to be a reciprocal quadratic. However, upon computing the damping ratio, we find that it is greater than unity. Consequently, the denominator must be factored. Then,

$$G(s) = \frac{1}{(s + 10)(s + 200)} \tag{2}$$

Thus, the given function $G(s)$ is equal to the product of two reciprocal first order factors. Referring to Fig. 8.4.5-1, the individual curves may be drawn, as shown in Fig. 8.6.2-1. Note that on the frequency response curve for $G_1(s)$, there is a break at frequency $\omega_1 = 10$, while there is a break at frequency

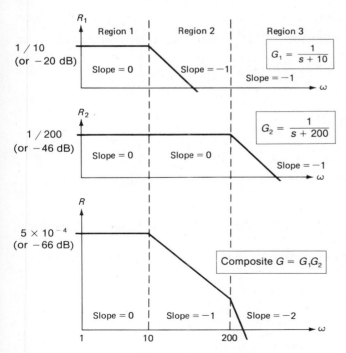

Figure 8.6.2-1   Magnitude curve for product of broken lines.

$\omega_2 = 200$ on the curve for $G_2(s)$. Hence, the frequency domain is divided into three regions, using the frequencies 10 and 200 as the boundaries. To assist in separating the regions, dotted vertical lines are used. It is suggested that the slope of each portion of each curve be noted in each region and that the intercept of each frequency response be determined at *one* frequency. For this problem, the frequency chosen is $\omega_0 = 1$. Region 1 extends from a frequency of 0 to 10. Summing the slopes in this region, we have

$$0 + 0 = 0 \tag{3}$$

Hence, in region 1, the composite will have a slope that is equal to 0. Region 2 extends from a frequency of 10 to 200. Summing slopes, we have

$$-1 + 0 = -1 \tag{4}$$

Hence, in region 2, the composite will have a slope that is equal to -1. Region 3 extends from a frequency of 200 to infinity. Summing the slopes, we have

$$-1 - 1 = -2 \tag{5}$$

We now have all the slopes of the composite. All that is needed is *one* point through which the resulting frequency response will pass. Multiplying the intercepts at the selected frequency of $\omega = 1$, we have

$$I_0 = \frac{1}{10} \times \frac{1}{200} = \frac{1}{2000} = 5 \times 10^{-4} \tag{6}$$

To check the numerical value, apply the alternative method to determine the composite intercept. Now, for this specific case, there is no exponential, hence, making use of eq. 8.6.2-5, the intercept is given by eq. 8.6.2-3.

$$I_c = I_0 = \lim_{s \to 0} \left[\frac{1}{s^2 + 210s + 2000} \right] \tag{7}$$

$$= 5 \times 10^{-4} \tag{8}$$

which agrees with eq. (6).

Using the point

$$\omega = 1 \text{ rad/s}, \qquad R = 5 \times 10^{-4}$$

and the three slopes $0, -1, -2$, the composite curve is drawn in Fig. 8.6.2-1.

∎

■ **EXAMPLE 8.6.2-2** Asymptotic Composite Magnitude Curve with Exponential Factor

Sketch the asymptotic composite magnitude curve for the following function:

$$G(s) = \frac{400}{s(s + 20)} = s^n G_1 \tag{1}$$

where

$$n = -1 \tag{2}$$

$$G_1 = \frac{400}{s + 20} \tag{3}$$

Solution. The composite intercept I_c is

$$I_c = I_e I_0 \tag{4}$$

where

$$I_e = 1 \quad \text{at a frequency } \omega = 1 \tag{5}$$

$$I_0 = \lim_{s \to 0} G_1(s) = \lim_{s \to 0} \frac{400}{s + 20} = 20 \tag{6}$$

Then

$$I_c = 1 \times 20 = 20 \quad \text{at frequency } \omega = 1 \tag{7}$$

The individual curves and the composite are sketched in Fig. 8.6.2-2. Note that

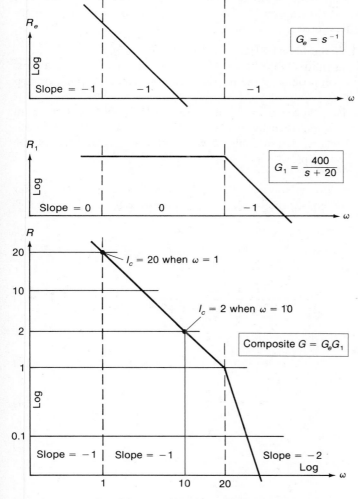

Figure 8.6.2-2 Composite magnitude with exponential factor.

the composite intercept I_c is located at frequency $\omega = 1$. If it is not convenient to locate the intercept at this frequency, make use of eq. 8.6.2-4. For example, at a frequency $\omega = 10$,

$$I_e = 10^n = 10^{-1} = 0.1 \tag{8}$$

and in that case,

$$I_c = 0.1 \times 20 = 2 \quad \text{at frequency } \omega = 10 \tag{9}$$

Note that the curve passes through both the points given by eqs. (7) and (9). ■

■ **EXAMPLE 8.6.2-3** Design of Suspension System for Motorcycle
Consider the motorcycle on a bumpy road as described in Ex. 7.6.2-1. Assume that the actions of front and rear suspension systems are independent. (The more accurate case is modeled in Ex. 12.3.1-3.) Consequently, we may assume that one-half the total mass is divided between front and rear suspension systems. Given the cruising speed V_c, design the suspension system to satisfy the following specifications:

(a) M = one-half the mass of frame and driver = 100 kg.
(b) The damping ratio should be chosen to meet the given specifications, but without compromising the transient response to an impulse or step input.
(c) At the cruising speed $V_c = 20$ m/s (about 50 mph) and with an average spacing between bumps $L = 3.75$ m (about 12 ft), the amplitude of vertical vibration shall be less than one-quarter that of the height of the bumps.

Solution.   There are four unknowns, D, K, ω_n, and ζ. The Laplace transfer function was given by eq. (2) of Ex. 7.6.2-1, repeated below.

$$\frac{y}{Y_a}(s) = \frac{Ds + K}{Ms^2 + Ds + K} \tag{1}$$

Normalize

$$\frac{y}{Y_a}(s) = \frac{D}{M} \frac{s + \alpha}{s^2 + 2\zeta\omega_n s + \omega_n^2} \tag{2}$$

where

$$\omega_n = \sqrt{\frac{K}{M}} \tag{3}$$

$$\zeta = \frac{D}{2M\omega_n} \tag{4}$$

$$\alpha = \frac{K}{D} = \frac{K}{M}\frac{M}{D} \tag{5}$$

$$= \frac{\omega_n^2}{2\zeta\omega_n} = \frac{\omega_n}{2\zeta} \tag{5a}$$

The cruising frequency ω_c may be found with the use of eq. (5) of Ex. 7.6.2-1, which is repeated below.

$$\omega_c = \frac{2\pi V_c}{L} \tag{6}$$

$$= \frac{2\pi 20}{3.75} = 33.5 \tag{6a}$$

Referring to eq. (2), it is clear that the composite curve will be given by the product of a first order factor with a reciprocal quadratic. The slope of each of these individual curves (refer to Figs. 8.4.4-1 and 8.4.7-1) is zero when $\omega = 0$. Hence, the slope of the composite will be zero when $\omega = 0$. Consequently, we may use eq. 8.6.2-3 to determine the intercept I_0. Applying this to eq. (1), we have

$$I_0 = \lim_{s \to 0} \frac{Ds + K}{Ms^2 + Ds + K} = 1 \tag{7}$$

It should be pointed out that we have determined the intercept I_0 without knowing any of the quantities to be found in this example.

At this point, we have a general idea of the shape of the composite curve. However, we do not know the corner frequencies (or break points). Also we do not know which of the two corner frequencies, α or ω_n, is the higher frequency. In order to span the many possibilities with only a few cases, consider the following four:

$$\zeta > \tfrac{1}{2} \quad \text{or} \quad \alpha < \omega_n \tag{8}$$

$$\zeta = \tfrac{1}{2} \quad \text{or} \quad \alpha = \omega_n \tag{9}$$

$$\zeta < \tfrac{1}{2} \quad \text{or} \quad \alpha > \omega_n \tag{10}$$

$$\zeta \ll \tfrac{1}{2} \quad \text{or} \quad \alpha \gg \omega_n \tag{11}$$

Each of the above four cases will be investigated.

(a) Damping Ratio Greater than One-Half

Let us refer to the corner frequency α as α_1, which, according to eq. (8), is less than the natural frequency. Thus, knowing only the relative frequencies, the two individual frequency response curves may be sketched as shown in Fig. 8.6.2-3. There are three regions separated by the two corner frequencies α_1 and ω_n. Region 1 covers the frequency range from 0 to α_1. Summing the slopes in this region, we have

$$0 + 0 = 0$$

Region 2 covers the frequency range from α_1 to ω_n. Summing the slopes in this region, we have

$$1 + 0 = 1$$

Region 3 covers the frequency range from ω_n to infinity. Summing the slopes

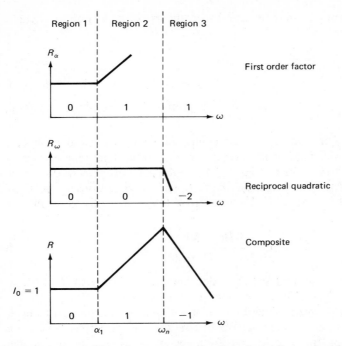

Figure 8.6.2-3   Frequency response for damping ratio greater than one-half.

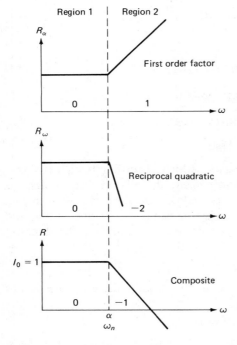

Figure 8.6.2-4   Frequency response for damping ratio equal to one-half.

in this region we have

$$+1 - 2 = -1$$

Using the intercept $I_0 = 1$, as found in eq. (7), and the three slopes, $0, -2, -1$, we may construct the composite as shown in Fig. 8.6.2-3.

(b) Damping Ratio Equal to One-Half

This case is covered by eq. (9), which indicates that the two corner frequencies are equal. Consequently, there will be only two regions since the two corner frequencies coincide. The individual curves may be sketched as shown in Fig. 8.6.2-4. Using an approach similar to that in part (a), sum the slopes in region 1 to obtain

$$0 + 0 = 0$$

and in region 2

$$1 - 2 = -1$$

Using the intercept $I_0 = 1$, found in eq. (7) and the two slopes, $0, -1$, construct the composite as shown in Fig. 8.6.2-4.

(c) Damping Ratio Less than One-Half

Refer to the corner frequency α as α_2 which, according to eq. (10), is greater than ω_n. The two individual curves are sketched in Fig. 8.6.2-5. There will be three regions separated by the two corner frequencies ω_n and α_2. Using an

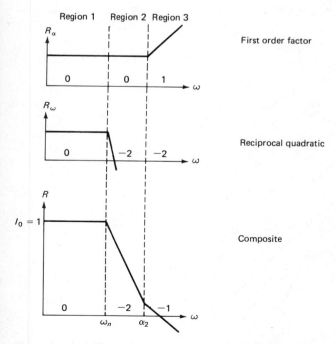

Figure 8.6.2-5   Frequency response for damping ratio less than one-half.

approach similar to that in part (a), sum the slopes in each region. In region 1 we have

$$0 + 0 = 0$$

In region 2 we have

$$0 - 2 = -2$$

In region 3 we have

$$1 - 2 = -1$$

Using the intercept $I_0 = 1$ and the above three slopes, $0, -2, -1$, construct the composite as shown in Fig. 8.6.2-5.

(d) Damping Ratio Much Less Than One-Half

According to eq. (11), the corner frequency α_3 will be much greater than ω_n. The curves for this case will be similar to those for part (c) as shown in Fig. 8.6.2-5. Summarizing, the composite curves for the above are placed on the same graph in Fig. 8.6.2-6.

We are now ready to make a selection that will satisfy specification (c). Note on Fig. 8.6.2-6 as the damping ratio is decreased, the output/input ratio R decreases up to a point. This point occurs when the corner frequency coincides with the cruising frequency ω_c. If the frequency α_3 is chosen higher than this, although the curve changes at frequencies higher than the cruising frequency, there is no further improvement for specification (c). Since raising the corner frequency α is tantamount to lowering the damping ratio, such a choice would compromise the transient response, in violation of specification

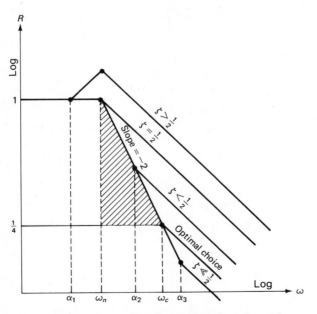

Figure 8.6.2-6 Composites for various damping ratios.

(b). Hence, the optimal choice is

$$\alpha = \omega_c \tag{12}$$

$$= 33.5 \text{ rad/s} \tag{13}$$

Now that we know α, let us employ log-log geometry to find ω_n. On Fig. 8.6.2-6, refer to the shaded triangle. Note that the slope of the hypotenuse is equal to -2, and that specification (c) requires a drop in R equal to 4. Then the corresponding horizontal distance (ratio of frequencies) is equal to the square root of 4 or 2. Consequently,

$$\frac{\omega_n}{\omega_c} = \frac{1}{2} \tag{14}$$

or

$$\boxed{\omega_n = \frac{33.5}{2} = 16.7 \text{ rad/s}} \tag{15}$$

The damping ratio is obtained from eq. (5a)

$$\zeta = \frac{\omega_n}{2\alpha} \tag{16}$$

Apply eq. (14)

$$\boxed{\zeta = \tfrac{1}{2}\left(\tfrac{1}{2}\right) = \tfrac{1}{4}} \tag{17}$$

The value of the spring K may be found with eq. (3)

$$K = M\omega_n^2$$

$$= 100(16.7)^2 \tag{18}$$

or

$$\boxed{K = 2.8 \times 10^4 \text{ N/m}} \tag{19}$$

Equation (5) may be employed to find D

$$D = \frac{K}{\alpha}$$

$$= \frac{2.8 \times 10^4}{33.5} \tag{20}$$

or

$$\boxed{D = 843 \text{ N} \cdot \text{s/m}} \tag{21}$$

The design has met all specifications and the design is complete.   ■

8.6.3 Phase Frequency Response Curves Using Products

Recall that the frequency response originated with the complex form

$$G(j\omega) = Re^{j\phi} \qquad [8.6.3\text{-}1]$$

where R = magnitude (which is a function of frequency ω)
   ϕ = phase (which is a function of frequency ω)

Consider a system transfer function that is equal to the product of two factors, G_1 and G_2. Then,

$$G(j\omega) = G_1(j\omega)G_2(j\omega) \qquad [8.6.3\text{-}2]$$

Writing these in polar form, we have

$$Re^{j\phi} = R_1 e^{j\phi_1} R_2 e^{j\phi_2}$$
$$= R_1 R_2 e^{j(\phi_1 + \phi_2)} \qquad [8.6.3\text{-}3]$$

Thus, while the magnitudes R_1 and R_2 are multiplied, the phase angles ϕ_1 and ϕ_2 are summed. Since the phase curve is plotted on a linear scale (rather than a log scale as is used for magnitude) then phase curves can be added algebraically. The *sum* of straight line curves on algebraic or *linear scales* requires that the *slopes* and the *intercepts* are *summed*.

The *phase* curves each require two corner frequencies, one at a frequency equal to $0.1\omega_a$ and the other at $10\omega_a$ (where ω_a is the corner frequency of the *magnitude* curve). As a result, the composite phase curve will require twice as many regions as did the magnitude composite.

■ EXAMPLE 8.6.3-1 Composite Phase Curve

Construct the phase frequency response curve for Ex. 8.6.2-1. Refer to the lower half of Fig. 8.4.5-1, where the phase curve is given. Note that for each decade (a decade is an order of magnitude change, for example, from 0.1 to 1) there is a phase angle shift of $-45°$ (phase lag = $45°$). Hence, the slope of the phase curve is equal to $-45°$ per decade. Knowing the slope and the corner frequencies, the phase curves for the two factors may be drawn, as show below. The corner frequencies for the first factor are

$$\omega_1 = 0.1 \times 10 = 1 \text{ rad/s}$$
$$\omega_2 = 10 \times 10 = 100 \text{ rad/s} \qquad (1)$$

For the second factor, we have

$$\omega_3 = 0.1 \times 200 = 20 \text{ rad/s}$$
$$\omega_4 = 10 \times 200 = 2000 \text{ rad/s} \qquad (2)$$

These four frequencies divide the frequency domain into five regions. Follow a procedure that is similar to that for summing magnitude curves, listing the slopes in each region. In order to determine the intercept, one may use either boundary of the individual curves. One is equal to zero, while the other is

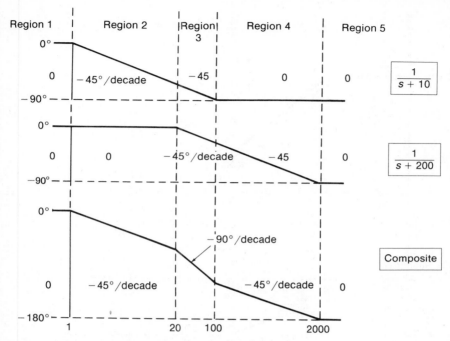

Figure 8.6.3-1 Phase curve for product of broken lines.

equal to $-90°$. Either will produce consistent results. The individual curves and the composite are sketched in Fig. 8.6.3-1.

8.6.4 Sketching the Actual Curve Using the Composite Asymptote

Once the composite asymptote is drawn, then locate the actual points above or below the *composite asymptote* just as they were shown for the individual curves. Care must be exercised to measure all distances *vertically* from the composite asymptote.

■ **EXAMPLE 8.6.4-1** Actual Magnitude Composite Curve
Refer to Ex. 8.6.2-1, where the magnitude asymptote was constructed. Check the intercept using eq. 8.6.2-3, and sketch the actual composite magnitude frequency response curve. Equation 8.6.2-3 applies either to the unfactored or to the factored form of the system transfer function. Using the former, apply eq. 8.6.2-3 to eq. (1) of Ex. 8.6.2-1.

$$I_0 = \lim_{s \to 0} \left[\frac{1}{s^2 + 210s + 2000} \right] = \frac{1}{2000} = 5 \times 10^{-4} \tag{1}$$

which agrees with eq. (6) of the example.

The individual curves for Ex. 8.6.2-1 are reciprocal first order factors. Refer to Fig. 8.4.5-1, where the asymptotes and actual curves are shown. Note

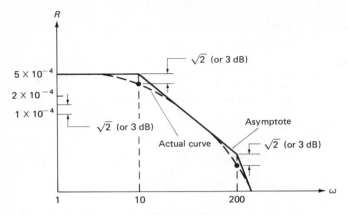

Figure 8.6.4-1 Actual magnitude frequency response curve.

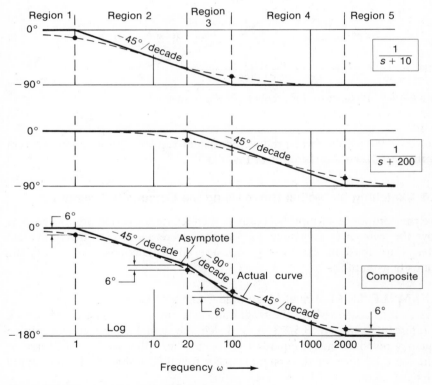

Figure 8.6.4-2 Actual composite phase curve.

that the actual curve passes through a point that is located a distance equal to $\sqrt{2}$ (or 3 dB) below the corner. Now we could either apply this correction factor to the individual curves or to the composite. The latter will be employed in this example. Note that the distance $\sqrt{2}$ can be constructed by taking one-half the distance 2. The resulting curve is shown in Fig. 8.6.4-1.   ■

■ **EXAMPLE 8.6.4-2** Actual Composite Phase Curve
Refer to Ex. 8.6.3-1, where the asymptotic phase curve was constructed. Refer to Fig. 8.4.5-1, where the general asymptotic phase curves are drawn. Note that for these general curves the actual phase curve passes through a point 6° below the corner at frequency = $0.1a$, and through a point 6° above the corner at frequency = $10a$.

Apply the general rule to this specific example. To construct the actual curve, let it pass through these points and also let it become asymptotic to the asymptotes. This procedure is applied to the individual and to the composite curves. The results are shown in Fig. 8.6.4-2.   ■

8.6.5 Magnitude and Phase Curves on Same Graph

The magnitude and phase curves can be drawn on the same graph. In order to reconcile the *log* scale required for the magnitude against the *linear* scale required for the phase curve, the following layout is suggested. Use the same log frequency scale for both curves. On the left-hand vertical border, use a log scale for the magnitude curve. On the right-hand vertical border, use a linear scale for the phase curve. Arrange the magnitude and phase scales so that magnitude = 1 lines up with phase = $-180°$.

■ **EXAMPLE 8.6.5-1** Magnitude and Phase on Same Graph
For the function given below, sketch the asymptotic magnitude and phase curves on the same graph:

$$G(s) = \frac{64 \times 10^6(s + 2000)}{s^2(s + 4000)} \tag{1}$$

Solution. The given function consists of an exponential s^{-2}, a first order factor $(s + 2000)$ in the numerator and a first order factor $(s + 4000)$ in the denominator. An appropriate location for the composite intercept is at

$$\omega = 1000 \tag{2}$$

which is reasonably below the lowest frequency in either the numerator or denominator.

Apply eq. 8.6.2-3 to eq. (1)

$$I_0 = \lim_{s \to 0} \frac{64 \times 10^6(s + 2000)}{(s + 4000)} = 32 \times 10^6 \tag{3}$$

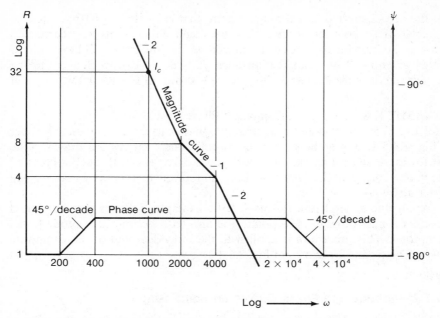

Figure 8.6.5-1 Magnitude and phase curves on the same graph.

Apply eq. 8.6.2-4 to the exponential

$$I_e = (1000)^{-2} = 10^{-6} \tag{4}$$

Apply eqs. (3) and (4) to eq. 8.6.2-2

$$I_c = (32 \times 10^6)(10^{-6}) = 32 \tag{5}$$

The composite intercept has been computed. It has a magnitude = 32 at a frequency $\omega = 1000$. Considering the three magnitude curves (not shown), the composite slope begins with -2. At a frequency $\omega = 2000$ the slope changes to -1, and at a frequency $\omega = 4000$ the slope is -2. The magnitude curve may be sketched as shown on Fig. 8.6.5-1.

The phase curve starts at $-180°$ and is horizontal. At a frequency $\omega = 200$, it starts to rise with a slope $= 45°/\text{decade}$ until it reaches frequency $\omega = 400$, at which the phase curve levels off. It remains a horizontal line until frequency $\omega = 2 \times 10^4$, after which it drops with a slope equal to $-45°/\text{decade}$ until frequency $\omega = 4 \times 10^4$, and levels off. At this point, if the phase curve was drawn accurately, it will be on the $-180°$ line. ∎

8.6.6 Frequency Response for Unfactored Function

In Sections 8.6.1 through 8.6.5, the method required that the function be in factored form. However, if the function is not in factored form, the frequency response can still be determined. Set $s = j\omega$ and apply Appendix 5.

■ **EXAMPLE 8.6.6-1** Frequency Response for Unfactored System
Plot the frequency response for the following function:

$$G(s) = \frac{10s^3 + 24s^2 + 60s + 103}{2s^4 + 3s^3 + 7s^2 + 19s + 2905} \tag{1}$$

Set

$$s = j\omega \tag{2}$$

Then

$$G(j\omega) = \frac{-j10\omega^3 - 24\omega^2 + j60\omega + 103}{2\omega^4 - j3\omega^3 - 7\omega^2 + j19\omega + 2905} \tag{3}$$

Separate real and imaginary terms in numerator and denominator.

$$G(j\omega) = \frac{(-24\omega^2 + 103) + j(-10\omega^3 + 60\omega)}{(2\omega^4 - 7\omega^2 + 2905) + j(-3\omega^3 + 19\omega)} \tag{4}$$

Collect the terms, first in rectangular form, then transform to polar form, shown symbolically as

$$G_1(j\omega) = \frac{A_1 + jB_1}{A_2 + jB_2} = \frac{R_1 e^{j\psi_1}}{R_2 e^{j\psi_2}} \tag{5}$$

$$\text{Magnitude} \quad R = \frac{R_1}{R_2} \tag{6}$$

$$\text{Phase} \quad \psi = \psi_1 - \psi_2 \tag{7}$$

Compute the quantities A_1, A_2, B_1, B_2, R_1, R_2, ψ_1, and ψ_2 or each frequency ω, and plot the magnitude and phase curves. As a check, the composite intercept can be found by using eqs. 8.6.2-2 through 8.6.2-5.

$$I_c = I_0 I_e \tag{8}$$

where

$$I_0 = \lim_{s \to 0} G(s) = 103/2905 \tag{9}$$

$$I_e = 1 \quad (\text{since } n = 0) \tag{10}$$

Then

$$I_c = 103/2905 \tag{11}$$

■

8.7 FILTERS AND RESONATORS

A chemical filter allows certain sized particles through while it restrains others. An optical filter is selective with color. Similarly, a mechanical filter delineates frequencies.

8.7.1 Classification by Frequency Response

The frequency response curve indicates how the ratio of output to input amplitude varies with input frequency. This ratio is not constant over the frequency range. The range over which the curve rises and falls with be used as means to classify systems.

The most common of all systems is the low pass filter shown in Fig. 8.7.1-1(a). At low frequencies, the output/input ratio R is equal to unity. This implies that at low frequencies, the system behaves as a rigid body. At high frequencies, the ratio R becomes less than unity. This implies that the system does not transmit, or pass, the high frequencies. Refer to Fig. 8.7.1-1(a). Note that the input shown consists of two frequencies, one low, one high. Also note

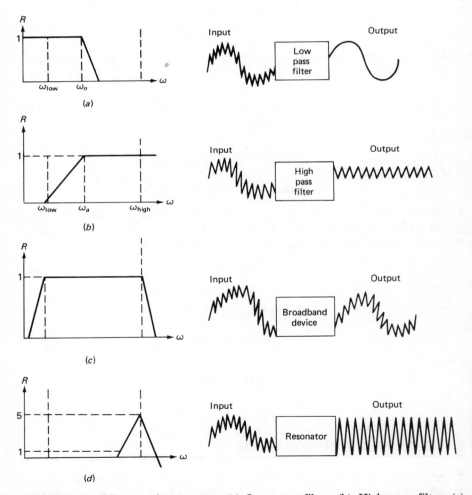

Figure 8.7.1-1 Filters and resonators. (a) Low pass filter. (b) High pass filter. (c) Broadband device. (d) Resonator

that the output contains only the low frequency. Since this system passes only the low frequencies, it is called a "low pass filter."

If the ratio is equal to unity in the high frequency range, then the system will transmit, or pass, the high frequencies. At the same time, the ratio is less than unity at the low frequencies, rendering the system unable to transmit, or pass, the low frequencies. Refer to Fig. 8.7.1-1(b). Note that of the two input frequencies, only the high frequency is passed. Since this system will pass only the high frequencies, it is referred to as a "high pass filter."

If the frequency response curve is approximately unity for a range of frequencies, then the system will pass all of these frequencies. It is called a "broadband device." Refer to Fig. 8.7.1-1(c). Note that both of the input frequencies are passed.

If the system produces a definite peak in the frequency response curve at one frequency, it is referred to as a resonator at that frequency. Refer to Fig. 8.7.1-1(d). Note that only one frequency will be passed and that it will be magnified or amplified.

8.7.2 Filters

Low Pass Filter   A low pass filter transmits, or passes, the low frequencies. The ideal low pass filter passes frequencies below ω_n and does not pass those above. In a real low pass filter, instead of totally cutting off the high frequencies, it attenuates them. Here the slope of the curve above ω_n is significant. If the slope is very steep, then frequencies somewhat distant from ω_n will experience a sizable reduction or attenuation. In Fig. 8.7.1-1(a), an input containing high and low frequencies enters a low pass filter. Note that only the *low* frequency component is passed.

■ **EXAMPLE 8.7.2-1** Typical Mechanism as Low Pass Filter
Show that the stepper motor in Ex. 8.5.2-2 is a low pass filter. There is a sinusoidal input. Since the motor may run at different speeds, different frequencies will be involved.

Refer to Fig. 8.5.2-2, the magnitude frequency response curve. Note that at low frequencies (below 100 rad/s) the magnitude is equal to unity. This implies that the input motion to the motor is transmitted properly. That is, at low frequencies, the motion is transmitted or passed. The frequency response curve also shows that at high frequencies (in excess of 1000 rad/s) the magnitude is much less than unity. Since this device *passes* the *low* frequencies, while attenuating the highs, it may be classified as a low pass filter.   ■

High Pass Filter   A high pass filter transmits or passes, the high frequencies. For the ideal high pass filter, the frequency response curve would drop vertically at frequency ω_a, not permitting any frequencies below this to pass. In a real high pass filter, there is a slope to the curve, resulting in an attenuation rather than a cutoff. In Fig. 8.7.1-1(b), an input containing high and low frequencies enters the high pass filter. Note that the high frequency is passed, while the low frequency is not.

■ **EXAMPLE 8.7.2-2** *RC* Circuit as High Pass Filter

Show that the circuit shown in Fig. 8.7.2-1 is a high pass filter, and discuss its use in an amplifier system.

Solution.   Apply Kirchhoff's law to the loop as shown in Section 3.3.1.

$$V_1 = \left[R + \frac{1}{Cs} \right] I = \left[\frac{RCs + 1}{Cs} \right] I \tag{1}$$

The output voltage V_2 is given by the potential drop across the resistor R. Then

$$V_2 = IR \tag{2}$$

Eliminate I between eqs. (1) and (2). Then

$$V_2 = \left[\frac{RCs}{RCs + 1} \right] V_1 \tag{3}$$

or

$$\frac{V_2}{V_1}(s) = \frac{s}{s + 1/RC} \tag{4}$$

Equation (4) is formed by the product of an exponential and a reciprocal first order factor. These individual curves and their composite are plotted in Fig. 8.7.2-2. Note that the composite has the form of Fig. 8.7.1-1(b). Consequently, the system is a high pass filter.

This circuit is often used as the coupling between stages of a low cost amplifier. Since the device is a high pass filter, only the high frequencies are passed. Consequently, the sound so produced is rich in highs but devoid of lows. As a result, the low cost amplifier sounds "tinny." In order to improve the performance, it is desirable to make the corner frequency $1/RC$ as low as possible. This requires that both R and C be large. Usually the resistance R is limited by the circuit that follows this one. Hence, the remaining choice is to use a large capacitor. This, in turn, often requires a large physical size, or the use of fairly expensive components.   ■

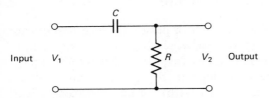

Figure 8.7.2-1   High pass electrical filter.

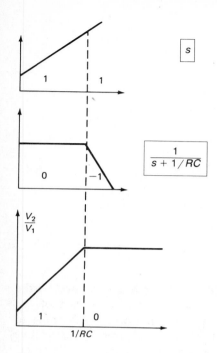

$$\boxed{s}$$

$$\boxed{\frac{1}{s + 1/RC}}$$

Figure 8.7.2-2 Frequency response of *RC* filter.

■ EXAMPLE 8.7.2-3 Torsiograph as a High Pass Filter

The torsiograph measures torsional vibration. The base is fastened to the machine whose torsional vibration is to be studied. The base has a stud or short shaft to which a clock spring and the stator of the readout device is fastened. The other end of the clock spring is attached to a hollow flywheel. The rotor of the readout device is made integral with the flywheel. The readout provides magnetic damping. Referring to Fig. 8.7.2-3, note that the readout device reads the difference between the input θ_a and the motion θ of the flywheel. Show that this device is a high pass filter. Show that all seismic devices are high pass filters.

Solution. First, model the system to establish that this device *is* a seismic instrument. In this device, it is easy to identify the mass, which is the flywheel. The flywheel is supported by a clock spring, which acts both as a spring and as a frictionless bearing. It may be assumed that the clock spring has significant mass. According to Section 2.1.4, one-third the mass of the clock spring is to be added to the mass of the flywheel, to determine the equivalent mass J. The spring rate K of the clock spring is given as item (5) in Table 2.2.3-4.* Note that one end of the clock spring is attached to the flywheel, while the other end is fastened to a stud which is part of the base whose motion is the input

---

*Appendix 2.

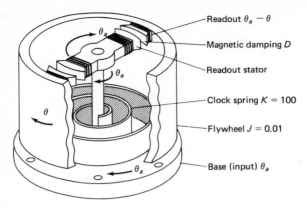

Readout $\theta_a - \theta$

Magnetic damping D

Readout stator

Clock spring $K = 100$

Flywheel $J = 0.01$

Base (input) θ_a

Figure 8.7.2-3   Torsiograph with electrical output.

θ_a. The output of the device is an electrical signal which is generated by the relative motion of the two parts of a magnetic device. One part of the magnetic device is fastened to the input θ_a, while the other is fastened to the flywheel whose motion is θ. Thus, the final output is given by the relative motion, or the difference $\theta_a - \theta$. The magnetic device has magnetic damping, as described in Section 2.4.4.

The magnetic damper D and the clock spring K both have the same boundaries. Hence, they are in parallel and can be represented by a single mechanical network $Ds + K$. This network connects the lumped mass J to the input θ_a which may be thought of as a moving base. The free body diagram is shown in Fig. 8.7.2-4. The output is obtained by the difference between the input θ_a and the output θ.

The output of this device is obtained by the difference of the output and input. Hence, this device is a seismic instrument. The out/in ratio is given by

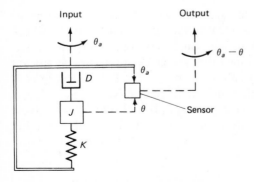

Figure 8.7.2-4   Free body diagram for torsiograph.

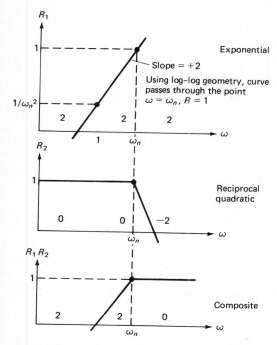

Figure 8.7.2-5   Frequency response for torsiograph.

eq. 7.8.1-4. Then

$$\frac{\theta_a - \theta}{\theta_a} = \frac{s^2}{s^2 + 2\zeta\omega_n s + \omega_n^2} \tag{1}$$

The Laplace transfer function of this system given in eq. (1) is formed by the product of an exponential and a reciprocal quadratic. Plots of the individual curves and their composite are shown in Fig. 8.7.2-5.

Referring to the composite, at frequencies below ω_n, the out/in ratio is attenuated (reduced) at low frequencies. Above ω_n, the out/in ratio is equal to unity. Thus, the torsiograph passes only high frequencies. It is a high pass filter.   ∎

8.7.3 Broadband Devices

A broadband device passes all frequencies within a specified range. See Fig. 8.7.1-1(c). Ideally, this device has a frequency response curve that is flat over the specified range. The ideal broadband device has a frequency response curve that is shaped like a rectangle, as shown in Fig. 8.7.3-1(a). In actual practice, if the frequency response curve wavers up and down within specified limits, the curve is still considered flat The *bandwidth* is defined as the range of frequencies over which the curve remains within the specified limits. The

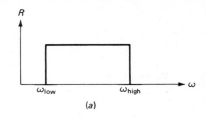

(a)

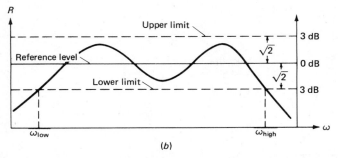

(b)

Figure 8.7.3-1   Frequency response of broadband devices. (a) Ideal broadband device. (b) Practical broadband device.

usual limits are taken as $\pm \sqrt{2}$. (Recall that on log-log scales, a distance up or down implies a multiplication or division by that quantity.) See Fig. 8.7.3-1(b).

■ **EXAMPLE 8.7.3-1** Design of Electrostatic Speaker

The electrostatic speaker uses electrostatic forces to drive the acoustical element. At light weight conducting plate is placed in close proximity with a similar plate. The plates are separated by a dielectric material that also is elastic. One plate is held fixed, while the other, under electrostatic forces, is

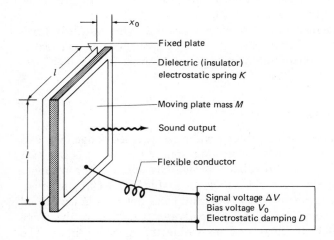

Figure 8.7.3-2   Electrostatic speaker.

capable of dynamic motion. The two plates form the plates of a parallel plate capacitor. Since the field is uniform, the moving plate has a uniform force distribution over its entire surface. As a result, the entire plate is translated perpendicular to its face. Consequently, there is no need for a cone as was the case for a magnetic speaker. The absence of a cone and its elaborate suspension offers some advantages for the electrostatic speaker. Such a speaker could be mounted in a picture frame and hung on a wall. See Fig. 8.7.3-2.

The "loudness" of sound is a function of the pressure variation P resulting from the mechanical motion X. If the mechanical displacement X of the speaker were maintained at constant amplitude for any applied frequency, the low frequencies would not sound as loud as the high frequencies. This is due to the requirement to move a greater volume of air at the low frequencies in order to produce the same pressure variation. One might refer to the situation as a mismatch in impedance between the speaker mass and the air mass. There is also a mismatch in spring rates of both bodies. At high frequencies, where little motion of air is required, the matching is good. Assume that the effect can be approximated by a high pass filter where the corner frequency ω_a is given by

$$\left(\frac{\omega_a}{\omega_n}\right)^2 = \frac{M}{\rho_a l^3} \tag{a}$$

where

$\qquad M$ = mass of acoustic element

$\qquad \rho_a$ = mass density of air $\left(\text{about } 1 \text{ kg/m}^3\right)$ $\hfill$ (b)

$\qquad l$ = maximum dimension of speaker acoustic element

Assume that the motion X of the moving plate is 10% of the plate spacing X_0. Consequently, we may neglect the change in capacitance C due to the displacement X. As a result, the system is cascaded. Design the speaker system for a frequency range 2×10^4 rad/s to 2×10^5 rad/s (about 32,000 Hz) with an input signal ± 150 V causing a displacement of $\pm 10^{-3}$ cm (about 4×10^{-4} in.).

Solution.

Cascaded System

The input to the system is a sinusoidal voltage V_1, which must pass through the electrostatic circuit. The resulting voltage V_2 produces a mechanical force F. This force, in accordance with the system dynamics, results in a mechanical displacement X of the acoustic element. The motion of the acoustic element sets up a modulated pressure P which is the sound that is heard. The overall behavior, from input voltage to sound output, is given by the product of the above four effects. See Fig. 8.7.3-3. Each of these will be discussed in turn.

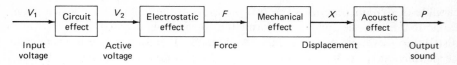

Figure 8.7.3-3   Cascaded electrostatic speaker.

Circuit Effect

The force of attraction of the two plates is independent of the polarity of the voltage across the plates. To prevent such ambiguity the voltage that is applied to the plates must not change sign even though the input signal does. The remedy is achieved with the inclusion of a bias voltage. The bias voltage also improves the linearity of the system. The circuit is shown in Fig. 8.7.3-4.

In studying the steady state response to a sinusoidal input for a linear system, the constant bias supply voltage V_0 may be ignored. Voltages V_1 and V_2 become

$$V_1 = \left(R + \frac{1}{Cs} \right) I = \left[\frac{RCs + 1}{Cs} \right] I$$

$$V_2 = \frac{1}{Cs} I$$

Then

$$\frac{V_2}{V_1} = \frac{I}{Cs} \times \frac{Cs/I}{RCs + 1} = \frac{\omega_c}{s + \omega_c} \tag{1}$$

where

$$\omega_c = \frac{1}{RC} \tag{2}$$

The frequency response for this effect is shown in Fig. 8.7.3-5(a).

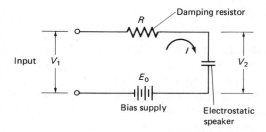

Figure 8.7.3-4   *RC* circuit for electrostatic speaker.

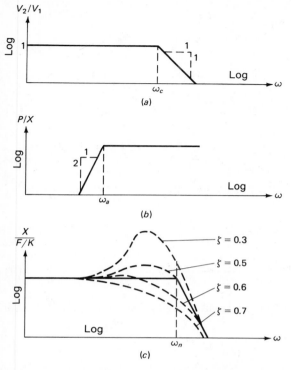

Figure 8.7.3-5   Frequency response for subsystems of electrostatic speaker. (a) Circuit effect. (b) Acoustic effect. (c) Mechanical effect.

Electrostatic Effect

The capacitance C, the electrostatic spring K, and the equivalent damping D, are found in Table 2.4.2-1,* item (1):

$$C = \frac{\epsilon A}{X_0} \tag{3}$$

$$K = -\frac{V_0^2 \epsilon A}{X_0^3} + K_1 \quad \text{(where } K_1 = \text{spring constant} \tag{4}$$
$$\text{for dielectric material)}$$

$$D = \frac{V_0^2 \epsilon^2 A^2 R}{X_0^4} = \frac{K \epsilon A R}{X_0} \tag{5}$$

Since the acoustic element is mechanical,

$$D = \frac{2 \zeta K}{\omega_n} \tag{6}$$

*Appendix 2.

Equate eqs. (5) and (6)

$$\frac{K\epsilon AR}{X_0} = \frac{2\zeta K}{\omega_n}$$

or

$$R = \frac{2\zeta X_0}{\epsilon A\omega_n} \tag{7}$$

Apply eqs. (3) and (7), to eq. (2)

$$\omega_c = \frac{\epsilon A\omega_n}{2\zeta X_0}\frac{X_0}{\epsilon A} = \frac{\omega_n}{2\zeta} \tag{8}$$

Small Changes for Linearity

From eq. (4) we have,

$$X_0^3 = \frac{V_0^2\epsilon A}{(K_1 - K)} \tag{9}$$

In order to determine the effects of small changes, take the derivative with respect to X. Then

$$3X^2 = 2V_0\frac{dV}{dX}\frac{\epsilon A}{K}$$

or

$$\frac{dX}{dV} = \frac{2V_0\epsilon A}{3X_0^2(K_1 - K)} \tag{10}$$

Substitute for $(K_1 - K)$

$$\frac{dX}{dV} = \frac{2V_0\epsilon AX_0^3}{3X_0^2V_0^2\epsilon A} = \frac{2X_0}{3V_3} \tag{11}$$

For small but finite changes, replace the differentials

$$\frac{\Delta X}{\Delta V} = \frac{2}{3}\frac{X_0}{V_0} \quad \text{or} \quad \frac{\Delta V}{V_0} = \frac{3\Delta X}{2X_0} \tag{12}$$

Plate Spacing X_0 and Bias Voltage V_0

Let the mechanical displacement ΔX be equal to 10% of X_0. Then

$$\frac{\Delta X}{X_0} = 10\% = 0.1$$

or

$$X_0 = \frac{\Delta X}{0.1} = \frac{10^{-3}}{0.1} = \boxed{10^{-2} \text{ cm}} \tag{13}$$

Then,

$$\frac{\Delta V}{V_0} = \frac{3}{2} 10\% = 0.15$$

or

$$V_0 = \frac{\Delta V}{0.15}$$

Given the signal voltage $\Delta V = 150$, then the bias voltage is

$$V_0 = \frac{150}{0.15} = \boxed{1000 \text{ V}} \tag{14}$$

Electrostatic Spring (ϵ for mylar* = 4×10^{-11})

In eq. (4), everything is known but the area A which is $= l^2$

$$(K_1 - K) = \frac{1000^2 \times 4 \times 10^{-11} l^2}{(10^{-4})^3} = 4 \times 10^7 l^2 \tag{15}$$

Mass M from mechanical aspects

$$M = \frac{(K_1 - K)}{\omega_n^2} = \frac{4 \times 10^7 l^2}{(2 \times 10^5)^2} = 10^{-3} l^2 \tag{16}$$

Acoustic Effect

The acoustic effect may be approximated as a high pass filter as shown in Fig. 8.7.3-5(b) and eq. (a). Apply eq. (16) to eq. (a) and use the specified frequency range.

$$\left(\frac{2 \times 10^4}{2 \times 10^5}\right)^2 = \frac{10^{-3} l^2}{l^3} \tag{17}$$

or

$$l = \boxed{0.10 \text{ m (about 4 in.)}} \tag{18}$$

Thus, the speaker has been sized to meet the low frequency requirement.

Mechanical Effect

The mechanical aspect of the system is a spring-mass with damping. The frequency response is given by a reciprocal quadratic, as shown in Fig. 8.7.3-5(c) using several values of damping ratio.

---

*See Appendix 6.

Return to eqs. (15) and (16) now that l is known.

$$(K_1 - K) = 4 \times 10^7 \times 0.10^2 = \boxed{4 \times 10^5 \text{ N/m}} \tag{19}$$

$$M = 10^{-3} \times 0.10^2 = \boxed{10^{-5} \text{ kg}} \tag{20}$$

Thickness of the plate t

$$M = \rho_p l^2 t$$

where ρ_p = mass density of plate material (aluminum film)

or

$$t = \frac{M}{\rho_p l^2} = \frac{10^{-2}}{3 \times 10^2}$$

$$= \boxed{3.3 \times 10^{-5} \text{ cm (about 13 microinch)}} \tag{21}$$

Damping

From the reciprocal quadratic curve and also from the composite curve, the optimum damping ratio is

$$\zeta = \tfrac{1}{2} \tag{22}$$

Apply this to eq. (8)

$$\omega_c = \frac{\omega_n}{2\zeta} = \omega_n = \boxed{2 \times 10^5 \text{ rad/s}} \tag{23}$$

Compute the required equivalent damping, using eq. (6)

$$D = 2 \times \tfrac{1}{2} \times 0^{-5} \times 2 \times 10^5 = \boxed{2 \text{ N} \cdot \text{s/m}} \tag{24}$$

The damping resistor is found by using eq. (5)

$$R = D \left(\frac{X_0^2}{V_0 \epsilon A} \right)^2 = 2 \left(\frac{(10^{-4})^2}{1000 \times 4 \times 10^{-11} \times 0.100^2} \right)^2$$

$$= \boxed{1250 \ \Omega} \tag{25}$$

Overall System

The system is cascaded, as shown by Fig. 8.7.3-3. The overall frequency response is given by the product as shown in Fig. 8.7.3-6. The individual curves were shown in Fig. 8.7.3-5. Note that eq. (23) indicates that the two corner frequencies ω_c and ω_n coincide.

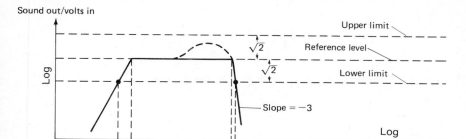

Figure 8.7.3-6   Frequency response for electrostatic speaker system.

Bandwidth

The bandwidth ranges from ω_{b_1} to ω_{b_2} which is slightly greater than the required range. The design is completed and it satisfies all requirements.   ■

8.7.4 Resonator

A resonator provides magnification to a small range of frequencies in the vicinity of the peak or resonance. Considering the peak as the reference, when the curve falls below the lower limit* ($\sqrt{2}$ below the peak), and this will happen at two points, these frequencies will be the *band limits* or *sidebands* ω_{b_1} and ω_{b_2}. Note that for any frequency ω_i eq. 8.4.7-4 indicates

$$\gamma_i = \frac{\omega_i}{\omega_n}$$

See Fig. 8.7.4-1.

Of particular interest are the band limits for the reciprocal quadratic. Divide eq. 8.5.1-9 by $\sqrt{2}$ and equate to eq. 8.5.1-1:

$$\frac{R_r}{\sqrt{2}} = \frac{1}{\sqrt{2}} \frac{1}{2\zeta} \sqrt{\frac{1}{1 - \zeta^2}} = \sqrt{\frac{1}{\left(1 - \gamma_b^2\right)^2 + \left(2\zeta\gamma_b\right)^2}} \qquad [8.7.4\text{-}1]$$

Square both sides and solve for γ_b^2

$$\gamma_b^2 = 1 - 2\zeta^2 \pm 2\zeta\sqrt{1 - \zeta^2} = \gamma_r^2 + 2\zeta\beta$$

or

$$\gamma_b = \sqrt{\gamma_r^2 \pm 2\zeta\beta} \qquad [8.7.4\text{-}2]^{\dagger}$$

---

*In systems where the power is proportional to the ordinate squared, at this point, the power level will be equal to $(1/\sqrt{2})^2$ or $\frac{1}{2}$. Hence, for such systems, the points at which this occurs are called "half-power" points.

†Numerical values appear in Table 8.7.4-1.

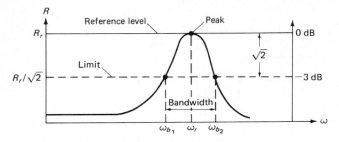

Figure 8.7.4-1   Band limits of a resonator.

TABLE 8.7.4-1   BAND LIMITS FOR RECIPROCAL QUADRATIC RESONATOR

ζ	γ_{b_1}	γ_{b_2}	Bandwidth	R_r	$R_r / \sqrt{2}$
0.0100	0.9898	1.0099	0.0200	50.0025	35.3571
0.0200	0.9794	1.0194	0.0400	25.0050	17.6812
0.0500	0.9461	1.0464	0.1003	10.0125	7.0799
0.1000	0.8837	1.0858	0.2021	5.0252	3.5533
0.1500	0.8114	1.1188	0.3073	3.3715	2.3840
0.2000	0.7267	1.1454	0.4187	2.5516	1.8042
0.2500	0.6252	1.1658	0.5406	2.0656	1.4606
0.3000	0.4976	1.1800	0.6824	1.7471	1.2354
0.3500	0.3151	1.1877	0.8727	1.5250	1.0784
0.3820	0.0457	1.1892	1.1435	1.4163	1.0015

When $\zeta \ll 1$, then eq. 8.7.4-2 simplifies

$$\gamma_b \approx \sqrt{1 \pm 2\zeta} \approx 1 \pm \zeta \quad \text{for } \zeta \ll 1 \qquad [8.7.4\text{-}3]$$

The bandwidth is given by

$$\gamma_{b_2} - \gamma_{b_1} \approx 1 + \zeta - (1 - \zeta) = 2\zeta \quad \text{for } \zeta \ll 1 \qquad [8.7.4\text{-}4]$$

■ **EXAMPLE 8.7.4-1** Design of Frahm Vibrating Reed Tachometer
The vibrating reed tachometer is used to measure the speed of a rotating machine where there is no accessible point at which the rotating shaft can be reached. The device consists of a number of small cantilever reeds, each having a different natural frequency. Each reed has very little damping. If the engine speed corresponds to the resonant frequency of one of the reeds, then that reed will vibrate noticeably. Numbers are printed at the base of each reed. Thus the operator notices which reed is moving at a resonant amplitude and thus determines the engine speed. It should be pointed out that the device depends upon the vibration of the engine. Obviously, it will not work on machines that are very well balanced. See Fig. 8.7.4-2. Given the geometry of one reed and its characteristics, design the remaining ten reeds so that their frequencies differ by the minimum detectable difference ($\sqrt{2}$).

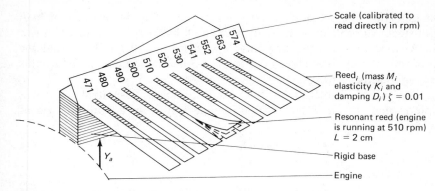

Figure 8.7.4-2   Frahm vibrating reed tachometer.

Solution.   First establish the set of frequencies. For very low damping ratio, the bandwidth is equal to 2ζ, with half of this on each side of the resonant frequency. Assume the damping ratio is 0.01. This means the sidebands are 0.01 times the resonant frequency. Note that for different frequencies, the sidebands will be somewhat different. Working to the nearest integers, the sequence of frequencies is as shown on the scale on Fig. 8.7.4-2. If the frequency response curves are drawn for each reed, it will become apparent that these curves will intersect at their respective sidebands (half-power points) as shown in Fig. 8.7.4-3.

The magnitude R_r of the peak for $\zeta = 0.01$ is found in Table 8.5.2-1, $R_r = 50$. This indicates that for an input frequency that corresponds to one of the 11 resonant frequencies there will be a magnification of 50, while the amplitudes of the two neighboring reeds will be much less than 50. On the other hand, if the input frequency should fall midway between two resonances, then the two reeds will vibrate at amplitudes that are 35.3 times the input amplitude (or $\sqrt{2}$ below that at resonance). This amplitude will still be detectable. Hence, the scheme permits interpolation between readings.

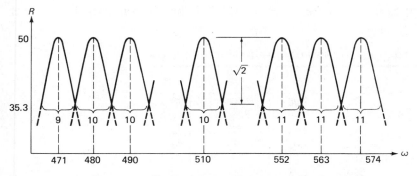

Figure 8.7.4-3   Frequency distribution for vibrating reed tachometer.

Assume that the set of reeds is stamped out of a single sheet of high strength steel. All reeds have the same width and thickness. They differ only in length. Assume that the damping ratio for all reeds is 0.01. Given that the length of the resonant reed shown in Fig. 8.7.4-3 is 2 cm, determine the lengths of the neighboring reeds.

In Appendix 2, Table 2.1.4-1, item (1) and Table 2.2.3-2, item (1), it was determined that the lumped mass and lumped spring for a cantilever are as follows:

$$M_{eq} = \tfrac{1}{4}M \quad \text{where} \quad M = \rho AL \tag{1}$$

$$K_{eq} = \frac{C_1}{L^3} \tag{2}$$

$$\rho = \text{mass density} \tag{3}$$

and

$$A = \text{cross sectional area} \tag{4}$$

The resonant frequency is almost equal to the undamped natural frequency for the given damping ratio of 0.01 (see Table 8.5.2-1).

$$\omega_r \approx \omega_n = \sqrt{\frac{K}{M}} = \sqrt{\frac{C_1/L^3}{AL/4}} \tag{5}$$

$$= \frac{1}{L^2}\sqrt{\frac{4C_1}{A}} = \frac{C_2}{L^2} \tag{6}$$

The lengths of the reeds may be determined by a ratio of frequencies. Given the length $L_0 = 2$ cm for the reed having a resonant frequency of 510 rpm, then the length L_i of the ith reed having a resonant frequency of ω_{n_i} is given by

$$L_i = L_0\sqrt{\frac{\omega_{n_0}}{\omega_{n_i}}} \tag{7}$$

Substituting the numerical values, Table 8.7.4-2 may be constructed.   ∎

TABLE 8.7.4-2   REED LENGTHS FOR FRAHM TACHOMETER

	ω_n	L	ω_n	L
	471	2.08	530	1.96
	480	2.06	541	1.94
	490	2.04	552	1.92
	500	2.02	563	1.90
Given	510	2	574	1.87
	520	1.98		

SUGGESTED REFERENCE READING FOR CHAPTER 8

Log-log plots and bode diagrams [7], p. 397–407; [8], p. 374; [12]; [15], p. 318; [17], p. 317–341; [26], p. 179–207; [40], p. 407–420; [42], p. 193–195; chap. 11; [44], p. 343–346.

Polar plots [7], p. 396–397, 409–412; [8], p. 364; [26], p. 179; [40], p. 407–441; [42], p. 190, chap. 10; [44], p. 341.

Root locus [7], p. 196–228; [8], p. 332–335; [17], p. 384, 401, chap. 11; [26], chap. 6; [40], p. 411–420; [42], chap. 12.

Filters [17], p. 316, 325, 329; [20]; [40], p. 404.

Broadband devices [20].

PROBLEMS FOR CHAPTER 8

8.1. For the system shown, $M = 1.0$ kg and $D = 2.0$ N $\cdot$ s/m.
 (a) Find the Laplace transfer function $DV(s)/F(s)$.
 (b) Sketch the asymptotic amplitude response for $DV(j\omega)/F(j\omega)$ on log-log paper. Select convenient scales.
 (c) Sketch to scale the asymptotic phase curve for the above transfer function.

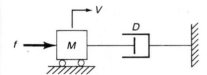

Figure P-8.1

8.2. Plot on log-log paper the asymptotes for the amplitude response of the transfer function

$$G(s) = \frac{1}{5}\frac{(s + 20)}{(s + 2)}$$

Also find actual values of $G(j\omega)$ at several frequencies and sketch the actual curve.

8.3. For the transfer function

$$G(s) = \frac{50}{s^2 + 2s + 25}$$

sketch the asymptotes for the amplitude and phase response on log-log and semilog paper, respectively.

8.4. (a) On log-log paper sketch the asymptotes for the amplitude response for the transfer function

$$G(s) = 5\frac{s + 2}{s + 10}$$

 (b) Find the actual values of $G(j\omega)$ at the corner frequencies for the numerator and the denominator. Sketch the actual curve.

8.5. Sketch the low frequency and high frequency asymptotes on log-log paper for each of the following transfer functions:

(a) $G(s) = \dfrac{500}{s^2 + 5s + 100}$

(b) $G(s) = 2(s + 4)$

Also compute $G(j\omega)$ at the corner frequency for each function, and sketch the actual amplitude response curves.

8.6. For the system shown, $K = 4000$ N/m, $D = 2000$ N $\cdot$ s/m, and $x_a = X_a e^{j\omega t}$.
(a) Find the transfer function $|X/X_a|$ as a function of ω.
(b) Plot on log-log paper $|X/X_a|$ versus ω.
(c) On semilog paper plot the phase ϕ versus ω for the above function.

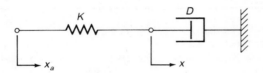

Figure P-8.6

8.7. Plot the asymptotes for the amplitude response $|G(j\omega)|$ on log-log paper for the following transfer function:

$$G(s) = \dfrac{100(s + 1)}{s^2 + 100}$$

Also sketch the actual amplitude response.

8.8. (a) Using two cycle log-log paper, and a strip of paper as a tape measure, show that the distance from 0.5 to 2 is the same as that from 1 to 4, and from 2 to 8. Explain.
(b) Show that the distance from A to B is the ratio B/A.
(c) Show that folding the strip provides the square root of B/A.
(d) Using the strip and the log-scales, find $(7/3)(0.6/0.8)\sqrt{8/0.5}$.
(e) On two cycle log-log paper, draw several right triangles whose hypotenuses are lines with a slope of -2. Show that the vertical side represents a ratio that is equal to the square of the horizontal side. Show how to obtain the square root of 2.

8.9. Plot the frequency response curves using four points as given by Table 8.5.2-1 for damping ratios $0, \frac{1}{4}, \frac{1}{2}$, and 1 for the system of Sec. 7.5.1.

8.10. Similar to Problem 8.9, but for Ex. 7.5.2-1 (Phono Pickup).

8.11. Similar to Problem 8.9, but for Ex. 7.8.5-1 (Vibrometer).

8.12. Similar to Problem 8.9, but for Ex. 7.9.1-1 (Reciprocating Pump for Air Conditioner). Why is plot similar to Problem 8.11?

8.13. Repeat Problem 8.12, for force transmitted to base.

8.14. Repeat Problem 7.22 (Suspension for Motorcycle) using log-log.

8.15. Repeat Problem 7.28 (Phono Pickup) using log-log.

8.16. Repeat Problem 7.32 (Pneumatic Valve) using log-log.

8.17. Rework Problem 8.14 by multiplying the intercepts at a different frequency and compare the results to the original. Explain.

8.18. In Fig. 8.4.4-1, prove that the two asymptotes intersect at the point $(a, 1)$. Verify the location of the intercept of the exponential in Ex. 8.7.2-3.

8.19. Plot the frequency response for Problem 7.39 (Air Controlled Turbine).

8.20. Plot the frequency response for Problem 4.45 (Cascaded Hydraulic Actuator).

8.21. List four methods by which the damping ratio can be determined experimentally.

8.22. Explain from a physical point of view why all seismic devices do not pass the low frequencies, but transmit the highs.

8.23. Rework Ex. 8.7.2-1 (Low Pass Filter) using damping ratios of 0.4, 0.5, 0.6, and 0.7. Select the optimum value and defend your choice.

8.24. *Asymptotic Magnitude Curves.* For each function listed below, plot the magnitude asymptote. *Note*: parts (a) through (f) may be plotted on the same two-cycle paper without interfering with each other. The same applies to parts (g) through (n).

(a) $40s$

(b) $\dfrac{200}{s + 20}$ $10\left(\dfrac{20}{s + 2}\right)$

(c) $\dfrac{s^2 + 20s + 400}{20}$

(d) $\dfrac{s + 30}{2}$ $\dfrac{s + 30}{30}(15)$

(e) $\dfrac{2}{s^2}$

(f) $\dfrac{500}{s^2 + 15s + 100}$ $\to 5\left(\dfrac{100}{s^2 + 15s + 100}\right)$

(g) $1000/s$

(h) $20(s + 2)$ 20

(i) $\dfrac{10}{s}$

(j) $2(s^2 + 2s + 16)$ 2

(k) $\dfrac{6000}{s^2 + 20s + 400}$ 15

(l) $\dfrac{8}{s^2}$ 8

(m) $\dfrac{400}{s + 20}$ 20

(n) $\dfrac{1200}{s^2 + 10s + 100}$ 12

8.25. *Actual Magnitude Curves.* For each function listed below, plot the actual magnitude curve. *Note*: Parts (a) through (f) may be plotted on the same two-cycle paper without interfering with each other. The same applies for parts (g) through (n).

(a) $\dfrac{4000}{s^2 + 10s + 400}$

(b) $\dfrac{10}{s + 5}$

(c) $20(s + 2)$

(d) $s^2 + 10s + 25$

(e) $\dfrac{2000}{s^2 + 40s + 400}$

(f) $\dfrac{1000}{s}$

(g) $21s$

(h) $\dfrac{7}{s}$

(i) $s + 20$

(j) $\dfrac{30}{s + 4}$

(k) $\dfrac{1000}{s^2 + 8s + 100}$

(l) $30s^2$

(m) $8 \times 10^{-4}s^2$

(n) $7.41 \times 10^{-3}(s^2 + 4.5s + 2025)$

8.26. *Asymptotic Magnitude and Phase Curves on Same Graph.* For each part of Problem 8.24 plot the asymptotic magnitude and phase curves on the same graph, using the same log frequency scale (along the horizontal). On the left-hand vertical border use a log scale for the magnitude curve. On the right-hand vertical border use a linear scale for the phase curve. Arrange the magnitude and phase scales so that magnitude = 1 lines up with phase = $-180°$. *Note*: the phase curves will require three-cycle paper. Also, don't put too many parts of the problem on the same sheet, since the phase curves may interfere with each other.

8.27. *Actual Magnitude and Phase Curves on Same Graph.* For each part of Problem 8.25 plot the actual magnitude and phase curves on the same graph, using the same log frequency scale. Arrange the scales as described in Problem 8.26.

8.28. *Composite Curves (Magnitude Curves, Asymptotes Only).* For each of the following, sketch the Bode plot (asymptotic magnitude curve) for each factor, and draw the composite on five-cycle log-log graph paper:

(a) $\dfrac{600(s + 30)}{(s + 2)(s + 450)s}$

(b) $\dfrac{30(s^2 + 30s + 900)}{(s^2 + 2s + 4)(s + 450)s}$

(c) $\dfrac{10(s + 30)}{(s^2 + 32s + 60)(s + 0.15)}$

(d) $\dfrac{(s + 30)40}{(s^2 + 2s + 4)s^2}$

(e) $\dfrac{400(s + 8)}{s(s^2 + 30s + 900)}$

(f) $\dfrac{210(s + 20)}{(s + 2)(s^2 + 64)}$

(g) $\dfrac{12(s^2 + 20s + 400)}{(s^2 + 2s + 2)(s + 200)s}$

(h) $\dfrac{40(s + 10)}{(s + 2)(s + 50)s}$

(i) $\dfrac{(s + 20)400}{(s^2 + 104s + 400)s^2}$

(j) $\dfrac{220(s + 10)}{(s^2 + 400)s}$

(k) $\dfrac{2100(s^2 + 100)}{(s^2 + 4)(s^2 + 2500)}$

(l) $\dfrac{1000(s + 30)}{s(s^2 + 4)(s + 450)}$

(m) $\dfrac{932(s + 4)}{(s + 2)(s^2 + 64)}$

(n) $\dfrac{6400(s + 1)(s + 4)}{(s + 2)(s^2 + 64)(s + 16)}$

(o) $\dfrac{20000}{(s + 5)(s + 10)(s + 20)}$

(p) $\dfrac{3000}{(s + 2)(s + 5)(s + 20)}$

8.29. *Composite Curves (Actual Magnitude Curves).* For each resulting curve in Problem 8.28, sketch the actual curve for each factor and the actual composite curve.

8.30. *Composite Curves (Phase Curves, Asymptotes Only).* For each transfer function in Problem 8.28, sketch the phase curve for each factor, and plot the composite on five-cycle log-log paper.

8.31. *Composite Curves (Actual Phase Curves)*. For each resulting curve in Problem 8.30, sketch the actual curve for each factor and for each composite curve.

8.32. *Asymptote Composite Curves on Same Paper*. Using the results of Problems 8.28 and 8.30, plot the composites on the same paper, lining up magnitude = 1 with phase = $-180°$.

8.33. *Actual Composite Curves on Same Paper*. Similar to Problem 8.32, but applied to Problems 8.29 and 8.31.

8.34. Design the circuit shown as a broadband device, where the bandwidth is equal to 10^6 rad/s. (See Fig. P-8.34.)

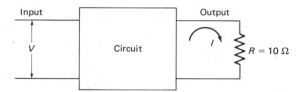

Figure P-8.34

8.35. Design the device in Problem 8.34 as a low pass filter, where the attenuation is greater than $\sqrt{2}$ at frequencies above 10^4 rad/s.

8.36. *Design of Stepper Motor*. Complete the design that was started in Ex. 8.5.2-2. Include the length and diameter of the wire for the coil, the wire diameter and radius of the mechanical spring, and the dimensions of the mass.

8.37. *Design of High Pass Filter*. Refer to Ex. 8.7.2-3 (Torsiograph). Design the system so that the lowest frequency passed (using limits equal to $\pm \sqrt{2}$) is 10 rad/s, where $J = 0.01$ N $\cdot$ m $\cdot$ s.

8.38. *Design and Build Electrostatic Speaker*. Design an electrostatic speaker using household materials, such as aluminum foil and plastic wrap. Stack the various sheets in the form of a pyramid so that there is about 1 cm (about $\frac{3}{8}$ in.) border between any two adjacent edges (to prevent arcing). To drive the speaker, use an output transformer in reverse to step up the voltage. (If possible use a half-wave rectifier in series with a resistor of sufficient size to eliminate "frequency doubling.") Estimate the frequency range and test the device to prove your estimate is reasonable.

8.39. Plot the frequency response for Ex. 8.6.6-1 (Frequency response for Unfactored System), $G_a(s)$.

8.40. Plot the frequency response for the following:
(a) $G(s) = 10s^3 G_a(s)$
(b) $G(s) = 10 G_a(s)/s$
(c) $G(s) = 10s^2(s + 10)G_a(s)$
(d) $G(s) = 10(s + 10)G_a(s)/s$
where the function $G_a(s)$ is given in Ex. 8.6.6-1 and where $G_a(j\omega)$ was plotted in Problem 8.39.

Chapter 9

Design Oriented Approach to Arbitrary Inputs

This chapter collects and presents assorted inputs that do not fit into any of the categories covered up to this point. The chapter begins with an introduction to periodic nonsinusoidal motion, and it sets the stage for the formal treatment of Fourier analysis. The formal approach is applied to a series circuit. In this system, the problem is to reproduce a triangular wave shape. Representing this wave shape by its Fourier series, one can readily comprehend how the system must perform. A second example involves the design of a resonating air pump whose input is pulsating. Representing the input by its Fourier series, the system is completely designed.

Another type of nonsinusoidal input is examined, the random distribution of frequencies. The example chosen is the design of a device that will filter aircraft vibration and yet pass the desirable signal.

9.1 PERIODIC NONSINUSOIDAL MOTION

When two or more sinusoids of different frequencies appear at the same time, the result, obtained by the algebraic sum, is periodic but not sinusoidal. Of particular interest is the case in which the sinusoids have frequencies that occur in integer multiples. Given a sinusoid whose frequency is ω, then a sinusoid with a frequency n times this is referred to as the nth *harmonic*.

9.1.1 Fourier Analysis

A periodic function can be expressed as a sine-cosine series. This is particularly useful in system analysis when the input to the system is periodic but is not sinusoidal. The sine-cosine series (Fourier series) can be constructed which

TABLE 9.1.1-1   NONSINUSOIDAL PERIODIC FUNCTIONS

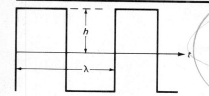

1. Square wave

$$\frac{4h}{\pi} \sum_{n=1}^{\infty} \frac{1}{n} \sin n\omega t \qquad n = 1, 3, 5, \ldots$$

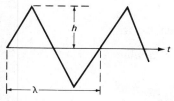

2. Saw tooth

$$\frac{2h}{\pi} \sum_{n=1}^{\infty} \frac{(-1)^{n+1}}{n} \sin n\omega t$$

$$n = 1, 2, 3, \ldots$$

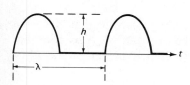

3. Triangular wave

$$\frac{8h}{\pi^2} \sum_{n=1}^{\infty} \frac{(-1)^{(N-1)/2}}{n^2} \sin n\omega t$$

$$n = 1, 3, 5, \ldots$$

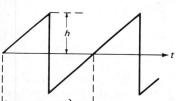

4. Half-wave rectification

$$\frac{h}{\pi} + \frac{h}{2} \sin \omega t - \frac{2h}{\pi} \sum_{n=1}^{\infty} \frac{1}{n^2 - 1} \cos n\omega t$$

$$n = 2, 4, 6, \ldots$$

5. Full wave rectification

$$\frac{2h}{\pi} - \frac{4h}{\pi} \sum_{n=1}^{\infty} \frac{1}{n^2 - 1} \cos n\omega t$$

$$n = 2, 4, 6, \ldots$$

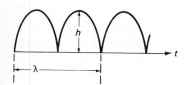

represents the nonsinusoidal periodic function as the sum of sinusoids. These can be used one at a time, and the results summed (assuming linear superposition).

The means by which one can construct the Fourier series is derived* in the standard texts on differential equations, and is given below

$$f(t) = C_0 + \sum_{n=1}^{\infty} A_n \cos n\omega t + \sum_{n=1}^{\infty} B_n \sin n\omega t \qquad [9.1.1\text{-}1]$$

*See reference [7] p. 556–558.

where

$$A_n = \frac{1}{\pi} \int_0^{2\pi} f(t)\cos n\omega t \, dt \qquad [9.1.1\text{-}2]$$

$$B_n = \frac{1}{\pi} \int_0^{2\pi} f(t)\sin n\omega t \, dt \qquad [9.1.1\text{-}3]$$

$$C_0 = \frac{1}{2\pi} \int_0^{2\pi} f(t) \, dt \qquad [9.1.1\text{-}4]$$

$$f(t) = \text{periodic but nonsinusoidal function} \qquad [9.1.1\text{-}5]$$

Some useful periodic nonsinusoidal functions are tabulated in Table 9.1.1-1, where

$$\lambda = \text{period of function} = \frac{2\pi}{\omega} \qquad [9.1.1\text{-}6]$$

$$h = \text{amplitude of function} \qquad [9.1.1\text{-}7]$$

■ EXAMPLE 9.1.1-1 Series For Triangular Wave

How many terms of the Fourier series for a triangular wave are required before the amplitude has dropped below 1% of the amplitude of the first sinusoidal term?

Solution. The series is given in Table 9.1.1-1, entry 3. Note that since the amplitude is an absolute, the (-1) term may be disregarded. Also note that the amplitude does not involve the $\sin n\omega t$ term. Hence, the amplitude A_n of any term is given by

$$A_n = \frac{8h}{\pi^2 n^2} \qquad (1)$$

For the first term of the series, $n = 1$, but for the rest of the terms, the term number is *not* equal to n.

To satisfy the requirement given in this example,

$$\frac{A_n}{A_1} \le 1\% = 0.01 \qquad (2)$$

Substitute numerical values and eq. (1)

$$\left(\frac{8h}{\pi^2 n^2}\right)\left(\frac{\pi^2 1^2}{8h}\right) \le 0.01 \qquad (3)$$

Simplify, invert and note the inequality sign

$$n^2 \ge 100 \qquad (4)$$

or

$$n \ge 10 \qquad (5)$$

Since n is an odd number, then let

$$n = 11 \qquad (6)$$

The quantity n has been found. Now determine which term this corresponds to. Note that n is an odd numbered series beginning with unity. Then, the mth term is

$$m = \frac{n - 1}{2} \qquad (7)$$

Substitute eq. (6)

$$m = \frac{11 - 1}{2} = 5 \text{ terms} \qquad (8) \blacksquare$$

9.2 DESIGN OF BROADBAND DEVICE WITH FOURIER SERIES INPUT

9.2.1 Need for Broadband Device

To reproduce the function that is expressed as a Fourier series, it is necessary for the system to transmit all frequencies equally. This requires that the system be a broadband device (see Section 8.7.3). If the device is not a broadband device, then those frequencies out of its range will not be reproduced properly. Consequently, the Fourier series will not be reproduced properly, and the resulting composite will not be exactly like the original input.

■ **EXAMPLE 9.2.1-1** Triangular Input to Series Circuit
A periodic function in the form of a triangular wave is applied to the series circuit shown in Fig. 9.2.1-1. Discuss the behavior of the system if the output is taken across the R-C series network. Design the system so that terms up to the fifth harmonic are neither amplified nor attenuated by a factor larger than $\sqrt{2}$. The following parameters are given:

$\omega = 10^4$ Hz   (fundamental frequency of triangular wave)

$C = 0.028 \; \mu F = 2.8 \times 10^{-8}$ F

$L = 1.5 \; mH = 1.4 \times 10^{-3}$ H

R_a and R_b to be determined

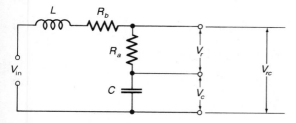

Figure 9.2.1-1   Series circuit.

Solution.   Sum the voltages around the loop, according to Kirchhoff's law.

$$V_{in}(s) - (Ls^2 + Rs + 1/C)Q(s) = 0 \tag{1}$$

where

$$R = R_a + R_b \tag{1a}$$

$$Q(s) = \frac{V_{in}(s)}{Ls^2 + Rs + 1/C} \tag{2}$$

Normalize

$$Q(s) = \frac{V_{in}/L}{s^2 + 2\zeta\omega_n s + \omega_n^2} \tag{3}$$

where

$$\omega_n^2 = \frac{1}{LC} \tag{4}$$

or

$$\omega_n = \sqrt{\frac{1}{LC}} = \sqrt{\frac{1}{(1.4 \times 10^{-3})(2.8 \times 10^{-8})}} = 160 \times 10^3 \tag{4a}$$

$$\zeta = \frac{R}{2L\omega_n} \tag{5}$$

or

$$R = 2\zeta L\omega_n \tag{5a}$$

The input V_{in} to this circuit is in the form of a Fourier series as given by item 3 in Table 9.1.1-1.

$$V_{in} = \frac{8h}{\pi^2}\left[\sin \omega t - \frac{1}{9}\sin 3\omega t + \frac{1}{25}\sin 5\omega t - \cdots\right] \tag{6}$$

where

$$\omega = 10^4 \text{ Hz} = 62.8 \times 10^3 \text{ rad/s*} \tag{7}$$

$$3\omega = 188 \times 10^3 \text{ rad/s} \tag{8}$$

$$5\omega = 312 \times 10^3 \text{ rad/s} \tag{9}$$

The voltage across the *R-C* series network is given by

$$V_{rc} = (R_a s + 1/C)Q(s) \tag{10}$$

---

*Fundamental frequency ω was given.

Apply eq. (3)

$$\frac{V_{rc}}{V_{in}}(s) = \frac{R_a}{L} \frac{s + \omega_a}{s^2 + 2\zeta\omega_n s + \omega_n^2} \tag{11}$$

where

$$\omega_a = \frac{1}{R_a C} \tag{12}$$

ω_n and ζ were defined in eqs. (4a) and (5). The denominator of eq. (11) is a quadratic whose corner frequency is ω_n, but whose numerator is a first order factor with a corner frequency ω_a. See Fig. 9.2.1-2. It will be shown that the location of the corner frequency ω_a is critical. If this frequency is low (ω_a') the frequency response curve rises above the upper limit resulting in a bandwidth ω_b', which is considerably below that (ω_{b_3}) of the reciprocal quadratic. If the corner frequency ω_a'' is somewhat higher than ω_a', but lower than ω_n, this results in a bandwidth ω_b'' that is *greater* than ω_{b_3}. The reason for this is twofold: the curve first climbed before descending (leaving some clearance before crossing the lower limit) and the frequency response curve for the R-C series system has a slope of -1 for frequencies above ω_n (while the slope for the reciprocal quadratic is -2). It is clear from Fig. 9.2.1-2 that the optimum location for the corner frequency ω_a would cause the frequency curve to rise and just touch the upper limit. Hence, the optimum location becomes

$$\frac{\omega_n}{\omega_a} = \sqrt{2} \tag{13}$$

and

$$\frac{\omega_b}{\omega_n} = 2 \tag{14}$$

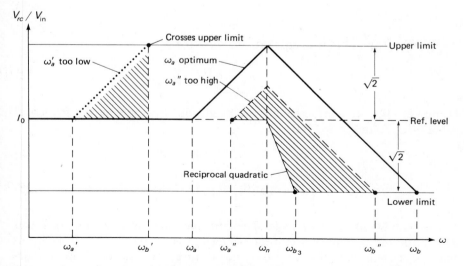

Figure 9.2.1-2   Output across R-C series network.

To make sure that the actual curve does not rise above nor fall below the limits [in Fig. 9.2.1-2], let

$$0.5 \leq \zeta \leq 0.707 \tag{15}$$

Apply these limits to eq. (5a)

$$224\Omega < R \leq 317\,\Omega \tag{16}$$

Equations (13) and (12) provide a means to determine resistor R_a

$$R_a = \frac{1}{\omega_a C} = \frac{\sqrt{2}}{\omega_n C} \tag{17}$$

Apply numerical values

$$R_a = \frac{\sqrt{2}}{160 \times 10^3 \times 2.8 \times 10^{-8}}$$

$$\boxed{R_a = 317\,\Omega} \tag{18}$$

Obviously, from eq. (1a) we conclude

$$R \geq R_a \tag{19}$$

This requires that we use the upper limit for R in eq. (16). Then,

$$\boxed{R = 317\,\Omega} \tag{20}$$

and this, in turn, requires that

$$\boxed{\zeta = 0.707} \tag{21}$$

Apply eqs. (18) and (20) to eq. (1a) and solve for R_b

$$\boxed{R_b = 0} \tag{22}$$

Using eq. (14), we can determine the bandwidth by applying the numerical value of ω_n found in eq. (4a).

$$\omega_b = 2\omega_n = 2 \times 160 \times 10^3$$

$$\boxed{\omega_b = 320 \times 10^3 \text{ rad/s}} \tag{23}$$

Is this bandwidth large enough? The fifth harmonic is given by eq. (9) as 312×10^3 rad/s. This is within the bandwidth given by eq. (23). This means that the system will effectively transmit all three frequencies. Thus we conclude that the R-C series network is an acceptable design.   ∎

9.3 DESIGN OF RESONATOR WITH FOURIER SERIES INPUT

There are instances when it is not desirable to transmit all the frequencies of the Fourier series. In certain cases, the input is unintentionally a nonsinusoidal periodic function. However, if one of the frequencies of the Fourier series offers some operational advantage, then it will pay to design a resonator that will resonate at that frequency.

■ **EXAMPLE 9.3.1-1** Design of Resonating Vibrator Pump
An inexpensive pump for moving small quantities of air can be powered by a resonating vibrator. The vibrator employs an electromagnetic actuator mounted on a curved cantilever spring. Two flapper valves alternately open and close ensuring unidirectional flow of air. See Fig. 9.3.1-1(a), where the known parameters are shown. Complete the design using cost-effective methods, and where the system operates from 60 Hz house current.

Solution.

System Modeling

This system does not have a lumped mass, but instead the mass is distributed. The spring, in turn, has appreciable mass. Following the approach in Chapter 2, this system can be closely approximated by lumped parameters. Since the excitation force $F(t)$ is located along the axis of the magnetic coil, all parameters will be "reflected" to this point. Consequently, the equivalent lumped mass M_{eq} will be located at the center of the armature. Now that the lumped mass has been identified, we can follow the methods in Chapter 3 to formulate the free body diagram.

The mass M_{eq} is connected to the ground through the cantilever spring K_c. The restriction of the air through the valves and pipe line provides damping D_a as computed in Chapter 2. The air damping force may be assumed to act against the moving side of the bellows, which is *not* located at the center of the armature (where all the other system parameters are located). It will be necessary to "reflect" this damping term using the methods in Section 2.1.5(b). The reflected damper D_{eq} is

$$D_{eq} = D_a \left(\frac{l_1 + l_2}{l_1} \right)^2 \tag{1}$$

The input is a forcing function $F(t)$ applied to the mass M_{eq}. Resisting this motion, there are an equivalent spring K_{eq} and an equivalent damper D_{eq}, each of which connect the mass M_{eq} to the ground. The free body diagram is shown in Fig. 9.3.1-1(b), for which (after normalizing) we have

$$\frac{X}{F/K}(s) = \frac{\omega_n^2}{S^2 + 2aS + \omega_n^2} \tag{1a}$$

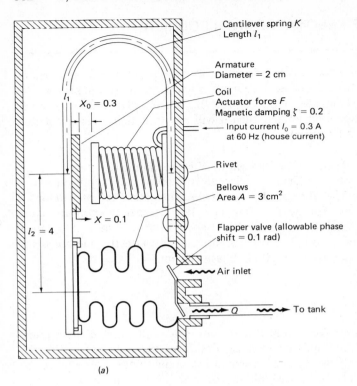

Cantilever spring K
Length l_1

Armature
Diameter = 2 cm

Coil
Actuator force F
Magnetic damping $\zeta = 0.2$

Input current $I_0 = 0.3$ A
at 60 Hz (house current)

Rivet

Bellows
Area $A = 3$ cm^2

Flapper valve (allowable phase
shift = 0.1 rad)

Air inlet

Q — To tank

$X_0 = 0.3$

$X = 0.1$

$l_2 = 4$

l_1

(a)

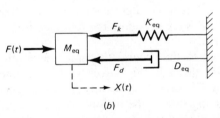

(b)

Figure 9.3.1-1   Resonating vibrator pump. (a) Actual system. (b) Free body diagram.

The electric current input to the coil is sinusoidal. The armature, which is fastened to the cantilever spring, is attracted to the coil no matter in which direction the current flows. Hence, the device *rectifies* the sinusoidal input, resulting in a fully rectified forcing function whose wave shape is described by entry 5 in Table 9.1.1-1. The Fourier series is given as

$$F(t) = \frac{2h}{\pi} - \frac{4h}{\pi}\left[\frac{1}{3}\cos 2\omega t + \frac{1}{15}\cos 4\omega t + \cdots\right] \qquad (2)$$

where

$$\omega = 60 \text{ Hz} = 60 \times 2\pi = 377 \text{ rad/s}$$

The constant term $2h/\pi$ will merely affect the central position of the arma-

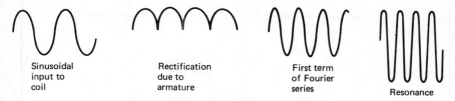

Sinusoidal
input to
coil

Rectification
due to
armature

First term
of Fourier
series

Resonance

Figure 9.3.1-2   Process flow for rectified vibrator.

ture. The device is to be designed to resonate at a frequency equal to the second harmonic 2ω. All harmonics above this are to be attenuated. The resonant frequency is

$$\omega_r = 2\omega = 2 \times 377 = 754 \text{ rad/s} \tag{2a}$$

The process flow diagram for this system is shown in Fig. 9.3.1-2. The input to the system is a sinusoid of frequency $\omega = 377$ into the coil. The armature rectifies this. Only the first term of the Fourier series (the second harmonic $2\omega = 754$) will be used by the device. The system resonates at this frequency, whose amplitude is magnified. Note that the rectification process has made this system behave as a "frequency doubler."

For maximum resonance, it is desirable to have very little damping. However, when the damping ratio falls below 0.2, there is the problem of system stability. Hence, selecting the lowest stable value, let

$$\zeta = \boxed{0.2}$$

Refer to Table 8.5.2-1, for $\zeta = 0.2$,

$$R_r = 2.5 \tag{2b}$$

$$\gamma_r = 0.959$$

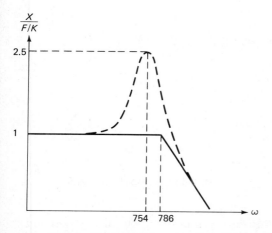

Figure 9.3.1-3   Frequency response for vibrator air pump.

Since the resonant frequency is $\omega_r = 754$, then the undamped natural frequency ω_n is

$$\omega_n = \frac{\omega_r}{\gamma_r} = \frac{754}{0.959} = \boxed{786 \text{ rad/s}} \tag{2c}$$

The frequency response curve is sketched in Fig. 9.3.1-3.

Design of Cantilever Spring

Refer to Appendix 2, Table 2.1.4-1 and 2.2.3-3. For a cantilever beam,

$$M_{eq} = 0.23M \approx \tfrac{1}{4}M$$

$$M = \rho l_1 bh$$

$$K_{eq} = \frac{3EI}{(l_1)^3} \tag{2d}$$

$$I = \frac{bh^3}{12}$$

Then, neglecting all other springs and masses,

$$\omega_n = \sqrt{\frac{K_{eq}}{M_{eq}}} = \sqrt{\frac{3Ebh^3/12(l_1)^3}{\rho l_1 bh/4}} = \frac{h}{(l_1)^2}\sqrt{\frac{E}{\rho}}$$

or

$$h = (l_1)^2 \omega_n \sqrt{\frac{\rho}{E}} \tag{3}$$

where

$$\omega_n = 786 \text{ rad/s}$$

$$E = 2 \times 10^{12} \text{ dyn/cm}^2 \text{ (about } 30 \times 10^6 \text{ psi)}$$

$$\rho = 8 \text{ g/cm}^3 \tag{3a}$$

Let

$$l_1 = 15 \text{ cm (about 6 in.)}$$

then

$$h = 15^2 \times 786 \sqrt{\frac{8}{(2 \times 10^{12})}} = 0.354 \text{ cm (about 0.14 in.)}$$

This is not a stock size. Since this device is based upon all low cost items, select the nearest stock dimension, and solve for the length l_1 that can be cut to any length. See stock below:

STOCK THICKNESS FOR SPRING STEEL BARS

cm	0.159	0.238	0.318	0.475	0.635
in.	1/16	3/32	1/8	3/16	1/4

Select $h = \boxed{0.318 \text{ cm } (1/8 \text{ in.})}$ and solve for l_1

$$l_1 = \boxed{14.2 \text{ cm } \left(\text{about } 5\tfrac{1}{2} \text{ in.}\right)} \tag{3b}$$

Note that the width b dropped out of the above computations.

Spring steel bars are made in widths that are integer multiplies of 1 cm as well as multiples of $\frac{1}{2}$ in. Then select the width b equal to the diameter of the armature, which is 2 cm. The actual values of the spring and mass can be determined:

$$K_{eq} = \frac{3EI}{(l_1)^3} = \frac{3Ebh^3}{12(l_1)^3} \tag{4}$$

$$= \frac{3 \times 2 \times 10^{12} \times 2 \times (0.318)^3}{12(14.2)^3}$$

$$= 11.2 \times 10^6 \text{ dyn/cm (about 62 lb/in.)} = \boxed{11.2 \times 10^3 \text{ N/m}}$$

$$M_{eq} = \frac{\rho l_1 bh}{4} = \frac{8 \times 14.2 \times 2 \times 0.318}{4} \tag{5}$$

$$= \boxed{18.1 \text{ g (about 0.04 lb)}}$$

Coil Design

Refer to the frequency response curve to determine the force required at resonance. At resonance we have

$$\frac{X}{F/K} = 2.5$$

or

$$F = \frac{KX}{2.5} = \frac{11.2 \times 10^6 \times 0.1}{2.5}$$

$$= 4.48 \times 10^5 \text{ dyn (about 1 lb)} = 4.48 \text{ N} \tag{6}$$

Area at the air gap is equal to the area of the armature

$$A_a = \frac{\pi b^2}{4} = \frac{\pi 2^2}{4} = 3.14 \text{ cm}^2 = 3.14 \times 10^{-4} \text{ m}^2 \tag{6a}$$

Refer to Table 2.4.1-1,* Item (4). The force exerted by a magnetic system whose major magnetic resistance is the air gap is

$$F = \frac{A_a N^2 I_0^2 \mu_0}{2 X_0^2}$$

or

$$N = \frac{X_0}{I_0} \sqrt{\frac{2F}{A_a \mu_0}} = \frac{0.3 \times 10^{-2}}{0.3} \sqrt{\frac{2 \times 4.48}{3.14 \times 10^{-4} \times 1.257 \times 10^{-6}}} \quad (7)$$

$$= \boxed{1510 \text{ turns}}$$

To provide the value of damping ratio selected, note that

$$\zeta = \frac{D_{eq}}{2 M_{eq} \omega_n}$$

or

$$D_{eq} = 2 \zeta \omega_n M_{eq} \quad (8)$$

$$= 2 \times 0.2 \times 786 \times 18.1 = 5690 \text{ dyn} \cdot \text{s/cm} = \boxed{5.69 \text{ N} \cdot \text{s/m}}$$

Bandwidth at Resonance

The band limits are given in Table 8.7.4-1. For $\zeta = 0.2$

$$\gamma_{b_1} = 0.727 \qquad \gamma_{b_2} = 1.14 \quad (9)$$

or

$$\omega_{b_1} = 0.727 \omega_n = 0.727 \times 786 = 571 \text{ rad/s}$$
$$\omega_{b_2} = 1.14 \omega_n = 1.14 \times 786 = 896 \text{ rad/s} \quad (10)$$

To prevent the output from dropping below a level equal to $\sqrt{\frac{1}{2}}$ times the peak level, the input frequency must lie between the band limits 565 to 886. For typical power sources, the frequency regulation is of the order of 10%. Then the input frequency lies in the range

$$\omega = \omega_r \pm 10\% = 754 \pm 75.4 = 679 \text{ to } 829 \quad (11)$$

which lies within the band limits, and is therefore suitable.

Design of Flapper Valve

Each of the flapper valves is a cantilever whose natural frequency must be selected high enough to open and close at the correct time. The allowable timing error is given in terms of allowable phase shift ψ, there $\psi = 0.1$ rad

---

*Appendix 2.

(about 5.7°). A cantilever has mass and a spring rate, and for typical materials it also has structural damping. Thus, the flapper valve is a reciprocal quadratic system whose frequency response curve is shown on Fig. 8.4.7-1. The phase angle ψ is given by eq. 8.4.7-3, repeated below.

$$\psi = -\tan^{-1} \frac{2\zeta\gamma}{1-\gamma^2}$$

or

$$2\zeta\gamma = (1-\gamma^2)\tan(-\psi) \tag{12}$$

where

$$-\psi = \sin^{-1}10\% \approx 0.1 \approx \tan\psi$$

$$\zeta = \text{structural damping in valve material} = \boxed{0.2}$$

$$\gamma = \frac{\omega}{\omega_n'}$$

$$\omega = \text{input frequency} = 754 \text{ rad/s}$$

$$\omega_n' = \text{natural frequency of flapper valve}$$

Apply numerical values

$$2 \times 0.2\gamma = (1-\gamma^2) \times 0.1$$

or

$$\gamma^2 + 4\gamma - 1 = 0$$

Rejecting the negative root, then

$$\gamma = 0.237 \tag{13}$$

Then

$$\omega_n' = \frac{\omega}{\gamma} = \frac{754}{0.237} = \boxed{3190 \text{ rad/s}} \tag{14}$$

The dimensions of the valve will be determined by use of eq. (3). The valve will be fabricated of high endurance sheet plastic.

$$\rho' = 3 \text{ g/cm}^3$$

$$E' = 10^{10} \text{ dyn/cm}^2$$

Let

$$l' = 1 \text{ cm} \tag{14a}$$

Then

$$h' = 1^2 \times 3190\sqrt{\frac{3}{10^{10}}} = 5.53 \times 10^{-2} \text{ cm}$$

Referring to a manufacturer's stock catalog, as simulated below,

STOCK HIGH ENDURANCE PLASTIC SHEET THICKNESS

cm	0.03	0.04	0.05	0.07	0.1
in.	0.0118	0.0157	0.0197	0.0275	0.0394

select $h' = \boxed{0.05}$ and solve for l'

$$l' = \boxed{1.01 \text{ cm (abouto 0.4 in.)}} \tag{15}$$

Pump Capacity

The pump capacity Q is given by the volumetric displacement per unit time. This quantity is reduced slightly by the timing error (phase shift) in the flapper values. The "lost motion" C_1 is

$$C_1 = \gamma \sin \psi = 0.237 \sin 0.1 = 0.024 = 2.4\% \tag{16}$$

Then

$$Q = (100\% - C_1) N 2 X \frac{l_1 + l_2}{l_1} A \tag{17}$$

where

N = number of strokes per unit time = 120 strokes/s

X = amplitude* of vibration of armature = 0.1 cm

l_1 = length of cantilever from coil to armature = 14.2 cm $\qquad$ (18)

l_2 = length from armature to bellows = 4 cm

A = effective pumping area = 3 cm^2

Then

$$Q = (.976)120 \times 2 \times 0.1 \times \frac{(14.2 + 4)}{14.2} \times 3 \tag{19}$$

$$= \boxed{90 \text{ cm}^3/\text{s (about 300 in.}^3/\text{min)}}$$

which is an amazing volume for such a small, simple device. ■

---

*Note that the stroke is twice the amplitude.

9.4 DESIGN OF LOW PASS FILTER OF RANDOM NOISE

A low pass filter passes only the low frequencies while it attenuates the highs. The high frequencies are usually a distribution of random frequencies with random amplitudes. The design of the proper filter could be complicated. For this reason, the following heuristic treatment of random noise is presented.

9.4.1 Heuristic Description of Random Noise

Section 9.1 was concerned with the case of nonsinusoidal periodic functions. In that case, the amplitude and frequency distribution was readily determined, and the periodic function could be represented as a series of discrete known frequencies each with known amplitudes. Random noise, on the other hand, is a composite of random frequencies each with random amplitudes. In this case, a Fourier series would not readily be determined. Instead, we may describe the situation in terms of a probable distribution of frequencies and amplitudes.

A common distribution is Gaussian distribution which appears as a bell-shaped curve on linear scales, and which may be approximated as a double broken line on log-log scales. See Fig. 9.4.1-1. In order to interpret the above plot, compare the random input to a choral group. Each singer may have either a small or large part in the overall effect. A large number of singers may sing softly so that their overall participation represents a certain sound level. On the other hand, occasionally, one singer has a solo which may be electronically amplified to maintain the sound level. Note that no one has all the volume all the time. Similarly, the various frequencies in the random distribution may occasionally have all the noise level for a very short time, or they may maintain a very low level all the time. Also this distribution need not be static, but like the choral group, may change in such a way to maintain a constant overall or average level.

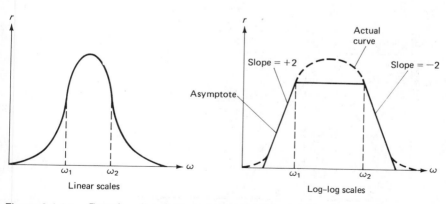

Figure 9.4.4-1 Gaussian distribution of random noise input.

9.4.2 Resulting Random Output

Given a random input to the system, how is the distribution altered by the system frequency response? Referring to the analogy to a choral group, say the momentary sound distribution is an equal mixture of all voice parts. Suppose the microphone reproduces only the high frequencies. The resulting mixture of random voices will favor the highs, but will still be a random mixture. However, the frequency distribution curve will no longer resemble the original one, but it will be altered by the system frequency response curve. Thus, the new random frequency distribution curve can be obtained by multiplying this input by the frequency response curve for the system. For illustrative purposes, let us use a typical system in the form of a reciprocal quadratic. The input, frequency response curve, and the resulting output are shown in Fig. 9.4.2-1.

9.4.3 Response at Resonant Frequency

A problem of particular concern is the response of the system at resonance. With a random input, within the random distribution of frequencies, the resonant frequency is likely to be present. If so, this input would be magnified as the system approaches steady state. The question then becomes one of how much opportunity does the system have to reach steady state with a random input.

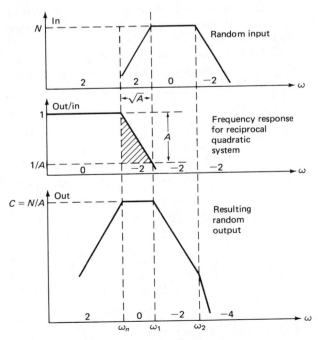

Figure 9.4.2-1   Random input to reciprocal quadratic.

A random distribution may be thought of either as a very brief burst at the full input level at the resonant frequency, or it may be considered as a continuous sinusoidal component with a very small amplitude at the resonant frequency. Either way, the system is *not* bombarded at its resonant frequency continuously at the *full* input level. Thus, the situation may be thought of in terms of an equivalent *continuous* input at a *reduced* level. Assuming a random but uniform distribution of frequencies, the reduction factor f_r may be determined by the ratio of the bandwidth of the system to the bandwidth of the noise input. Then,

$$f_r = \frac{\omega_{b_2} - \omega_{b_1}}{\omega_2 - \omega_1} \qquad [9.4.3\text{-}1]$$

Equation 9.4.3-1 reflects the probable opportunity of any frequency (including the resonant frequency) to pass through the "window" whose width is the system bandwidth. The smaller this window, and the broader the noise spectrum, the smaller will be this factor. The equivalent amplitude of any frequency is then given by the product of the reduction factor f_r by the full input level N. The output or response to this reduced input is given by the product of the input by the resonant peak R_r. Then the output is

$$\text{out} = \frac{\omega_{b_2} - \omega_{b_1}}{\omega_2 - \omega_1} N R_r \qquad [9.4.3\text{-}2]$$

The system bandwidth and the resonant peak are tabulated in Table 8.7.4-1 for various damping ratios. Note that as ζ *decreases*, the resonant peak R_r *increases*, but the bandwidth $\omega_{b_2} - \omega_{b_1}$ decreases. In the limit, when $\zeta \to 0$, then

$$\lim_{\zeta \to 0} \left(\omega_{b_2} - \omega_{b_1} \right) = 2\zeta\omega_n$$

and

$$R_r = \frac{1}{2\zeta} \qquad [9.4.3\text{-}3]$$

then

$$\lim_{\zeta \to 0} \text{out} = \frac{2\zeta\omega_n}{\omega_2 - \omega_1} \frac{N}{2\zeta} = \frac{\omega_n}{\omega_2 - \omega_1} N \qquad [9.4.3\text{-}4]$$

Thus, the resonant response to random input for near zero damping approaches a limit, as opposed to the infinite response of the same system to a nonrandom input at the resonant frequency.

9.4.4 Design Criteria

Given the random input noise level N and the required random output level C, then the attenuation A is

$$A = \frac{N}{C} \qquad [9.4.4\text{-}1]$$

For the log-log triangle shown shaded in Fig. 9.4.2-1 the vertical side represents the attenuation (the reduction in amplitude on the system frequency response curve due to the drop in the curve beyond the corner frequency ω_n) and the horizontal side of the triangle represents the corresponding change in frequency. Since the slope of the hypotenuse of the triangle is equal to -2 then a *reduction* in amplitude (attenuation) of A corresponds to an *increase* in frequency of $\sqrt{A}$. From the triangle we have,

$$\frac{\omega_1}{\omega_n} = \sqrt{A} \qquad\qquad [9.4.4\text{-}2]$$

Applying eq. 9.4.4-1

$$\frac{\omega_1}{\omega_n} = \sqrt{\frac{N}{C}} \qquad\qquad [9.4.4\text{-}3]$$

or

$$\omega_n = \omega_1 \sqrt{\frac{C}{N}} \qquad\qquad [9.4.4\text{-}4]$$

Thus, to design a low pass filter with random input noise, the necessary natural frequency is given by eq. 9.4.4-4. Once this frequency is so determined, the system is designed so that this is its undamped natural frequency.

■ **EXAMPLE 9.4.4-1** Design of Low Pass Filter

An aircraft instrument must work accurately in spite of aircraft vibration. The objective is to design a low pass filter that will pass the low frequency signals and yet filter (diminish) the objectionable vibration. Design the filter to meet the following specifications and also determine the bandwidth (highest frequency signal that the filter will pass).

(a) Aircraft vibration shall be reduced to 1.81% of its original value. ($CN = 0.0181$.)

(b) Vibration is random noise with Gaussian distribution from 600 to 6000 rad/s.

(c) Pass the signals with error no greater than 4%.

(d) Instrument is a spring-mass system with damping, where $J = 11.7$ g $\cdot$ cm^2.

Solution. The system will be designed as a low pass filter where the corner frequency is chosen to pass the signal, and then drop as sharply as possible. The natural frequency that will result in the specified attenuation of aircraft noise is given by eq. 9.4.4-4, which is repeated below.

$$\omega_n = \omega_1 \sqrt{\frac{C}{N}} \qquad\qquad (1)$$

$$= 600\sqrt{0.0181} = \boxed{80.7 \text{ rad/s}} \qquad\qquad (2)$$

For the specific problem at hand, to meet the accuracy specification, the curve cannot rise above the upper limit of 1.04. This implies that the peak at resonance cannot rise above 4% of the reference level.

In Table 8.5.2-1, note if the damping ratio is equal to or greater than 0.6, the resonant peak will not exceed the specified value of 4%. However, when the damping ratio exceeds 0.7, the table indicates that the curve will rapidly drop below the lower limit of 0.96. Hence, the optimum value of damping lies between 0.6 and 0.7 (and probably lies closer to 0.6). Thus, let

$$\boxed{\zeta = 0.6} \tag{3}$$

Using Table 8.5.2-1, the resonant frequency is 42.4 rad/s. The frequency response curve is shown plotted in Fig. 9.4.4-1 The frequency response curve indicates that the out/in ratio is unity for frequencies from 0 to about 20 rad/s. Above 20 rad/s, the ratio differs only slightly from unity with the maximum deviation being 1.04 at the resonant frequency 42.4 rad/s. This represents only a 4% departure from unity. At a frequency of 60 rad/s, the ratio is 0.96. Beyond this frequency, the ratio drops off rapidly. Thus, for the signal frequency range of 0 to 60 rad/s, the maximum error is 4%, which meets the specifications. Hence, the bandwidth is given by

$$\boxed{\omega_b = 60 \ \text{rad/s}} \tag{4}$$

The frequency response curve also indicates how well the device acts as a low pass filter to aircraft vibration. Aircraft vibration, at a frequency of 600 rad/s, is greatly attenuated. The out/in amplitude ratio is 0.0181 at this frequency. Thus, only 1.81% of the aircraft vibration will affect the reading of the instrument.

In order to determine the spring K, recall that

$$\omega_n = \sqrt{\frac{K}{J}} \tag{5}$$

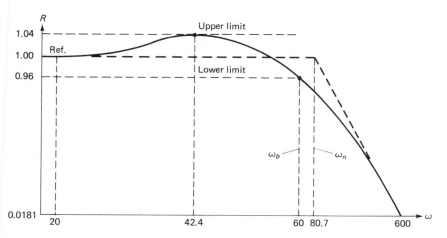

Figure 9.4.4-1 Frequency response of filter.

or

$$K = J\omega_n^2$$

$$= 11.7(80.7)^2 = \boxed{7.62 \times 10^4 \, \text{dyn} \cdot \text{cm/rad}} \tag{6}$$

In order to determine the damping constant D, recall that

$$D = 2J\zeta\omega_n = 2 \times 10.0 \times 0.6 \times 80.7 = \boxed{968 \, \text{dyn} \cdot \text{s} \cdot \text{cm}} \tag{7}$$

∎

SUGGESTED REFERENCE READING FOR CHAPTER 9

Fourier series [1], p. 148–161; [4], p. 56–58; [5], p. 333; [8], p. 273–276; [17], p. 529–534; [44], p. 90, 91.
Random inputs [4], p. 79; [5], chap. 10; [17], p. 267.
Impact and shock [4], p. 79; [5], p. 105–109; [44], p. 121–125, 136–139.

PROBLEMS FOR CHAPTER 9

9.1. How many terms of the respective Fourier series are required before the amplitude drops below 1% of the first sinusoidal term for each of the following:
   (a) Square wave.
   (b) Sawtooth wave.
   (c) Half-wave rectification.
   (d) Full wave rectification.

9.2. A system has a frequency response curve in the form of a reciprocal quadratic. How many terms of the respective Fourier series are required before the output amplitude drops below 10^{-4} that of the first sinusoidal term for each of the following inputs:
   (a) Sawtooth wave.
   (b) Triangular wave.
   (c) Half-wave rectification.
   (d) Full wave rectification.

9.3. *Nonsinusoidal Inputs to Reciprocal Quadratic.* A system has a frequency response curve in the form of a reciprocal quadratic with $\omega_n = 100$ and $\zeta = 0.5$. For each function shown in Table 9.1.1-1, sketch the resulting output and explain. The fundamental input frequency is equal to 50 rad/s.

9.4. *Use of Cascade Compensator (Cascade Network).* The object of this problem is to alter the response found in the previous problem. This will be accomplished without changing the system, but instead by using a suitably designed additional system (referred to as a compensator system or network) placed in cascade with the original system. Sketch the frequency response of the compensator, plot the overall response of the new cascaded system, compare the results to the original one, and briefly explain, using each of the following compensators:
   (a) Compensator is identical to the system in previous problem.
   (b) Compensator is a reciprocal quadratic with $\omega_n = 110$.
   (c) Compensator is a first order factor with corner at 50.

(d) Compensator is a first order factor with corner at 70.

(e) Compensator is a first order factor with corner at 85.

(f) Select a compensator that will provide maximum bandwidth.

(g) Select a compensator that will pass only the first term of the series, but will filter all the others.

9.5. In Ex. 9.3.1-1 (Resonating Vibrator Pump) explain how the magnetic actuator can be a frequency doubler. Can an electrostatic device also be a frequency doubler? Explain.

9.6. Justify the use of only the second harmonic in Ex. 9.3.1-1.

9.7. In Ex. 9.3.1-1 (Vibrator Pump) how does "lost motion" occur due to the phase shift in the flapper valve?

9.8. Redesign the system in Ex. 9.3.1-1 for damping ratio equal to 0.3.

9.9. In Ex. 9.3.1-1, the displacement X was equal to $X_0/3$. Redesign the system using a smaller displacement (say, $X = X_0/5$).

9.10. In Ex. 7.6.2-1 (Motorcycle) approximate the bumpy road as a sequence of half-wave sinusoids.

9.11. Rework Ex. 9.2.1-1, Triangular Input to Series Circuit, for half-wave rectified input Consider only the first three terms (the second to the sixth harmonic) where the fundamental frequency is equal to 52.0 rad/s.

9.12. *Embroidery Attachment for Sewing Machine.* A plastic cam is used as part of an attachment for a sewing machine that enables it to make an embroidery stitch.

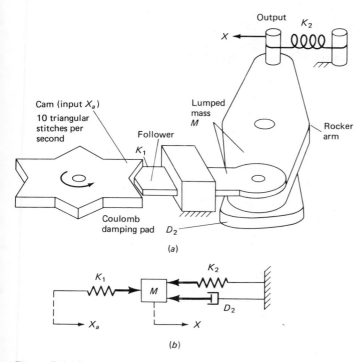

(a)

(b)

Figure P-9.12

Assuming that the cam provides a triangular input, determine how closely this device follows the triangular shape. See Fig. P-9.12. The following parameters are given:

$$M = 100 \text{ dyn} \cdot \text{s}^2/\text{cm} \qquad D_2 = 1.28 \times 10^4 \text{ dyn} \cdot \text{s/cm}$$

$$K_1 = 2 \times 10^6 \text{ dyn/cm} \qquad K_2 = 0.56 \times 10^6 \text{ dyn/cm}$$

9.13. *Phonograph Pickup with Triangular Input.* A phono pickup is a mechanical device that follows a groove cut in a record and actuates an electrical sensor to produce signals in accordance with vibrations cut in the groove. A sharp needle follows the groove, which wavers from side to side, appearing like a displacement input θ_a. It is assumed that the needle is a rigid but massless body that follows the groove perfectly. The motion is transmitted through an input link Z_a that is to be

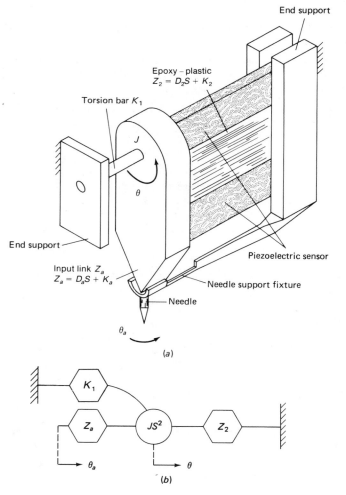

Figure P-9.13

synthesized. The final motion θ twists a piezoelectric sensor. In the process, a torsion bar, which supports the free end of the device, is also twisted. The sensor and its epoxy-plastic binder will be assumed to be a spring-damper $K_2 + D_2 s$ parallel network. (a) Design the device for a bandwidth of 10^4 Hz (6.28×10^4 rad/s). See Fig. P-9.13. (b) Show how the device transmits the first three terms of the Fourier series for a triangular wave with a fundamental frequency equal to 2,000 Hz.

9.14. Using as many sketches as is appropriate, demonstrate random noise as depicted by Fig. 9.4.1-1. For simplicity, use only four frequencies: $\omega_1 = 600$, $\omega_2 = 6000$, $\omega_{low} = 300$, and $\omega_{high} = 12,000$ rad/s.

9.15. Why does a system *not* resonate even though the resonant frequency is present in the random noise input?

9.16. Redesign the low pass filter of random noise in Ex. 9.4.4-1 by changing specification (a) from 1.81% to 4%.

9.17. Similar to Problem 9.16, but changing specification (b) from 600–6000 to 1200–12,000.

9.18. Similar to Problem 9.16, but changing specification (c) from 4% to 10%.

9.19. Redesign the circuit in Ex. 9.2.1-1, Triangular Input to Series Circuit, treating the triangular input as noise to be filtered (attenuated) to 1.81% of its original value. The system must pass signals with an error no greater than 4%. Inductance $L = 1.4$ mH. All other system parameters may be changed. Determine the bandwidth.

9.20. *Design of Vibrator Powered Tool Using Sawtooth Wave.* Many power tools, like the saw shown, require power during the cutting stroke, and prefer a quick return. There are several low-cost electronic circuits that can produce a sawtooth current. Making use of frequency doubling, design the system for use with house current (fundamental = 60 Hz). See Fig. P-9.20.

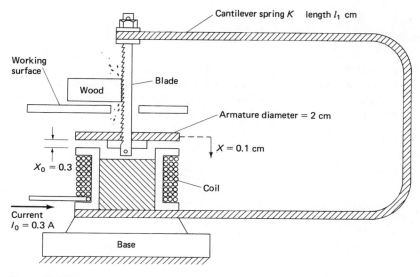

Figure P-9.20

9.21. Redesign the power saw in Problem 9.20 using a permanent magnet to prevent frequency doubling.

9.22. *Design of Vibrator Powered Electric Shaver.* (See Fig. P-9.22.) A low-cost electric shaver uses a vibrator to operate on ordinary house current. Make use of the following to design the system:

    frequency doubler operation
    amplitude of blade vibration = 3 mm = 0.003 m
    cantilever beam width = 0.03 m
    damping ratio = 0.3
    nominal gap = $5X$, where X is displacement of the armature

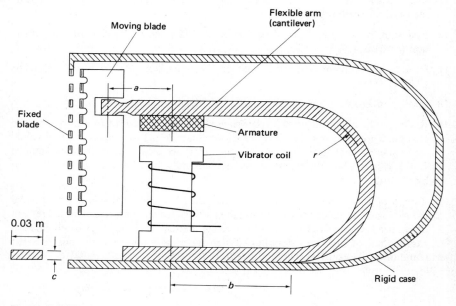

Figure P-9.22

9.23. Refer to Problem 7.53, Carrier Current Transmission. In that problem, the signal through the relay (inductor) was assumed sinusoidal. However, the operation of the relay makes use of a half-wave rectifier. Rework the problem using the series for half-wave rectification.

Chapter **10**

Design Oriented Approach to Microsystems

Entire electrical circuits can be etched into a micro–silicon chip. The entire circuit is so small, it is invisible to the naked eye. An entire computer, containing over a million components, can be etched on a silicon chip smaller than the head of a thumbtack.

There is no doubt that electrical circuits are successfully microminiaturized. These techniques are now being applied to microminiaturizing components in other disciplines. There now exist microcomponents (such as valves, capillary columns, pumps, nozzles, diaphragms, tuning forks, cantilevers, and springs) that are smaller than the period at the end of this sentence.

10.1 MICROLITHOGRAPHY

The *chip* is a slice (or a wafer) of some crystalline substance (usually silicon because of its unusual properties). The direction of the cut is chosen along certain crystal lattice lines in order to take advantage of the different properties of silicon that are a function of crystal structure.*

10.1.1 The Etching Process

The chip is coated with photosensitive material, and microphotography (equivalent to using a microscope in reverse to reduce the size of a picture) is used to cause changes in the coating. Certain chemicals are added that will etch away

---

*See references [51] and [52].

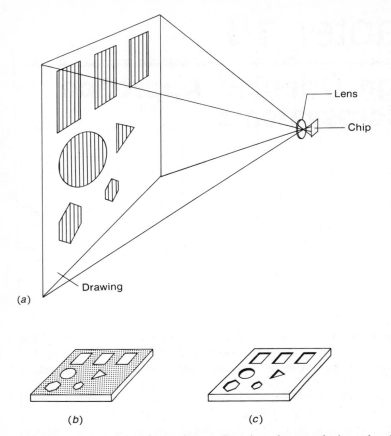

Figure 10.1.1-1   Microlithography. (a) Drawing photographed on the chip. (b) Mask. (c) Etched chip.

the coating where the photograph is black, or others that will etch away where it is white, and yet others that will change the chemical nature of the coating according to the image that has been photographed upon it. The resulting chip is coated again, and there may be further microphotographs or other chemical agents, and this may be accompanied with heat, pressure, or electrical fields. After a number of such processes, there is a three-dimensional cavity on the chip, resembling a contour map. During the processes, certain impurities (such as boron or phosphorus) are added (called doping) to further change the properties of either the coating or the silicon. See Fig. 10.1.1-1.

The process starts with a drawing, usually about one meter by one meter (about 40 inches). See Fig. 10.1.1-1(a). Areas to be etched into the chip are drawn in black. An optical reducing system produces a sharp image that is one-thousandth the size of the drawing, one millimeter by one millimeter. The front side of the chip is coated with a photosensitive film.

By photochemical means, holes are produced in the film corresponding to the black areas on the original drawing. See Fig. 10.1.1-1(b); the chip is shown enlarged to show detail. The film with holes so produced is called a *mask*.

An etchant is poured over the chip. Where there are holes in the mask, the etchant eats into the chip, leaving a shallow cavity. See Fig. 10.1.1-1(c). The mask acts like a stencil, producing a contour map on the surface of the silicon chip.

Additional processes may be superimposed, resulting in a composite configuration. It should be pointed out that the lens system produces an extremely sharp image, resulting in shapes of high tolerances, on the order of fractions of a micron (millionth of a meter) with straight or curved lines and round or sharp corners.

Transistors are made this way, as are other electrical components, such as resistors, capacitors, inductances, photoresistors, thermisters, and so on. The fabrication process also includes all the wiring and interconnections. Since the process operates on the whole surface of the chip, any number of components can be made at the same time. The process is called *batch fabrication* for this reason. Over the years, efforts have succeeded in reducing the size of the components until an entire microprocessor can be put on a single chip about a quarter-inch square. Since the process is basically photography with a few added dippings, there is no added expense in making more than one item at a time. For example, it costs the same to take a photo of a blank wall as it does to photograph a colony of ants. Also, there are techniques for repeating certain parts of the photo. It isn't necessary for an artist to draw one thousand resistors, for example.

The same batch fabrication techniques that have been developed for electrical components have been applied to mechanical ones. However, there still are many fabrication difficulties. But in time, we can expect to see batch fabricated mechanical components; for example, there could be a thousand valves, a thousand nozzles, accelerometers, pumps, diaphragms, and so on. These could be connected by tiny pipelines, producing the mechanical equivalent to ultracomplicated systems. There could be a microcosm of micromechanical devices that are integrally connected.

10.1.2 Isotropic and Anisotropic Etching

Silicon has very interesting electrical, mechanical, and chemical properties. Also, silicon can be chemically etched with microscopic precision. A chemical protective coating called a mask will prevent the etchant from reaching portions of the silicon, leaving a cavity in the unprotected areas. By appropriate sequences of masking and etching, various configurations can be etched in the surface.

There are two types of etchants, isotropic and anisotropic. *Isotropic etchants* eat away the material at the same rate in all directions. This produces a round-bottom cavity. See Fig. 10.1.2-1. Common isotropic etchants are hydrofluoric, nitric, and acetic acids.

Anisotropic etchants, such as potassium hydroxide, sodium hydroxide, and other caustic agents, etch at different rates along different crystalline planes. To appreciate this effect, consider three major crystalline planes defined as 110, 100, and 111 (called the Miller coordinate indexes). See Fig. 10.1.2-2. If we think of the crystal of silicon as a cube, the 110 plane passes diagonally

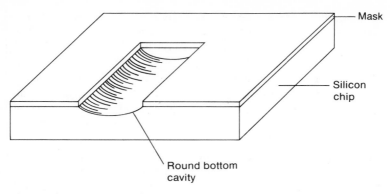

Figure 10.1.2-1   Isotropic etching.

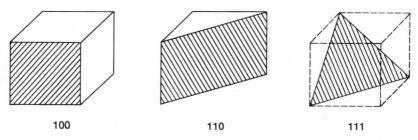

Figure 10.1.2-2   Crystalline planes.

through two diagonally opposite edges. The 100 plane is one of the faces of the cube. The 111 plane passes through three corners of the cube and through none of the edges.

The importance of these planes is that anisotropic etchants progress rapidly in a direction perpendicular to the 110 plane, slowly perpendicular to the 100 plane, and very slowly or not at all perpendicular to the 111 plane. Thus, various shapes* can be produced by a choice of the crystalline plane along which the chip has been cut, the shape of the opening in the mask, and the time allowed for the etching process. For example, a chip cut along the 111 plane would not etch at all, since this plane is highly resistant to etching perpendicular to the plane. As another example, etching a chip cut along the 110 plane would yield a cavity with vertical (perpendicular) sides. See Fig. 10.1.2-3.

In addition to isotropic and anisotropic etching, the process may be modified by "doping" (adding chemical impurities) that will alter the speed of or even halt the etching in specified directions. With a combination of all of the processes, a large number of configurations can be attained. All of these

*See references [51] and [52].

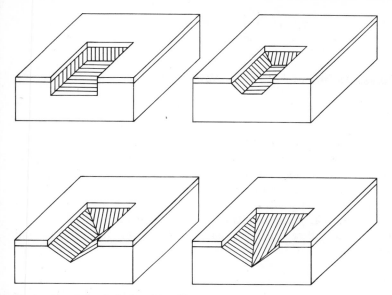

Figure 10.1.2-3   Anisotropic etching.

involve etching (removal of material). Material (not necessarily silicon) can also be deposited by, for example, chemical, electrochemical, and thermal processes. Silicon can be added in the form of a gas, silane (SiH_4), and the layer deposited is called an epitaxial layer.

Some exotic materials are coming into common use, such as piezoelectric, strain sensitive resistance, and temperature sensitive resistance.*

10.1.3 Categories of Chip Manufacture

Complex shapes can be formed either by multiple etching processes or by assembling separate chips. However, these are not all accomplished with the same ease or at the same manufacturing costs. For this reason, it will prove convenient to establish four categories of chip manufacture that involve successively more complicated methods.

(1). *All processes on one side of chip.* This is the lowest cost, since neither the chip nor the optical system is moved.

(2). *Work on both sides of chip.* This involves a little more cost than category 1, since the chip must be turned over and then must be realigned.

(3). *Stacked chips.* In this case, several chips are fabricated separately and then must be stacked. This requires an alignment procedure at assembly, which is somewhat more expensive than category 2.

*See reference [56].

(4). *Assembly of microparts*. This is the most involved and expensive procedure. It requires the actual manipulation of microscopic parts.

10.1.4 Types of Movement

The categories of mechanical movement will be done with the means of fabrication in mind.

(1). *Materials that either elongate or twist*. This provides very limited movement. There are materials, such as piezoelectric and magnetostrictive, that change dimensions (slightly) due to voltage in the first case and a magnetic field in the second.

(2). *Structure is flexible*. This provides limited movement, but which can be sufficient for many applications. There are many flexible structures, but the three most popular are cantilever, torsion bar, and diaphragm.

(3). *Sliding members*. This provides large movement. Of necessity, this involves at least two separate chips, which might at first glance suggest the need for category 4 of chip manufacture. However, it is possible to start with a silicon substrate, deposit a dissolvable layer, deposit silicon over that, and then dissolve the layer.

(4). *Bearings and pivots*. This provides unlimited movement. The means of fabrication is similar to item 3 but employing a wheel and axle machine instead of sliding members. The problem with today's techniques is the poor tolerances available.

10.2 FABRICATION OF BASIC MICROCONFIGURATIONS

10.2.1 Some Rules for Batch Fabrication

A single crystal of silicon can be larger than a one-gallon bottle. (A boule, a super-large crystal, can be one meter long.) A single crystal can be stronger than a polycrystalline metal, since intercrystal bonds are usually weaker than the material itself. Hence, a single crystal of silicon is stronger than steel. Also, because of the monocrystalline nature of silicon, it has a higher endurance limit than most metals. Thus, silicon can be bent back and forth, resulting in repeated cycles of tension and compression, many more times than most metals. These two features make silicon an ideal mechanical structural material.

The next question is, "Can silicon be batch fabricated in the complicated shapes required for mechanical devices?" The answer is a qualified "yes." One of the problems is that the choice of crystal plane needed for one shape may conflict with that for another. While a variety of shapes is possible, there is a limit when considering the restraint of batch fabrication. It is not practical to use screws or rivets, or to weld or solder. It is not feasible to position tiny pieces in place as is customary when working in the macro-world. The bonding of more than one chip to another becomes a slow and expensive process, and surely, one does not cut, drill holes, saw, mill, or otherwise machine the microscopic parts as has been customary in working with normally sized devices. Batch fabrication doesn't permit the worker to handle the parts.

In Section 10.1.2 it was shown that various materials can be deposited (in batch fabrication) producing raised geometric shapes. In that section, only silicon dioxide and metal were mentioned, but there are many other materials, including silicon itself. Using silane gas (SiH_4), it is possible to deposit silicon on the chip. What is important is that the deposited silicon copies the crystal structure and orientation of the silicon chip. This is called an *epitaxial layer* and may be from 5 to 20 μm thick.

In Section 10.1.2 it was shown that certain chemical ingredients dope or alter the electrical, mechanical, and chemical properties of the silicon chip. Dopants like boron, phosphorus, antimony, and arsenic can either retard or accelerate the etching process. Some can stop the etching altogether and are called *stop-etch dopants*. This provides yet another means to shape the silicon chip.

Considering all of the aforementioned limits, before micromechanical devices can reach the status of present-day electrical chips, it will require a combination of innovative etching sequences and reconceptualization of the mechanical functions so that the required shapes can be batch fabricated. With this in mind, let us review and summarize the mechanical configurations that are currently available, with an eye to employing them in the fabrication of micromechanical devices.

10.2.2 Fabrication of a Transistor

A silicon crystal is sliced along planes that will produce etching of cavities with near vertical sides. These thin slices, called wafers, will serve as the base, or *substrate*, of the chip. The top surface is oxidized (using superheated steam) to a thickness of 0.5–0.6 μm [1 μm (micrometer) is equal to 10^{-6} m, or 39.4 μin.]. This will be the first of several oxide coatings. The coating will be covered with a mask (which is the first of several) using a microlithographic process. The oxide layer #1 is etched away (where not protected by the mask), leaving exposed silicon. See Fig. 10.2.2-1(a). Now the etched oxide coating #1 will be used as a mask for a doping operation. Impurities such as boron and phosphorus are applied to the surface and are isotropically absorbed into the exposed silicon. Note the rounded corners and that some doping material does manage to "crawl" under the mask at its edges. This has been allowed for in the original shape of the first mask. Incidentally, the first mask would not have successfully restrained the doping agent, hence, it was used to create a tougher mask (made of silicon dioxide, a type of glass). The doping agent that has been absorbed into the silicon will give it properties that will enable it to serve as the P element of the transistor. The undoped regions (pure silicon) will serve as the N element. See Fig. 10.2.2-1(b).

Having done their respective jobs, mask #1 and oxide coating #1 are removed by appropriate chemical solvents. The entire top surface is reoxidized

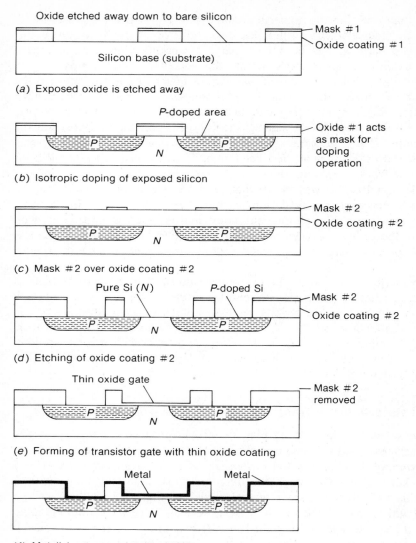

Oxide etched away down to bare silicon

Mask #1
Oxide coating #1

Silicon base (substrate)

(a) Exposed oxide is etched away

P-doped area

Oxide #1 acts
as mask for
doping
operation

P   P   N

(b) Isotropic doping of exposed silicon

Mask #2
Oxide coating #2

P   P   N

(c) Mask #2 over oxide coating #2

Pure Si (N)   P-doped Si

Mask #2
Oxide coating #2

P   P   N

(d) Etching of oxide coating #2

Thin oxide gate

Mask #2
removed

P   P   N

(e) Forming of transistor gate with thin oxide coating

Metal   Metal

P   P   N

(f) Metalizing to complete the MOS

Figure 10.2.2-1   Fabrication of a transistor.

with a thick coating (about 1–1.5 μm). Mask #2 is microlithographed over this. See Fig. 10.2.2-1(c). By chemical etching, oxide coat #2 is removed and, consequently, it exposes most of the P-doped regions as well as the small undoped area. See Fig. 10.2.2-1(d).

Using a suitable solvent, mask #2 is removed, while oxide coating #2 remains. Using appropriate masking #3, oxidizing #3, and etching #3, a very thin oxide is formed in the well over the undoped region. Note that this thin film of oxide contacts part of each P-doped area as well as the undoped area. See Fig. 10.2.2-1(e). This will become the gate of the final transistor.

For the final operation, a metal such as aluminum is deposited over appropriately masked areas to complete the fabrication of the transistor, a MOS (metal oxide semiconductor). See Fig. 10.2.2-1(f).

Upon reviewing the overall sequence of processes, note that all were performed from one side of the chip. This demonstrates category 1 and represents the lowest cost of manufacture.

Note that in the fabrication of a transistor, it was never necessary to handle the chip, to reposition the fixture that held the chip, or to manipulate tiny parts or particles individually. (What kind of tweezers would the machinist use for such a manipulation, and would it have to be through a microscope?) All processes were accomplished by means of photographic methods and chemistry. Such processes are called *micromachining*. Also note that there could have been thousands of transistors on the same chip that would have

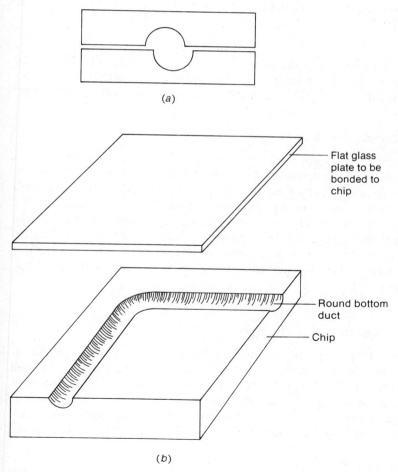

Figure 10.2.3-1   Duct. (a) Alignment problem using stacked chips. (b) Use of flat plate produces half-round duct.

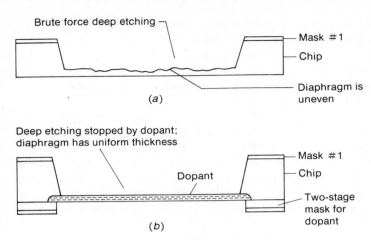

Figure 10.2.4-1 Fabricating a diaphragm. (a) Brute force deep etching. (b) Use of stop-etch dopant.

been micromachined all at the same time—and at no extra cost. This is the magnificent advantage of micromachining, the ability to fabricate any number of devices at the same time and at no added cost—so unlike macromachining.*

10.2.3 Fabrication of a Duct

Isotropic etching does not depend upon crystalline planes and produces a half-round cavity. If the mask is a long line, this will produce a round bottom duct, which is suitable for a fluid conduit. See Fig. 10.2.3-1. Since this doesn't place any restrictions upon the orientation of the crystalline planes, such a requirement may be saved for some other configuration that *does* need it. The novice might recommend forming a matching half (a) that would produce a duct with a circular cross section. However, this would require precise align-ment of the two halves. Instead, use a flat glass plate. A half-round duct will serve adequately (b).

10.2.4 Fabrication of a Diaphragm

Consider the fabrication of a diaphragm. It is possible to use a pyramidal cavity† and to stop the etching process before the etchant eats entirely through the chip. See Fig. 10.2.4-1(a). In this "brute force" approach, the bottom of the cavity, the diaphragm, is not uniform in thickness due to imperfections in the etchant and the surfaces of the chip. Instead, an appropriate mask (usually a two-stage one like that in Fig. 10.2.2-1) is employed for doping the

---

*See references [51], [52], and [54].

†See Fig. 10.1.2-3.

back of the chip. It is much easier to control a thin penetration of dopant than to etch for almost the thickness of the chip. Using a stop-etch dopant, the silicon is conditioned to stop the etch on the normal side at the appropriate depth. As a result the diaphragm has uniform thickness. See Fig. 10.2.4-1(b).

Note that the fabrication process represents category 2. However, in this case, the extra cost (above that for category 1) is negligible, since there is no precise alignment required. The dopant occupies an area that is larger than the diaphragm, the overlap being greater than typical manufacturing tolerances.

10.2.5 Fabrication of a Nozzle

A nozzle can be fabricated in much the same way as a diaphragm. Instead of stopping the etch, let the etchant eat all the way through the chip. While this is conceptually possible, it is not accurate enough, for the same reasons as those given for the diaphragm. See Fig. 10.2.5-1(a). Due to accumulation of errors in the speed of etching, the final nozzle opening N would not be accurately determined. Instead, using the stop-etch dopant method, create a small diaphragm area at the bottom of the cavity. Then etch an orifice in the diaphragm. See Fig. 10.2.5-1(b). This becomes quite a game in chemistry. The dopant must stop the etching of the cavity but not the etching of the orifice. Alternate methods make use of positive and negative masking.

Similar to the case of fabricating a diaphragm, the fabrication of a nozzle represents category 2. Ordinarily, the added cost of category 2 can possibly

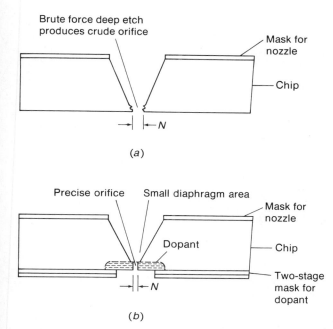

(a)

(b)

Figure 10.2.5-1   Fabricating a nozzle. (a) Brute force approach. (b) Use of stop-etch dopant.

offset the advantages of the technique. However, in this case, no precision alignment is required, making the cost just slightly more than that for category 1. Thus, this technique is practical.

10.2.6 Fabrication of a Cantilever

Another very useful shape for micromechanical devices is the cantilever. By a sequence of clever steps, this complicated shape can be batch fabricated. The most dramatic step is the undercutting of the cantilever leaf. This uses a combination of all of the aforementioned techniques. See Fig. 10.2.6-1(a).

An alternate fabrication technique makes use of stacked chips. The difficulties are means to align the chips and the chance that the bonding material will fall into the device. To minimize alignment problems, the chips are made with exactly the same size and shape and, in addition, they make use of alignment notches. See Fig. 10.2.6-1(b). The bonding problem is avoided by not using any type of cement. Instead, the mating surfaces of the chips are

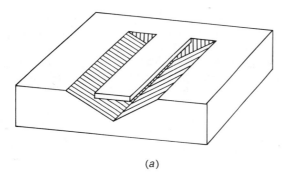

(a)

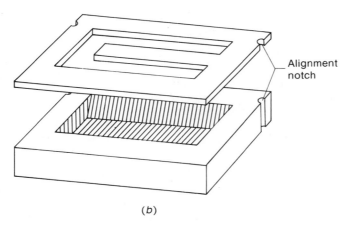

(b)

Figure 10.2.6-1   Fabricating a cantilever. (a) Fabricated from one side of chip. (b) Stacked chips.

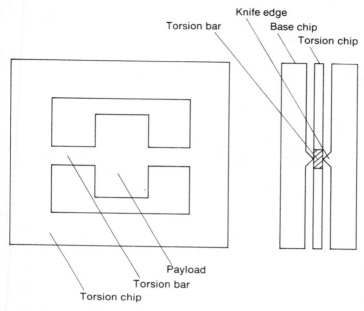

Figure 10.2.7-1   Torsion bar suspension.

oxidized with a jet of steam (forming silicon dioxide, which is glass). Then the chips are assembled and subjected to high temperature and pressure, which melts the glass, making the chips self-adhering—the perfect weld.

Note that in fabricating the cantilever, two different categories are represented. Fig. 10.2.6-1(a) demonstrates category 1, where the entire configuration is fabricated on one side of the chip. This configuration is particularly useful when the system electronics are to be integrated on the same chip. Fig. 10.2.6-1(b) portrays category 3, making use of stacked chips.

10.2.7 Fabrication of a Torsion Bar

A torsion bar (or a pair of them) with rectangular cross section may be successfully employed to support a payload, permitting significant rotation, with little bending. See Fig. 10.2.7-1. The cross section of the torsion bar is selective in its bending strength, resisting bending in its plane, but weak out of its plane. It is necessary for the torsion bar to be weak in some plane in order to twist easily. To compensate for the weakness out of its plane, knife edges on two additional stacked chips limit the movement.

10.2.8 Fabrication of a Bearing

The engineer is well aware of the need and applications of the wheel and axle as a basic machine. However, in the twenty or so years that micromechanical

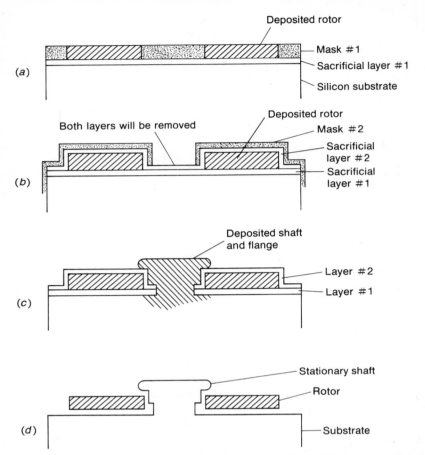

Figure 10.2.8-1 Bearing fabricated by dissolving sacrificial layer. (a) Rotor is supported by layer #1. (b) Preparation for shaft. (c) Deposition of shaft. (d) Final product, wheel and axle.

engineering has been in existence, the microbearing is only just beginning* to become practical. The major problem is to fabricate the shaft and rotor without handling the microparts. That is, it is desirable to avoid category 4 of chip manufacture.

There are a number of facilities currently developing various means to fabricate the microbearing. To typify these, consider Fig. 10.2.8-1. Starting with a silicon substrate, a sacrificial layer (a material that can easily be dissolved away later) is deposited over the top surface. See Fig. 10.2.8-1(a). Over this layer, mask #1 is laid. This mask defines the contours of the rotor. The rotor is made by either an epitaxial layer of silicon or other suitable depositable material, and the mask is removed. The rotor adheres to the

*See reference [56].

sacrificial layer and lies a small distance above the silicon substrate. This distance is equal to the thickness of the layer.

A second sacrificial layer is deposited over the entire top surface. See Fig. 10.2.8-1(b). Over this, mask #2 is laid, which exposes a central opening that will ultimately be the stationary shaft. A solvent is used to dissolve the sacrificial layer and thus exposes a portion of the silicon substrate.

Mask #2 is removed and silicon (or other suitable material) is deposited in the opening. It is allowed to overflow (or other suitable means are employed) to create a flange. See Fig. 10.2.8-1(c). The material that fills this area may be either solid or hollow, depending upon the method employed.

The sacrificial layers are dissolved, leaving the rotor free of its surroundings. See Fig. 10.2.8-1(d). There is a gap surrounding the rotor. This gap is equal to the thickness of the sacrificial layer. It was necessary to make the layer thick enough to permit the solvent to enter and flow freely in order to dissolve away the layer. Here, we note the conflict: The layer was necessary to provide clearance between the rotor and the stationary shaft, yet this layer had to be thick enough for its subsequent removal. Consequently, the clearances are relatively large. This problem, among others, is currently under investigation. One solution is to deposit material in the gaps in an effort to close up or reduce the clearances.

Note that all processes were accomplished from the same side of the chip. The fabrication of the bearing by the above method constitutes category 1.

10.3 SOME MICROMECHANICAL DEVICES

10.3.1 Ink Jet Printer*

Using thousands of micro-nozzles, ink can be ejected onto paper to form a screened picture. The droplet size is of the order of those in a spray. Hence, high resolution is possible. Using several banks of micro-nozzles, the picture may be printed in color. See Fig. 10.3.1-1(a).

Each droplet is ejected from the nozzles and is propelled onto the paper by electrostatic forces. The cone of each nozzle is coated with insulating material (silicon dioxide) and this is coated by depositing a metal film to form the cathode of the electrostatic drive system. The anode is installed at each nozzle exit.

To assist in propelling the ink droplets, a vibrator pump agitates the ink supply, but not enough to eject the ink. The final thrust is provided by the ink jet nozzles, each of which is electrically charged according to the picture plan. The system works as if the pump provides the major thrust and the electrostatic force is the booster, but the boost is applied only where an ink drop is scheduled to appear. See Fig. 10.3.1-1(b).

---

*See reference [49].

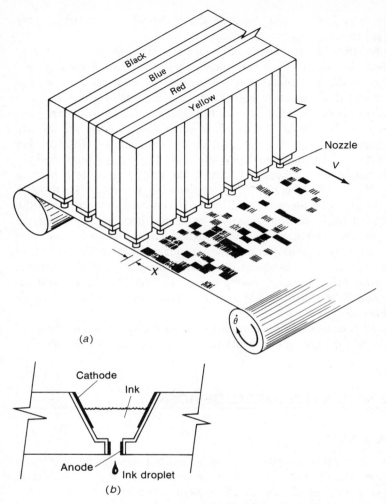

(a)

(b)

Figure 10.3.1-1   Ink jet printer. (a) Complete system. (b) Nozzle.

10.3.2 Check Valve

A check valve constrains the fluid to flow in only one direction. Complying with the design* restriction to use only bending and twisting to provide motion, the check valve to be discussed here uses a cantilever beam to accomplish the task.[†] See Fig. 10.3.2-1(a).

The check valve makes use of chip fabrication category 3, stacked chips. Normally, the cantilever lies against the valve seat, thus closing the port to

---

*See Section 10.1.3, item 4.

[†]See reference [58].

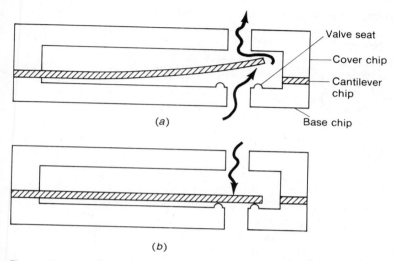

Figure 10.3.2-1   Cantilever check valve. (a) Flow is allowed upward. (b) Flow down is checked.

fluid flow. In operation, when the fluid flow is upward, it exerts pressure against the cantilever. The cantilever, acting like a spring, is deflected, uncovering the port, allowing the fluid to flow upward.

When the fluid tries to flow in the opposite direction, downward, it presses against the cantilever, which is pressed against the seat, closing the port, preventing the fluid from flowing downward. See Fig. 10.3.2-1(b).

10.3.3 Diaphragm Valve

This valve uses a diaphragm* to provide the motion. From a structure point of view, the diaphragm is treated as a disk with clamped edges. See Fig. 10.3.3-1(a).

The valve shown is normally open. Air pressure acting against the diaphragm closes the valve. Note that the air supply does not mix with the fluid the valve is handling.

In Fig. 10.3.3-1(b) the valve shown is normally closed. The valve is opened by electrostatic forces on the plate attached to the diaphragm. The diaphragm is bonded to the base and thus does not allow the two fluids to mix. Since the electronics could be integrated in the top plate, it was unnecessary to etch a hole in the top wafer. A similar valve could be placed surgically under the

*See reference [58].

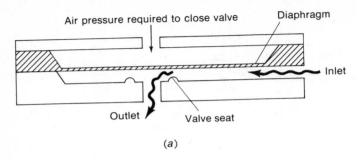

(a)

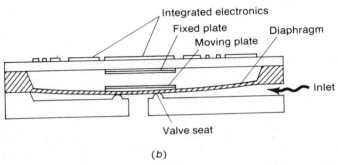

(b)

Figure 10.3.3-1 Diaphragm valves. (a) Pressure actuated normally open valve. (b) Electrostatically actuated normally closed valve.

skin, but it would be actuated externally by magnetic means, where the magnetic device need not be micro-sized, since it is outside the body.

10.3.4 Diaphragm Pump

A diaphragm is set into vibration by means of an electrostatic driver. The driver consists of a pair of plates, one affixed to the center of the diaphragm, the other on the top plate. As the diaphragm vibrates, it alternately applies pressure or vacuum to the chamber. See Fig. 10.3.4-1(a). On the upstroke, the vacuum in the chamber draws the inlet check valve away from its seat (below the cantilever chip), opening the inlet port, permitting the fluid to enter the chamber. The cantilever chip is shown in Fig. 10.3.4-1(b) with the inlet valve (cantilever) raised off its seat, allowing the fluid to flow into the chamber.

On the downstroke of the diaphragm, the pressure so produced presses the inlet valve down, closing the inlet port. At the same time, the pressure forces the outlet valve away from its seat (above the cantilever chip; not shown). This permits the fluid to flow out of the pump. See Fig. 10.3.4-1(c).

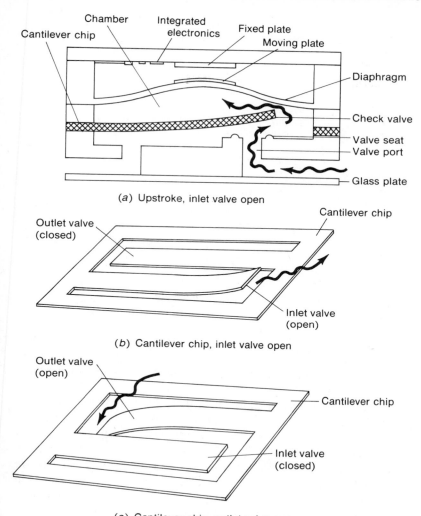

(a) Upstroke, inlet valve open

(b) Cantilever chip, inlet valve open

(c) Cantilever chip, outlet valve open

Figure 10.3.4-1 Vibrating diaphragm pump.

10.3.5 Torsional Mirror*

If a light beam is directed at a mirror, a very small rotation of the mirror will produce a substantial deflection of the reflected beam. This system behaves like an optical-mechanical amplifier. For the microdevice, the rotation is not achieved by means of a shaft in bearings. Instead, the mirror is supported on a pair of torsion bars. This proves to be exactly what the optical system that we

---

*See reference [50].

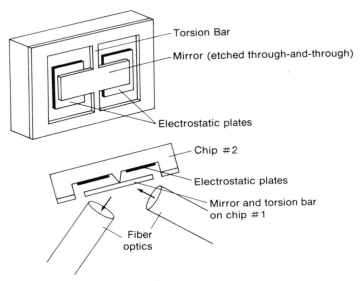

Figure 10.3.5-1   Torsional mirror for optical communication.

are interested in needs, anyway. The optical system requires that the mirror oscillate at very high frequencies, pulsing and modulating the light beam. Thus, it is used as a communication device, where thousands of such beams and mirrors fit in a small tube.

One such mirror assembly is shown in Fig. 10.3.5-1. A large number of these are fabricated on a single chip, thus greatly reducing the cost of assembling the many microdevices. The fabrication uses category 3, stacked chips. The mirrors are etched by a through-and-through process that leaves a multitude of mirrors, each supported on torsion bars. Electrostatic plates are deposited on a second chip. An oscillating voltage applied to these (opposite voltage to each plate) will produce a torque on each mirror, oscillating it through a proportional angle. (This can be employed in a number of ways. One has fiber optics to and from the mirror, as part of a fiber optics communications system.) A knife edge is etched (or deposited) on the second chip. This keeps the mirror from being pulled onto the electrostatic plates.

10.3.6 Capillary Column*

A very long capillary tube is often required in chemical analysis. The walls of the tube at certain points have a conductive coating, making it possible to measure the electrical resistance of the fluid as it passes through the tube. Note that to fit a very long tube into microspace, the tube is wound in a spiral (either double or quadruple). The set of either double or quadruple spiral

---

*See reference [52].

Beginning of quadruple spiral

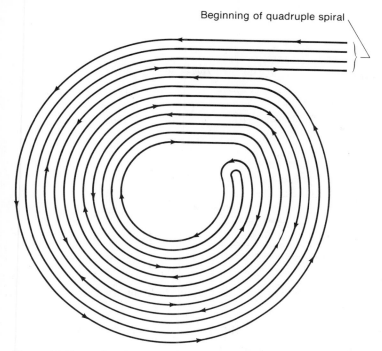

Figure 10.3.6-1   Quadruple spiral capillary column.

grooves lies in a plane, requiring that at certain points the paths turn around, double back on themselves, finally tracing the entire surface. Such a system of grooves is shown in Fig. 10.3.6-1. The arrangement shown permits the use of either one very long bending and winding path or two parallel paths. At the upper right-hand side of the diagram, the four tube ends may be spliced in a variety of ways depending upon the requirements at hand. Such fluid switching may be accomplished by means of microvalves (not shown).

The grooves are isotropically etched and thus have round bottoms. The electrical contacts are deposited in the grooves. Thus fabrication follows category 1. The chip is covered with a glass plate that is bonded by using heat and pressure. Although two pieces are employed, this still does not qualify as a stacked chip technique. It ranks as one of the lowest cost micromechanical devices to produce.

10.3.7 Heat Exchanger

A heat exchanger requires that two fluids pass freely in immediate proximity with each other without contaminating one another. There are a number of ways in which this can be done. One way is to use the quadruple set of grooves shown in Fig. 10.3.6-1. The hot fluid proceeds through one double (folded) tube while the cold fluid passes through the other folded groove, which is in

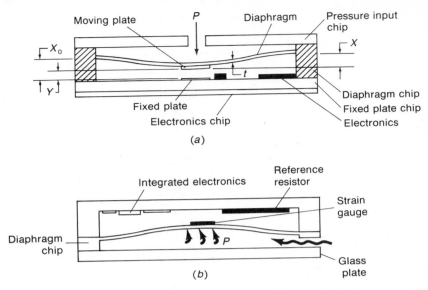

Figure 10.3.8-1 Pressure sensor. (a) Capacitive output. (b) Strain gauge output.

close proximity. The fluid may flow either in the same or opposite directions, resulting in either parallel- or counter-flow.

An alternate method is to use both sides of a diaphragm. Silicon is a fair conductor of heat, making it suitable for this application. As an added sophistication, raised lines may be deposited on each side of the diaphragm to act as baffles, and micropumps may be employed to circulate the fluid to increase its heat transfer characteristics.

10.3.8 Pressure Sensor*

Using a diaphragm, pressure can be measured. See Fig. 10.3.8-1. The diaphragm is deflected by the pressure and this deflection, in turn, is measured either by a strain gauge or by a capacitor. Since the strain gauge (or capacitor) is located on the other side of the diaphragm, it will not be contaminated by the fluid.

The fabrication process makes use of stacked chips, where no precise alignment is required. Each chip is coated with a film of glass so that under high temperature and pressure the chips become self-bonding.

10.3.9 Temperature Sensor

There are two common methods for measuring temperature: One utilizes a thermistor (an element whose resistance changes with temperature), the other a bimetallic strip. In the former, the thermistor is placed in a Wheatstone

---

*See references [54], [55], and [56].

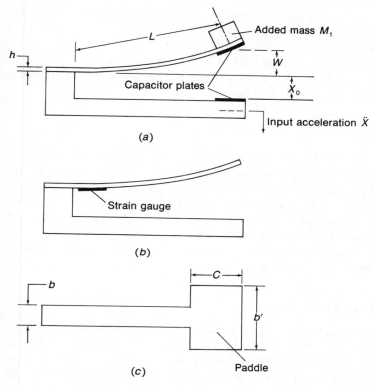

Figure 10.3.10-1   Vib-accelerometer. (a) Lumped mass and capacitive output. (b) Strain gauge output. (c) Paddle.

bridge that is calibrated to transform change in resistance to change in temperature.

The bimetallic strip is fabricated by electro-depositing a metal with high thermal expansion on a thin cantilever that has low thermal expansion. A temperature change causes one material to expand faster than the other, causing the bimaterial strip to curl. The resulting displacement of the strip is measured either by strain gauge or by capacitive means.

10.3.10 Accelerometer* (Vibration Sensor)

The accelerometer and vibrometer measure vibration. As discussed in Sections 7.8.3 and 7.8.5, the difference between these two instruments is the natural frequency. For micro-sized devices, the natural frequency is, of necessity, very high, thus making the microdevice a candidate for an accelerometer. The microdevice makes use of a cantilever. See Fig. 10.3.10-1(a). Vibration sets the

---

*See reference [53].

cantilever vibrating at the appropriate amplitude, which is measured either by a strain gauge or by capacitive means.

To increase the sensitivity and also to lower the natural frequency of the instrument, an additional mass is applied to the end of the cantilever. This is done either by depositing gold or by shaping the cantilever so that it has a large paddle. See Figs. 10.3.10-1(b) and (c).

10.3.11 Tunister*

The tunister is an electromechanical transformer. It resembles an accelerometer, but uses an electrical input instead of mechanical vibration. The cantilever is usually very thin and operates in an evacuated chamber (so that it has very low damping). As a result, when vibrated at its resonant frequency, the output amplitude is very large. Since the device has extremely low damping, it has a very narrow bandwidth. Thus, the device behaves like a high "Q" system that competes with purely electrical ones. It is used to transmit a single frequency for which it has excellent noise suppression.

10.3.12 Gears and Linkages

With the success of a good bearing, the use of gears and linkages is possible. At the present time, the major problems are large clearances (see Section 10.2.8) and short life. However, there is a huge amount of research going on at the time of this printing. When these two major problems are overcome, the bearing will take its place in the set of reliable building blocks of microsystems.

10.3.13 Electrostatic Motor

An electrostatic motor has been built and run. While there still are problems (in addition to the ones cited in Section 10.3.12) there is an optimistic view of its future. The device uses alternate blades that are charged and produce electrostatic forces.

10.3.14 Turbine

A jet of air was aimed at a rotor with flat blades. It ran at very high speed and, surprisingly, its life was much longer than expected. It is possible that, with the ultrahigh finishes achievable in microdevices, the rotor became an air bearing. A thin film of air supported the rotor and a thin cylinder of air protected the shaft. This device may suggest approaches of research to provide general improvement in the microbearing.

---

*See reference [64].

10.4 DESIGN OF MICRODEVICES

For convenience, a device will be defined as micro-sized if its largest dimension is 200 μm. One μm (micron) is a millionth of a meter (or about forty millionths of an inch). Then 200 μm is about 0.008 inch.

What techniques are available to the designer of a microdevice? Would it be practical to start with a macro (normally sized) device, like the vibrator pump in Fig. 9.3.1-1 and divide by one hundred and scale down all of its dimensions. If we were to divide all of its dimensions by 100, the pump would be small enough to qualify as a microdevice. In scaling down each dimension, for example, its mass would become $(1/100)^3$ or 10^{-6} times its original mass, as expected. However, none of the other properties would diminish by the same ratio. As a result, its natural frequency would be in the millions and the amplitude of vibration would be less than a micron. Note that these do not follow the original 1/100 scaling. What we are experiencing is typical of large-scale miniaturization.

10.4.1 Design of Cantilever Devices

A number of microdevices rely upon a cantilever for their operation. Such devices are check valve, tunister, bimetallic temperature sensor, vib-accelerometer, vibrating reed tachometer, and dc accelerometer. The generalized model of these devices is shown in Fig. 10.4.1-1,

where L = effective length of cantilever
h = thickness of cantilever
b = width of cantilever
X = displacement when valve is fully open
F = force to fully bend the cantilever spring
r = radius of half-round duct
d = diameter of valve port
X_0 = nominal spacing of electrostatic plates
r_0 = radius of electrostatic plates
V_0 = voltage applied to plates
F_0 = electrostatic force
l = length of fluid path through valve

Equivalent Spring [cgs units (cm, g, s, dyn)]   See entry 1, Table 2.2.3-3, Appendix 2.

$$K_c = 3EI/L^3 \qquad [10.4.1\text{-}1]$$

where

$$I = bh^3/12 \qquad [10.4.1\text{-}2]$$

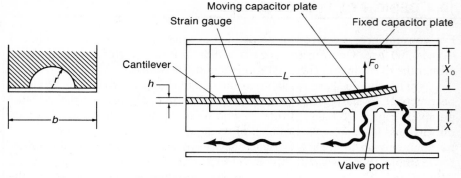

Figure 10.4.1-1   Generalized cantilever device.

Then

$$K_c = \frac{Ebh^3}{4L^3}$$ [10.4.1-3]

$$= \frac{Eb}{4}\left(\frac{h}{L}\right)^3$$ [10.4.1-4]

Force to Fully Bend Spring (cgs units)

$$F = K_c X$$ [10.4.1-5]

Bending Stress (cm g, s, dyn)

$$S = \frac{C_f 3EhX}{2L^2}$$

$$= 1.5EC_f \frac{hX}{L^2}$$ [10.4.1-7]

where C_f = stress concentration factor

Beam Aspect Ratios   To be sure that the beam covers the valve seat, let

$$\frac{b}{d} \geq 1.25$$ [10.4.1-8]

To minimize twisting of the beam, let

$$\frac{b}{h} \geq 3$$ [10.4.1-9]

To be sure the beam bends properly, let

$$\frac{L}{X} \geq 24$$ [10.4.1-10]

and

$$\frac{L}{h} \geq 12$$ [10.4.1-11]

and

$$\frac{L}{b} \geq 8$$
[10.4.1-12]

Design Considerations for Flow Areas   Through the valve port:

$$A_p = \pi \left(\frac{d}{2}\right)^2$$
[10.4.1-13]

Through the half-round duct:

$$A_d = \tfrac{1}{2}\pi r^2$$
[10.4.1-14]

Through fully opened valve: Note that this area is approximately a cylinder bounded by the cantilever and the substrate.

$$A_v = \pi d X$$
[10.4.1-15]

In order that all three areas be equal (for efficient fluid flow through the entire device), equate eqs. 10.4.1-13 and 10.4.1-15 and simplify.

$$d = 4X$$
[10.4.1-16]

Equate eqs. (10.4.1-13 and 10.4.1-14 and simplify.

$$d = r\sqrt{2}$$
[10.4.1-17]

Effect of Pressure   Force F_p against cantilever due to pressure drop ΔP,

$$F_p = A_p \Delta P$$
[10.4.1-18]

$$= \frac{\pi d^2 \Delta P}{4}$$
[10.4.1-19]

Output Sensor*   Three output sensors will be discussed in this section: strain gauge, piezoelectric element, and capacitive device. A strain gauge is a resistance element whose resistance changes with strain. A piezoelectric element develops a voltage due to strain. A parallel plate capacitor is one whose capacitance changes with the spacing of the plates.

For the strain gauge and piezoelectric sensors, the output is not linear and usually is calibrated empirically. The general relationships may be given respectively by

$$R_g = R(x)$$
[10.4.1-20]

$$V_z = V(x)$$
[10.4.1-21]

For a capacitor, refer to entry 1, Table 2.4.2-1, Appendix 2.

$$C = \frac{\epsilon A}{X_0}$$
[10.4.1-22]

---

*See references [54] and [56].

where

$$A = \pi r_0^2 \qquad [10.4.1\text{-}23]$$

Static Actuator [mks units (m, kg, s, N)]   There are a number of ways to actuate the cantilever. However, only two will be considered here, piezoelectric and capacitive. The piezoelectric device produces a large force for a small voltage. However, there are fabrication problems in attempting to deposit a piezoelectric material. In addition a piezoelectric sensor is nonlinear and responds poorly to very high frequencies. For low frequencies, it is calibrated empirically, where the general relationship is given by

$$F_z = f(V_0) \qquad [10.4.1\text{-}24]$$

For a capacitive actuator, the electrostatic force F_0 is given in entry 1, Table 2.4.2-1, Appendix 2.

$$F_0 = \frac{V_0^2 \epsilon A}{2 X_0^2} \qquad [10.4.1\text{-}25]$$

where

$$A = \pi r_0^2 \qquad [10.4.1\text{-}26]$$

Solve for the voltage V_0.

$$V_0 = \sqrt{\frac{2 X_0^2 F_0}{\epsilon \pi r_0^2}} \qquad [10.4.1\text{-}27]$$

$$= \frac{X_0}{r_0} \sqrt{\frac{2 F_0}{\pi \epsilon}} \qquad [10.4.1\text{-}28]$$

To avoid corona, arcing, and other electrical problems, it is necessary to check that the voltage gradient γ does not exceed the allowable value. The gradient is

$$\gamma = \frac{V_0}{X_0} \qquad [10.4.1\text{-}29]$$

The allowable gradient γ_{allow} is listed in Appendix 6, Dielectric Properties. Note that γ_{allow} is a function of the medium (dielectric) between the plates and not of the plates themselves. Also note that

$$10^6 \text{ V/m} = 1 \text{ V}/\mu\text{m} \qquad [10.4.1\text{-}30]$$

Note that eq. 10.4.1-25 will be a fair approximation if the valve displacement X is much less than the nominal plate spacing X_0. This requires that X_0 be large, but that in turn requires a high driving voltage V_0. As a reasonable compromise, let

$$X_0 \geq 5X \qquad [10.4.1\text{-}31]$$

In order to have as large an electrostatic force as possible, the plates should be as large as possible. Let plate diameter $2r_0$ approach the beam width b.

$$2r_0 \leq b \qquad [10.4.1\text{-}32]$$

Equations (like eq. 10.4.1-25) that involve a capacitor are accurate only if the plate diameter $2r_0$ is much larger than the plate spacing X_0. Then let

$$\frac{2r_0}{X_0} \geq 10 \qquad [10.4.1\text{-}33]$$

Equation 10.4.1-25 assumes that the plates remain parallel. This approximation is reasonably true if the plate diameter is much smaller than the cantilever length L. The condition is satisfied if we let

$$\frac{L}{2r_0} \geq 8 \quad \text{or} \quad \frac{L}{r_0} \geq 16 \qquad [10.4.1\text{-}34]$$

Vibrator Actuator   Using a vibrator actuator, the system will be frequency sensitive, requiring the use of frequency response methods as treated in Chapters 7, 8, and 9. These considerations all begin with the natural frequency, which in this case, is the natural frequency of the cantilever. Refer to Ex. 9.3.1-1, eq. (3),

$$\omega_n = \sqrt{K_{eq}/M_{eq}} \qquad [10.4.1\text{-}35]$$

$$= \frac{h}{L^2} \sqrt{\frac{E}{\rho}} \qquad [10.4.1\text{-}36]$$

The advantage of using a vibrator actuator is that, at resonance, the dynamic force F_d required to produce a displacement X is less than the static force F_s.

$$F_d = F_s/R_r \qquad [10.4.1\text{-}37]$$

where R_r is given by eq. 8.5.1-9, repeated below.

$$R_r = \frac{1}{2\zeta} \sqrt{\frac{1}{1 - \zeta^2}} \qquad [10.4.1\text{-}38]$$

The cantilever will be driven at its resonant (rather than at its natural) frequency in order to take advantage of the dynamic magnification factor. The resonant frequency is given by eq. 8.5.1-8, repeated below.

$$\omega_r = \omega_n\sqrt{1 - 2\zeta^2} \qquad [10.4.1\text{-}39]$$

where ζ = damping ratio as described in Section 6.2.4. This is the total effect of all damping sources, such as structural damping, fluid resistance, electrostatic damping, and so on.

Vibration Measurement   There are four basic seismic instruments, as covered in Section 7.8: seismograph, vibrometer, torsiograph, and accelerometer. The first three have very low natural frequencies and measure the amplitude of vibration directly. The accelerometer has a high natural frequency and measures the acceleration of vibration. It will be seen that the natural frequency of micro-sized devices is very high, making it necessary to restrict micro-sized seismic instruments to the accelerometer only.

The output of an accelerometer is given by eq. 7.8.5-3, repeated below.

$$R = (\omega/\omega_n)^2 \qquad [10.4.1\text{-}40]$$

Acceleration Input Those devices that respond to the vibratory movement of the foundation actually respond to acceleration. For a uniform cantilever, the acceleration of its distributed mass is equivalent to a uniformly distributed load. See reference [10], p. 98, case 2a, where $a = 0$.

$$X = \frac{3WL^3}{24EI} = \frac{WL^3}{8EI} \qquad [10.4.1\text{-}41]$$

where

$$W = \text{load per unit length}$$

$$= \ddot{X}\rho bh \qquad [10.4.1\text{-}42]$$

$\ddot{X} = $ acceleration perpendicular to cantilever

■ **EXAMPLE 10.4.1-1** Design of Cantilever Check Valve
Design a micro–check valve whose port diameter $d = 10$ μm. Assume $C_f = 8$. Determine the force to fully open the valve and find the natural frequency. See Fig. 10.3.2-1.

Solution.   Use eqs. 10.4.1-16, 10.4.1-17, 10.4.1-8, 10.4.1-9 and 10.4.1-10.

$$X = \frac{d}{4} = \frac{10}{4} = \boxed{2.5\ \mu\text{m}} \qquad (1)$$

$$r = \frac{d}{\sqrt{2}} = \frac{10}{\sqrt{2}} = \boxed{7.07\ \mu\text{m}} \qquad (2)$$

$$b \geq 1.25d = 1.25 \times 10 = \boxed{12.5\ \mu\text{m}} \qquad (3)$$

$$h \leq \frac{b}{3} = \frac{12.5}{3} = 4.17\ \mu\text{m} \qquad (4)$$

$$L \geq 24X = 24 \times 2.5 = 60\ \mu\text{m} \qquad (5)$$

Check dimensions by using eqs. 10.4.1-10 through 10.4.1-12

$$\frac{L}{X} = \frac{60}{2.5} = 24 \qquad \text{OK} \qquad (6)$$

$$\frac{L}{h} = \frac{60}{4.7} = 14.4 > 12 \quad \text{OK} \qquad (7)$$

$$\frac{L}{b} = \frac{60}{12.5} = 4.8 < 8 \quad \text{NG} \qquad (8)$$

Note that eq. (5) is an inequality; L could be larger. Solve for L in eq.

10.4.1-12

$$L \geq 8b = 8 \times 12.5 = 100 \ \mu\text{m} \qquad (9)$$

The allowable stress is found in Appendix 6.

$$S_{\text{allow}} = 4.2 \times 10^9 \ \text{dyn/cm}^2 \qquad (10)$$

Determine the actual stress by applying eq. 10.4.1-7.

$$S = \frac{1.5 \times 1.9 \times 10^{12} \times 8 \times 4.17 \times 2.5}{100^2}$$

$$= 23.8 \times 10^9 > 4.2 \times 10^9 \quad \text{NG} \qquad (11)$$

Solve for the value of L that would permit the system to meet the stress requirement. Set $S = S_{\text{allow}}$ in eq. 10.4.1-7 and solve for L.

$$L = \sqrt{\frac{1.5EC_f hX}{S_{\text{allow}}}} \qquad (12)$$

$$= \sqrt{\frac{1.5 \times 1.9 \times 10^{12} \times 8 \times 4.17 \times 2.5}{4.2 \times 10^9}}$$

$$= 238 \times 10^{-4} \ \text{cm} = 238 \ \mu\text{m} \qquad (13)$$

Equation (13) indicates that the length has exceeded the limiting size for a microdevice. Examine eq. (12) to determine how L can be reduced. The only negotiable parameter is h, which must be made smaller to solve the immediate problem. The only limitation expressed upon h is in eq. (4), which fortunately permits h to be smaller. Then, let

$$L = \boxed{200 \ \mu\text{m}} \qquad (14)$$

and solve for h in eq. (12).

$$h = \frac{L^2 S_{\text{allow}}}{1.5EC_f X} \qquad (15)$$

$$= \frac{200^2 \times 4.2 \times 10^9}{1.5 \times 1.9 \times 10^{12} \times 8 \times 2.5}$$

$$= \boxed{2.95 \ \mu\text{m}} \qquad (16)$$

Use eq. 10.4.1-4 to determine the equivalent spring rate.

$$K_c = \left(\frac{19 \times 10^{11} \times 12.5 \times 10^{-4}}{4} \right) \left(\frac{2.95}{200} \right)^3$$

$$= 1.91 \times 10^3 \ \text{dyn/cm} \qquad (17)$$

Use eq. 10.4.1-5 to determine the force to fully open the valve.

$$F = 1.91 \times 10^3 \times 2.5 \times 10^{-4}$$

$$= 0.477 \text{ dyn} = \boxed{4.77 \times 10^{-6} \text{ N}} \qquad (18)$$

Use eq. 10.4.1-36 to determine the natural frequency.

$$\omega_n = \frac{2.95 \times 10^{-4}}{(200 \times 10^{-4})^2} \sqrt{\frac{1.9 \times 10^{12}}{2.3}}$$

$$= \boxed{6.70 \times 10^5 \text{ rad/s}} \qquad (19)$$

All important dimensions have been determined. All design requirements have been met. The design is complete. ■

■ **EXAMPLE 10.4.1-2** Design of Tunister*
Design the largest micro-sized tunister whose resonant frequency is 5×10^5 rad/s. The device uses an electrostatic actuator. About midway along the length of the cantilever, it acts as the gate of a transistor (not to be designed in this example) that is essentially the output. The device is in an evacuated chamber, permitting the damping ratio to be as low as 0.002. The shape of a tunister is similar to that of an accelerometer, but has both the plates (Fig. 10.3.10-1a) and the strain gauge (Fig. 10.3.10-1b).

Solution. The largest element is the cantilever. Let its length be equal to the size limit for microdevices.

$$L = \boxed{200 \ \mu\text{m}} = 2 \times 10^{-2} \text{ cm} \qquad (1)$$

Use eq. 10.4.1-34 to determine plate radius.

$$r_0 \le L/16 = 200/16 = \boxed{12.5 \ \mu\text{m}} \qquad (2)$$

Use eq. 10.4.1-12 to determine beam width.

$$b < L/8 = 200/8 = \boxed{25 \ \mu\text{m}} \qquad (3)$$

Check plate radius with eq. 10.4.1-32.

$$r_0 < b/2 = 25/2 = 12.5 \ \mu\text{m} \quad \text{OK} \qquad (4)$$

---

*See reference [64].

Use eq. 10.4.1-33 to determine nominal gap.

$$X_0 < 2r_0/10 = 2 \times 12.5/10 = \boxed{2.5 \; \mu\text{m}} \tag{5}$$

Use eq. 10.4.1-31 to determine displacement.

$$X < X_0/5 = 2.2/5 = 0.5 \; \mu\text{m} \tag{6}$$

Use eq. 10.4.1-10 to check

$$L/X = 200/0.5 = 400 > 24 \quad \text{OK} \tag{7}$$

The relationship between natural and resonant frequencies is given by eq. 10.4.1-39. Solve for ω_n.

$$\omega_n = \frac{\omega_r}{\sqrt{1 - 2\zeta^2}} = \frac{5 \times 10^5}{\sqrt{1 - 2(0.002)^2}}$$

$$= \boxed{5.00 \times 10^5 \; \text{rad/s}} \tag{8}$$

Solve eq. 10.4.1-36 for h.

$$h = \frac{\omega_n L^2}{1}\sqrt{\frac{\rho}{E}} \tag{9}$$

$$= \frac{5 \times 10^5 \times (2 \times 10^{-2} \; \text{cm})^2}{1}\sqrt{\frac{2.3}{1.9 \times 10^{12}}}$$

$$= 2.20 \times 10^{-4} \; \text{cm} = \boxed{2.20 \; \mu\text{m}} \tag{10}$$

Using the allowable stress given for silicon in Appendix 6, check the actual stress with eq. 10.4.1-7. Note that the factor hX/L^2 is dimensionless. Hence, any consistent set of units may be used.

$$S = \frac{1.5 \times 1.9 \times 10^{12} \times 8 \times 2.20 \; \mu\text{m} \times 0.5 \; \mu\text{m}}{(200 \; \mu\text{m})^2} \tag{11}$$

$$= 6.27 \times 10^8 \; \text{dyn/cm} < 4.2 \times 10^9 \quad \text{OK} \tag{12}$$

Check beam thickness with eq. 10.4.1-11.

$$L/h = 200/2.2 = 90.9 > 12 \quad \text{OK} \tag{13}$$

Check beam width with eq. 10.4.1-9.

$$b/h = 25/2.2 = 11.4 > 3 \quad \text{OK} \tag{14}$$

The equivalent spring is given by eq. 10.4.1-4.

$$K_c = \frac{19 \times 10^{11}}{4} \text{ dyn/cm}^2 \times 25 \times 10^{-4} \text{ cm} \times (2.20/200)^3 \qquad (15)$$

$$= 1.58 \times 10^3 \text{ dyn/cm} \qquad (16)$$

The static force required to bend the spring through a displacement X is

$$F_s = K_c X \qquad (17)$$

$$= 1.58 \times 10^3 \text{ dyn/cm} \times 0.5 \times 10^{-4} \text{ cm}$$

$$= 0.079 \text{ dyn} = 0.79 \times 10^{-6} \text{ N} \qquad (18)$$

The dynamic force required to bend the spring by the same amount in eq. (17) is reduced by the magnification ratio R_r, which is given by eq. 10.4.1-38.

$$R_r = 1/(2 \times 0.002) \qquad (19)$$

$$= 250 \qquad (20)$$

The dynamic force is given by eq. 10.4.1-37.

$$F_d = F_s/R_r \qquad (21)$$

$$= 0.79 \times 10^{-6} \text{ N}/250 = 3.16 \times 10^{-9} \text{ N} \qquad (22)$$

This force will be produced by electrostatic means, where

$$F_0 = F_d = 3.16 \times 10^{-9} \text{ N} \qquad (23)$$

The voltage required to produce this force is given by eq. 10.4.1-27.

$$V_0 = \frac{2.5 \ \mu\text{m}}{12.5 \ \mu\text{m}} \sqrt{\frac{0.316 \times 10^{-8} \times 2}{\pi 0.8854 \times 10^{-11}}} \qquad (24)$$

$$= 3.01 \text{ V} \qquad (25)$$

To find the allowable voltage gradient for vacuum (free space) see Appendix 6, Dielectric Properties.

$$\gamma_{\text{allow}} = 10^6 \text{ V/m} = 1 \text{ V}/\mu\text{m} \qquad (26)$$

Then, the allowable voltage is found by solving eq. 10.4.1-29 for V.

$$V_{\text{allow}} = \gamma_{\text{allow}} X_0 \qquad (27)$$

$$= 1 \text{ V}/\mu\text{m} \times 2.5 \ \mu\text{m} = 2.5 \text{ V} < 3.01 \text{ V} \quad \text{NG} \qquad (28)$$

Equation (28) indicates that the voltage required to bend the cantilever is greater than the allowable voltage. This requires a redesign. Let us start by meeting the allowable voltage and working backwards. Then let

$$V_0 = V_{\text{allow}} = 2.5 \text{ V} \qquad (29)$$

Note that many of the quantities that have been determined will not be affected by the voltage requirement. First, solve for F_0 in eq. 10.4.1-28.

$$F_0 = 2.19 \times 10^{-9} \text{ N} = F_d \tag{30}$$

$$F_s = F_d R_r = 0.549 \times 10^{-6} \text{ N} = 0.0549 \text{ dyn} \tag{31}$$

$$X = F_s/K_c = \boxed{0.347 \,\mu\text{m}} \tag{32}$$

which is smaller than that found in eq. (6). Consequently, this will satisfy all the design ratio inequalities. The design is complete.   ■

10.4.2 Design of Torsion Bar Devices*

There are microdevices that depend upon a torsion bar for their operation. Typically, the cross section of the bar is rectangular. The torsion bar supports a mass (the payload) and will twist through an appropriate angle. The generalized model of a torsionally supported system is shown in Fig. 10.4.2-1,

where a = half the width of the torsion bar
$\quad\quad b$ = half the depth of the torsion bar
$\quad\quad L$ = length of the torsion bar
$\quad\quad \theta$ = angle of twist of torsion bar
$\quad\quad T$ = torque required to twist the torsion bar
$\quad\quad G$ = modulus of elasticity in shear
$\quad\quad \ell$ = length of mass
$\quad\quad w$ = width of mass
$\quad\quad h$ = thickness of mass
$\quad\quad r$ = moment arm for capacitor plates
$\quad\quad r_0$ = radius of capacitor plates
$\quad\quad X_0$ = nominal spacing of capacitor plates
$\quad\quad V_0$ = voltage applied to capacitor plates
$\quad\quad F_0$ = force developed by electrostatic means on each capacitor plate
$\quad\quad T_0$ = algebraic sum of all torques produced by electrostatic forces

Equivalent spring[†]

$$K_t = \frac{T}{\theta} = \frac{CG}{L} \tag{10.4.2-1}$$

where

$$C = ab^3 \left[\frac{16}{3} - 3.36 \frac{b}{a} \left(1 - \frac{b^4}{12a^4} \right) \right] \tag{10.4.2-2}$$

---

*See reference [50].

[†]Reference [10], p. 290, case 4.

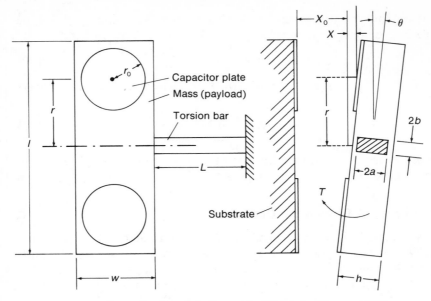

Figure 10.4.2-1   Generalized model of system with torsion bar suspension.

Using eqs. 10.4.2-1 and 10.4.2-2, torsional stiffness and bending stiffness can be determined. The following conclusions can be made:

$$a = b \quad \text{stiffest torsional spring} \qquad [10.4.2\text{-}3]$$

$$a \gg b \quad \text{weakest torsional spring} \qquad [10.4.2\text{-}4]$$

$$a/b = 3 \quad \text{optimum aspect ratio}$$

$$\text{(for combined torsion and bending)} \qquad [10.4.2\text{-}5]$$

Shear stress in torsion*

$$\tau_s = \frac{T(3a + 1.8b)}{8a^2b^2} \qquad [10.4.2\text{-}6]$$

Equivalent Mass   The moment of inertia about the center of mass is given in entry 3, Table 2.1.3-1, Appendix 2.

$$J = J_c = \frac{\rho h w \ell (h^2 + \ell^2)}{12} \qquad [10.4.2\text{-}7]$$

Torque to Twist Torsion Bar

$$T = K_t \theta \qquad [10.4.2\text{-}8]$$

---

*Reference [10], p. 290, case 4.

Aspect Ratios   The dimension a of the torsion bar is chosen larger than dimension b. The dimension $2a$ should not be larger than the mass,

$$2a \leq h \qquad [10.4.2\text{-}9]$$

and should be parallel to dimension h. To be sure the mass acts like a rigid body, let

$$\ell \geq w \geq \ell/5 \qquad [10.4.2\text{-}10]$$

and

$$w \geq h \geq w/5 \qquad [10.4.2\text{-}11]$$

Output Sensor*   Similar to eqs. 10.4.1-20 through 10.4.1-23, the torsional system may employ the following output sensors: a torsional strain gauge, a torsional piezoelectric element, or a parallel plate capacitor.

Static Actuator*   Similar to eq. 10.4.1-24, a piezoelectric element may be used to twist the torsion (bar) spring. There may be as many as four, one on each side of the torsion bar. For a capacitor actuator, there may be as many as four pairs of capacitor plates, two on each side of the h-t face of the mass. The electrostatic force F_0, the voltage V_0, and the allowable gradient γ_{allow} are given by eqs. 10.4.1-25 through 10.4.1-30.

The torque T_0 developed by the electrostatic forces is given by

$$T_0 = NrF_0 \qquad [10.4.2\text{-}12]$$

where

$$N = \text{number of forces} \qquad [10.4.2\text{-}13]$$

In order to determine the maximum plate size, refer to Fig. 10.4.2-1. This is a matter of geometry.

$$2r_0 \leq w \qquad [10.4.2\text{-}14]$$

The nominal plate spacing X_0 should be large enough that the rotation through an angle θ with a moment arm r will not appreciably upset the basic relationships for a parallel plate capacitor. The distance X traveled by the center of each plate is

$$X = r \sin \theta$$

and this should be smaller than $\frac{1}{5}$ of the nominal plate spacing. Then,

$$X = r \sin \theta \leq \tfrac{1}{5} X_0 \qquad [10.4.2\text{-}15]$$

Since the nature of the device limits the travel,

$$\theta \ll 1 \qquad [10.4.2\text{-}16]$$

---

*See references [54] and [56].

and

$$\sin \theta \simeq \theta \qquad [10.4.2\text{-}17]$$

Then

$$\frac{X_0}{r\theta} \geq 5 \qquad [10.4.2\text{-}18]$$

Vibrator Actuator The natural frequency for a torsionally oscillating mass is given by

$$\omega_n = \sqrt{K_t/J} \qquad [10.4.2\text{-}19]$$

where K_t and J are given by eqs. 10.4.2-1 and 10.4.2-7, respectively. The advantages and equations relevant to resonant conditions are given by eqs. 10.4.1-37 through 10.4.1-39.

10.4.3 Design of Diaphragm Devices

A number of microdevices depend upon a diaphragm for their operation. Such devices are valve, pump, and pressure sensor. The generalized model of these devices is shown in Fig. 10.4.3-1,

where a = radius of diaphragm
  t = thickness of diaphragm
  Y = displacement of center of diaphragm
  P = pressure due to fluid in flow area
  d = port diameter
  r = radius of half-round duct
  X_0 = nominal spacing of capacitor plates
  r_0 = radius of capacitor plates
  V_0 = electrostatic voltage
  P_0 = electrostatic pressure

Displacement and Stress (any units) For micro-sized devices, the diaphragm usually is very thin compared to the displacement. Consequently, the membrane stress is appreciable. For this reason, the formulae* for displacement and for stress include both flexural and membrane stresses. The displacement is given by

$$\frac{P}{E}\left(\frac{a}{t}\right)^4 = \frac{5.33}{1-\nu^2}\left(\frac{Y}{t}\right) + \frac{2.6}{1-\nu^2}\left(\frac{Y}{t}\right)^3 \qquad [10.4.3\text{-}1]$$

where ν = Poisson's ratio
  E = Young's modulus

---

*See reference [10], p. 407, case 3.

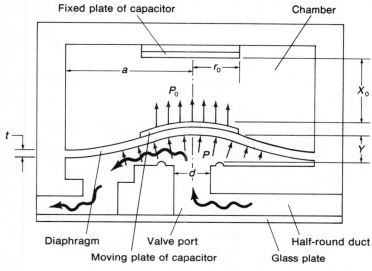

Figure 10.4.3-1   Generalized diaphragm device.

and the stress by

$$\frac{s}{E}\left(\frac{a}{t}\right)^2 = \frac{2}{1-\nu}\left(\frac{Y}{t}\right) + 0.976\left(\frac{Y}{t}\right)^2 \qquad [10.4.3\text{-}2]$$

Diaphragm Aspect Ratios   In order for the diaphragm to deflect properly, both the thickness t and the displacement Y should be very much smaller than the diaphragm radius a.

$$\frac{a}{t} \geq 100 \qquad [10.4.3\text{-}3]$$

$$\frac{a}{Y} \geq 100 \qquad [10.4.3\text{-}4]$$

Flow Areas   For design of flow areas, see eqs. 10.4.1-13 through 10.4.11-7, but where Y is used in place of X.

Output Sensor   For the output sensors, refer to eqs. 10.4.1-20 through 10.4.1-23. In order to produce the largest output possible, the designer is tempted to make the capacitor plates as large as possible. However, there is a trade-off: Only the *change* in capacitance produces an output. This suggests the use of only the center of the diaphragm. As a fair compromise, let

$$\frac{a}{r_0} \geq 4 \quad \text{(for sensor)} \qquad [10.4.3\text{-}5]$$

In order to minimize the influence of the displacement upon the capacitive effects, let

$$\frac{X_0}{Y} \geq 10 \qquad \qquad [10.4.3\text{-}6]$$

Static Actuator For audio frequencies (200–5000 Hz) a piezoelectric element is effective. For a small size and a low voltage, it produces a comparably large force. See eq. 10.4.1-24.

For a capacitive actuator, it is recommended that the entire surface of the diaphragm be used. This is not overindulgence, but a means to treat the diaphragm like a membrane. Since the diaphragm will most likely be very thin, a uniform pressure over its entire surface is preferable to a load concentrated in a small area. In this case, let

$$r_0 = a \quad \text{(for actuator)} \qquad \qquad [10.4.3\text{-}7]$$

The entire surface was used because of the membrane nature of the diaphragm. However, this will result in poor capacitive performance unless the nominal plate spacing is large. Then, let

$$\frac{2r_0}{X_0} \geq 20 \qquad \qquad [10.4.3\text{-}8]$$

The electrostatic forces and voltage gradient follow in the same manner as eqs. 10.4.1-25 through 10.4.1-30. Since the electrostatic force is uniformly distributed over the entire surface of the diaphragm, it may be considered an electrostatic pressure P_0, where

$$P_0 = F_0/\pi r_0^2 \qquad \qquad [10.4.3\text{-}9]$$

for which eq. 10.4.1-25 becomes

$$P_0 = \frac{F_0}{A} = \frac{V_0^2 \epsilon}{2X_0^2} = \frac{\epsilon}{2}\left(\frac{V_0}{X_0}\right)^2 \qquad \qquad [10.4.3\text{-}10]$$

Apply the allowable voltage gradient, eq. 10.4.1-29.

$$P_0 = \frac{F_0}{A} = \frac{\epsilon}{2}(\gamma_{\text{allow}})^2 \qquad \qquad [10.4.3\text{-}11]$$

Vibrator Actuator (units—microns) Since the displacement of a thin diaphragm is nonlinear, there will be some difficulty in computing the natural frequency. Hence, only a rough approximation* will be offered. (The actual value should be determined empirically.)

$$\omega_n \simeq 2 \times 10^6 \, t/a^2 \qquad \qquad [10.4.3\text{-}12]$$

The conditions at resonance are similar to those covered in eqs. 10.4.1-37 through 10.4.1-39.

---

*Approximately correct for flexure.

Design Details for Vibrator Pump   A vibrator pump can be fabricated from a vibrating diaphragm and two cantilever check valves. The diaphragm is vibrated at its resonant frequency. See Fig. 10.3.4-1. The check valves are opened and closed by the pumping action of the diaphragm. To be sure that each valve opens and closes quickly and at the right time, it is necessary to minimize phase lag (which produces lost motion). To keep the phase lag below 10%, the natural frequency of each check valve ω_{nc} should be higher than that of the diaphragm ω_{nd}. However, high frequency requires a stiff cantilever spring, which will introduce other problems. As a compromise, let,

$$\frac{\omega_{nc}}{\omega_{nd}} \geq 4 \qquad [10.4.3\text{-}13]$$

To minimize lost motion due to the fluid volume expended in opening and closing each valve, let the displaced volume of each valve be less than 10% of that of the diaphragm. This provides a means to establish aspect ratios between diaphragm dimensions and cantilever dimensions as follows:

$$a \geq L \qquad [10.4.3\text{-}14]$$

$$\frac{Y}{X} \geq 2 \qquad [10.4.3\text{-}15]$$

$$\frac{L}{b} \gg 1 \qquad [10.4.3\text{-}16]$$

Pump Flow Rate [cgs units (cm, g, s)]   The volumetric flow rate is equal to the displaced volume of the diaphragm per unit time. The volumetric displacement of a diaphragm is *not* equal to the area multiplied by X, nor is it the volume of a cone, but somewhere between the two limits. There will be losses due to timing (or phase) errors due to the valves not opening or closing on time. There will be losses due to the volumetric requirement to direct some of the flowing fluid to open and close the valves. Considering all of the above, the pump flow rate becomes

$$Q \simeq 0.20 Y a^2 \omega_r \qquad [10.4.3\text{-}17]$$

Flow velocity through the valve port is given by

$$V_p = Q/A_p = 4Q/\pi d^2 \qquad [10.4.3\text{-}18]$$

■ **EXAMPLE 10.4.3-1** Design of Pressure Sensor*
Design the largest micro–pressure sensor for a pressure equal to 3.67×10^{-5} dyn/μm². Determine sensitivity. See Fig. 10.3.8-1a.

Solution.   A diaphragm will be used. Let the diameter be equal to to the size limit.

$$2a = 200 \ \mu m \qquad (1)$$

---

*See references [54], [55] and [56].

or

$$a = \boxed{100 \ \mu m} \qquad (2)$$

The plate radius for a sensor is given by eq. 10.4.3-5.

$$r_0 < \frac{a}{4} = \frac{100}{4} = \boxed{25 \ \mu m} \qquad (3)$$

Nominal plate spacing is given by eq. 10.4.3-8.

$$X_0 < \frac{2r_0}{20} = \frac{2 \times 25}{20} = \boxed{2.5 \ \mu m} \qquad (4)$$

The displacement of the diaphragm is given by eq. 10.4.3-6.

$$Y < \frac{X_0}{10} = \frac{2.5}{10} = \boxed{0.25 \ \mu m} \qquad (5)$$

The relationship between pressure and diaphragm thickness is given by eq. 10.4.3-1. Using the given pressure, solve for t by iteration.

$$\boxed{t \simeq 0.5 \ \mu m} \qquad (6)$$

The stress is given by eq. 10.4.3-2.

$$S = \frac{19 \times 10^{11}}{1} \left(\frac{0.5}{100} \right)^2 \left[\frac{2}{(1 - 0.18)} \left(\frac{0.25}{0.5} \right) + \frac{0.976}{1} \left(\frac{0.25}{0.5} \right)^2 \right] \qquad (7)$$

$$= 6.26 \times 10^7 < 4.2 \times 10^9 \quad \text{OK} \qquad (8)$$

The nominal capacitance is given by eq. 10.4.1-22.

$$C = \frac{\epsilon A}{X_0} \qquad (9)$$

where

$$A = \pi r_0^2 \qquad (10)$$

$$C = \frac{(0.885 \times 10^{-11} \ \text{F/m}) \times \pi (25 \times 10^{-6} \ \text{m})^2}{2.5 \times 10^{-6} \ \text{m}} \qquad (11)$$

$$\boxed{= 0.695 \times 10^{-14} \ \mu m} \qquad (12)$$

The capacitance change for full pressure (sensitivity)

$$\frac{\Delta C}{C} = \frac{X_0}{X_0 - Y} - 1 = \frac{2.5}{2.5 - 0.25} - 1 = 0.11 = 11\% \qquad (13)$$

For port and duct, use eqs. 10.4.1-16 and 10.4.1-17, respectively.

$$d = 4Y = 4 \times 0.25 = \boxed{1.0 \ \mu\text{m}} \tag{14}$$

$$r = d/\sqrt{2} = \boxed{0.707 \ \mu\text{m}} \tag{15}$$

∎

■ **EXAMPLE 10.4.3-2** Design of Diaphragm Control Valve
Design the largest micro–control valve. Assume the chamber is filled with ether. Also assume fluid pressure $P_f = 1.33 \times 10^{-5}$ dyn/μm². See Fig. 10.3.3-1b.

Solution. Let the diaphragm diameter be equal to the size limit for microdevices. Then the diaphragm radius is

$$\boxed{a = 100 \ \mu\text{m}} \tag{1}$$

The plate radius is given by eq. 10.4.3-7.

$$r_0 = a = \boxed{100 \ \mu\text{m}} \tag{2}$$

Plate spacing is given by eq. 10.4.3-8.

$$X_0 \le \frac{2r_0}{20} = \frac{2 \times 100}{20} = \boxed{10 \ \mu\text{m}} \tag{3}$$

Diaphragm displacement is given by eq. 10.4.3-6.

$$Y \le \frac{X_0}{10} = \frac{10}{10} = 1 \ \mu\text{m} \tag{4}$$

It will subsequently develop that this exceeds the diaphragm thickness t, which will introduce a design difficulty. Observing the inequality, let

$$\boxed{Y = 0.25 \ \mu\text{m}} \tag{5}$$

Electrostatic pressure is given by eq. 10.4.3-11. For ether,

$$P_0 = \frac{4.0 \times 10^{-11} \ \text{F/m}}{2} \left(5.0 \times 10^6 \ \text{V/m}\right)^2 \tag{6}$$

$$= 500 \ \text{N/m}^2 = 5.0 \times 10^{-5} \ \text{dyn}/\mu\text{m}^2 \tag{7}$$

Electrostatic voltage is given by eq. 10.4.1-29.

$$V_0 \leq X_0 \gamma_{\text{allow}} = 10 \ \mu\text{m} \times 5 \ \text{V}/\mu\text{m} = \boxed{50 \ \text{V}} \tag{8}$$

Net pressure P against diaphragm

$$P = P_0 - P_f = 3.67 \times 10^{-5} \, \text{dyn}/\mu\text{m}^2 \tag{9}$$

and solve eq. 10.4.3-1 for t by iteration.

$$\boxed{t = 0.5 \ \mu\text{m}} \tag{10}$$

Determine the stress by using eq. 10.4.3-2.

$$S = 19 \times 10^{11} \left(\frac{0.5}{100}\right)^2 \left[\frac{2}{1 - 0.18}\left(\frac{0.25}{0.5}\right) + 0.976\left(\frac{0.25}{0.5}\right)^2\right] \tag{11}$$

$$= 6.26 \times 10^7 \, \text{dyn}/\text{cm}^2 < 4.2 \times 10^9 \quad \text{OK} \tag{12}$$

Port and duct sizes are given by eqs. 10.4.1-16 and 10.4.1-17, respectively.

$$d = 4Y = 4 \times 0.5 = \boxed{2.0 \ \mu\text{m}} \tag{13}$$

$$r = 2.0/\sqrt{2} = \boxed{1.42 \ \mu\text{m}} \tag{14}$$

■

SUGGESTED REFERENCE READING FOR CHAPTER 10

Silicon structure [51]; [52].
Sensors [54]; [56].
Micromachining [51]; [52]; [54] [56].
Ink jet printer [49].
Cantilever [58].
Diaphragm [58].
Torsional mirror [50].
Capillary [52].
Pressure sensor [54]; [55]; [56].
Vib-accelerometer [53].
Tunister [64].

PROBLEMS FOR CHAPTER 10

General note for all problems in this chapter: Briefly describe how to fabricate the device, using only those fabrication techniques currently in use today. If the technique is not described in this chapter, then cite the literary source (and indicate why it should be included in this chapter).

10.1. *Creative Engineering*. Make a simple layout for each of the following devices and explain how it operates including a few important dimensions:
   (a) Clamp.
   (b) Two opposing fingers to grab microparticles.
   (c) Latch.

10.2. *Self-Transporter*. Make a simple layout of a device that is self-propelled and can carry a payload equal to its own size. (Some suggestions: Use a set of vibrating pins set at 45° angle on a moderately soft surface. On down stroke, pins dig into surface. On upstroke, they slide, thus providing half-wave rectification.)

10.3. Similar to Problem 10.2, but for travel in liquids.

10.4. Continuing with either Problem 10.2 or 10.3, make a simple layout of a radio control to steer the self-propelled device.

10.5. Show by a simple layout how an ink jet printer can serve as a graphics printer. Indicate the size of the matrix needed to produce pictures of newspaper quality.

10.6. Make a layout of a set of check valves, where each one opens at a different pressure. Suggest some uses.

10.7. Make a simple layout of a pressure regulator system that uses a pressure sensor to open and close the switch to a vibrator pump.

10.8. Make a simple layout showing several of a one-thousand-fiber optic communication system. Show how fibers and mirrors are supported.

10.9. Make a simple layout of screws that can be hollowed out, without depreciating their strength significantly, to house the following sensors, making it possible to insert such sensors in a machine without drilling holes:
   (a) Pressure sensor.
   (b) Temperature sensor.
   (c) Accelerometer.
   (d) Vibrating tachometer (like Frahm device in Ex. 8.7.4-1).
   (e) Fuel-air analyzer.

10.10. Make a simple layout of a temperature control system for either heating or cooling, with appropriate source. (Some suggestions: Use a temperature sensor that opens and closes a valve from a heat exchanger.)

10.11. *Micro–Pressure Regulator*. Make the micro-sized counterpart of the regulator shown in Fig. P-16.25. (Suggestion: Use a diaphragm that reacts to pressure and closes a valve in the flow path, thus offering a restriction to flow when pressure is too high.)

10.12. *Automotive Devices*. Make a detailed layout of devices to accomplish the following tasks:
   (a) One thousand valves to serve as a carburetor.
   (b) Air-fuel mixer and supercharger.

(c) Computer for "smart ignition timer." Adjusts spark advance to suit temperature, pressure, engine speed, and so on.

(d) Computer for "smart carburetor." Adjusts air-fuel mixture for maximum efficiency, depending upon several variables: temperature of engine, air, and fuel; pressure of engine and air; engine speed; humidity; engine load, and so on.

10.13. *BioMedical Applications.* Make a detailed layout showing how each of the following may be measured:
   (a) Pulse.
   (b) Respiration.
   (c) Skin temperature.
   (d) Blood pressure.

10.14. *Heart Attack Alert.* Make a detailed layout of a vital signs monitor that also makes an electrocardiogram via radio link to a nearby recording machine. Device sounds an alert either by beeper or by radio communication to a medical attendant.

10.15. *Insulin Dispenser for Diabetics.* Make a detailed layout of a system that checks the diabetic's blood chemistry and dispenses the appropriate amount of insulin. This is preferable to the "one shot" arrangement currently in use when such automation is not available.

10.16. *Timed Medicine Dispenser.* Similar to the previous problem, but where blood chemistry is not performed. Instead, the medicine is dispensed at a predetermined rate.

10.17. Using a uniform cantilever, determine the lowest natural frequency obtainable if the beam thickness is limited (to prevent buckling) to $L/h < 240$.

10.18. Continuing with the previous problem, let the paddle [shown in Fig. 10.3.10-1(c)] be four times wider and four times thicker than the cantilever, and one-fourth the length of L.

10.19. *Design of Cantilever Check Valve.* Design the largest micro-sized cantilever check valve for which $C_f = 8$. Determine the force to fully open the valve, and find the natural frequency. See Fig. 10.3.2-1.

10.20. *Design of Cantilever Control Valve.* Design the largest micro-sized cantilever control valve, assuming the fluid is air or some material that will not interfere with the operation of the capacitive actuator. (Suggestion: The numerical value of h will be in conflict several times. Continue, each time reducing the value until all inequality constraints are met.)

10.21. Redesign the valve in the previous problem if the fluid pressure is 0.7×10^{-4} dyn/μm^2.

10.22. *Design of Vib-Accelerometer.* Design the largest micro-sized accelerometer to measure vibration in the frequency range 40,000–100,000 rad/s.

10.23. Similar to the previous problem, but for the frequency range 1000–40,000 rad/s.

10.24. *Design of Tunister.* Redesign the tunister of Ex. 10.4.1-2 for a resonant frequency of 10^6 rad/s.

10.25. *Design of Torsional Mirror for Optical Communication.* Design the device discussed in Section 10.3.5 for amplitude equal to 0.01 rad, $r = 2r_0$ and damping ratio = 0.6. (For the layout, add electrostatic plates to Fig. 10.3.2-1.) Find K_t, J_c, W_n and the bandwidth.

10.26. *Design of Pressure Sensor.* Redesign the device in Ex. 10.4.3-1 for a pressure equal to 3.0×10^{-5} dyn/s.

10.27. *Design of Diaphragm Control Valve.* Redesign the device in Ex. 10.4.3-2 where the electrostatic drive must act against pressure and also against the diaphragm as a spring. Assume fluid pressure $P_f = 2.0 \times 10^{-5}$ dyn/μm^2.

10.28. In eq. 10.4.3-1 the first term corresponds to flexure and the second to membrane effects. Determine the numerical value of Y/t for which:
(a) Both terms are equal.
(b) The second term is less than 5% of the first. (Hence flexure rules and membrane may be neglected.)
(c) The first term is less than 5% of the second. (Hence membrane rules and flexure may be neglected.)

10.29. Show that when $Y/t \ll 1$, eq. 10.4.3-12 is a good approximation of natural frequency.

10.30. *Design of Vibrating Pump.* Design the largest microsized diaphragm pump for which the amplitude of vibration is equal to 0.3 μ and damping ratio is equal to 0.2. Be sure to check beam width b for all four criteria. Determine the resonant frequency and pump flow rate. See Fig. 10.3.4-1.

Chapter 11

Synthesis Using Laplace and Logical Theorems

11.1 DOMINANCE

11.1.1 Definition

The extensive table of Laplace transform pairs in Appendix 3 points out that many different terms affect the solution. Upon close examination, it becomes clear that not all terms have an equal effect upon the solution; some seem to overrule the dynamic stage. Such a term is called the *dominant* one. When there is clearly a dominant term, the properties of the entire system can be estimated by observing only those of the dominant term.

The opposite of dominance is the case for which no subsystem exerts the major influence upon the overall system performance. In that case, no one subsystem prevails and it becomes necessary to study the system as a whole. This leads to two questions: "How can we ascertain if dominance exists?" and "How can we identify which is the dominant factor?"

These questions and their answers are covered in the subsequent sections.

11.1.2 Dominant Factor and the Transient Solution in Time Domain

In order to demonstrate how a dominant factor influences the system solution, assume that the characteristic determinant has been factored. Using the method of partial fractions in Section 4.5, the result takes the form of a sum of low order subsystems. Each of these can be transformed—one at a time—to the time domain and plotted. Here, it is convenient to compare the coefficients and the decay constants of each solution. This provides a quantitative means by which to establish if there is dominance and which term dominates.

For example, consider the result in the time domain as follows:

$$y(t) = Ae^{-ct} + Be^{-at}\sin(\omega t + \phi) \qquad [11.1.2\text{-}1]$$

If the coefficient

$$A \gg B \qquad\qquad [11.1.2\text{-}2]$$

then the magnitude of the exponential factor dominates. If

$$A \ll B \qquad\qquad [11.1.2\text{-}3]$$

the quadratic factor dominates. See Fig. 11.1.2-1. Now examine the decay coefficients, a and c. These define which dies out faster.

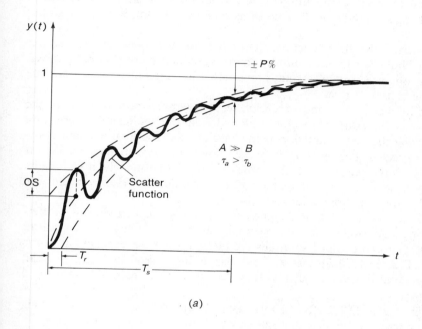

(a)

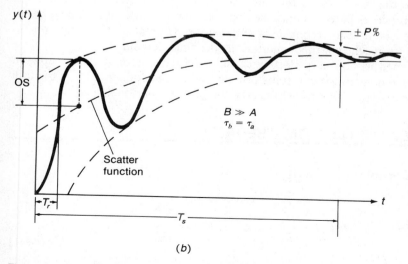

(b)

Figure 11.1.2-1   Dominance. (a) Exponential rules. (b) Sinusoid rules.

An alternate means to make the same measurements can be facilitated by the scatter function, covered in the next section.

11.1.3 Scatter Function

Suppose one or more functions of time $f_1(t)$, $f_2(t)$, $f_3(t)$, ..., $f_n(t)$ are summed with a function $f_0(t)$ that has some special properties. Suppose the same set of functions $f_1(t), ..., f_n(t)$ is summed with a different function $g_0(t)$ that also has special properties, but different ones from those of $f_0(t)$. The functions $f_0(t)$ and $g_0(t)$ may themselves be sums of functions and are called *scatter functions*.

Ordinarily, the time functions (both the summed ones and the scatter function) would be summed point by point to produce the final plot. However, this often obscures certain system properties. Also, the objective can become lost in the clutter of terms.

To minimize clutter and obscurity, let us examine the properties of the system when the scatter function is plotted first and the summed terms are plotted using the scatter function as the base or axis. The final outcome is the sum, as expected, but the sum is broken down into two major components where the relative weights of each can be ascertained. Ultimately, it will not be necessary to actually make these plots in the time domain. However, they are shown this way now for explanation purposes.

Refer to Fig. 11.1.2-1, where one exponential is summed with two different damped sinusoids. The exponential has been selected as the scatter function. The two damped sinusoids are plotted using the exponential as the axis. In Fig. 11.1.2-1(a) the sinusoid has a high frequency and a small amplitude, and it dies out quickly. It is clear that in this case the exponential dominates.

Refer to Fig. 11.1.2-1(b). In this case, the sinusoid has a large amplitude and a low frequency, and it dies out slowly. Here the sinusoid dominates.

If we did not have all the information and were forced to estimate the system properties based on only the dominant term in each of the above cases, the result would be a reasonable approximation. Ultimately, this same conclusion will be reached without plotting in the time domain. In fact, it will be accomplished without any plotting at all, and what's more, it will be done in the Laplace domain without transforming to the time domain. This is the task for the scatter function, to help handle the problem of dominance.

11.2 SCATTER LOGICAL THEOREM

11.2.1 Derivation

Given

$$L_1(s) = \frac{N_1}{D_1} \quad \text{and} \quad L_2(s) = \frac{N_2}{D_2} \qquad\qquad \text{[11.2.1-1]}$$

if

$$D_1 = F(s)D_2 \quad \text{and} \quad N_1 + FN_2 = cD_2$$

where $F(s)$ will be called the scatter or separation function. Then

$$\mathcal{L}^{-1}L_1(s) + \mathcal{L}^{-1}L_2(s) = \mathcal{L}^{-1}\frac{c}{F(s)} \qquad [11.2.1\text{-}2]$$

Proof.

$$L_3(s) = L_1(s) + L_2(s) = \frac{N_1}{D_1} + \frac{N_2}{D_2}$$

$$= \frac{N_1}{FD_2} + \frac{N_2}{D_2} = \frac{N_1 + FN_2}{FD_2} = \frac{cD_2}{FD_2} = \frac{c}{F}$$

$$L_3(t) = \mathcal{L}^{-1}L_3(s) = \mathcal{L}^{-1}\frac{c}{F(s)} \qquad [11.2.1\text{-}3]$$

The value of the scatter logical theorem is that it serves several roles: It predicts the separation or scatter of the sum in the time domain; it provides a method to select a scatter function; it determines dominance.

■ **EXAMPLE 11.2.1-1** Determination of Scatter

Given entries 412, 422, 432, and 442, Appendix 3, determine which to use as the base function (the one to which each of the other three will be added) and determine the scatter of each of the other three functions with respect to it. The three functions will be referred to as *scatter mates* of the base function. (Note that the base function is *not* the scatter function which will be determined subsequently.)

Solution.   Refer to the solution of each function in the time domain in Appendix 3. By inspection, it is clear that these have a number of similar terms. Upon closer inspection, we note that, of the terms they all have in common, the only one with opposite signs is entry 412. This means that summing with entry 412 is the only way to cancel terms, which suggests that it be used as the base function.

Since entry 412 has a high power of s in the denominator, it will serve in some cases as function #1, and as #2 in other cases. Since there are so many functions to consider, a shortcut is in order. The strategy is to sum numerators (multiplied by the appropriate function) to take the form of the denominator. Upon inspection of the numerator of entry 412, it will be convenient to let

$$\frac{\omega_n^4}{s^3} = \frac{\omega_n^2}{s^3}(\omega_n^2) \qquad (1)$$

Entry 422 has s^2 in the denominator. To make it compatible with entry 412 and the form of eq. (1), let

$$\frac{\omega_n^2(s + 2a)}{s^2} = \frac{\omega_n^2}{s^3}(s^2 + 2as) \qquad (2)$$

Now when entries 412 and 422 are summed, the result is

$$\frac{\omega_n^2\left(\omega_n^2 + s^2 + 2as\right)}{s^3\left(s^2 + 2as + \omega_n^2\right)} = \frac{\omega_n^2}{s^3} \tag{3}$$

Take the inverse Laplace transform of eq. (3) with the aid of entry 143.

$$\mathscr{L}^{-1}\frac{\omega_n^2}{s^3} = \frac{\omega_n^2 t^2}{2!} \tag{4}$$

Equation (4) is the scatter function between entries 412 and 422. To check the result, sum the inverses given for each in Appendix 3.

Next, consider entry 432. Use the same strategy as before—complete the numerator so it takes the form of the denominator. In order to do this, note that in the Laplace domain, entry 432 has the term ω_n^2 in its numerator. But entry 412 also has this term in its numerator. Only one will do. The one in the numerator of entry 432 is not negotiable since it is intimately tied by a summing sign, while the one in the numerator of entry 412 can be considered a multiplier [in fact this was done in eq. (1)]. It will be convenient to start with entry 432 and to factor the term

$$\frac{\omega_n^4}{2as^4} \tag{5}$$

To make entry 412 compatible with this factor, modify entry 412 as follows:

$$\frac{\omega_n^4}{s^3} = \frac{\omega_n^4}{2as^4}(2as) \tag{6}$$

Now sum entries 412 and 432.

$$\frac{\omega_n^4\left(2as + s^2 + \omega_n^2\right)}{2as^4\left(s^2 + 2as + \omega_n^2\right)} = \frac{\omega_n^4}{2as^4} \tag{7}$$

Take the inverse Laplace transform of eq. (7) with the help of entry 142.

$$\mathscr{L}^{-1}\frac{\omega_n^4}{2as^4} = \frac{\omega_n^4 t^3}{(2a)3!} \tag{8}$$

Equation (8) is the scatter function between entries 412 and 432. To check the result, sum the inverses given for each entry. The remaining function will be left as a problem in the problem section of this chapter.   ∎

■ **EXAMPLE 11.2.1-2** Synthesis Using Unknown Scatter Function
For the function given below, determine the mating function that will produce the lowest scatter, and check result.

$$\frac{s^4}{(s^2 + \omega^2)(s + \alpha^2)} = F(s) \tag{1}$$

Solution.   In general, the lowest possible scatter is zero. Let us determine if zero is the scatter for this specific example. To accomplish the task, complete the numerator. To do this, first multiply out the denominator.

$$D = s^4 + (\alpha^2 + \omega^2)s^2 + \alpha^2\omega^2 \tag{2}$$

To work with minimum scatter, let the given function supply the highest derivative term. In this case it is ω^4. The scatter function will then complete the numerator.

$$F_s(s) = \frac{(\alpha^2 + \omega^2)s^2 + \alpha^2\omega^2}{(s^2 + \omega^2)(s^2 + \alpha^2)} \tag{3}$$

Sum functions.

$$F(s) + F_s(s) = \frac{s^4 + (\alpha^2 + \omega^2)s^2 + \alpha^2\omega^2}{(s^2 + \omega^2)(s^2 + \alpha^2)} = 1 \tag{4}$$

Upon taking the inverse Laplace transform, the result is the δ function. Checking this result will be more of a challenge, since the result is not an entry in the extensive table of Appendix 3. (We might entertain the idea of adding this entry upon its completion. Many entries in the table were derived by using this method.) In order to check the result, make use of the following:

$$\alpha^2 \text{ times entry 325} \quad \text{(in time domain)} \tag{5}$$

$$\omega^2 \text{ times entry 325} \quad \text{(in time domain)} \tag{6}$$

$$\alpha\omega \text{ times entry 323} \quad \text{(in time domain)} \tag{7}$$

$$\text{entry 327} \quad \text{(in time domain)} \tag{8}$$

Sum all of the above terms, noting that there will be eight trigonometric terms. These will all cancel out, leaving only the δ function which at time $t = 0^+$ is equal to 0. Thus, the result has been verified.

On many occasions, it will not be convenient (sometimes not even possible) to check the results. If we trust the theorem, we can nevertheless add a new entry to the table of Laplace transform pairs. If the inverse Laplace transform is known for *one* function, and if we either know or desire some specific scatter function, work backwards and derive the scatter mate. In this case the inverse is known for the original function. The scatter is known to be the δ function. Then subtract the scatter from the known result to obtain the inverse of the unknown, which in this case is eq. (4).

The inverse for eq. (1), from Appendix 3, is given by entry 327. Then the inverse for eq. (4) becomes

$$\mathcal{L}^{-1}F_s(s) = \mathcal{L}^{-1}F(s) - \delta \tag{9}$$

$$= \frac{\omega^3 \sin \omega t - \alpha^3 \sin \alpha t}{\alpha^2 - \omega^2} \tag{10}$$

Equations (3) and (10) comprise a Laplace transform pair. This pair came about by using the original function as the first term in eq. (2), the expanded denominator. It is conceivable to utilize the original function for each term of the denominator, creating a whole family of Laplace transform pairs. This is done in Problems 11.18 through 11.27. ∎

11.3 LAPLACE OPERATORS AND THEOREMS

11.3.1 Shifting Theorem

The shifting theorem is given by entry 114, Appendix 3, repeated below.

$$\mathcal{L}^{-1}F(s + a) = e^{-at}f(t) \qquad [11.3.1\text{-}1]$$

∎ **EXAMPLE 11.3.1-1** Use of Shifting Theorem
Starting with entry 315, Appendix 3, apply the shifting theorem and derive entry 415, Appendix 3.

$$F_{315} = \frac{\omega}{s^2 + \omega^2}, \qquad f_{315}(t) = \sin \omega t \qquad (1)$$

Apply the shifting theorem to the Laplace function

$$F_{315}(s + a) = \frac{\omega}{(s + a)^2 + \omega^2} = \frac{\omega}{s^2 + 2as + a^2 + \omega^2} \qquad (2)$$

Refer to entry 6, Appendix 1,

$$a^2 + \omega^2 = \omega_n^2 \qquad (3)$$

Apply to eq. (2) and multiply by ω_n/ω.

$$F_{415} = F_{315}(s + a) = \frac{\omega_n}{s^2 + 2as + \omega_n^2} \qquad (4)$$

Apply the shifting theorem to the time function and multiply by ω_n/ω.

$$f_{415}\left(\frac{\omega_n}{\omega}\right)e^{-at}f_{315}(t) = \frac{\omega_n}{\omega}e^{-at}\sin \omega t \qquad (5)$$

Equations (4) and (5) comprise the Laplace transform pair, which can be checked in Appendix 3. ∎

11.3.2 Derivative in Time Domain

The derivative in the time domain is given by entry 126, Appendix 3, repeated below:

$$\mathcal{L}\frac{df(t)}{dt} = sF - f(0^-) \qquad [11.3.2\text{-}1]$$

or

$$\mathcal{L}^{-1}sF = \frac{df(t)}{dt} + \mathcal{L}^{-1}f(0^-) \qquad [11.3.2\text{-}2]$$

■ **EXAMPLE 11.3.2-1** Use of Time Derivative
Derive entry 227, Appendix 3, from entry 226, Appendix 3, by taking the time derivative.

$$F_{227} = sF_{226} = \frac{s^2}{(s+a)(s+b)} \tag{1}$$

$$f_{227} = \mathscr{L}^{-1}f_{226}(0) + \frac{df_{226}(t)}{dt} \tag{2}$$

$$\mathscr{L}^{-1}f_{226}(0) = \mathscr{L}^{-1} - \frac{a-b}{b-a} = \delta \tag{3}$$

$$f_{227} = \delta + \frac{a^2e^{-at} - b^2e^{-bt}}{b-a} \tag{4}$$

Equations (1) and (4) comprise the Laplace transform pair, which can be checked in Appendix 3.   ■

11.3.3 Integral in Time Domain

The integral in the time domain is given by entry 124, Appendix 3, repeated below.

$$\frac{F}{s} = \int_{0^+}^{t} f(T)\, dT \tag{11.3.3-1}$$

■ **EXAMPLE 11.3.3-1** Use of the Time Integral
Derive entry 215, Appendix 3, from entry 216, Appendix 3.
    In the Laplace domain, multiply by a/s.

$$F_{215} = a\frac{F_{216}}{s} = \frac{a}{s(s+a)} \tag{1}$$

In the time domain, multiply by a and take the integral.

$$f_{215} = a\int_{0^+}^{t} f_{216}(T)\, dT = a\int_{0}^{t} e^{-aT}\, dT \tag{2}$$

$$= 1 - e^{-at} \tag{3}$$

Equations (1) and (3) comprise the Laplace transform pair, which can be checked in Appendix 3.   ■

11.3.4 Derivative in the Laplace Domain

The derivative in the Laplace domain is given by entry 136, Appendix 3, repeated below.

$$\frac{dF}{ds} = -tf(t) \tag{11.3.4-1}$$

■ **EXAMPLE 11.3.4-1** Use of Derivative in Laplace Domain
Derive entry 335, Appendix 3, from entry 315, Appendix 3.
    Apply the theorem in the Laplace domain and multiply by $-\omega$.

$$F_{335} = -\omega \frac{dF_{315}}{ds} = -\omega \frac{d}{ds}\left(\frac{\omega}{s^2 + \omega^2}\right) \tag{1}$$

$$= \frac{2s\omega^2}{\left(s^2 + \omega^2\right)^2} \tag{2}$$

Apply the theorem in the time domain and multiply by $-\omega$.

$$f_{335} = \omega t \sin \omega t \tag{3}$$

Equations (2) and (3) comprise the Laplace transform pair, which can be
checked in Appendix 3.   ■

11.3.5 Scatter Logical Theorem

The scatter logical theorem is given by eqs. 11.2.1-1 and 11.2.1-2.

■ **EXAMPLE 11.3.5-1** Use of Logical Theorem
Derive entry 524, Appendix 3, from 514, Appendix 3.
    The correct result can be found in the Laplace domain by completing the
numerator. Note that entry 524 is missing the ω_n^2 in the numerator. Hence this
will be a factor in modifying entry 514. Multiply numerator and denominator
of entry 514 by ω_n. Also, to make entry 524 compatible with entry 514,
multiply numerator and denominator of entry 524 by s. The two terms are
prepared for the summing operation.

$$L_1 + L_2 = \frac{r^2\omega_n^2}{\omega_n s(s + c)(s^2 + 2as + \omega_n^2)} + \frac{r^2(s^2 + 2as)}{\omega_n s(s + c)(s^2 + 2as + \omega_n^2)} \tag{1}$$

$$= \frac{r^2}{\omega_n s(s + c)} \tag{2}$$

Take the inverse Laplace transform of eq. (2), using entry 215, Appendix 3,

$$\mathcal{L}^{-1} \frac{r^2}{\omega_n s(s + c)} = \frac{r^2}{c\omega_n}(1 - e^{-ct}) \tag{3}$$

Add eq. (3) to $-f_{514}$, according to eq. 11.2.1-2, and solve for f_{524}.

$$f_{524} = \frac{2r^2}{c\omega_n} - \frac{(\omega_n^2 + r^2)}{c\omega_n}e^{-ct} - R\sin(\omega t - \phi_r + \phi) \tag{4}$$

The second term of eq. (1) and eq. (4) comprise the Laplace transform pair,
which can be checked in Appendix 3.   ■

11.4 TRANSIENT RESPONSE FOR SECOND ORDER SYSTEMS

11.4.1 Purpose

The transient response of underdamped second order systems is a decaying sinusoid. The output approaches steady state after some time, but until then there are several transient properties to be considered.

11.4.2 Generalized Model of Second Order Systems

Considering the large sampling of different systems in Appendix 3, the following represents a generalized model.

$$X(s) = \frac{(Es + B)s^n}{s^2 + 2as + \omega_n^2} \qquad [11.4.2\text{-}1]$$

The solution in the time domain is

$$x(t) = C + De^{-at}\sin(\omega_d t + n\phi) \qquad [11.4.2\text{-}2]$$

A plot of the generalized solution is shown in Fig. 11.4.2-1. Note that the curve is a decaying sinusoid that is oscillating about the steady state value C. When a scatter function is involved, it is used in place of the steady state value. The scatter function makes it possible to determine the transient properties in the presence of other functions. See Fig. 11.1.2-1.

The scatter function has an additional role. The scatter logical theorem and the many examples in Section 11.2.2 indicate that groups of functions have the same transient with respect to a common scatter function. This permits the computation of *the* transient property that applies for a large number of different systems. In particular, note that in the table of Laplace transformation pairs in Appendix 3, entries 412, 422, 432, and 442 all have identical sinusoidal terms. They differ only by scatter functions. Consequently, it will suffice to compute a transient property for the simplest case (time domain solution has no summed terms). For each specific case, add the numerical value of its scatter function. The same applies for entries 413, 423, 433, and 443. Incidentally, it now becomes clear how the Laplace table of Appendix 3 was organized.

When the scatter function is employed, it is possible to delete all terms but the sinusoid. In this case, a more useful form of eq. 11.4.2-2 is suggested:

$$x(t) = \frac{\omega_n}{\omega_d}e^{-at}\sin(\omega_d t + n\phi) \qquad [11.4.2\text{-}3]$$

where

$$n = \text{any positive or negative integer} \qquad [11.4.2\text{-}4]$$

Note that in eq. 11.4.2-3,

$$\text{phase angle} = n\phi \qquad [11.4.2\text{-}5]$$

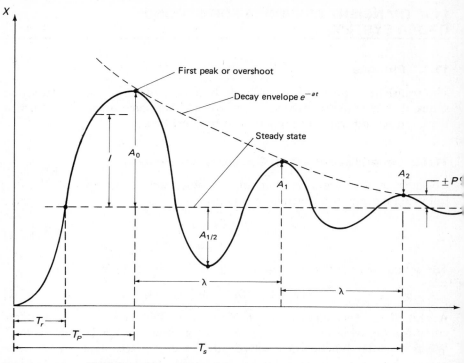

Figure 11.4.2-1   Transients for second order systems.

11.4.3 Initial Values of Transient I_0

The initial *value* of the transient is the value at time* $t = 0^+$. This is not to be confused with the response due to initial *conditions*.

The initial value I_0 can be computed by setting

$$t = 0 \qquad\qquad\qquad [11.4.3\text{-}1]$$

in eq. 11.4.2-3.

$$I_0 = x(0) = \frac{\omega_n}{\omega_d}\sin(n\phi) \qquad\qquad [11.4.3\text{-}2]$$

Since n is a positive or negative integer, the numerical value can easily be obtained by using entries 15 through 18, Appendix 1. Also, apply entry 4, Appendix 1. Table 11.4.3-1 summarizes the results. Note that the results in the table are a function of ζ only.

---

*Time an infinitesimal time after $t = 0$.

TABLE 11.4.3-1   INITIAL VALUE OF TRANSIENTS

n	I_0
± 4	$\pm 4\zeta(2\zeta^2 - 1)$
± 3	$\pm 4\zeta^2 - 1$
± 2	$\pm 2\zeta$
± 1	± 1
0	0

11.4.4 Time for First Crossing of Steady State Value, Rise Time T_r

The rise time T_r is defined as the first time the curve crosses the steady state value or the first time the sinusoid crosses the scatter function. The rise time is an index of how quickly the system crosses the steady state value, but of course, unless the system is critically damped, it will continue past the point, only to return and cross again in the opposite direction. Consequently, the rise time by itself is insufficient. Nevertheless, it does indicate the speed of system response.

In order to compute the rise time, let

$$x(t) = 0 \qquad [11.4.4\text{-}1]$$

in eq. 11.4.2-3 and solve for t, calling this T_r.

$$\sin(\omega_d t + n\phi) = 0 \qquad [11.4.4\text{-}2]$$

or

$$t = T_r = (m\pi - n\phi)/\omega_d \qquad [11.4.4\text{-}3]$$

where

$$m = \text{smallest integer for } T_r > 0 \qquad [11.4.4\text{-}4]$$

In order to obtain dimensionless time, multiply both sides by ω_n.

$$\omega_n T_r = (m\pi - n\phi)/\beta \qquad [11.4.4\text{-}5]$$

where

$$\beta = \sqrt{1 - \zeta^2} = \omega_d/\omega_n \qquad [11.4.4\text{-}6]$$

$$\phi = \sin^{-1}\beta \qquad [11.4.4\text{-}7]$$

Note that the result is a function of ζ, m, and n, where the latter two terms depend upon the entry number.

■ **EXAMPLE 11.4.4-1** Rise Time
Determine the rise time for entries 414, 424, and 434, Appendix 3.

Solution.   Refer to Appendix 3 for the time domain terms for the three entries. By inspection, entry 424 has the simplest form. For this,

$$m = 1, \qquad n = 1 \qquad (1)$$

The dimensionless rise time is given by eq. 11.4.4-5. Apply the appropriate numerical values,

$$\omega_n T_r = \frac{\pi - \phi}{\beta} \tag{2}$$

For entry 424, this is the time to cross the zero line. For entry 414, it is the time to cross the $+1$ line. For entry 434, this is the time to cross the scatter function

$$\frac{\omega_n^2 t}{2a} - 1 \tag{3}$$

■

11.4.5 Peak Time T_p

Peak time T_p is the time to each maximum value. It can be determined by setting the derivative (with respect to time) equal to zero, solving for time, and testing for maximum or minimum. [The derivative of a sinusoid (with phase shift θ) is given in Problem 11.2, where $\theta = n\phi$.] Apply this to eq. 11.4.2-3 and divide both sides by ω_n.

$$\frac{\dot{x}}{\omega_n} = \frac{dx}{dt} = -\frac{\omega_n}{\omega_d} e^{-at} \sin\left[\omega_d t + (n - 1)\phi\right] = 0 \qquad [11.4.5\text{-}1]$$

which is satisfied if

$$\sin\left[\omega_d t + (n - 1)\phi\right] = 0 \qquad [11.4.5\text{-}2]$$

Solve for time t and call this peak time, T_p.

$$t = T_p = \left[m\pi - (n - 1)\phi\right]/\omega_d \qquad [11.4.5\text{-}3]$$

where

$$m = \text{smallest integer for } T_p > 0 \qquad [11.4.5\text{-}4]$$

To obtain dimensionless time, multiply both sides by ω_n as was done for eq. 11.4.4-5.

$$\omega_n T_p = \left[m\pi - (n - 1)\phi\right]/\beta \qquad [11.4.5\text{-}5]$$

Note that the result is a function of ζ, m, and n, where the latter two depend upon entry number.

■ EXAMPLE 11.4.5-1 Peak Time

Determine peak time for entry 414, Appendix 3, where $m = 1$, $n = 1$. Using eq. 11.4.5-5, dimensionless peak time is

$$\omega_n T_p = \pi/\beta \tag{1}$$

■

11.4.6 Peak, or Overshoot, A_0

The overshoot is the amount of travel past the steady state (or the scatter function) value the first time. It occurs at the peak time. In order to determine the overshoot, apply the peak time T_p of eq. 11.4.5-5 to eq. 11.4.2-3. The displacement that results is called overshoot A_0.

$$A_0 = \frac{\omega_n}{\omega_d} e^{-aT_p} \sin(\omega_n T_p + n\phi) \qquad [11.4.6-1]$$

The exponent becomes upon expansion

$$-aT_p = -a[m\pi + (1-n)\phi]/\omega_d \qquad [11.4.6-2]$$

The sinusoidal term becomes upon expansion

$$\sin[m\pi - (1-n)\phi + n\phi] = \sin\phi = \beta \qquad [11.4.6-3]$$

according to entry 15, Appendix 1,

$$\sin\phi = \beta = \omega_d/\omega_n \qquad [11.4.6-4]$$

and entry 3, Appendix 1,

$$a = \zeta\omega_n \qquad [11.4.6-5]$$

Apply the appropriate equations to eq. 11.4.6-1.

$$A_0 = e^{-\zeta[m\pi+(1-n)\phi]/\beta} \qquad [11.4.6-6]$$

Note that the result is a function of ζ, m, and n, where the latter two depend upon the entry number.

■ **EXAMPLE 11.4.6-1** Overshoot and Maximum Value
Determine the overshoot and maximum (or peak) value for entries 414, 424, 434, and 444, Appendix 3.

Solution. Referring to Appendix 3, in the time domain, the simplest form is for entry 424. Let this be the base function, for which the others will be its scatter mates. The overshoot is given by eq. 11.4.6-6. In the time domain. solution for entry 424, note that

$$m = 1, \qquad n = 1 \qquad (1)$$

Applying appropriate numerical values, overshoot becomes

$$A_0 = e^{-\zeta\pi/\beta} \qquad (2)$$

Note that the final result is a function of damping ratio ζ only, where the term β is defined in Appendix 1. The result in eq. (2) specifies the overshoot or travel past the scatter function. The peak, or maximum value, is obtained by adding the scatter function to this. The scatter for entry 414 is unity. The scatter for entry 434 is given in Ex. 11.4.4-1, eq. (3). The scatter function for entry 444 is

$$\frac{\omega_n^2 t^2}{2!} - 1 \tag{3}$$

Note that for entries 434 and 444, the time t is required. This is the peak time T_p (actual, not dimensionless) as found in the previous example.   ∎

11.4.7 Time Constant and Settling Time, τ and T_s

The *time constant* is an index to the rate of decay where *time* is the reference. The *settling time* relates the rate of decay in terms of the *number of time constants*.

The amplitude of damped oscillation is given approximately by the decay envelope. See Fig. 11.4.2-1. The decay curve was discussed in Section 5.5. Since all second order systems have the same decay envelope, then the time constant τ and the settling time T_s are common to all second order systems. Thus,

$$\tau = \frac{1}{a} = \frac{1}{\zeta\omega_n} \tag{11.4.7-1}$$

$$T_s = N\tau = \frac{N}{a} = \frac{N}{\zeta\omega_n} \tag{11.4.7-2}$$

where N is given by eq. 5.5.2-3, repeated below.

$$P = e^{-N} \times 100\% \tag{11.4.7-3}$$

where

$$P = \text{percent transient residue} \tag{11.4.7-4}$$

Equation 11.4.7-3 appears in tabulated form in Table 5.5.2-1.

■ **EXAMPLE 11.4.7-1** Quick-Acting Circuit
Refer to the circuit in Fig. 11.4.7-1. Design the system, using a step input, to meet the following specifications:

$$T_r \leq 10^{-6} \text{ s} \qquad \text{rise time} \tag{1}$$

$$OS \leq 36.4\% \qquad \text{overshoot} \tag{2}$$

$$T_s \leq 5 \times 10^{-6} \text{ s} \quad \text{settling time for 5\% transient residue} \tag{3}$$

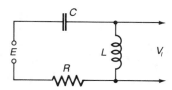

Figure 11.4.7-1   Quick-acting circuit.

Solution.   The out/in ratio is given by

$$\frac{V_l}{E}(s) = \frac{s^2\omega_n^2}{s^2 + 2as + \omega_n^2} \tag{4}$$

For a step input of magnitude U,

$$E(s) = \frac{U}{s}$$

Then

$$V_l(s) = \frac{U\omega_n^2 s}{s^2 + 2as + \omega_n^2} \tag{5}$$

Refer to entry 416, Appendix 3, for which $n = -1$.

Apply the overshoot specification, eq. (2), to eq. 11.4.6-6, where β and ϕ are given by eqs. 11.4.4-6 and 11.4.4-7, respectively. Let $m = 0$. Then

$$A_0 = e^{-\zeta 2\phi/\beta} = 36.4\% = 0.364 \tag{6}$$

Solve for ζ by iteration.

$$\boxed{\zeta = 0.4} \tag{7}$$

Apply this and specification (1) to eq. 11.4.4-5.

$$\omega_n T_r = \frac{0 - (-1.16)}{0.917} = 1.26 \tag{7}$$

Solve for ω_n and apply specification (1).

$$\omega_n \geq 1.26/T_r = 1.26/10^{-6} = 1.26 \times 10^6 \text{ rad/s} \tag{8}$$

For settling time, use specification (3) and Table 5.5.2-1 for

$$P = 5\%, \qquad N = 3 \tag{9}$$

Then

$$T_s = 3\tau = \frac{3}{\zeta\omega_n} \tag{10}$$

Solve for ω_n.

$$\omega_n \geq \frac{3}{\zeta T_s} = \frac{3}{0.4 \times 5 \times 10^{-6}} = 1.5 \times 10^6 \text{ rad/s} \tag{11}$$

Noting the inequality between eqs. (8) and (11), select the larger one.

$$\boxed{\omega_n = 1.5 \times 10^6 \text{ rad/s}} \tag{12}$$

To conclude the design, note that there are three unknowns, L, R, and C, but there are only two equations. Hence, one is arbitrary. Choose

$$L = 0.001 \text{ H} \tag{13}$$

Then

$$C = \frac{1}{L\omega_n^2} = \frac{1}{0.001 \times (5 \times 10^6)^2} = \boxed{4.44 \times 10^{-10} \text{ F}} \tag{14}$$

and

$$R = 2L\zeta\omega_n = 2 \times 0.001 \times 0.4 \times 1.5 \times 10^6 \tag{15}$$

$$\boxed{= 1200 \ \Omega} \tag{16}$$

∎

■ **EXAMPLE 11.4.7-2** Dynamics of Pneumatic Control Valve
A control valve regulates or controls the amount of fluid that flows in a particular process. The amount of flow Q is proportional to the opening Y of the valve. The opening of the valve is adjusted remotely by means of air pressure P applied to the diaphragm. This pressure times the area A of the diaphragm applies a force F upon the valve stem. If there were no dynamics involved, this force would act against the various springs and the resultant force would determine the opening of the valve. Thus, the flow rate of the process fluid is controlled by pneumatic means. See Fig. 11.4.7-2. The valve will always be partly open, responding to step changes in Y, the travel of the valve stem. Design the system to meet the following specifications:

$$T_r \leq 0.01 \text{ s} \quad \text{rise time} \tag{1}$$

$$\text{OS} = 16.3\% \quad \text{overshoot} \tag{2}$$

Solution.   The system model includes air pressure P acting upon diaphragm area A, fluid pressure p acting on valve plug area a, viscous drag D on the valve plug due to fluid velocity $\dot{Y}_b$, and initial displacement y_0. The displacement Y of the valve stem becomes

$$Y(s) = \frac{AP(s) - DsY_b(s) - ap(s) + (Ms + D)y_0}{Ms^2 + Ds + K} \tag{3}$$

where

$$K = K_1 + K_2 + K_3 \tag{4}$$

Neglect* all terms but $PA(s)$ and normalize.

$$Y(s) = \frac{AP}{s(s^2 + 2as + \omega_n^2)} \tag{5}$$

which is the portion of entry 414 in the Laplace domain in Appendix 3. Using the response in the time domain, note that $n = 1$.

Overshoot is given by eq. 11.4.6-6, where β and ϕ are given by eqs. 11.4.4-6 and 11.4.4-7, respectively, and let $m = 1$. Then,

$$\text{OS} = e^{-\zeta\pi/\beta} = 16.3\% = 0.163 \tag{6}$$

---

*A better treatment is covered in Problem 11.34.

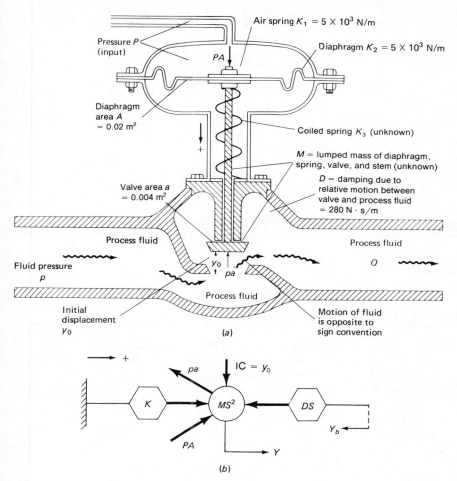

Figure 11.4.7-2   Pneumatically operated control valve. (a) Actual system. (b) Free body diagram.

Solve for ζ by iteration

$$\boxed{\zeta = 0.5} \tag{7}$$

Dimensionless rise time is given by eq. 11.4.4-5. Apply numerical values and let $m = 1$.

$$\omega_n T_r = (\pi - \phi)/\beta = 2.41 \tag{8}$$

Apply specification (1) and solve for ω_n.

$$\omega_n = 2.41/0.01 = \boxed{241 \text{ rad/s}} \tag{9}$$

Using appropriate numerical values, determine the system parameters.

$$M = \frac{D}{2\zeta\omega_n} = \frac{280}{2 \times 0.5 \times 241} = \boxed{1.16 \text{ kg}} \tag{10}$$

$$K = M\omega_n^2 = 1.16 \times (241)^2 = 6.75 \times 10^5 \text{ N/m} \tag{11}$$

$$K_3 = K - K_1 - K_2 = (6.75 - 0.5 - 0.5) \times 10^4$$

$$\boxed{= 5.75 \times 10^4 \text{ N/m}} \tag{12}$$

∎

SUGGESTED REFERENCE READING FOR CHAPTER 11

System properties [26], p. 87–93; [7], p. 221, 271.
Dominance [26], p. 146, 198.

PROBLEMS FOR CHAPTER 11

11.1. Given the transient responses of two different systems, and if the difference of these transient responses in the time domain is $f(t)$, prove that the difference of their steady state responses will also be $f(t)$ when
(a) $f(t) = 0$
(b) $f(t) = 1$
(c) $f(t) = $ any function of time

11.2. Prove

$$\frac{d}{dt}\left[e^{-at}\sin(\omega_d t + \theta)\right] = -\omega_n e^{-at}\sin(\omega_d t + \theta - \phi)$$

11.3. Rework Problem 11.2 with $\theta = n\phi$.

11.4. Rework Problem 11.2 for cosine instead of sine.

11.5. Perform the following operations on the jump* functions;
(a) Start with the unit doublet and take successive integrals.
(b) Start with the unit parabola and take successive derivatives.

11.6. Set the appropriate term equal to zero and derive:
(a) 220 group† (entries 222 through 229) from the 240 group.
(b) 210 group (entries 211 through 219) from the 220 group.

11.7. Use the Shifting Theorem to derive:
(a) Entry 265 from entry 144.
(b) Entry 465 from entry 316.

---

*Entries 141 through 148 in Appendix 3.

†All group and entry numbers in these problems refer to the table of Laplace transform pairs in Appendix 3.

11.8. *Derivatives in the Laplace Domain.*
   (a) Derive entry 265 from entry 216.
   (b) Find the inverse Laplace transform of

$$F_b(s) = \frac{2\omega_n^2(s + a)}{\left(s^2 + 2as + \omega_n^2\right)^2}$$

(Suggestion: Apply entry 136 to entry 415.)

11.9. *Derivatives in the Time Domain.* For each entry* listed below, take successive derivatives in the time domain and derive all the Laplace transform pairs in the rest of the group.[†]

(a) 211	(b) 222	(c) 243
(d) 261	(e) 285	(f) 311
(g) 321	(h) 332	(i) 341
(j) 412	(k) 422	(l) 432
(m) 442	(n) 462	(o) 472
(p) 493	(q) 513	(r) 524
(s) 535	(t) 546	

11.10. Similar to Problem 10.9, but take successive integrals:*

(a) 219	(b) 229	(c) 247
(d) 269	(e) 286	(f) 319
(g) 329	(h) 338	(i) 348
(j) 419	(k) 427	(l) 438
(m) 449	(n) 468	(o) 478
(p) 497	(q) 518	(r) 526
(s) 536	(t) 546	

11.11. Rewrite the scatter logical theorem in an alternate form where the statement includes the notion of completing the numerator. This is the practical method of using the theorem as treated in Exs. 11.2.1-1 and 11.2.1-2, where completing the numerator was used as a short cut.

11.12. Show that consecutive entries in the 210 group[†] (entries 211 through 219) obey the scatter logical theorem.

11.13. Show that consecutive entries in the 220 group[†] (entries 222 through 229) obey the scatter logical theorem.

11.14. Refer to the 260 group[†] (entries 261 through 269). Do consecutive or do alternate entries follow the scatter logical theorem?

11.15. Show that alternate entries of the 310 group[†] (entries 311 through 319) obey the scatter logical theorem.

11.16. Show that alternate entries of the 320 group[†] (entries 321 through 329) obey the scatter logical theorem.

11.17. Refer to the 330 group[†] (entries 332 through 338). Do alternate entries obey the scatter logical theorem?

---

*If any entry does not appear in the table of Laplace transform pairs in Appendix 3, discuss the value of adding or omitting it.

[†]See footnote on page 444.

11.18. Below, there are a number of sets of four entries.* Determine the scatter for the first entry of each set with each of the other three in the set.
- (a) 411: 421, 431, 441.
- (b) 412: 422, 432, 442.
- (c) 413: 423, 433, 443.
- (d) 414: 424, 434, 444.
- (e) 415: 425, 435, 445.
- (f) 416: 426, 436, 446.
- (g) 417: 427, 437, 447.
- (h) 418: 428, 438, 448.
- (i) 419: 429, 439, 449.

11.19. Show that it was predictable that there would be three scatter logical mates for each entry in the previous problem. What other conclusions can be drawn?

11.20. Similar to Problem 11.18, but applied to the following sets of four:*
- (a) 513: 523, 533, 543.
- (b) 514: 524, 534, 544.
- (c) 515: 525, 535, 545.
- (d) 516: 526, 536, 546.
- (e) 517: 527, 537, 547.
- (f) 518: 528, 538, 548.

11.21. Similar to Problem 11.19, but applied to Problem 11.20.

11.22. Similar to Problem 11.18, but applied to the following sets of two:*
- (a) 462: 472.   (b) 463: 473.
- (c) 464: 474.   (d) 465: 475.
- (e) 466: 476.   (f) 467: 477.
- (g) 468: 478.

11.23. Similar to Problem 11.19, but applied to Problem 11.22, and where the prediction involves only one mate.

11.24. The following entries* do not appear in the table of Laplace transform pairs in Appendix 3. Write the function in the Laplace domain (by extrapolating the entries in the same group) and derive the function in the time domain, using two methods where possible.

221	241	242	248	249
331	339	349	411	421
428	429	431	439	441
461	469	471	479	491
492	498	499	511	512
519	523	527	533	537
543	547			

11.25. Complete Ex. 11.2.1-1.

_____

*See footnote on page 445.

11.26. In Ex. 11.2.1-2, there are four terms in eq. (2), the expanded denominator. The given function $F(s)$ could conceivably be associated with any one of these four terms. In the example, $F(s)$ was associated with the first term of eq. (2). Associate $F(s)$ with each term of eq. (2) and, for each one, determine the scatter mate* and the scatter function.

11.27. *Creation of a Table of Laplace Transform Pairs.* Starting with the solution to a differential equation, such as eq. 6.3.4-8i, and using only the entries in the 100 family (entries 111 through 157) and the scatter logical theorem, derive all the entries* from 411 through 449.

11.28. Using the format of that in Section 11.4, derive generalized expressions for rise time, peak time, and overshoot, for the 490 group (entries 493 through 497).

11.29. Similar to Problem 11.28, but applied to the 460 group (entries 462 through 468).

11.30. Similar to Problem 11.28, but applied to the 470 group (entries 472 through 478).

11.31. Show how the scatter logical theorem permits the generalized expressions of eqs. 11.4.4-5, 11.4.5-5, and 11.4.6-6 to include all the scatter mates without rewriting the expression. Can this concept be extended to include the results of Problems 11.28 through 11.30?

11.32. Similar to Problem 11.28, but applied to the 510 group (entries 513 through 518).

11.33. Show that the circuit in Fig. 11.4.7-1 has better response to a parabolic input than any other series circuit.

11.34. *Design of Balanced (Compensated) Control Valve.* See Fig. P-11.34. A control valve regulates or controls the flow of fluid. A pneumatically operated valve permits remote operation of the valve. The position of the valve stem should ideally be proportional to the air pressure P against the diaphragm. In a simple pneumatically controlled valve, like those shown in Figs. 11-4.7-2, and P-4.43, this is not true. These are subject to errors due to the motion and pressure of the fluid, which affect the position of the valve. A pneumatically operated valve should be positioned only by the pneumatic pressure. Hence, to eliminate the obvious offenders, an improved design (the one in great use today) makes use of a balanced pair of valves, placed so that the effects of the fluid motion and pressure cancel, thus making the valve self-compensating.

The valve must be quick-acting. However, if the valve stem moves too fast, it will impact against the seat upon closing. To provide the optimum compromise, a porous material[†] is installed in the air feedline. Design the system to meet the following specifications:

(1) Time to open or close is less than 0.05 s
(2) Impact stress upon suddenly closing is less than 5×10^8 N/m^2
   (about 70,000 psi)

---

*See footnote on page 445.

[†]In designing the porous damper D_4, use $C = 800$ in Appendix 2, Table 2.3.4-1, item (1).

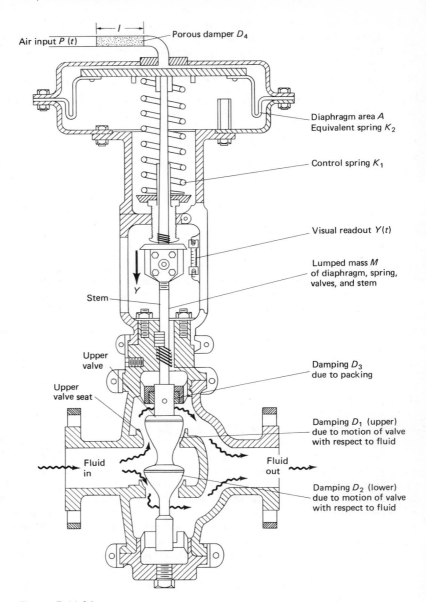

Figure P-11.34

(3) Total travel or stroke $L = 10^{-2}$ m (about 0.4 in.)
(4) $M_v \geq 1.4$ kg
(5) $K_1 \geq 10K_2$
(6) $l \leq 0.2$ m

For dimensions, refer to Fig. P-11.34,

where M_v = mass of valve stem and twin valve assembly (to be determined)
  M_c = mass of diaphragm center = 0.17 kg (about 0.4 lb)
  M_f = mass of folds of diaphragm = 0.1 kg (about 0.2 lb)

M_k = mass of control spring K_1 = unknown
(assume = 0.7 kg and check later)
M_s = mass of each valve seat = 0.1 kg
K_1 = control spring (to be determined)
K_2 = diaphragm spring = 5×10^3 N/m (about 25 lb/in.)
K_s = elasticity of each valve seat = 5×10^7 N/m
$D_1 = D_2$ = damping due to valve in fluid = 70 N · s/m
D_3 = damping due to packing = 6 N · s/m
D_4 = damping due to porous material (to be determined)
D_s = structural damping in each valve seat = 2500 N · s/m
A = area of diaphragm = 0.02 m²
A_c = contact area of each valve = 1.6×10^{-5} m²

11.35. *Design of Rotary Conveyor.* A bottling plant injects a fairly viscous material into bottles using a positive displacement pump-type of device. A crank, which is synchronized to the conveyer through appropriate gearing, goes through one cycle during the time it takes for one bottle to pass beneath the dispenser. On the intake stroke, the inlet valve opens admitting the correct amount of fluid. At the same time, the outlet valve is closed. On the injection stroke, the volume is ejected, opening the outlet valve and closing the inlet valve. See Fig. P-11.35.

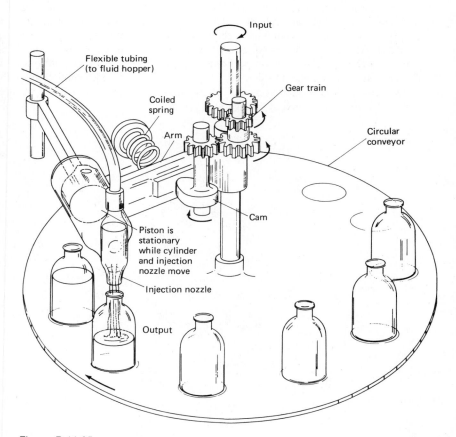

Figure P-11.35

In some installations, the conveyer is stopped for each bottle (waiting for it to be filled) and then started again. In order to develop a high speed plant, the stop-start procedure is not applied to the fragile bottles but is applied to the rugged injection system instead. The bottles travel on a circular conveyer that turns at constant speed while the injection system travels along the small arc following the bottle while it is being filled. A model of the system has the form of an out/in ratio equal to entry 447. The input is a ramp. Design the system to meet the following specifications:

$$J \geq 0.5 \text{ N} \cdot \text{s/m} \quad \text{(for structural rigidity)} \tag{1}$$

$$T_r \leq 0.01 \text{ s} \tag{2}$$

$$\text{OS} \leq 25\% \tag{3}$$

$$T_s \leq 0.05 \text{ s} \tag{4}$$

11.36. *Magnetic Torquer with Pulse Width Modulation.* The device shown in Fig. P-11.36 is referred to as a magnetic torquer. It is essentially an electric motor that rotates through a limited angle. Like many heavy duty magnetic devices, the torque produced by the device is not proportional to current. However, it has been established that for a particular value of current, the resulting torque is highly repeatable and reliable. Hence, to obtain a variable torque that is proportional to some parameter, the current is held constant, but is applied in pulses whose width is varied. Since the pulse width can be changed easily and accurately, the torque is proportional to pulse width. As such the system is referred to as a pulse torquer, or a torquer with pulse width modulation. The system with pulse input is modeled with entry 515. Design the system to meet the specifications listed in the previous problem.

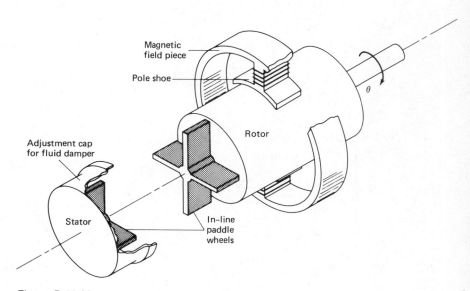

Figure P-11.36

11.37. *In-Depth Study of Undamped Second Order System*. For the systems shown in Fig. 3.6.2-1, where damping ratio = 0, determine the transient and steady state responses and find the first and second derivatives of the response, for each of the following:

(a) Initial velocity.
(b) Initial displacement.
(c) Impulse input.
(d) Step input.
(e) Ramp input.

11.38. *In-Depth Study of Underdamped Second Order System*. Similar to Problem 11.37, but for the underdamped case.

11.39. *In-Depth Study of Critically Damped Second Order System*. Similar to Problem 11.37, but where the damping ratio is equal to unity.

11.40. *In-Depth Study of Overdamped Second Order System*. Similar to Problem 11.37, but where the damping ratio is greater than unity.

11.41. A certain underdamped system has a response given by

$$\frac{X}{X_{in}}(s) = \frac{s^2 + \omega_n^2}{s^2 + 2as + \omega_n^2}, \qquad IC = 0$$

Find $x(t)$, $\dot{x}(t)$, $\ddot{x}$, and the steady state response for each of the following inputs:

(a) impulse;   (b) step;   (c) ramp;   (d) parabola.

11.42. Similar to Problem 11.41, but for the following response:

$$\frac{X}{X_{in}}(s) = \frac{2as + \omega_n^2}{\left(s^2 + 2as + \omega_n^2\right)\omega_n}$$

11.43. Similar to Problem 11.41, but for the following response:

$$\frac{X}{X_{in}}(s) = \frac{s + a}{s^2 + 2as + \omega_n^2}$$

11.44. Similar to Problem 11.41, but for the following response:

$$\frac{X}{X_{in}}(s) = \frac{as + \omega_n^2}{s^2 + 2as + \omega_n^2}$$

11.45. Similar to Problem 11.41, but for the following response:

$$\frac{X}{X_{in}}(s) = \frac{s + c}{s^2 + 2as + \omega_n^2}$$

11.46. Similar to Problem 11.41, but for the following response:

$$\frac{X}{X_{in}}(s) = \frac{r^2}{(s + c)\left(s^2 + 2as + \omega_n^2\right)}$$

Chapter **12**

Modeling of High Order Systems

12.1 DEFINITION

High order systems will be defined as those that have more than one variable. In most cases, this will require the use of simultaneous equations.

12.2 DYNAMIC SYSTEMS

The differential equations of a system with more than one variable can be written in the Laplace domain and then can be treated like simultaneous linear algebraic equations.

12.2.1 Simultaneous Algebraic Equations

Consider a set of simultaneous linear algebraic equations in several variables, say, X_1, X_2, and X_3 with constant coefficients A through I, and inputs L through N, as follows:

$$AX_1 + DX_2 + GX_3 = L$$

$$BX_1 + EX_2 + HX_3 = M \qquad [12.2.1\text{-}1]$$

$$CX_1 + FX_2 + IX_3 = N$$

While there are several ways to handle the subsequent computations for solving these equations, the following is suggested since it will later prove convenient for analyzing and designing. Rewrite the equations in a tabulated form or system array as covered in the next section.

12.2.2 System Array

The system array may be considered as a bookkeeping scheme to assist in and to organize the computation of the solution. This is done as follows:

1. Tabulate the constant coefficients in the same order in which they were given. (For the illustration at hand, these will be arranged in three columns of three entries.) This will be referred to as the *system characteristic*.
2. Place the *inputs* in a *column* in the same order in which they were given, and locate this column as an additional column to the right of the system characteristic.
3. Place the outputs (variables or unknowns) at the head of each column. This is referred to as the *output row*.

For the illustration at hand, the system array becomes

Output row

X_1	X_2	X_3	Inputs
A	D	G	L
B	E	H	M
C	F	I	N

System characteristic — Input column

[12.2.2-1]

While there are several computational schemes for solving the above set of equations (it is both a table and an array) it is suggested that Cramer's rule be used. This will prove useful in factoring the resulting expressions, and it will also retain the system characteristic in a useful form.

12.2.3 Cramer's Rule

Cramer's rule is an algorithm for solving simultaneous linear algebraic equations with constant coefficients. Each of the outputs (variables or unknowns) is given by a fraction whose numerator and denominator are determinants which are shown below:

(a) **Denominator**   The denominator Δ (also called the *characteristic determinant*) is given by the determinant of the system characteristic.

(b) **Numerator**   The numerator (also called the *input determinant*) is a determinant formed in a manner similar to the denominator, except that the "input column" is substituted for the column under the output thus determined. For the illustration at hand, the solution becomes

$$X_1 = \begin{vmatrix} L & D & G \\ M & E & H \\ N & F & I \end{vmatrix} + \Delta \qquad\qquad [12.2.3\text{-}1]$$

$$X_2 = \begin{vmatrix} A & L & G \\ B & M & H \\ C & N & I \end{vmatrix} + \Delta \qquad\qquad [12.2.3\text{-}2]$$

$$X_3 = \begin{vmatrix} A & D & L \\ B & E & M \\ C & F & N \end{vmatrix} + \Delta \qquad\qquad [12.2.3\text{-}3]$$

where the characteristic determinant is

$$\Delta = \begin{vmatrix} A & D & G \\ B & E & H \\ C & F & I \end{vmatrix} \qquad\qquad [12.2.3\text{-}4]$$

12.2.4 Algebra Of Arrays

Some simple rules of the algebra of arrays may be formulated by inspection of eqs. 12.2.1-1, 12.2.2-1, and 12.2.3-1 through 12.2.3-4. These rules will permit manipulations of the array that will transform the original system into an equivalent one.

(a) **Multiplying a Row by a Constant or by an Operator** $G(s)$ Since each row is an equation, then this is equivalent to multiplying both sides of the equation by the same quantity, or operating on both sides the same way. The characteristic determinant is multiplied by $G(s)$.

(b) **Summing Rows**   Since each row is an equation, then adding rows is tantamount to adding equals to equals. Inputs are included in this transformation. Hence, the rule may be stated. If one row of the system array is multiplied by $G(s)$ and added (element by element) to another row, then the characteristic determinant is unchanged and the system is transformed into an equivalent system whose new inputs are automatically accounted for in the operation.

(c) **Multiplying a Column by a Constant or by an Operator** $G(s)$ If a column is multiplied by $G(s)$, then the characteristic determinant is multiplied by $G(s)$, and the heading must be divided by $G(s)$.

(d) **Summing Columns**   If column n is multiplied by $G(s)$ and is added to column p then the characteristic determinant is unchanged and the system is transformed into an equivalent system whose outputs are transformed as follows; all outputs except that for column n remain the same, and the output for column n is X_n'. The heading (new variable X_n') is given by

$$X_n' = X_n - X_p G(s) \qquad\qquad [12.2.4\text{-}1]$$

where

$$G(s) = \text{constant or function of } s \qquad [12.2.4\text{-}2]$$
$$X_n = \text{original heading (output)} \qquad [12.2.4\text{-}3]$$
$$X_n' = \text{transformed heading (output)} \qquad [12.2.4\text{-}4]$$

12.3 ANALYSIS OF COUPLING BETWEEN EQUATIONS OF MOTION

12.3.1 Type of Coupling

The tabular form or system array for a system with two degrees of freedom has a characteristic that is two by two. For the moment let us ignore the input column and the output row, in order to concentrate upon the characteristic.

$M_1 S^2 + Z_1 + \cdots$	$-C$
$-C$	$M_2 S^2 + Z_2 + \cdots$

[12.3.1-1]

The off-diagonal terms C represent the coupling.

■ **EXAMPLE 12.3.1-1** Motor and Pump

In Fig. 12.3.1-1, a motor drives a fluid pump. The connecting shaft K is somewhat elastic, and the fluid offers viscous damping D to the pump. An input torque T_a is applied to mass J_1, the twisted shaft exerts a torque T_k in opposite directions on both masses J_1 and J_2, and the damping D is applied to mass J_2. The actual system and its free body diagram appear in Fig. 12.3.1-1. Model the system.

Solution. Note that torque T_k is applied in a negative direction to J_1 but in a positive direction to J_2. Sum the torques on each mass

$$\sum T_1 = J_1 s^2 \theta_1 = T_a - T_k = T_a - K(\theta_1 - \theta_2)$$
$$\sum T_2 = J_2 s^2 \theta_2 = T_k - T_d = K(\theta_1 - \theta_2) - Ds(\theta_2 - 0) \tag{1}$$

Separating the variables

$$(J_1 s^2 + K)\theta_1 - K\theta_2 = T_a(s)$$
$$-K\theta_1 + (J_2 s^2 + Ds + K)\theta_2 = 0 \tag{2}$$

Using eqs. (2), formulate the system array.

θ_1	θ_2	Inputs
$J_1 s^2 + K$	$-K$	$T_a(s)$
$-K$	$J_2 s^2 + Ds + K$	0

(3)

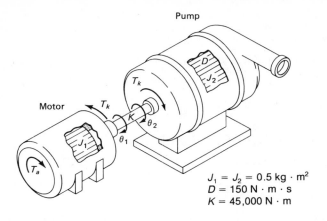

$$J_1 = J_2 = 0.5 \text{ kg} \cdot \text{m}^2$$
$$D = 150 \text{ N} \cdot \text{m} \cdot \text{s}$$
$$K = 45,000 \text{ N} \cdot \text{m}$$

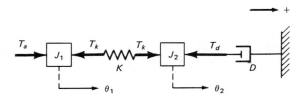

Figure 12.3.1-1   Motor and pump (two mass system).

Noting that the off-diagonal terms are springs, we conclude the system is elastically coupled. Use Cramer's rule to solve for each unknown

$$\theta_1(s) = \frac{\begin{vmatrix} T_a(s) & -K \\ 0 & J_2 s^2 + Ds + K \end{vmatrix}}{\begin{vmatrix} J_1 s^2 + K & -K \\ -K & J_2 s^2 + Ds + K \end{vmatrix}} \tag{4}$$

$$\theta_2(s) = \frac{\begin{vmatrix} J_1 s^2 + K & T_a(s) \\ -K & 0 \end{vmatrix}}{\begin{vmatrix} J_1 s^2 + K & -K \\ -K & J_2 s^2 + Ds + K \end{vmatrix}} \tag{5}$$

■ **EXAMPLE 12.3.1-2** Superheterodyne Interstage Amplifier
A superheterodyne system is used in a radio receiver to produce a constant frequency for the amplifier stages. This scheme offers two advantages: optimum amplifier gain, and effective means to reject random noise. A typical interstage amplifier is approximated in Fig. 12.3.1-2. Formulate the system array in terms of networks Z_i, where $Z_i = D_i s + K_i$.

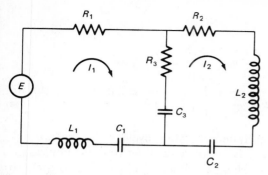

Figure 12.3.1-2 Superheterodyne interstage amplifier.

Solution. Sum the voltage drops according to Kirchhoff's laws in Section 3.3.1. Note that currents I_1 and I_2 flow in opposite directions through R_3 and C_3. Then,

$$\Delta V = \left(R_1 + \frac{1}{C_1 s} \right)(I_1 - I_2) \tag{1}$$

Apply Kirchhoff's law to each loop, noting the sign convention for the currents.

$$E = L_1 s I_1 + R_1 I_1 + \frac{I_1}{C_1 s} + R_3(I_1 - I_2) + \frac{1}{C_3 s}(I_1 - I_2) \tag{2}$$

$$0 = L_2 s I_2 + R_2 I_2 + \frac{I_2}{C_2 s} - R_3(I_1 - I_2) - \frac{1}{C_3 s}(I_1 - I_2) \tag{3}$$

Rearranging terms in eqs. (2) and (3), the system array becomes

I_1	I_2	Inputs	
$L_1 s + R_1 + R_3 + \dfrac{1}{C_1 s} + \dfrac{1}{C_3 s}$	$-\left(R_3 + \dfrac{1}{C_3 s} \right)$	$E(s)$	(4)
$-\left(R_3 + \dfrac{1}{C_3 s} \right)$	$L_2 s + R_2 + R_3 + \dfrac{1}{C_2 s} + \dfrac{1}{C_3 s}$	0	

Use rule (a) of Section 12.2.4; multiply each row by s and regroup the terms.

I_1	I_2	Inputs	
$L_1 s^2 + Z_1 + Z_3$	$-Z_3$	$sE(s)$	(5)
$-Z_3$	$L_2 s^2 + Z_2 + Z_3$	0	

where

$$Z_i = R_i s + \frac{1}{C_i} = D_i s + K_i \qquad (6)$$

Noting the off-diagonal terms in eq. (5) and their representation in eq. (6), we may conclude that the system is R-C series network coupled. ∎

■ **EXAMPLE 12.3.1-3** Suspension System for Motorcycle

The objective of a suspension system is to isolate the rider from the unpleasant vibrations due to road conditions. These conditions cause the motorcycle to translate vertically and to rotate about a horizontal axis that is perpendicular to the plane of the paper. See Fig. 12.3.1-3. Neglecting the mass of the tires, this system consists of a single mass (the vehicle frame plus driver) that has two degrees of freedom. Model the system.

Solution. A motorcycle suspension system was analyzed for one degree of freedom in Ex. 6.5.1-2. The motorcycle will now be considered as a system with two degrees of freedom. There are several possible coordinate systems that could be used to express these two degrees of freedom. Let us choose vertical translation of the center of mass, and the rotation about the center of mass. Refer to Fig. 12.3.1-3.

For the free body diagram, consider the frame of the motorcycle to be a rigid body with mass M and moment of inertia J. The input displacement Y_i at each wheel represents the road conditions. Assume that the suspension system for each wheel can be expressed in terms of a polynomial $Z_i(s)$ operating on the input displacement $Y_i(s)$. In this way it will be possible to model any type of suspension system. With this assumption we may write the

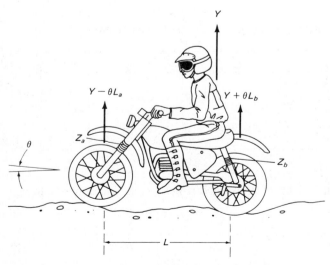

Figure 12.3.1-3 Motorcycle with two degrees of freedom.

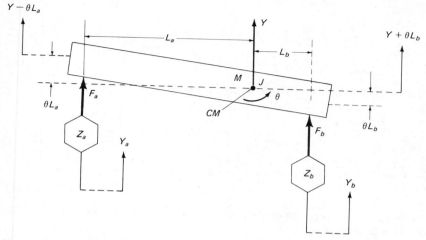

Figure 12.3.1-4   Free body diagram for motorcycle.

force exerted by each suspension system upon the frame, as follows

$$F_i = Z_i(s)Y_i(s) \tag{1}$$

Thus, if we decide that the suspension system consists of only a spring, then

$$Z_i = K_i$$

and

$$F_i = K_i Y_i \tag{2}$$

The free body diagram is shown in Fig. 12.3.1-4. Including the translation and the rotation, the forces F_a and F_b at each wheel become, respectively

$$F_a = Z_a[Y_a - (Y - \theta L_a)]$$
$$F_b = Z_b[Y_b - (Y + \theta L_b)] \tag{3}$$

Sum vertical forces according to Newton's law, assuming IC = 0

$$\sum F = M\ddot{Y}$$

$$F_a + F_b = Ms^2Y$$

or

$$(Ms^2 + Z_a + Z_b)Y - (Z_aL_a - Z_bL_b)\theta = Z_aY_a + Z_bY_b \tag{4}$$

Sum the torques about the center of mass according to Newton's law, assuming IC = 0

$$\sum T = J\ddot{\theta}$$

$$F_bL_b - F_aL_a = Js^2\theta$$

or

$$\left(J s^2 + Z_b L_b^2 + Z_a L_a^2 \right) \theta - \left(Z_a L_a - Z_b L_b \right) Y$$

$$= Z_b L_b Y_b - Z_a L_a Y_a \qquad (5)$$

Using eqs. (4) and (5), the system array becomes

Y	θ	Inputs
$Ms^2 + Z_a + Z_b$	$-(Z_a L_a - Z_b L_b)$	$Z_a Y_a + Z_b Y_b$
$-(Z_a L_a - Z_b L_b)$	$Js^2 + Z_a L_a^2 + Z_b L_b^2$	$Z_b L_b Y_b - Z_a L_a Y_a$

(6)

The terms L_a and L_b in the off-diagonal entries relate rotation and translation. Hence, the system is coordinate coupled.

$L_a = 1$ m $L_b = L_a/2$

$K_a = 1.3 \times 10^4$ N $\cdot$ m $K_b = 2K_a$

$M = 200$ kg $J = 22.5$ kg $\cdot$ m² ∎

12.4 LAGRANGE'S EQUATIONS

12.4.1 Formulation of Equations of Motion

Lagrange's equations result from an energy method with which one can write the differential equations of motion for fairly complex systems. While the method does not assist in solving these equations nor does it expose any system properties, it is extremely simple to use and it usually avoids any complicated coordinate transformation (for instance one may mix polar and Cartesian coordinates, or velocity and voltage, and so forth). The fundamental form of Lagrange's equations is given as follows:

$$\frac{d}{dt} \frac{\partial (\text{KE})}{\partial \dot{q}_i} - \frac{\partial (\text{KE})}{\partial q_i} + \frac{\partial (\text{PE})}{\partial q_i} + \frac{\partial (\text{DF})}{\partial \dot{q}_i} = Q_i \qquad [12.4.1\text{-}1]$$

where   KE = total kinetic energy of the system
   PE = total potential energy of the system
   DF = dissipation function = one-half the total power dissipated by viscous damping
   Q_i = total external force acting along coordinate q_i
   q_i = generalized coordinate for axis i
   $\dot{q}_i$ = time derivative of coordinate q_i, and it is assumed that $\dot{q}_i$ is independent of q_i

Note that one differential equation is generated for each coordinate axis i. Hence, Lagrange's equations provide n simultaneous equations.

■ **EXAMPLE 12.4.1-1** Lagrange's Equations for Damped Double Compound
Pendulum

Consider two distributed masses in the form of links or compound pendulums.
The link that is pivoted with respect to the ground is damped, and is forced by
a forcing function $T_1(t)$. See Fig. 12.4.1-1. Model the system.

Solution.   For an energy method, the velocities are required.

$$V_1 = L_1\dot{\theta}_1 \quad \text{(velocity of mass center } M_1)$$

$$V_a = L_a\dot{\theta}_1 \quad \text{(velocity of pivot of the two links)}$$

$$_aV_2 = L_2\dot{\theta}_2 \quad \text{(relative velocity of mass center } M_2$$
$$\text{with respect to pivot point } A)$$

For velocity V_2 use the law of cosines

$$V_2^2 = V_a^2 + {_aV_2^2} - 2V_a\,{_aV_2}\cos\phi$$

where

$$\cos\phi = \cos[\pi - (\theta_2 - \theta_1)] = -\cos(\theta_2 - \theta_1)$$

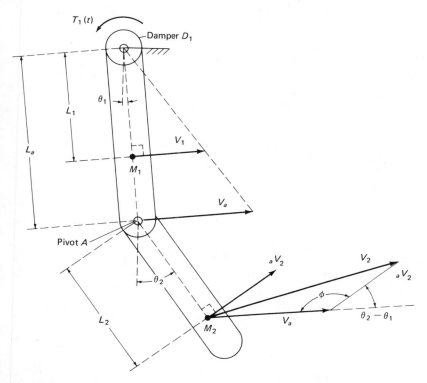

Figure 12.4.1-1   Damped and forced double compound pendulum.

Then

$$V_2^2 = \left(L_a \dot{\theta}_1 \right)^2 + \left(L_2 \dot{\theta}_2 \right)^2 + 2\left(L_a \dot{\theta}_1 \right)\left(L_2 \dot{\theta}_2 \right)\cos(\theta_2 - \theta_1)$$

The kinetic energy KE is given by the translatory plus rotary terms

$$\begin{aligned}
\text{KE} &= \tfrac{1}{2}M_1 V_1^2 + \tfrac{1}{2}M_2 V_2^2 + \tfrac{1}{2}J_1 \dot{\theta}_1^2 + \tfrac{1}{2}J_2 \dot{\theta}_2^2 \\
&= \tfrac{1}{2}\left(M_1 L_1^2 + M_2 L_a^2 + J_1 \right)\dot{\theta}_1^2 + \tfrac{1}{2}\left(M_2 L_2^2 + J_2 \right)\dot{\theta}_2^2 \\
&\quad + M_2 L_a L_2 \dot{\theta}_1 \dot{\theta}_2 \cos(\theta_2 - \theta_1)
\end{aligned} \tag{1}$$

The potential energy PE is equal to the work done in lifting each pendulum.

$$\begin{aligned}
\text{PE} &= M_1 g h_1 + M_2 g h_2 \\
&= \left(M_1 g L_1 + M_2 g L_a \right)(1 - \cos\theta_1) + M_2 g L_2 (1 - \cos\theta_2)
\end{aligned} \tag{2}$$

The dissipation function is equal to one half the damping power

$$\text{DF} = \tfrac{1}{2}\left(D_1 \dot{\theta}_1 \right)\dot{\theta}_1 = \tfrac{1}{2}D_1 \dot{\theta}_1^2 \tag{3}$$

The forcing function is applied about axis θ_1 only. Lagrange's equations will be used to formulate the differential equations. For the θ_1 coordinate

$$\frac{\partial(\text{KE})}{\partial \dot{\theta}_1} = \left(M_1 L_1^2 + M_2 L_a^2 + J_1 \right)\dot{\theta}_1 + M_2 L_a L_2 \dot{\theta}_2 \cos(\theta_2 - \theta_1)$$

$$\frac{d}{dt}\frac{\partial \text{KE}}{\partial \dot{\theta}_1} = \left(M_1 L_1^2 + M_2 L_a^2 + J_1 \right)\ddot{\theta}_1 + M_2 L_a L_2 \ddot{\theta}_2 \cos(\theta_2 - \theta_1)$$

$$\frac{\partial \text{KE}}{\partial \theta_1} = M_2 L_a L_2 \dot{\theta}_1 \dot{\theta}_2 \sin(\theta_2 - \theta_1)$$

$$\frac{\partial \text{PE}}{\partial \theta_1} = \left(M_1 g L_1 + M_2 g L_a \right)\sin\theta_1$$

$$\frac{\partial \text{DF}}{\partial \dot{\theta}_1} = D_1 \dot{\theta}_1$$

$$Q_1 = T_1$$

Apply Lagrange's equations in order to formulate the differential equation for the θ_1 coordinate. Note that this equation contains terms in θ_2 as well as θ_1. Since the oscillation can be assumed to produce very small angles, make the following small angle approximations:

$$\cos\theta = 1 - \tfrac{1}{2}\theta^2$$
$$\sin\theta = \theta \tag{4}$$

Assuming that the oscillations are small, then terms of the form θ, $\dot{\theta}$, and $\ddot{\theta}$ may be considered small. Hence, products of these (particularly products involving three terms) can be neglected. Then, the differential equation for the θ_1 coordinate becomes

$$
\left(M_1 L_1^2 + M_2 L_a^2 + J_1\right)\ddot{\theta}_1 + M_2 L_a L_2 \ddot{\theta}_2
$$
$$
+ \left(M_1 g L_1 + M_2 g L_a\right)\theta_1 + D_1 \dot{\theta}_1 = T_1(t) \tag{5}
$$

Transform eq. (5) to the Laplace domain, and include IC.

$$
\left[\left(M_1 L_1^2 + M_2 L_a^2 + J_1\right)s^2 + D_1 s + \left(M_1 g L_1 + M_2 g L_a\right)\right]\theta_1
$$
$$
+ \left(M_2 L_a L_2 s^2\right)\theta_2 = T_1(s) + \left(M_1 L_1^2 + M_2 L_a^2 + J_1\right)s\theta_{1_0}
$$
$$
+ M_2 L_a L_2 s\theta_{2_0} + D_1 \dot{\theta}_{1_0} \tag{6}
$$

Similarly for the θ_2 coordinate,

$$
\left(M_2 L_a L_2 s^2\right)\theta_1 + \left[\left(M_2 L_2^2 + J_2\right)s^2 + M_2 g L_2\right]\theta_2
$$
$$
= M_2 L_a L_2 s\theta_{1_0} + \left(M_2 L_2^2 + J_2\right)s\theta_{2_0} \tag{7}
$$

Using eqs. (6) and (7), the system array becomes

θ_1	θ_2	Inputs and IC
$\left(M_1 L_1^2 + M_2 L_a^2 + J_1\right)s^2$ $+ D_1 s + \left(M_1 g L_1 + M_2 g L_a\right)$	$\left(M_2 L_a L_2\right)s^2$	$T_1(s) + \left(M_2 L_a L_2\right)s\theta_{2_0} + D_1 \dot{\theta}_{1_0}$ $+\left(M_1 L_1^2 + M_2 L_2^2 + J_1\right)s\theta_{1_0}$
$\left(M_2 L_a L_2\right)s^2$	$\left(M_2 L_2^2 + J_2\right)s^2$ $+ M_2 g L_2$	$M_2 L_a L_2 s\theta_{1_0}$ $+\left(M_2 L_2^2 + J_2\right)s\theta_{2_0}$

$$\tag{8}$$

The off-diagonal terms are of the form $ML^2 s^2$, which is equivalent to the moment of inertia term $J^2 s^2$. Since this is a rotational system, the inertia terms should be of this form. Hence, the system has inertia coupling. ∎

12.5 THE GYROSCOPE

12.5.1 Conservation of Angular Momentum of the Gyroscope

The gyroscope in its simplest form consists of a flywheel that is spinning at high speed. The spin axis is supported in a way that permits complete rotation freedom. If the gyro is undisturbed, the spin axis will remain parallel to its original orientation. This property is due to the conservation of angular momentum.* If the supports (usually gimbals) are frictionless, then the vehicle

---

*Some individuals in the gyro manufacturing industry refer to this property as the "spatial memory" of the gyroscope.

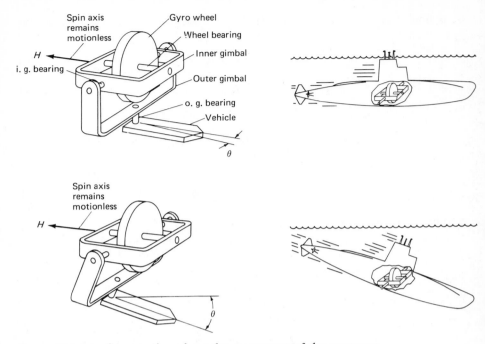

Figure 12.5.1-1   Conservation of angular momentum of the gyroscope.

may tilt through various angles while the gyro retains its orientation. See Fig.
12.5.1-1.

12.5.2 Angular Momentum

The sum of the moments of momentum of all the particles of a system is
referred to as angular momentum **H**:

$$\mathbf{H} = \int \mathbf{r} \times \mathbf{V} \, dm \qquad [12.5.2\text{-}1]$$

For simplicity, consider the specialized case of a body of revolution
rotating about its centroidal axis:

$$H = \int \omega r^2 \, dm \qquad [12.5.2\text{-}2]$$

Assume the system of particles is a rigid body; the angular velocity ω is
the same for all particles and, therefore, is constant with respect to m (the
angular velocity may vary with time, but at any instant all particles have the
same angular velocity). Consequently, the angular velocity may be taken out of
the integral sign when integrating with respect to m, thus,

$$H = \omega \int r^2 \, dm \qquad [12.5.2\text{-}3]$$

The integral $\int r^2\,dm$ is defined as the moment of inertia of the body of revolution. Thus, for the special case of a solid of revolution rotating about its centroidal axis:

$$\mathbf{H} = \omega \mathbf{I}$$ [12.5.2-4]

12.5.3 Gyroscopic Torque

If the gyro is not free, but is forced to turn, its spin axis is forced to change its orientation. Refer to Fig. 12.5.3-1. Let the initial position of the spin axis be represented by the vector ω_1 and the position after time Δt be the vector ω_2. The vector $\Delta \omega$ represents the change in angular velocity.

$$\Delta \omega = \omega \, \Delta \phi$$ [12.5.3-1]

where $\Delta \phi$ is the angle turned during time Δt.

Likewise, the angular momentum vector $\mathbf{H}$ may be shown initially and after time Δt, as indicated in Fig. 12.5.3-1(b)

$$\mathbf{H} = I\omega$$ [12.5.3-2]

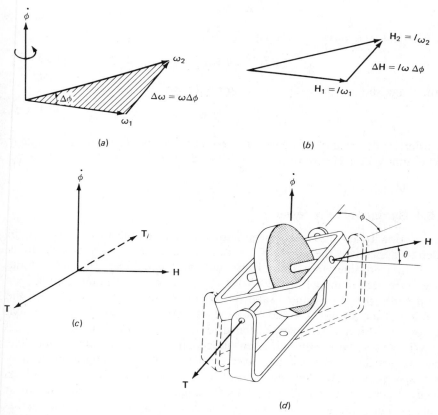

Figure 12.5.3-1   Gyroscopic torque.

where I = mass moment of inertia of wheel about the spin axis
       ω = angular velocity of wheel.

The change in magnitude of angular momentum is

$$\Delta H = I \Delta \omega \qquad [12.5.3\text{-}3]$$

The torque required to change the angular momentum is T_i, and the equal and opposite torque T felt by the structure is

$$T = \frac{\Delta H}{\Delta t} \qquad [12.5.3\text{-}4]$$

Apply eqs. 12.5.3-1 through 12.5.3-3 to eq. 12.5.3-4.

$$T = H \frac{\Delta \phi}{\Delta t} \qquad [12.5.3\text{-}5]$$

In the limit, $\Delta\phi/\Delta t$ approaches $\dot{\phi}$. If the turn rate vector $\dot{\phi}$ makes an angle ψ with the plane containing the vectors $\mathbf{H}_1$ and $\mathbf{H}_2$, then the perpendicular component is

$$\dot{\phi} \sin \psi$$

Applying the above concepts to eq. 12.5.3-5, we have

$$T = H \dot{\phi} \sin \psi \qquad [12.5.3\text{-}6]$$

as the magnitude of the torque. In vector notation, this becomes

$$\mathbf{T} = \mathbf{H} \times \dot{\phi} \qquad [12.5.3\text{-}7]$$

In order to determine the sense of the torque vector $\mathbf{T}$, rotate the angular momentum vector $\mathbf{H}$ toward the input rate vector $\dot{\phi}$ and apply the right-hand rule.

12.5.4 Gyroscopic Precession

This section treats the effects of a torque applied to the gyro. However, in order to appreciate the effect of a torque applied to the axis of a spinning wheel, first consider the same disturbance on a nongyroscopic body. If the wheel were *not* spinning [Fig. 12.5.4-1(a)], a torque about the inner gimbal axis would cause the inner gimbal (and its motionless wheel) to accelerate in the direction of the torque. It would not be at all surprising to see the inner gimbal tilt downward if a weight were fastened to one edge of the gimbal.

    When the wheel is spinning, an altogether different picture results. The same weight does not cause the inner gimbal to tilt at all, appearing to defy the law of gravity. Actually, no contradiction of the universal law takes place. The gyro develops a counter-torque by turning about the outer gimbal axis at a rate $\dot{\phi}$. This induced turning rate, $\dot{\phi}$, referred to as *gyroscopic precession*, is equal in magnitude but opposite in sign to the relationship discussed in the

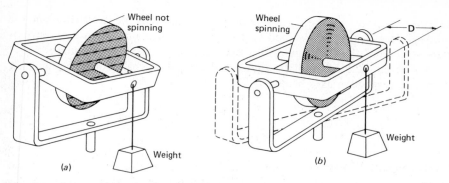

Figure 12.5.4-1   Gyroscopic precession.

preceding section. Then,

$$\dot{\phi} = \frac{T_i}{H}$$ [12.5.4-1]

12.5.5 Transient Response for Free Gyro

If the gyro is unrestrained (free to turn about any axis) it is called a *free gyro*. For the practical gyro, this is close to the actual case. Precision low-friction bearings are used. The free gyro is shown in Fig. 12.5.5-1.

In order to study the dynamics of a free gyro, concentrate upon the movement of the spin axis. It will be assumed that the gimbals provide the

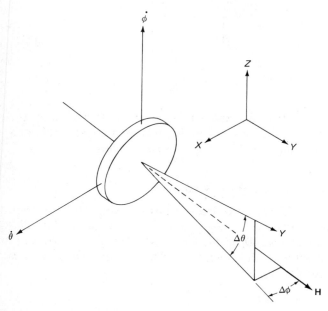

Figure 12.5.5-1   Free gyroscope.

necessary support of the spin axis and that they may be omitted from the analysis. It will also be assumed that the gimbal angles θ and ϕ are approximately 90° and that all angular displacements will be extremely small.

Using eq. 12.5.3-6 and the right-hand rule, the gyro torques T_{xh} and T_{zh} about the respective axes become

$$T_{xh} = H\dot{\phi}$$

$$T_{zh} = -H\dot{\theta}$$

[12.5.5-1]

Apply Newton's law about the x and z axes, noting that there are no restraints (springs, dampers, friction). The net torques T_x and T_z become

$$J_x\ddot{\theta} + H\dot{\phi} = T_x$$

$$J_z\ddot{\phi} - H\dot{\theta} = T_z$$

[12.5.5-2]

Transform eqs. 12.5.5-2 to the Laplace domain, where the initial conditions are given by

$$\theta(0) = \dot{\theta}(0) = \phi(0) = \dot{\phi}(0) = 0$$

[12.5.5-3]

Then

$$J_x s^2 \theta(s) + Hs\phi(s) = T_x(s)$$

$$- Hs\theta(s) + J_z s^2 \theta(s) = T_z(s)$$

[12.5.5-4]

Formulate the system array, and normalize.

θ	ϕ	Input
s^2	sH/J_x	T_x/J_x
$-sH/J_z$	s^2	T_z/J_z

[12.5.5-5]

The off-diagonal terms contain the angular momentum H and also the derivative operator s. Hence, the system coupling involves angular rate and angular momentum, which is referred to as *gyroscopic coupling*.

12.5.6 Gyro with Restraints, The Rate Gyro

In the previous sections, the gyro was free or unrestrained. Here, we will consider two restraints, restriction to one degree of freedom, and the installation of a spring K and a damper D on the free axis. See Fig. 12.5.6-1. Since the device has only one degree of freedom, it is called a *single axis gyro*. The equation of motion becomes

$$T_{in} = H\dot{\phi} = J\ddot{\theta} + D\dot{\theta} + K\theta$$

[12.5.6-1]

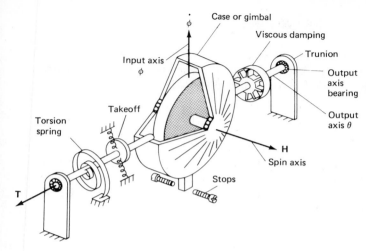

Figure 12.5.6-1   Rate gyroscope.

where θ = angular displacement of gimbal
J = moment of inertia of the gimbal and wheel about the output axis
D = viscous damping coefficient
K = coefficient of constraint (due to elastic restraint of the flex leads plus take-off-reaction torque) or torsion spring of a rate gyro
$\dot{\phi}$ = input velocity
H = angular momentum about spin axis

Note that the input is $\dot{\phi}$, the output is θ, and the two variables are separated. Thus, this is a system whose solution is of the form of a reciprocal quadratic.

$$\theta(s) = \frac{H\dot{\phi}}{Js^2 + Ds + K} \qquad [12.5.6\text{-}2]$$

The steady state response to a step input is

$$\theta_{ss} = \lim_{s \to 0} \frac{sH\dot{\phi}/s}{Js^2 + Ds + K} \qquad [12.5.6\text{-}3]$$

$$= \frac{H\dot{\phi}}{K} \qquad [12.5.6\text{-}4]$$

Since the output θ is proportional to the input rate $\dot{\phi}$, the device is called a rate gyro. Obviously, the instrument can be made very sensitive by choosing the ratio H/K (called the gyro gain)

$$H/K \gg 1 \qquad [12.5.6\text{-}5]$$

12.6 MECHANICAL NETWORKS

This technique makes use of LMNOP (linear mechanical network operators), which are operators in the Laplace domain. There are two types of operators, one for rigid bodies, the other for deformable bodies.

12.6.1 Rigid Body or Mass M

If all points of a body (or a group of bodies) maintain a fixed spatial relationship with respect to one another, the body (or group of bodies) will be treated as a rigid body. In a lumped system, a rigid body can be considered as a point mass. In such a case, we may employ Newton's law; the sum of forces $\sum F$ on a mass M will produce an acceleration $\ddot{x}$ of the mass. In the time domain

$$\sum F = M\ddot{x} \qquad [12.6.1\text{-}1]$$

In the Laplace domain

$$\sum F(s) = Ms^2X - Msx_0 - M\dot{x}_0 \qquad [12.6.1\text{-}2]$$

where x_0 and $\dot{x}_0$ are initial displacement and velocity, respectively. Transposing the initial condition terms to the left-hand side of the equation, the above equation becomes

$$\sum F(s) + IC = Ms^2X \qquad [12.6.1\text{-}3]$$

Diagrammatically, this will be shown as a circle with the term Ms^2 written within it, as shown in Fig. 12.6.1-1(a). The treatment of IC (initial conditions) will be deferred until Section 12.6.4.

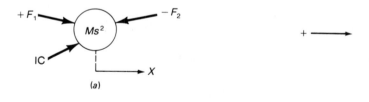

(a)

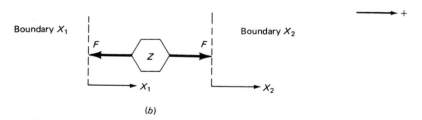

(b)

Figure 12.6.1-1 Linear mechanical networks. (a) Rigid body or mass. (b) Deformable body or network Z.

12.6.2 Deformable Body or Network Z

If a body or group of bodies is deformable, it will be referred to as a mechanical network. If the deformable body resists such deformation (by exerting reaction forces upon its boundaries) and if this resistance can be expressed in terms of a linear differential equation (or a linear operator Z) then the body will be referred to as a linear mechanical network (LMN). See Fig. 12.6.1-1(b). Examples of such devices are springs, dampers, and combinations of these. These may be alone, or in groups, and they may be connected directly, through levers, gears, fluids, and so forth.

The network will be located between boundaries, one or both of which move. These boundaries may be fixed walls (ground), moving walls (displacement input), or masses. No matter what the boundary is, the network will exert its forces against these boundaries. In general, consider that both boundaries move. A fixed boundary can be set equal to zero later. If both boundaries move the *same* amount (same magnitude and the same direction), the body obviously will *not* deform. Hence, it is the *relative* motion of these boundaries that affects the network, resulting in the following conventions:

1. A force (or torque) is exerted *out* of both ends against each boundary
2. The force (or torque) is given by the linear operator Z operating upon the difference of the boundaries. Use the algorithm for sign convention in Section 2.2.5(d).

Obviously in the Laplace domain, where X_1 and X_2 are the moving boundaries,

$$F = Z(s)(X_1 - X_2) \qquad [12.6.2\text{-}1]$$

To derive the network for a spring and a damper, transform eqs. 2.2.5-4 and 2.3.5-3, respectively, to the Laplace domain. Thus, the notation becomes:

$$Z_k = K \quad \text{for a spring} \qquad [12.6.2\text{-}2]$$

$$Z_d = Ds \quad \text{for a damper} \qquad [12.6.2\text{-}3]$$

12.6.3 Groups of Networks

(a) **Networks in Parallel** Networks are in parallel if *both* ends share common boundaries. They can be replaced by a single equivalent network Z_p, where

$$Z_p = Z_a + Z_b \qquad [12.6.3\text{-}1]$$

See Fig. 12.6.3-1(a).

(b) **Networks in Series** Networks are in series if they have but a single interface which is not in contact with nor affected by any outside force or

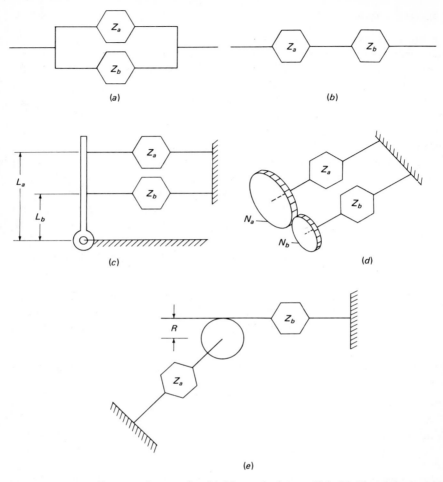

Figure 12.6.3-1   Groups of networks. (a) Networks in parallel. (b) Networks in series. (c) Networks connected by levers. (d) Networks connected by gears. (e) Networks connected by rack and pinion.

torque. They can be replaced by a single equivalent network Z_s, where

$$Z_s = \frac{Z_a Z_b}{Z_a + Z_b} \qquad [12.6.3\text{-}2]$$

See Fig. 12.6.3-1(b).

(c) Networks Connected by Levers   If two networks are connected by levers, they can be replaced by a single equivalent network that can be located anywhere on the lever. For example, if network Z_b is to be located at (reflected to) the site of network Z_a, the equivalent network Z_1 becomes

$$Z_l = Z_a + Z_b \left(\frac{L_b}{L_a}\right)^2 \qquad [12.6.3\text{-}3]$$

See Fig. 12.6.3-1(c).

(d) Networks Connected by Gears If two networks are connected by gears, they can be replaced by a single equivalent network that can be located on either shaft. If the chosen location is the site of network Z_a, then network Z_b must be "reflected" to that site. The equivalent network Z_g becomes

$$Z_g = Z_a + Z_b \left(\frac{N_a}{N_b}\right)^2 \qquad [12.6.3\text{-}4]$$

See Fig. 12.6.3-1(d).

(e) Networks Connected by a Rack and Pinion If a translatory network Z_t is fastened to a rack that engages a pinion, which is fastened to a rotary network Z_r, these can be replaced by a single network that is either a translatory equivalent one or a rotary equivalent one. If the equivalent network is to be a rotary one, then the translatory network Z_t must be "reflected" to the rotating shaft. The equivalent rotary network Z_{eq} becomes

$$Z_{eq} = Z_a + Z_b R^2 \qquad [12.6.3\text{-}5]$$

See Fig. 12.6.3-1(e).

12.6.4 One Mass Rule

The effects considered in the previous sections will now be applied simultaneously in order to summarize these effects. Consider a mass that is connected through networks to inputs and the ground in all possible ways as shown in Fig. 12.6.4-1. Transform all terms (including initial conditions) to the Laplace domain and solve for the displacement $X(s)$.

$$X = \frac{F' - F'' + Z_c X_c + Z_d X_d - Z_a X_a - Z_b X_b + IC}{MS^2 + Z_a + Z_1 + Z_c + Z_b + Z_2 + Z_d} \qquad [12.6.4\text{-}1]$$

where IC represents the initial conditions.

Upon comparison of the above equation against the figure to which it relates, the following conclusions can be drawn and can be stated as the one mass rule. The output or displacement of the mass in a system having one degree of freedom is given by a proper fraction, where:

1. The *numerator* is equal to the *sum* of each *input* multiplied by its connection to the mass. (Since forces are applied directly to the mass, the connection is unity.) In selecting the sign for each term, only the

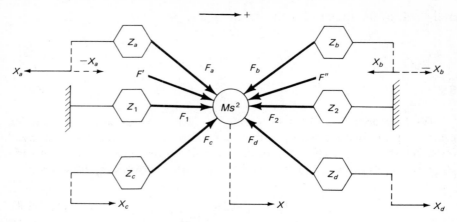

Figure 12.6.4-1   One mass rule.

direction of the *input* is taken, with *no* regard to which side of the mass the input is applied.

2. The *denominator* is given by the *sum* of Ms^2 and *all* the networks that are connected to the mass with *no* regard to which side of the mass they are applied. [Note that each network that connects an input to the mass will appear in the numerator as indicated by rule (1) and it will also be added to the Ms^2 in the denominator.]

3. Initial conditions IC. These develop in the ordinary procedure for taking the Laplace transform of derivatives. Note that all such terms appear only in the numerator.

12.6.5 Multimass Rule

When the system consists of more than one mass, the one mass rule could be applied to each mass. However, since an array will be formed, the term that was called the denominator of the one mass rule is placed as an entry on the diagonal. The term that was called the numerator of the one mass rule will be set as an entry in the input column. Those networks that connect the masses will be set as off-diagonal entries in the appropriate locations. Initial conditions, which were added to the numerator in the one mass rule, will be added to inputs in the input column. See Fig. 12.6.5-1.

These conclusions can be summarized and formalized in a form that will be termed the multimass rule as follows:

1. Each diagonal element is equal to $M_i s^2$ plus all the networks that touch mass M_i no matter where the other ends are attached and no matter in which direction they lie (where M_i is the ith mass of the system).

2. Each off-diagonal element is equal to the negative $(-Z_{ij})$ of the network that connects mass M_i to mass M_j. This quantity is entered in two places, row i, column j, and in row j, column i.

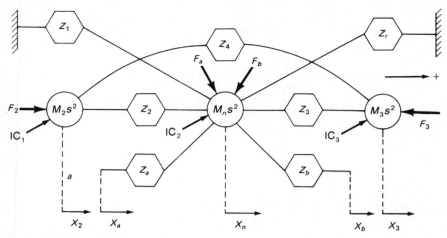

Figure 12.6.5-1   Multimass rule.

3. The input entry for row i is equal to the sum of each input (to the mass M_i) multiplied by the network that connects the input to the mass.
4. Initial conditions will be summed with inputs in the input column.

■ EXAMPLE 12.6.5-1 Houdaille Damper

Machines with long shafts and heavy masses often vibrate severely. In the actual machine, if there is no place to insert a damping device, one solution to this problem is to place a wheel within a wheel. Between the wheels, a viscous fluid is used. The oscillation, should it become excessive, will result in relative motion of the two wheels, with the attendant viscous damping. On the other hand, when the system is rotating without vibrating, there is no relative motion, and therefore, no damping. Thus, this device is adaptive in producing damping only when needed. See Fig. 12.6.5-2. There is an initial velocity, but no initial twisting of the shaft. Model the system.

Solution.   The elastic shaft K connects the input θ_a to mass J_1 while the damper D connects the two masses J_1 and J_2. The load and other torques T are applied to mass J_1. See Fig. 12.6.5-2(b).

The free body diagram can be constructed in the Laplace domain, as shown in Fig. 12.6.5-2(c). The following can be used to formulate the system array.

(a) *Diagonal entries*:
   Networks K and Ds are connected to J_1
   Network Ds is connected to J_2
(b) *Off-diagonal entries*:
   Network Ds connects J_1 with J_2

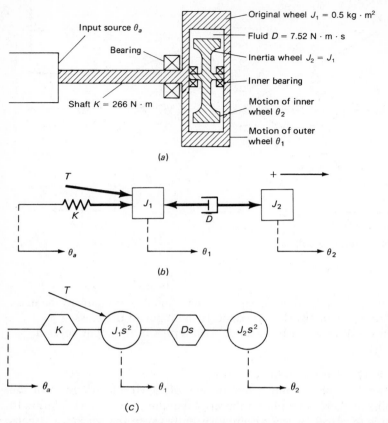

Figure 12.6.5-2 Houdaille damper. (a) Actual system. (b) Free body diagram. (c) Network diagram.

The array, thus far, becomes

θ_1	θ_2	
$J_1 s^2 + K + Ds$	$-Ds$	
$-Ds$	$J_2 s^2 + Ds$	

(1)

(c) *Inputs*:
  Input θ_a through K to J_1 (hence, place in row 1)
  Input T directly to J_1 (hence, add to row 1)
  No inputs to J_2 (hence, place 0 in row 2)
(d) *Initial conditions*: Since there is no initial twisting of the shaft, then $\theta_1(0) = \theta_2(0) = 0$. Initial speed implies $\dot{\theta}_1(0) = \dot{\theta}_2(0) = \dot{\theta}_0$.

Note that since initial displacements are equal to zero, then the IC will appear only for second derivative terms. This applies to J_1 in row 1 column 1, and also for J_2 in row 2 column 2. Then,

$$IC_1 = J_1 \dot{\theta}_0$$

$$IC_2 = J_2 \dot{\theta}_0$$

The final array becomes

θ_1	θ_2	Inputs and IC	
$J_1 s^2 + Ds + K$	$-Ds$	$K\theta_a + T + J_1\dot{\theta}_0$	(2)
$-Ds$	$J_2 s^2 + Ds$	$0 + J_2\dot{\theta}_0$	

∎

12.7 ELECTRICAL NETWORKS

12.7.1 Modeling of Electrical Systems

The basic relationship for electrical networks is Kirchhoff's laws. The sum of voltages around a closed loop is zero, and the sum of currents at a junction is zero. The basic electrical elements are inductance, resistance, and capacitance. These are all two port elements, like springs and dashpots, but unlike the mass which is a one port element. The port may be regarded as the boundaries of the element. It is assumed that current I passes through the element, and voltage is measured across the element. All elements are assumed to be "pure," that is, an inductance has zero resistance, a resistance has zero inductance, and so forth. It is also assumed that all elements are ideal, and that there are no leakages or losses, and that all elements may be lumped. It is also assumed that any oscillatory currents reach all parts of every element simultaneously. This is reasonably true if the dimensions or path lengths are less than 1% of the wavelength of the oscillation.

Kirchhoff's Laws

1. The algebraic sum of all the potential drops around a closed loop or closed circuit is zero (*loop analysis*).
2. The algebraic sum of all the currents flowing into a junction is zero (*node or admittance analysis*).

12.7.2 Mutual Inductance

If two inductors share a magnetic path, there will be a contribution in the change of magnetic flux in each inductor due to the other one. The effect is mutual and is ascribed to a property term *mutual inductance* L_{ij}. This produces a potential drop ΔV_i in loop i given by

$$\Delta V_i = L_{ij} L_j \qquad [12.7.2\text{-}1]$$

A similar effect (with subscripts changed) applies to loop j.

12.7.3 Amplifier Gain

It will be assumed that an amplifier can be represented by a set of "pure" lumped elements. The gain will be treated as a constant (with no phase shift or frequency dependence), while the actual amplifier will be treated as a set of appropriately located elements. Thus, the gain will be treated as a constant between two voltages (or currents for a current amplifier).

12.7.4 Loop Analysis

Using the above concepts, the system array can be constructed by using the following rules:

1. Each closed loop or closed circuit represents a subsystem which appears on the diagonal.
2. Using either current or charge, these become the headings or variables.
3. The off-diagonal elements are the electrical elements that are common to any two circuits either having currents flowing through that element or having mutual inductance.
4. The input column contains all the voltage sources.

■ **EXAMPLE 12.7.4-1** Circuits with Mutual Inductance
Use loop analysis on the circuit in Fig. 12.7.4-1 and formulate the system array.

Solution. The first diagonal entry, for current I_1, is obtained by traversing a path around the loop. In terms of current, using eq. 3.3.1-4 in the Laplace domain, we have

$$L_1 s + R_1 + \frac{1}{C_1 s} \tag{1}$$

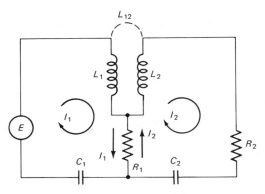

Figure 12.7.4-1   Coupled two loop electrical network.

For the other loop we have, for the second diagonal entry,

$$R_1 + L_2 s + R_2 + \frac{1}{C_2 s} \tag{2}$$

Both the common resistance R_1 and mutual inductance L_{12} are shared by the two loops. Hence the negative of the sum is placed in the off-diagonal entries for row 1 column 2, and vice-versa. This becomes

$$-R_1 - L_{12} s \tag{3}$$

There is only one input, E in the first loop. Place this in the input column, row 1. The array becomes

I_1	I_2	Inputs	
$L_1 s + R_1 + \dfrac{1}{C_1 s}$	$-R_1 - L_{12} s$	E	(4)
$-R_1 - L_{12} s$	$R_1 + L_2 s + R_2 + \dfrac{1}{C_2 s}$	0	

■

12.7.5 Node Analysis

Kirchhoff's second law applies to summing currents at a node, or junction. In this case, it will prove convenient to use admittances (reciprocals). The rules for constructing the system array are as follows:

1. Each node, or junction, represents a subsystem which appears on the diagonal.
2. The voltage at each node is used as the heading for each column.
3. The off-diagonal entries are the elements that connect two nodes. However, care must be exercised to avoid passing through a node on the way. If the ground is used as a node, this, too, must be avoided as a path to another node. Also, the laws of series and parallel elements must be observed (particularly if there is more than one valid path).
4. The input column contains all the current sources.

■ EXAMPLE 12.7.5-1 Transistor Amplifier

A single transistor amplifier is shown in Fig. 12.7.5-1(a). The transistor can be approximated as an R-C network as follows: The path from V_1 to V_4 is a resistor R_T in series with a capacitor C_T, where their junction is labeled V_2. The path from V_4 to V_3 is reciprocal resistance GM. See Fig. 12.7.5-1(b). Assume the power supply is well regulated permitting it to be treated as a fixed quantity. Formulate the system array.

Solution. There are five junctions, or nodes. The system order can be reduced by one if we eliminate node V_3 and replace the series resistors $1/GM$ and R with R_m.

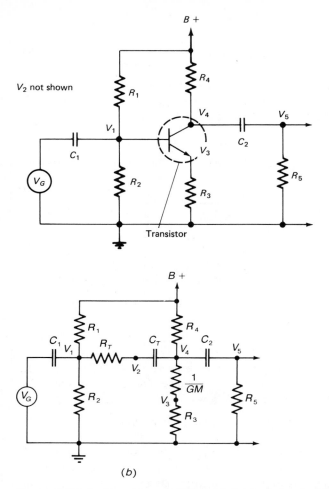

Figure 12.7.5-1 Transistor amplifier. (a) Original circuit. (b) Equivalent circuit.

At node V_1 there are four elements. Summing the admittances (the reciprocals), we have, for the first diagonal entry,

$$C_1 s + \frac{1}{R_1} + \frac{1}{R_2} + \frac{1}{R_T} \tag{1}$$

At node V_2 there are two elements. Then the second diagonal entry becomes

$$C_T s + \frac{1}{R_T} \tag{2}$$

A similar treatment applies to nodes V_4 and V_5. Hence, the remaining diagonal entries become

$$\frac{1}{R_m} + C_T s + C_2 s + \frac{1}{R_4} \tag{3}$$

and

$$C_2 s + \frac{1}{R_5} \tag{4}$$

Now for the off-diagonal entries. Taking care to avoid passing through other nodes or through ground, the connection from node V_1 to node V_2 is R_T. This is a bilateral operator (as are most electrical elements). Hence, in row 1 column 2 and vice-versa, place

$$-1/R_T \tag{5}$$

From node V_1 to node V_4 there is no path that doesn't pass through ground or through the well regulated power supply. Hence, in row 1 column 4 and vice-versa (since there is no valid path) place

$$0 \tag{6}$$

There is no valid path from node V_1 to node V_5. Hence, in row 1 column 5 and vice-versa place

$$0 \tag{7}$$

The path from node V_2 to V_4 passes through capacitor C_T. Hence, in row 2 column 4 and vice-versa place

$$-C_T s \tag{8}$$

The rest of the off-diagonal entries follow in a similar manner.

Finally, consider the two inputs. Input V_g passes through C_1 to V_1. Input B passes through R_1 to V_1 and also through R_4 to V_4. This completes the array, which becomes

V_1	V_2	V_4	V_5	Inputs
$C_1 s + \dfrac{1}{R_1} + \dfrac{1}{R_2} + \dfrac{1}{R_T}$	$-\dfrac{1}{R_T}$	0	0	$V_g C_1 s + \dfrac{B}{R_1}$
$-\dfrac{1}{R_T}$	$C_T s + \dfrac{1}{R_T}$	$-C_T s$	0	0
0	$-C_T s$	$\dfrac{1}{R_m} + C_T s + C_2 s + \dfrac{1}{R_4}$	$-C_2 s$	$\dfrac{B}{R_4}$
0	0	$-C_2 s$	$C_2 s + \dfrac{1}{R_5}$	0

$$(9) \ \blacksquare$$

SUGGESTED REFERENCE READING FOR CHAPTER 12

Modeling multiple degrees of freedom [1] p. 222–223, 229, 300; [4], p. 109–112; [5], p. 159, 211; [8], p. 341, 342; [9], p. 103–106; [16], p. 94, 95; [17], p. 419, 457, 458; [44], p. 163–168.

PROBLEMS FOR CHAPTER 12

12.1. Use rule (a) in Section 12.2.4 and multiply row 1 of the array, eq. 12.2.2-1, by $G(s)$ and compare the result to eqs. 12.2.3-1 through 12.2.3-4, where
(a) $G(s) = k$ (constant)
(b) $G(s) = s$
(c) $G(s) = 1/s$
(d) $G(s) = s^n$

12.2. Similar to Problem 12.1, but use rule (b) and multiply row 1 by $G(s)$ and add this to row 2.

12.3. Similar to Problem 12.1, but use rule (c) and multiply column 1 by $G(s)$ and treat the column head accordingly.

12.4. Similar to Problem 12.1, but use rule (d) and multiply column 1 by $G(s)$ and add this to column 2, treating the column heads accordingly.

12.5. Determine displacements $\theta_1(s)$ and $\theta_2(s)$ is the Laplace domain by expanding the appropriate determinants in Ex. 12.3.1-1 (Motor Pump). Explain why there is a cubic (rather than a quartic) in the denominator.

12.6. Similar to Problem 12.5, but applied to Ex. 12.6.5-1 (Houdaille Damper).

12.7. In Ex. 12.3.1-3 (Motorcycle), determine the displacement Y and rotation θ by expanding the appropriate determinants. Neglect damping.

12.8. For Ex. 12.3.1-3 (Motorcycle), choose the coordinates as displacements Y_a and Y_b. Write the equations of motion or formulate the system array.

12.9. Use Lagrange's equations to derive the equations of motion for the following:
(a) Ex. 12.3.1-1 (Motor Pump).
(b) Ex. 12.6.5-1 (Houdaille Damper).
(c) Ex. 12.3.1-3 (Suspension System for Motorcycle).

12.10. *Circuit With No Capacitance.* In Fig. 12.3.1-2, replace each capacitor C_i with a resistor R_{3i}. What does this do to the order of the system?

12.11. *Circuit With No Inductance.* In Fig. 12.3.1-2, replace each inductance L_i with a resistor R_{3i}. What does this do to the order of the system?

12.12. *Single Axis Gyro.* See Fig. 12.5.4-1. Compute the angular momentum H of a gyro wheel spinning at 24,000 rpm, having a mean rim diameter of 4 cm, and a mass $M = 1$ kg. Assume all the mass to be concentrated in the rim.

12.13. Find the magnitude and direction of the torque produced by an angular rate applied to the gyro in Problem 12.12 of
(a) 1 rad/s about the spin H axis
(b) 1 rad/s about the input ϕ axis
(c) 1 rad/s about the output θ axis

12.14. Find the precession due to a 1-kg mass hung at the end of a 4-cm arm attached to the gyro in Problem 12.12.

12.15. Find precession (drift) due to the wheel shifting 10^{-7} m (about 4 millionths of an inch) along its axis due to minute error in the wheel axis bearings for the gyro in Problem 12.12.

12.16. *Microphone.* Refer to Fig. P-12.16. Sound waves (pressure waves) strike a membrane, exerting forces against it. Although the membrane is slightly deformable, we will consider it rigid compared to the other members of the system. A diaphragm that has several folds (which make it very elastic) deforms readily, allowing the membrane to move. The subsequent motion of the membrane is transmitted through a rubber (deformable) block to an electrical sensing (deformable) element. Draw the free body diagram. Determine the output in the Laplace domain for a general input. Find the steady state response to a step input force of 5 dyn.

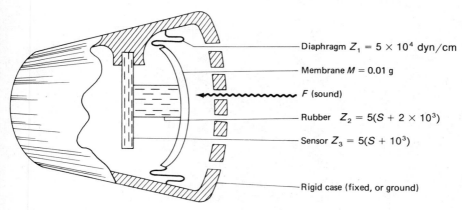

Diaphragm $Z_1 = 5 \times 10^4$ dyn/cm

Membrane $M = 0.01$ g

F (sound)

Rubber $Z_2 = 5(S + 2 \times 10^3)$

Sensor $Z_3 = 5(S + 10^3)$

Rigid case (fixed, or ground)

Figure P-12.16

12.17. *Compactor.* Refer to Fig P-12.17. The compactor is a pneumatic press that is used to crush bulky material. The mass of the piston, rod and press is lumped into M. The mass of the material and that of the platform is negligible. The bulky material is elastic and the crushing action absorbs energy. Hence, the material may be represented by a spring K_c and a damper D_c in parallel.

(a) Draw the network diagram.
(b) Determine the response to a general input.
(c) Determine the steady state response to a suddenly applied pressure P.

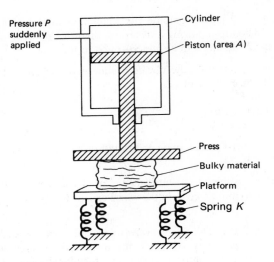

Figure P-12.17

12.18. *Recording Thermograph.* The device in Fig. P-12.18 makes a recording of temperature on a disk of paper. A bimetallic strip, wound in the shape of a spiral, curls as the temperature rises. This causes a shaft and arm to rotate, which produces a line on the paper. The paper is rotated slowly, thus providing a plot of temperature versus time.

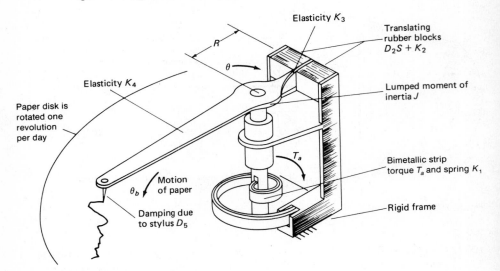

Figure P-12.18

(a) Draw the block diagram of the system
(b) Discuss which elements are in series, parallel, or combinations.
(c) Represent those networks that can be simplified (in their simplest form) by networks Z_i and write the differential equation. Assume IC = 0.
(d) Model the system for a constant temperature rise.
(e) Determine the steady state solution.

12.19. For each system in Fig. P-12.19 formulate the system array. Assume all motions are small.

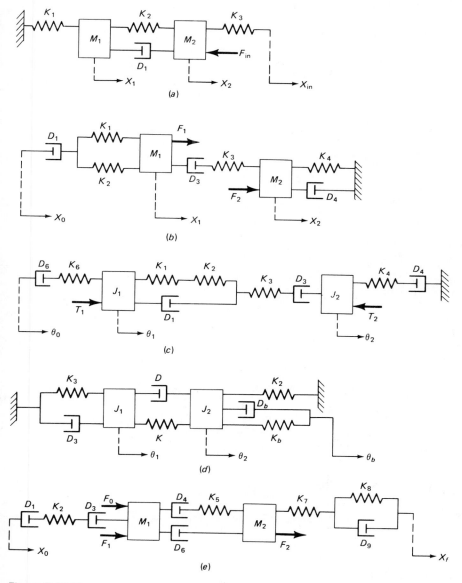

Figure P-12.19

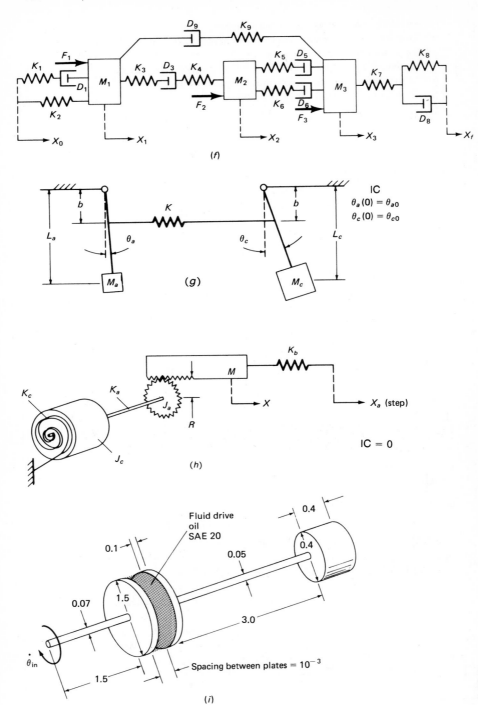

Figure P-12.19   (*Continued.*)

12.20. *Transformer Coupled Amplifier.* A transformer is used in an electrical circuit to match impedances of the two circuits that are coupled. In the amplifier shown in Fig. P-12.20, the value of R_3 is about 10^{-4} that of R_4. The mutual inductance in the transformer is L_{23}. Formulate the system array.

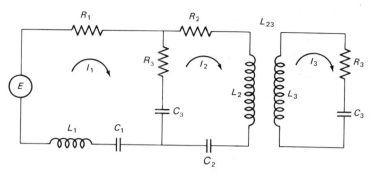

Figure P-12.20

12.21. In Problem 12.20 (Transformer Coupled Amplifier), insert a feedback transformer whose primary coil L_4 is in loop no. 3, and whose secondary coil is L_1 already shown in loop no. 1. Construct the system array.

12.22. A more exact model of the transistor amplifier of Ex. 12.7.5-1 is shown in Fig. P-12.22. Note that the quantity GM represents the output $V_2 - V_3$ of the transistor. Formulate the system array.

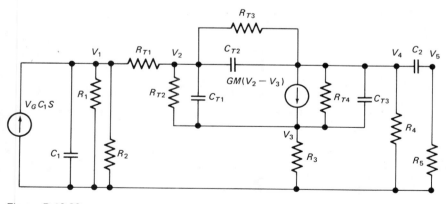

Figure P-12.22

12.23. *Intercom System.* Figure P-12.23 represents a simplified model of a four-way communication system. Formulate the system array. Briefly explain how each input affects all four circuits.

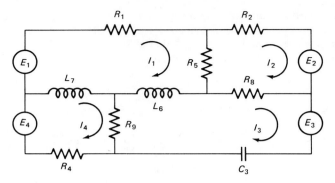

Figure P-12.23

12.24. *Transistor With Signal and Noise.* A transistor with signal and noise is modeled approximately in Fig. P-12.24. The signal is ac, hence the dc supply $B+$ may be neglected.

(a) Show that the output V_4 is of the form

$$V_4 = \frac{(M_1 s^2 + D_1 s + K_1)V_a + (D_3 s + K_3)V_b}{M_1 s^2 + (D_1 + D_2 + D_3)s + K_1 + K_2 + K_3}$$

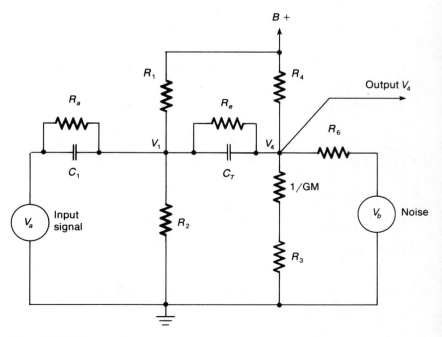

Figure P-12.24

where

$$M_1 = C_1 C_T, \qquad D_1 = \frac{C_1}{R_e} + \frac{C_T}{R_a}, \qquad K_1 = \frac{1}{R_e R_a}$$

(b) Find D_2, D_3, K_2, and K_3.

12.25. Formulate the system array for the circuit in Fig. P-12.25 using:
(a) loop analysis
(b) node analysis

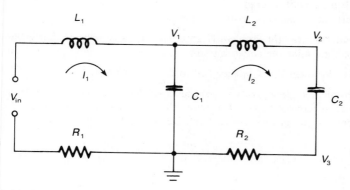

Figure P-12.25

12.26. Similar to Problem 12.25, but applied to Fig. P-12.26.

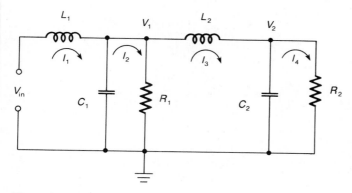

Figure P-12.26

12.27. *Two-Stage Transistor Amplifier With Signal and Noise.* Refer to Problem 12.11 (Circuit With No Inductance), Problem 12.24 (Transistor With Signal and Noise), and Ex. 12.3.1-2 (Superheterodyne Circuit), and use these as background.

(a) Show that a two-stage transistor amplifier with signal V_a and noise V_b has a system array of the form

V_1	V_2	Inputs
$Ms^2 + Z_3 + Z_4 + Z_5$	$-Z_3$	$Z_4 V_a(s)$
$-Z_3$	$aMs^2 + Z_3 + Z_6 + Z_7$	$Z_6 V_b(s)$

where

$$Z_i = D_i s + K_i \quad \text{(network)}$$

(b) Draw the circuit diagram.

(c) Determine all network terms and the constant a.

12.28. Show that the array for Problem 12.25(a) has the same form as that in Problem 12.27(a). Determine the network terms.

12.29. Similar to Problem 12.28, but applied to Problem 12.26(b).

12.30. Treat Problem 12.26(a) as two subsystems so that each one has the form of a quadratic, appropriately coupled. Show that the array for this system has the same form as that in Problem 12.27(a). Determine the networks terms.

Chapter **13**

Modeling with Block Diagrams

A block diagram can be used to express an entire differential equation, either in its original form or inverted so as to solve for the output. The block diagram technique makes it possible to apply the output of one differential equation as the input to another.

Refer to the brief introduction to block diagrams in Section 3.7. Also refer to Chapters 4, 5, 6, and 7 for the use of block diagrams for TUTSIM simulation.

13.1 SIGNAL FLOW DIAGRAMS

13.1.1 Modeling in Flow Diagram Form

Each subsystem of a complex system can be modeled in terms of building blocks. The overall system can be modeled by a network of such terms referred to as a *signal flow diagram*. There are several types of notation in common practice, two of which will be used in this text. The first of these is called the *flow graph*, which makes use of lines that connect circles or *nodes* in which the variable is written. The second type, called the *block diagram*, places the variables above lines which interconnect rectangles in which the appropriate functions are written.

13.1.2 Linear Operator

The linear operator operates upon its input to produce its output. Consequently, the linear operator is also the transfer function. The transfer function

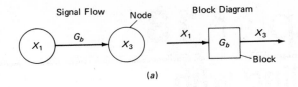

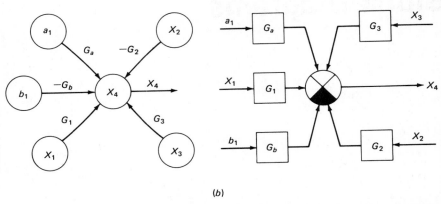

(b)

Figure 13.1.2-1   Signal flow diagram operators. (a) Linear operator or transfer function. (b) Summing operation.

or linear operator G_b is shown in both forms in Fig. 13.1.2-1(a), where

$$X_3 = G_b X_1 \qquad\qquad [13.1.2\text{-}1]$$

13.1.3 Sum

The sum of several quantities is attained only at a node. The quantity representing the sum is written either within the node or else it is written over the line drawn from the node. The quantities entering the sum are represented by lines with arrows in both notations. See Fig. 13.1.2-1(b).

13.1.4 Combinations

The arrow leaving the sum does not enter into this computation but may be used somewhere else. The equation of the operations shown in the figure is given as

$$X_4 = G_a a_1 - G_b b_1 + G_1 X_1 - G_2 X_2 + G_3 X_3 \qquad [13.1.4\text{-}1]$$

Note that the negative sign is accounted for in the transfer function on the flow graph. In the block diagram, the negative sign may be either written near the arrowhead entering the node, or one sector of the node may be blackened in.

13.1.5 Modeling of Some Mechanical Subsystems

First Order Angular Velocity Device   Refer to Fig. 3.6.1-1(d) whose differential equation is repeated below

$$J\dot{\omega} + D\omega = T \qquad\qquad [13.1.5\text{-}1]$$

Transform to the Laplace domain (noting IC = 0)

$$(Js + D)\omega(s) = T(s) \qquad\qquad [13.1.5\text{-}2]$$

The transfer function is

$$\frac{\omega}{T}(s) = \frac{1}{Js + D} \qquad\qquad [13.1.5\text{-}3]$$

The resulting block is shown in Fig. 13.1.5-1(a).

Second Order Translatory Device   Refer to Fig. 3.6.2-1(a) whose differential equation is repeated below

$$M\ddot{y} + D\dot{y} + Ky = f(t) \qquad\qquad [13.1.5\text{-}4]$$

Assume IC = 0, transform to the Laplace domain, and determine the transfer function.

$$\frac{Y}{F}(s) = \frac{1}{Ms^2 + Ds + K} \qquad\qquad [13.1.5\text{-}5]$$

The resulting block is shown in Fig. 13.1.5-1(b).

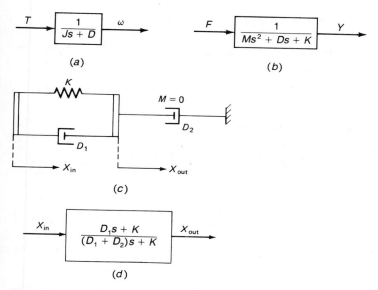

Figure 13.1.5-1   Block diagrams for mechanical subsystems. (a) First order. (b) Second order. (c) Mechanical lead-lag network. (d) Block diagram for lead-lag network.

Lead-Lag Mechanical Network   Refer to Fig. 13.1.5-1(c). Note that the output X_{out} appears at the interface of two parts of the network. Place an imaginary mass M_i at the interface, apply the one mass rule, and then take the limit as M_i approaches zero.

$$X_{out} = \frac{X_{in}(D_1 s + K)}{0 + D_1 s + K + D_2 s}$$ [13.1.5-6]

The transfer function is

$$\frac{X_{out}}{X_{in}}(s) = \frac{D_1 s + K}{(D_1 + D_2)s + K}$$ [13.1.5-7]

The resulting block is shown in Fig. 13.1.5-1(d).

13.1.6 Modeling of Some Electrical Subsystems

Second Order Series Circuit   Refer to Fig. 3.6.2-1(c) whose differential equation is repeated below

$$L\ddot{q} + R\dot{q} + \frac{q}{C} = V$$ [13.1.6-1]

Assume IC = 0, transform to the Laplace domain, and determine the transfer function

$$\frac{q}{V}(s) = \frac{1}{Ls^2 + Rs + 1/C}$$ [13.1.6-2]

The resulting block is shown in Fig. 13.1.6-1(a).

DC Generator (Tachometer)   A dc generator produces a voltage E_g that is proportional to shaft speed $\dot{\theta}$, where the constant of proportionality is K_g.

$$E_g = K_g \dot{\theta}$$ [13.1.6-3]

DC Motor with Fixed Field   Refer to Fig. 13.1.6-1(b). It is assumed that the armature has inductance L and resistance R. As the rotor turns at velocity $\dot{\theta}$, a voltage E_g is generated as given by eq. 13.1.6-3. This voltage, known as "back EMF," opposes the impressed voltage E. Apply Kirchhoff's law around the loop

$$E - E_g = L\frac{di}{dt} + Ri$$ [13.1.6-4]

Apply eq. 13.1.6-3, assume IC = 0, transfer to the Laplace domain, and solve for current I

$$I(s) = \frac{E(s) - K_g s\theta}{Ls + R}$$ [13.1.6-5]

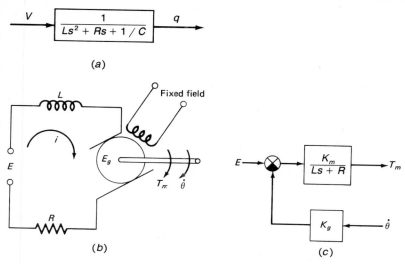

Figure 13.1.6-1   Block diagrams for electrical subsystems. (a) Series circuit. (b) DC motor with fixed field. (c) Block diagram for dc motor with fixed field.

For a linear motor, the output torque T_m is proportional to current I, where the constant of proportionality is K_m.

$$T_m = K_m I = \frac{K_m(E - K_g s\theta)}{Ls + R} \qquad [13.1.6\text{-}6]$$

The block diagram for eq. 13.1.6-6 is shown in Fig. 13.1.6-1(c).

Lead-Lag Electrical Circuit   Refer to Fig. 13.1.6-2(a). Apply Kirchhoff's law to each loop, assuming IC = 0, and transform to the Laplace domain

$$E_{\text{out}} = \left(R_2 + \frac{1}{Cs} \right) I \qquad \text{across } R_2 C$$

$$\qquad\qquad\qquad\qquad\qquad\qquad\qquad\qquad [13.1.6\text{-}7]$$

$$E_{\text{in}} = \left(R_1 + R_2 + \frac{1}{Cs} \right) I \quad \text{around loop}$$

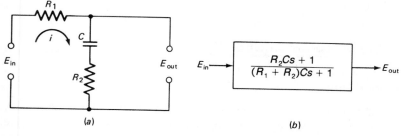

Figure 13.1.6-2   Electrical lead-lag circuit. (a) Electrical circuit. (b) Block diagram.

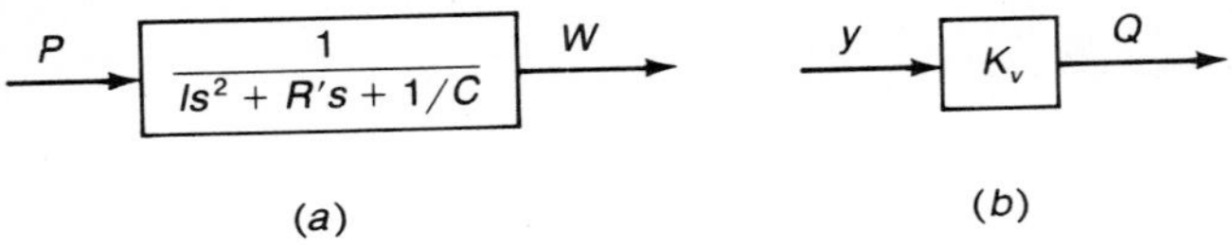

Figure 13.1.7-1 Block diagrams for fluidic subsystems. (a) Series fluidic circuit. (b) Linear valve.

The transfer function is

$$\frac{E_{out}}{E_{in}}(s) = \frac{(R_2 + 1/Cs)I}{(R_1 + R_2 + 1/Cs)I}$$

$$= \frac{R_2 Cs + 1}{(R_1 + R_2)Cs + 1} \qquad [13.1.6\text{-}8]$$

The block diagram is shown in Fig. 13.1.6-2(b).

13.1.7 Modeling of Some Fluidic Subsystems

Second Order Fluidic Circuit   Refer to Fig. 3.6.2-1(e), whose differential equation is repeated below

$$I\ddot{W} + R'\dot{W} + \frac{W}{C} = P \qquad [13.1.7\text{-}1]$$

Assume IC = 0, transform to the Laplace domain, and solve for the transfer function

$$\frac{W}{P}(s) = \frac{1}{Is^2 + R's + 1/C} \qquad [13.1.7\text{-}2]$$

The resulting block is shown in Fig. 13.1.7-1(a).

Linear Valve   Refer to Figs. 3.5.4-1 and 3.5.4-2. If the valve is designed for constant pressure drop, then the flow rate Q is proportional to the valve opening y, where K_v is the constant of proportionality. Then

$$Q = K_v y \qquad [13.1.7\text{-}3]$$

The resulting block is shown in Fig. 13.1.7-1(b).

13.2 BLOCK DIAGRAM ALGEBRA

13.2.1 General Rule

If a system of blocks can be confined to lie within a single closed boundary, and where there is only *one* input crossing *into* the boundary, and where there is only *one* output crossing *out* of the boundary, then the system can be

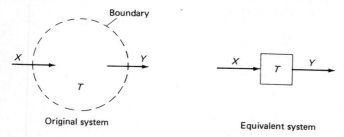

Figure 13.2.1-1   General rule for equivalent block.

expressed as a single equivalent block using the same single input and the same single output. See Fig. 13.2.1-1.

13.2.2 Open Loop Operations

Cascaded Blocks (Product)   Cascaded blocks can be lumped into a single block, by multiplying the cascades. See Fig. 13.2.2-1(a).

Summed Blocks   Summed blocks involving *one* input obviously satisfies the definition for the general rule in Section 13.2.1, and therefore can be lumped into a single equivalent block. See Fig. 13.2.2-1(b). Note that in each of the subsystems in Fig. 13.2.2-1 we cannot return to the starting point. Such subsystems are called open loops.

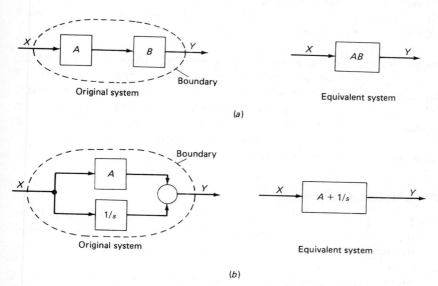

Figure 13.2.2-1   Open loop operations. (a) Cascaded blocks. (b) Summed blocks.

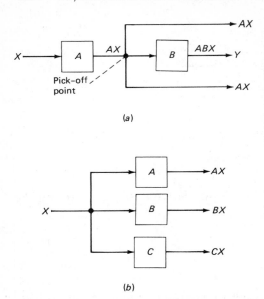

(a)

(b)

Figure 13.2.3-1   Common points. (a) Signal picked off at internal point. (b) Common input serves any number of blocks.

13.2.3 Common Points

Signals Picked off at Internal Point   Electrically speaking, one could use a (high impedance) voltmeter to measure the signal at any point in the circuit and not affect the circuit. Several such instruments could be connected to the same point and not affect the circuit. In such a system, the circuit appears like an "infinite source" that can feed any number of readout or pick-off devices without affecting its own behavior. See Fig. 13.2.3-1(a).

Common Input   Analogous to the infinite signal source described in the previous paragraph, there is an infinite supply or infinite input. For a physical example, consider a fluid system in steady state. The pressure is transmitted equally everywhere, no matter how many points are fed by this pressure. In this case, all points have a common input. It is assumed that the input can supply as many systems as needed without "loading" the input. The block diagram for such a system is shown in Fig. 13.2.3-1(b).

13.2.4 Closed Loop Rule

If a system of blocks forms a closed loop, being summed with a single input, and if there is *one* output, then a single block representation can be found. The equivalent block is also referred to as the closed loop rule. See Fig. 13.2.4-1.

   In order to derive the equivalent single block, we shall derive the *closed loop rule*. To do this, note that the output of block H is equal to HY. Let us

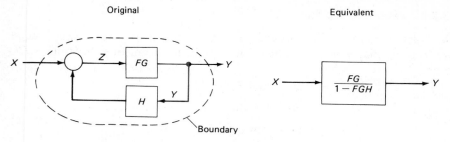

Figure 13.2.4-1   Closed loop rule.

call the output of the summing point Z.

$$Z = X + HY \qquad\qquad [13.2.4\text{-}1]$$

Note that the input and output of block FG are Z and Y, then,

$$Y = FGZ \qquad\qquad [13.2.4\text{-}2]$$

Eliminate Z between eqs. 13.2.4-1 and 13.2.4-2

$$FG(X + HY) = Y$$

or

$$FGX = Y - FGHY = (1 - FGH)Y \qquad\qquad [13.2.4\text{-}3]$$

In eq. 13.2.4-3, solve for the transfer function, which is given by the output/input ratio Y/X

$$\frac{\text{out}}{\text{in}} = \frac{Y}{X} = \frac{FG}{1 - FGH} \qquad\qquad [13.2.4\text{-}4]$$

The result shown in eq. 13.2.4-4 is called the *closed loop transfer function*. The quantity FG is called the *forward gain*, and H is called *feedback*. The product FGH is obtained by multiplying all the blocks proceeding around the loop. This product is called the *open loop gain L* and it includes the algebraic sign of the summing point. Thus, the open loop gain* is

$$L = +FGH \qquad\qquad [13.2.4\text{-}5]$$

and the closed loop transfer function becomes

$$\frac{\text{out}}{\text{in}} = \frac{FG}{1 - L} \qquad\qquad [13.2.4\text{-}6]$$

---

*Note that in Fig. 13.2.4-1, the path through the summing point is positive. In an earlier example, Fig. 13.1.2-1, each of two paths experienced a change in sign.

13.2.5 Fraction Loop Rule

The fraction loop rule is a special case of the closed loop rule. The *fraction loop rule* applies when the forward gain is in the form of a fraction N/B, where

$$N = \text{numerator}$$

$$B = \text{denominator}$$

$$FG = \text{forward gain} = N/B \qquad [13.2.5\text{-}1]$$

$$H = \text{feedback}^{\dagger}$$

Apply eqs. 13.2.5-1 to the closed loop rule, eq. 13.2.4-4

$$\frac{\text{out}}{\text{in}} = \frac{N/B}{1 - HN/B} = \frac{N}{B - HN} \qquad [13.2.5\text{-}2]$$

■ **EXAMPLE 13.2.5-1** Use of Closed Loop Rules

For the system in Fig. 13.2.5-1, find the open loop gain L and the closed loop transfer function $\theta_{\text{out}}/\theta_{\text{in}}$.

Solution.

Open Loop Gain L

Note that in making the circuit around the loop, the path through the summing point is negative.

$$L = -(10s + 9)\left(\frac{4}{s^2}\right)(2s) = \frac{-8(10s + 9)}{s} \qquad (1)$$

Closed Loop Transfer Function $\theta_{\text{out}} / \theta_{\text{in}}$

The forward gain FG from input θ_{in} to the output θ_{out} is

$$FG = (10s + 9)\left(\frac{4}{s^2}\right) \qquad (2)$$

Using the closed loop rule,

$$\frac{\theta_{\text{out}}}{\theta_{\text{in}}} = \frac{FG}{1 - L} = \frac{(10s + 9)(4/s^2)}{1 + 8(10s + 9)/s} = \frac{4(10s + 9)}{s^2 + 8s(10s + 9)} \qquad (3)$$

As an alternate method, use the fraction loop rule. Refer to eq. (2). The numerator N and denominator B are

$$N = 4(10s + 9)$$

$$B = s^2 \qquad (4)$$

---

†Including the algebraic sign of the summing point.

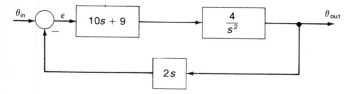

Figure 13.2.5-1   Closed loop system.

The feedback term is formed by completing the circuit after the forward gain. Note that in so doing, in this case, there is a change of sign in passing through the summing point.

$$H = -2s$$

then

$$HN = -2s(4)(10s + 9) = -8s(10s + 9) \tag{5}$$

Use the fraction loop rule

$$\frac{\theta_{out}}{\theta_{in}} = \frac{N}{B - HN} = \frac{4(10s + 9)}{s^2 + 8s(10s + 9)} \tag{6}$$

which agrees with eq. (3).   ■

13.2.6 Shifting Terms To Satisfy Boundary Requirements

When there is more than one input or more than one output crossing the boundary, it becomes necessary to rearrange certain terms. The quantity so shifted must be appropriately altered in order to retain the equivalent expression.

13.2.7 System with More than One Input and More than One Output

Consider a system with two inputs θ_{in} and T_d as shown in Fig. 13.2.7-1. If the system is linear, then the two inputs may be treated one at a time, and the results may be superposed.

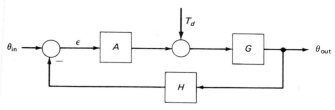

Figure 13.2.7-1   Closed loop system with more than one input.

Suppose we wish to find the output θ and also the error ϵ. The output can be found by the usual closed loop means. In order to find the error, ϵ will have to be treated as another output. Each of the two outputs will be found in terms of each of the two inputs.

Input = θ_{in} While $T_d = 0$ Feed forward (from input θ_{in} to output θ)

$$FG_\theta = AG \qquad [13.2.7\text{-}1]$$

Open loop gain

$$L = -AGH \qquad [13.2.7\text{-}2]$$

Closed loop transfer function

$$\frac{\theta_\theta}{\theta_{in}} = \frac{FG_\theta}{1 - L} = \frac{AG}{1 + AGH} \qquad [13.2.7\text{-}3]$$

or

$$\theta_\theta = \frac{AG\theta_{in}}{1 + AGH} \qquad [13.2.7\text{-}4]$$

Input = T_d While $\theta_{in} = 0$ Forward gain (from input T_d to output θ)

$$FG_t = G \qquad [13.2.7\text{-}5]$$

Closed loop transfer function

$$\frac{\theta_t}{T_d} = \frac{FG_t}{1 - L} = \frac{G}{1 + AGH} \qquad [13.2.7\text{-}6]$$

or

$$\theta_t = \frac{GT_d}{1 + AGH} \qquad [13.2.7\text{-}7]$$

Superposing the two results

$$\theta = \theta_\theta + \theta_t = \frac{G(A\theta_{in} + T_d)}{1 + AGH} \qquad [13.2.7\text{-}8]$$

Now consider ϵ as the output.

Input = θ_{in} While $T_d = 0$ The forward gain is found by tracing a path from the input θ_{in} to the output ϵ, which in this case is a very short path of magnitude unity with no change of sign. Then

$$FG = +1 \qquad [13.2.7\text{-}9]$$

The open loop gain L is given by eq. 13.2.7-2. Apply these to the closed loop rule.

$$\frac{\epsilon_\theta}{\theta_{in}} = \frac{FG}{1 - L} = \frac{1}{1 + AGH} \qquad [13.2.7\text{-}10]$$

Input $= T_d$ While $\theta_{in} = 0$ The forward gain is found by tracing a path from the input T_d to the output ϵ. Note that in passing through the summing point, the sign changes. Then

$$FG = -GH \qquad [13.2.7\text{-}11]$$

Applying the above to the closed loop rule,

$$\frac{\epsilon_t}{T_d} = \frac{FG}{1 - L} = \frac{-GH}{1 + AGH} \qquad [13.2.7\text{-}12]$$

In order to superpose the two results, solve eqs. 13.2.7-10 and 13.2.7-12, respectively, for ϵ_θ and ϵ_t. Sum the results.

$$\epsilon = \epsilon_\theta + \epsilon_t = \frac{\theta_{in} - GHT_d}{1 + AGH} \qquad [13.2.7\text{-}13]$$

■ **EXAMPLE 13.2.7-1** Loop Within a Loop with Several Inputs and Outputs For the system* shown in Fig. 13.2.7-2, determine the following:

$$\frac{X_{out}}{X_{in}}, \frac{\epsilon}{X_{in}}, \frac{\epsilon}{Q_d}, \frac{\epsilon}{F_d}$$

Solution.   Note that the system block diagram as given consists of a loop within a loop. The first step is to reduce the inner loop to a single block. However, when a boundary is drawn around the inner loop, three inputs Q_{in}, Q_d, and F_d cross this boundary. See Fig. 13.2.7-3(a). It will be necessary to shift terms. First shift F_d out of the inner loop to the summing point. This will require division by K_m to maintain equivalence. Now there are three inputs, all at the one summing point. Define a new term Ψ that is equal to this sum.

$$\Psi = Q_{in} + Q_d + F_d/K_m \qquad (1)$$

See Fig. 13.2.7-3(b). Now, only one input Ψ crosses the boundary, and the subsystem qualifies for reduction to a single block. Its closed loop transfer

---

*This is a hydraulic servo system that will be derived in Chapter 16, Ex. 16.2.3-1.

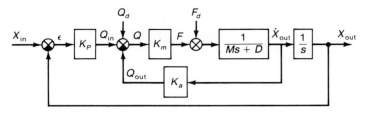

Figure 13.2.7-2   Loop within a loop.

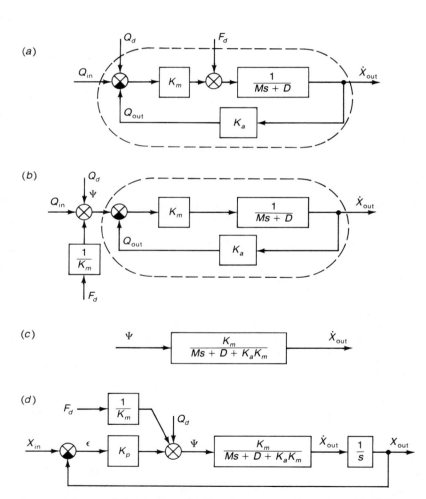

Figure 13.2.7-3   Shifting terms. (a) Two inputs cross boundary. (b) Shift one term. (c) Inner loop reduced to one block. (d) Complete system.

function will be determined. Trace a path from the input Ψ to the output $\dot{X}_{out}$ to find the forward gain FG_i. Note that there is no change of sign.

$$FG_i = \frac{K_m}{Ms + D} \tag{2}$$

Since this is a fraction, the fraction loop rule will be used. According to the rule multiply the numerator K_m by the feedback K_a and add this to the denominator. Thus, the inner loop becomes a single block whose transfer function is

$$\frac{\dot{X}_{out}}{\Psi} = \frac{K_m}{Ms + D + K_a K_m} \tag{3}$$

The inner loop has been reduced to a single block, as given by eq. (3) and shown in Fig. 13.2.7-3(c). Thus, the final system can be simplified as shown in Fig. 13.2.7-3(d).

In order to determine the overall closed loop transfer function, the forward gain FG_0 is found by tracing a path from X_{in} to X_{out}. Note that there is no change of sign.

$$FG_0 = \frac{K_p K_m}{(Ms + D + K_a K_m)s} \tag{4}$$

Since this is a fraction, the fraction loop rule will be used. According to the rule multiply the numerator $K_p K_m$ by the feedback (which is unity) and add to the denominator. The closed loop transfer function becomes

$$\frac{X_{out}}{X_{in}}(s) = \frac{K_p K_m}{Ms^2 + (D + K_a K_m)s + K_p K_m} \tag{5}$$

To determine ϵ/X_{in}, trace a path from X_{in} to ϵ. This is a direct path with no change of sign.

$$FG_x = +1 \tag{6}$$

Since this is not a fraction, the fraction loop rule will not be convenient. Instead, use the closed loop rule. For this the open loop gain L is required.

$$L = -\frac{K_p K_m}{s(Ms + D + K_a K_m)} \tag{7}$$

Apply eqs. (6) and (7) to the closed loop rule.

$$\frac{\epsilon}{X_{in}}(s) = \frac{FG_x}{1 - L} = \frac{1}{1 + \dfrac{K_p K_m}{s(Ms + D + K_a K_m)}} \tag{8}$$

Rationalize.

$$\frac{\epsilon}{X_{in}}(s) = \frac{s(Ms + D + K_a K_m)}{s(Ms + D + K_a K_m) + K_p K_m} \tag{9}$$

For ϵ/Q_d, the forward path is a fraction that changes sign

$$FG_q = \frac{-K_m}{s(Ms + D + K_a K_m)} \tag{10}$$

Since this is a fraction, it will pay to use the fraction loop rule. For this, the feedback is required. To find this, continue along the path until you return to the starting point Q_d.

$$H_q = +K_p \tag{11}$$

According to the rule, the feedback is multiplied by the numerator (which in this case is negative) and is subtracted from the denominator. Then

$$\frac{\epsilon}{Q_d}(s) = \frac{-K_m}{s(Ms + D + K_a K_m) + K_p K_m} \tag{12}$$

It should be pointed out that the negative sign of the entire term is not important. It merely states that the quantity ϵ has a sign that is opposite to Q_d. However, in practice, the sign of Q_d is seldom known.

The last quantity to be determined, ϵ/F_d, can be determined readily, since input F_d in Fig. 13.2.7-3(d) appeared at the same point as the input Q_d. Then replace Q_d in eq. (12) by F_d/K_m and divide both sides by K_m.

$$\frac{\epsilon}{F_d}(s) = \frac{-1}{s(Ms + D + K_a K_m) + K_p K_m} \tag{13}$$

∎

13.3 BLOCK DIAGRAMS IN ARRAY FORM

By inspection, the rule for formulating the system array for the block diagram becomes

1. Each diagonal entry is equal to unity.
2. Each off-diagonal entry is equal to minus the connection from node X_i to node X_j. Since these are unilateral connections, they appear in only *one* location, column i, row j.
3. The input column contains the inputs multiplied by their connection to the summing point.

■ **EXAMPLE 13.3.1-1** Loop Within a Loop
For the system in Fig. 13.2.7-2, find the error ϵ due to each input.

Solution. In this system, there are a number of unknowns. However, the problem can be solved by considering only three,

$$\epsilon, Q, \text{ and } \dot{X}_{\text{out}} \tag{1}$$

The remaining terms may be found as follows:

$$Q_{\text{in}} = K_p \epsilon$$

$$Q_{\text{out}} = \dot{X}_{\text{out}} K_a$$

$$F = Q K_m \tag{2}$$

$$X_{\text{out}} = \dot{X}_{\text{out}}/s$$

Using the variables in eq. (1), the array may be formed.

ϵ	Q	$\dot{X}_{\text{out}}$	Inputs
1	0	$1/s$	X_{in}
$-K_p$	1	K_a	Q_d
0	$\dfrac{-K_m}{Ms+D}$	1	$\dfrac{F_d}{Ms+D}$

$$\tag{3}$$

Use rule (a) of Section 12.2.4 to multiply row 1 by s and row 3 by $Ms + D$.

ϵ	Q	$\dot{X}_{\text{out}}$	Inputs
s	0	1	sX_{in}
$-K_p$	1	K_a	Q_d
0	$-K_m$	$Ms+D$	F_d

$$\tag{4}$$

Solve for ϵ using Cramer's rule.

$$\epsilon(s) = \frac{\begin{vmatrix} sX_{\text{in}} & 0 & 1 \\ Q_d & 1 & K_a \\ F_d & -K_m & Ms+D \end{vmatrix}}{\begin{vmatrix} s & 0 & 1 \\ -K_p & 1 & K_a \\ 0 & -K_m & Ms+D \end{vmatrix}} \tag{5}$$

Expand the determinants

$$\epsilon(s) = \frac{(Ms+D+K_aK_m)sX_{\text{in}} - K_mQ_d - F_d}{s(Ms+D+K_aK_m) + K_pK_m} \tag{6}$$

which agrees with eqs. (9), (12), and (13) of Ex. 13.2.7-1. ∎

SUGGESTED REFERENCE READING FOR CHAPTER 13

Block diagrams [7], p. 3–5, 71–79, 270–277; [17], p. 8, 9, 275, 276, 495–499; [26], p. 41–51; [40], p. 194–202; [41]; [42], p. 124–140; [46], p. 303–314; [47], p. 480–489.

State variables [7], p. 260–270; [15], p. 87–98, 293–323; [16], p. 326; [17], p. 360–369; [26], p. 260–265; [27], p. 157–214; [40], p. 42, 212, 216; [42], p. 351; [46], p. 327–335.

Computer methods [12]; [28]; [40], p. 525–542; [47], p. 272–287.

Taylor series [22], p. 102; [37], p. 348–350; [40], p. 121.

Perturbation [7], p. 80; [16], p. 327–362; [17], p. 94, 104–106, 112, 120, 358–360; [22], p. 387–391; [27], p. 144–173; [47], p. 195–219.

Linearization [7], p. 43–54; [15], p. 134–135, 313–318; [16], p. 334; [17], p. 356–360; [26], p. 23; [27], p. 144–155; [28], p. 139–144; [40], p. 121, 134, 462; [42], p. 320, 329; [46], p. 57, 62, 63; [47], p. 195–200.

Hamilton's principle [15], p. 30, 31, 128–135, 336–351, 433–440; [18], p. 230, 231.

Mechanical networks [40], p. 76–101.

Electrical networks [46], p. 21–24.

Arrays [22], p. 302.

Determinants [16], p. 155–158; [21]; [22], p. 302–311; [23]; [25]; [26], p. 48, 365.

Linear algebra [7], p. 277–296; [16], p. 469–483; [21]; [22]; [23].

Matrices [7], p. 298–307; [15], p. 55–62, 425–429; [16], p. 129–132, 159–165; [21]; [22], p. 302–331; [23]; [26], p. 361–374; [28], p. 23–30, 141–167, 225–228; [42], p. 352–358.

PROBLEMS FOR CHAPTER 13

13.1. For each system in Fig. 3.6.2-1, determine the transfer function and draw the block diagram.

13.2. For each of the following, draw the block diagram:
   (a) Ex. 7.6.1-1, eq. (6).   (b) Eq. 7.6.3-3.
   (c) Eq. 7.7.3-2.   (d) Eq. 7.8.1-3.
   (e) Ex. 4.3.1-3, eqs. (3) and (4). (f) Eq. 12.2.1-1.

13.3. Draw the block diagram for each of the following:
   (a) Fig. 12.3.1-1.   (b) Fig. 12.3.1-2.
   (c) Fig. 12.3.1-4.   (d) Eq. 12.5.5-4.
   (e) Fig. 12.6.4-1.   (f) Fig. 12.6.5-1.
   (g) Fig. 12.6.5-2.   (h) Fig. 12.7.4-1.
   (i) Fig. 12.7.5-1.   (j) Fig. P-12.20.
   (k) Eq. 2.5.3-2.   (l) Eq. 3.4.1-1.
   (m) Ex. 3.4.2-1, eq. (2).   (n) Ex. 3.4.2-2, eq. (1).
   (o) Eq. 3.5.2-6.   (p) Eq. 3.3.1-4.
   (q) Eq. 3.3.1-5.

13.4. Determine the closed loop transfer function for each system shown in Fig. P-13.4.

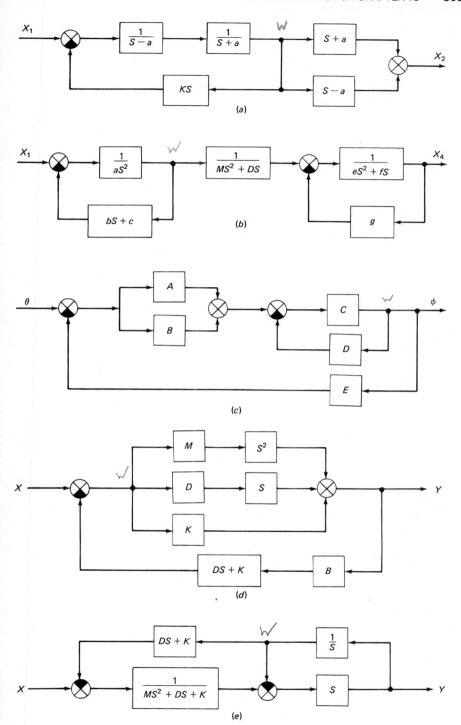

Figure P-13.4

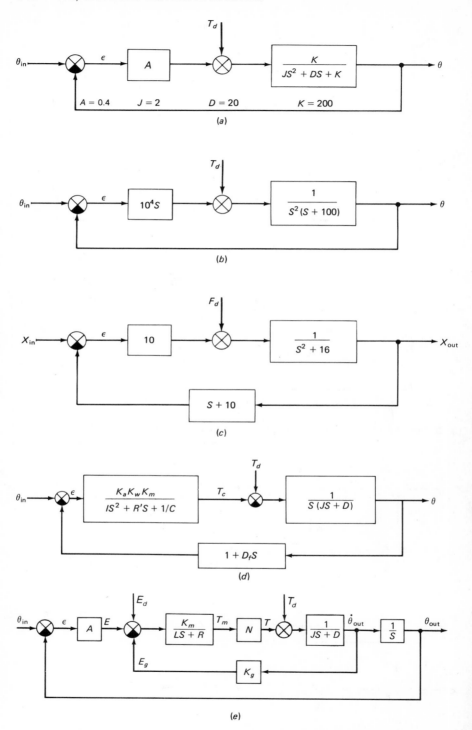

Figure P-13.5 The value $A = 0.4$ in (a) is the nominal value; for design purposes, it may be varied over a large range.

13.5. For each block diagram in Fig. P-13.5, determine the closed loop transfer function and the error due to each input.

13.6. For the block diagram shown in Fig. P-13.6, determine the following:
(a) $X(s)$ due to each input, θ, ϕ, and ψ.
(b) $\epsilon(s)$ due to each input, θ, ϕ, and ψ.

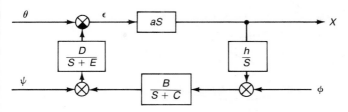

Figure P-13.6

13.7. For each part of Problem 13.5, determine the steady state error, where each input is a step.

13.8. For part (a) of Problem 13.6, determine the steady state output X_{ss} for each of the inputs, where each is a step.

13.9. For part (b) of Problem 13.6, determine the steady state error ϵ_{ss} for each of the inputs, where each is a step.

13.10. For the block diagram in Fig. P-13.10, find the steady state error ϵ_{ss} due to a step disturbance T_d, where the function $G(s)$ is each of the following:

(a) $G(s) = a$

(b) $G(s) = as$

(c) $G(s) = \dfrac{a}{s}$

(d) $G(s) = as^2$

(e) $G(s) = \dfrac{a}{Js + D}$

(f) $G(s) = as^3$

(g) $G(s) = \dfrac{a}{s(Js + D)}$

(h) $G(s) = as^n$

(i) $G(s) = s(s + a)$

(j) $G(s) = 1$

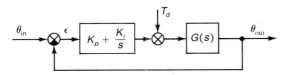

Figure P-13.10

13.11. For the system shown in Fig. P-13.11, solve for all the unknowns in terms of all of the inputs. *Suggestion*: Formulate the system array and apply Cramer's rule.

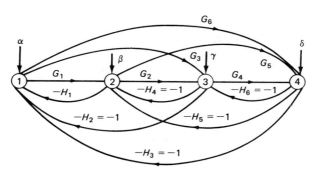

Figure P-13.11

13.12. If Problem 13.11 were solved by using the system array, perform the following, using rule (b) of Section 12.2.4: Subtract row 2 from row 1. Next, subtract row 3 from row 2. Finally, subtract row 3 from row 4. Will the result be easy to expand?

13.13. For the signal flow diagram in Fig. P-13.13, formulate the system array.

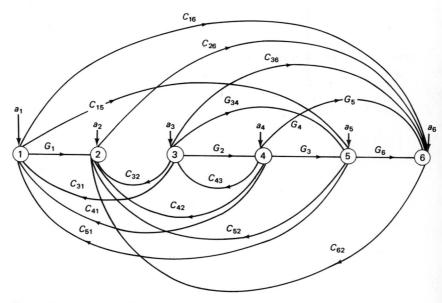

Figure P-13.13

Chapter **14**

Design Oriented Approach to High Order Systems

In Chapter 12, it was stated that a high order system was one with more than one variable, generally requiring the use of simultaneous equations. This chapter is not concerned with how such equations are formulated, formulation was covered in Chapters 12 and 13. This chapter concentrates upon the next step—the solution of those equations and the subsequent system design.

14.1 DOMINANT FACTORS IN HIGH ORDER SYSTEMS

14.1.1 Dominance

The high order system can be considered a coupled set of subsystems. This becomes quite apparent when the system equations are written in the form of a system array. We may consider the high order system as a set of diagonal entries or subsystems that are coupled by the off-diagonal terms. As such, it is clear that each subsystem exerts its influence, tugging and shoving the solution this way and that.

If there is a dominant factor, then it is a reasonable approximation to estimate the system properties by observing only those of the dominant factor.

14.1.2 Time Domain Solution

The relative effect of each term in the time domain solution is a fair means by which to judge dominance. In Chapter 11, this problem is dealt with directly, assisted by a comprehensive table of Laplace transform pairs.

14.1.3 Scatter Function

If there are too many terms to estimate clearly which one dominates or not, Chapter 11 offers one means to reduce the number of terms. This is done by selecting one or more terms to serve as the axis upon which the other terms are plotted. This axis is called the scatter function.

14.2 SYSTEMS WITH SMALL AMOUNTS OF DAMPING AND THE BAND ELIMINATOR

14.2.1 Approximation to System with No Damping

While no real system has zero damping, this model will apply to systems with a small amount of damping. Keep in mind that a system with no damping never reaches steady state. Let there be very little damping, little enough to neglect it in order to factor the quartic, but still enough to approach the steady state value.

14.2.2 System Array

The array for a fourth order system is given by

X_1	X_2	Input
$M_1 s^2 + K_1 + K_2$	$-K_2$	F_1
$-K_2$	$M_2 s^2 + K_2$	0

[14.2.2-1]

The characteristic determinant is given by

$$\Delta = M_1 M_2 \left\{ s^4 + \left[\omega_1^2 + \omega_2^2 (\mu + 1) \right] s^2 + \omega_1^2 \omega_2^2 \right\} \qquad [14.2.2\text{-}2]$$

where

$$\mu = \frac{M_2}{M_1}, \qquad \omega_1^2 = \frac{K_1}{M_1}, \qquad \omega_2^2 = \frac{K_2}{M_2} \qquad [14.2.2\text{-}3]$$

Equation 14.2.2-2 is a quadratic in s^2, which readily can be factored

$$\Delta = M_1 M_2 \left[s^2 + (r_1 \omega_2)^2 \right] \left[s^2 + (r_2 \omega_2)^2 \right] \qquad [14.2.2\text{-}4]$$

where

$$r_1^2 = \frac{b}{2} - \sqrt{\left(\frac{b}{2} \right)^2 - c} \qquad [14.2.2\text{-}5]$$

$$r_2^2 = \frac{b}{2} + \sqrt{\left(\frac{b}{2} \right)^2 - c} \qquad [14.2.2\text{-}6]$$

$$c = \left(\frac{\omega_1}{\omega_2} \right)^2 \qquad [14.2.2\text{-}7]$$

$$b = c + \mu + 1 \qquad [14.2.2\text{-}8]$$

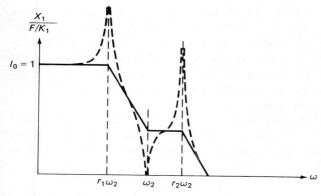

Figure 14.2.3-1   Frequency response in trap.

14.2.3 Frequency Response

Use Cramer's rule to solve for X_1 and X_2.

$$\frac{X_1}{F/K_1}(s) = \frac{(s^2 + \omega_2^2)\omega_1^2}{[s^2 + (r_1\omega_2)^2][s^2 + (r_2\omega_2)^2]} \qquad [14.2.3\text{-}1]$$

The frequency response is shown in Fig. 14.2.3-1. Note that when the input frequency is equal to ω_2 then

$$X_1 = 0 \qquad [14.2.3\text{-}2]$$

$$\frac{X_2}{F/K_1}(s) = \frac{\omega_1^2\omega_2^2}{[s^2 + (r_1\omega_2)^2][s^2 + (r_2\omega_2)^2]} \qquad [14.2.3\text{-}3]$$

The frequency response is given in Fig. 14.2.3-2. Note that when the input frequency is equal to ω_2, then

$$X_2 = \frac{F}{K_2} \qquad [14.2.3\text{-}4]$$

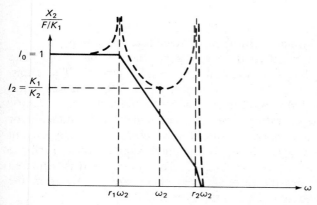

Figure 14.2.3-2   Frequency response in secondary subsystem.

Several observations can be made of the frequency plots and the above equations. Both curves approach infinity at frequencies $r_1\omega_2$ and $r_2\omega_2$. This means the system must be installed where the input frequency cannot reach either of these values. At frequency ω_2 the output X_1 approaches zero, while X_2 becomes F/K_2. The fact that the output X_1 approaches zero is an important property. This means at that frequency the input is totally absorbed —frequency ω_2 has been trapped.

The price paid for the frequency trap is the two resonant peaks that approach infinity. But the solution is fairly simple. Arrange for the ratios r_1 and r_2 to be large enough to set these troublesome peaks sufficiently remote from the trap frequency.

14.2.4 Design Approach

The design approach makes use of the following:

$$\omega_2 = \text{trapped frequency for } X_1 \qquad\qquad [14.2.4\text{-}1]$$

$$r_1 \text{ and } r_2 \text{ are the resonant frequency ratios} \qquad [14.2.4\text{-}2]$$

$$c \text{ and } \mu \text{ are tools of the design process} \qquad [14.2.4\text{-}3]$$

$$\text{Fig. 14.2.4-1 presents design curves} \qquad [14.2.4\text{-}4]$$

The design curves are plotted in Fig. 14.2.4-1 in terms of the two constants, c and μ, and present the two resonant frequency ratios r_1 and r_2.

Use c and one of the resonant frequency ratios (either r_1 or r_2) on the design curves, Fig. 14.2.4-1, to find μ and check the other resonant frequency ratio. Then, using the values thus found, complete the design:

$$M_2 = \mu M_1 \qquad K_2 = M_2\omega_2^2 \qquad [14.2.4\text{-}5]$$

$$X_2 = \frac{cF_1}{\mu K_1} \qquad X_1 = 0 \qquad [14.2.4\text{-}6]$$

■ **EXAMPLE 14.2.4-1** Dynamic Absorber for Reciprocating Pump
A reciprocating pump for an air conditioner is supported on four springs like that shown in Fig. 7.9.1-1, but where there are no dampers. The pump operating speed is 600 rpm (62.8 rad/s). However, in Ex. 7.9.1-1, it was found in eq. (4) that the undamped natural frequency of the support is 52.92 rad/s, which is too close to the operating speed. A mechanical band eliminator or frequency trap is required. A mass M_2 will be supported by a spring K_2 and will be placed under the base (mass M_1) of the pump support. Such a scheme is called a dynamic absorber. See Fig. 14.2.4-2. The motor speed is regulated to 10%. For safety, be sure the resonant frequencies are remote from the operating speed by at least 20%. Design the system.

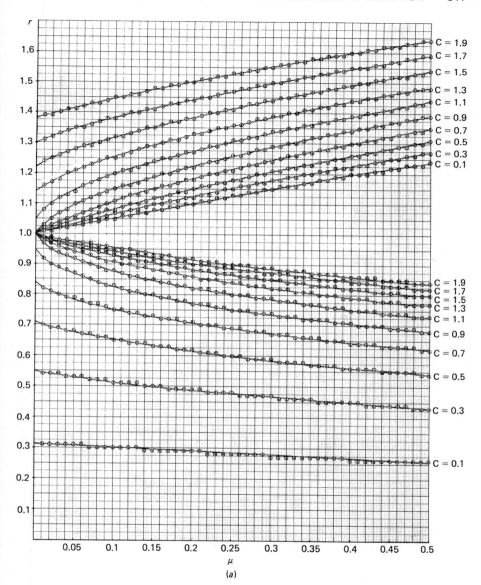

Figure 14.2.4-1   Design curves for band eliminator. (a) Values of μ from 0.0 to 0.5. (b) Values of μ from 0.5 to 2.0.

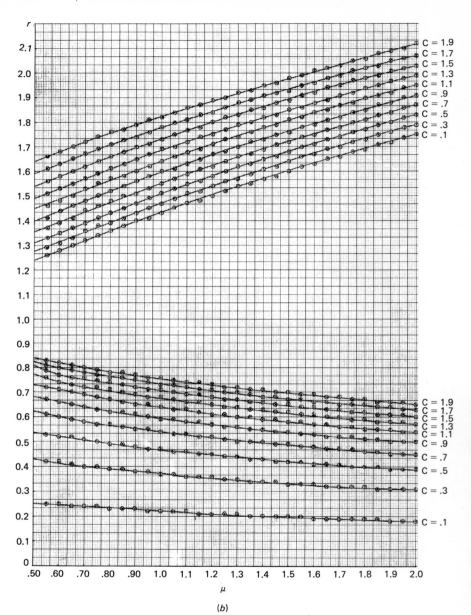

Figure 14.2.4-1   (*Continued*).

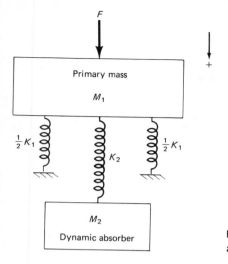

Figure 14.2.4-2 Undamped dynamic absorber.

Solution.   The system array for Fig. 14.2.4-2 is given by

	X_1	X_2	Inputs	
	$M_1 s^2 + K_1 + K_2$	$-K_2$	F_1	(1)
	$-K_2$	$M_2 s^2 + K_2$	0	

Note that eq. (1) is identical to eq. 14.2.2-1. Hence, the design symbols will be the same as those in Section 14.2.4.

$$\omega_1 = \sqrt{\frac{K_1}{M}} = \sqrt{\frac{2.8 \times 10^4}{10}} = 52.9 \text{ rad/s} \tag{2}$$

$$c = \left(\frac{\omega_1}{\omega_2}\right)^2 = \left(\frac{52.9}{62.8}\right)^2 = 0.71 \tag{3}$$

Refer to Fig. 14.2.4-1 and first try,

$$c = 0.71 \qquad r_1 = 100\% - 20\% = 0.8 \tag{4}$$

then

$$\mu = 0.05 \qquad r_2 = 1.05 = 100\% + 5\% \tag{5}$$

The upper resonant frequency $r_2\omega_2$ will be only 5% remote from the operating frequency and, hence, is unsafe. Try the other,

$$c = 0.71 \qquad r_2 = 100\% + 20\% = 1.2 \tag{6}$$

then

$$\mu = 0.22 \qquad r_1 = 0.69 = 100\% - 31\% \tag{7}$$

This is safe. Complete the design.

$$M_2 = \mu M_1 = 0.22 \times 10 = 2.2 \text{ kg} \tag{8}$$

$$K_2 = M_2\omega_2^2 = 2.2 \times 62.8^2 = 8.7 \times 10^3 \text{ N/m} \tag{9}$$

Apply the input speed to eqs. (2) and (3) of Ex. 7.9.1-1.

$$F_1 = F_a = M_p e \omega_2^2 = 1 \times 0.02 \times 62.8^2 = 79 \text{ N} \tag{10}$$

$$X_2 = \frac{cF_1}{\mu K_1} = \frac{0.71 \times 79}{0.22 \times 2.8 \times 10^4} = 0.0091 \text{ m} \quad \text{(about 0.36 in.)} \tag{11}$$

$$X_1 = 0 \quad \text{(the nature of the dynamic absorber)} \tag{12}$$

■

■ **EXAMPLE 14.2.4-2** Band Eliminator for Alarm System
The circuit for the sensing element of an alarm system consists of an inductance-capacitor $L_1 - C_1$ in series with negligible resistance. The alarm system operates trouble-free at 110 V at a frequency of 60 Hz (377 rad/s). In the same building, a remote control system* was installed on the same 60-Hz line. The control system operates at a voltage $V_c = 9.0$ V and a frequency $\omega_2 = 10^5$ Hz (6.28 × 10⁵ rad/s) with a frequency variation of ±10%. It was assumed that the high frequency would not interfere with the alarm system. Given the following circuit parameters, will there be interference? If so, design the band eliminator.

$$L_1 = 10 \ \mu\text{H} = 1.0 \times 10^{-5} \text{ H}$$

$$C_1 = 0.357 \ \mu\text{F} = 3.57 \times 10^{-7} \text{ F}$$

Solution.   Compute the natural frequency of the undamped alarm circuit.

$$\omega_1 = \sqrt{\frac{1}{L_1 C_1}} = \sqrt{\frac{1}{10^{-5} \times 3.57 \times 10^{-7}}} = 5.29 \times 10^5 \text{ rad/s} \tag{1}$$

The alarm circuit will resonate at a frequency of 5.29 × 10⁵ rad/s, which is too close to the remote control disturbance frequency of 6.28 × 10⁵ rad/s. Hence, there will be interference and a frequency trap is required. The circuit is shown in Fig. 14.2.4-3. The system array is formulated by node analysis and then each row is multiplied by s in accordance with rule (a) of Section 12.2.4,

---

*A remote control system of this type superimposes a 100-kHz signal on the 60-Hz, 110-V line. Resonators, appropriately located, will respond to this signal (without requiring additional wiring) by merely plugging into the line. Such a system is called carrier current transmission and is employed in the "bedside device" that turns lights on or off anywhere in the house. See Problem 7.54.

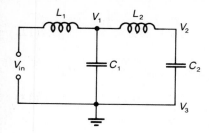

Figure 14.2.4-3 Band eliminator for alarm system.

giving

V_1		V_2	Inputs
$C_1 s^2 + \dfrac{1}{L_1} + \dfrac{1}{L_2}$		$-\dfrac{1}{L_2}$	$\dfrac{V_c}{L_1}$
$-\dfrac{1}{L_2}$		$C_2 s^2 + \dfrac{1}{L_2}$	0

$$(2)$$

Note that eq. (2) has the form of eq. 14.2.2-1, permitting us to use the design approach in Section 14.2.4, where:

$$X_1 = V_1 \qquad X_2 = V_2 \tag{3}$$
$$M_1 = C_1 \qquad M_2 = C_2 \tag{4}$$
$$K_1 = 1/L_1 \qquad K_2 = 1/L_2 \tag{5}$$
$$F_1 = V_c/L_1 \tag{6}$$

Since the frequency variation given for the disturbance is $\pm 10\%$, let the resonant frequencies be at least 20% remote from the trapped frequency ω_2.

Apply eq. (5) to eq. (6)

$$\frac{F_1}{K_1} = \frac{V_c/L_1}{1/L_1} = V_c \tag{7}$$

$$c = \left(\frac{\omega_1}{\omega_2}\right)^2 = \left(\frac{5.29 \times 10^5}{6.28 \times 10^5}\right)^2 = 0.71 \tag{8}$$

Refer to Fig. 14.2.4-1 and first try,

$$c = 0.71 \qquad r_1 = 100\% - 20\% = 0.8 \tag{9}$$

then

$$\mu = 0.05 \qquad r_2 = 1.05 = 100\% + 5\% \tag{10}$$

The upper resonant frequency $r_2\omega_2$ will be only 5% remote from the operating frequency and, hence, is unsafe. Try the other,

$$c = 0.71 \qquad r_2 = 100\% + 20\% = 1.2 \tag{11}$$

then

$$\mu = 0.22 \qquad r_1 = 0.69 = 100\% - 31\% \tag{12}$$

This is safe. Complete the design.

$$C_2 = \mu C_1 = 0.22 \times 0.357\,\mu\text{F}$$

$$= 0.0785\,\mu\text{F} \tag{13}$$

$$L_2 = \frac{1}{C_2 \omega_2^2} = \frac{1}{0.0785 \times 10^{-6} \times (6.28 \times 10^5)^2}$$

$$= 3.23 \times 10^{-5}\,\text{H} \tag{14}$$

$$V_2 = \frac{cF_1}{\mu K_1} = \frac{cV_c}{\mu} = \frac{0.71 \times 9.0}{0.22}$$

$$= 29\,\text{V} \tag{15}$$

Note that voltage V_2 does not affect the alarm system. Hence, 29 V presents no problem. However, the voltage V_1 is important and

$$V_1 = 0 \quad \text{(the nature of a frequency trap)} \tag{16}$$

Thus the disturbance of 9.0 V at a frequency of 10^5 Hz from the remote control system will not affect the alarm system. ∎

14.3 PROPORTIONAL SYSTEMS

14.3.1 Comparison to Systems with No Damping

In Section 14.2, it was shown that an undamped system produces a characteristic determinant Δ that is a quartic with only the *even* powers of the operator s. This is tantamount to a quadratic in s^2 which is factored readily (see eq. 14.2.2-4). A nonconservative system, one with either resistance or damping, produces a characteristic determinant that contains all powers of s. This would normally be a deterrent to factoring. However, if the K_i terms of a conservative system were replaced with network terms Z_i, then the determinant could be factored in the same way as that in Section 14.2.

14.3.2 Proportional Networks

If all the networks (all must be of the same form) Z_i in a lumped linear system are proportional to each other, then the system will be referred to as a *proportional system*. For such a system, all networks Z_i can be expressed as products of constants C_i with a common network Z. Then

$$Z_i = C_i Z \tag{14.3.2-1}$$

For a further convenience, let

$$M_1 = M \quad \text{and} \quad M_2 = \mu M \tag{14.3.2-2}$$

The array for a proportional system is given by

X_1	X_2	Inputs
$Ms^2 + (c + 1)Z$	$-Z$	F
$-Z$	$\mu Ms^2 + Z$	0

[14.3.2-3]

where, for example,

$$Z = Ds + K \qquad [14.3.2-4]$$

Use Cramer's rule to solve for X_2

$$X_2 = \frac{FZ}{\Delta} = \frac{F(Ds + K)}{\Delta} \qquad [14.3.2-5]$$

The denominator can be factored by the method used in eq. 14.2.2-4.

$$\Delta = \mu M^2 \left(s^2 + r_1^2 \frac{Z}{M}\right)\left(s^2 + r_2^2 \frac{Z}{M}\right) \qquad [14.3.2-6]$$

where the roots

r_1 and r_2 are given by eqs. 14.2.2-5 and 14.2.2-6, and by Fig. 14.2.4-1

[14.3.2-7]

Then

$$\frac{X_2}{F/K} = \frac{KD(s + \omega_a)}{\mu M^2(s^2 + 2\zeta_1\omega_1 s + \omega_1^2)(s^2 + 2\zeta_2\omega_2 s + \omega_2^2)} \qquad [14.3.2-8]$$

where

$\omega_1 = r_1\omega_n$	$\omega_2 = r_2\omega_n$	[14.3.2-9]
$\zeta_1 = r_1\zeta$	$\zeta_2 = r_2\zeta$	[14.3.2-10]
$\omega_n^2 = \dfrac{K}{M}$	$\zeta = \dfrac{D}{2M\omega_n}$	[14.3.2-11]
$\omega_a = \dfrac{K}{D}$	$Z_1 = cZ$	[14.3.2-12]

$$I_0 = \frac{KD\omega_a}{\mu M^2 \omega_1^2 \omega_2^2} = \frac{1}{\mu r_1 r_2} \qquad [14.3.2-13]$$

∎

14.4 SYMMETRIC SYSTEMS AND THE LOW PASS FILTER

14.4.1 Definitions

If the characteristic determinant of the system array is point symmetrical, then the system is referred to as symmetric. A point symmetric determinant is $2n$ by $2n$ where all entries have an equal counterpart located on the opposite end of a line through the center of the determinant. A point symmetric six by six determinant is given by eq. 14.4.1-1. A symbolic notation is given by eq. 14.4.1-2, which applies to the general $2n$ by $2n$ determinant.

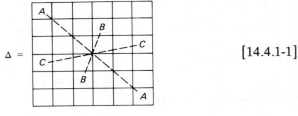

$$\Delta = \qquad [14.4.1\text{-}1]$$

$$\Delta = \qquad [14.4.1\text{-}2]$$

14.4.2 Folding* a Point Symmetric Determinant

The following theorem applies to a point symmetric determinant of any (even) order. It should be pointed out that the notation in the previous section implies partitioning the matrix into n by n submatrices. The subsequent folding operations will involve the folding of the entire submatrices. In the process, note that the image left by the folding process is like an imprint (mirror image or inverted image). First, fold the right-hand side of the determinant about the Y axis onto the left-hand side. Note that the symbol ᴌ becomes ⅃ and that the symbol ⅂ becomes ⌐. Add. The result is

$$\Delta = \begin{array}{|c|c|} \hline \mathsf{L+⅃} & \mathsf{ᴌ} \\ \hline \mathsf{⅂+⌐} & \mathsf{⅂} \\ \hline \end{array} X \qquad [14.4.2\text{-}1]$$

Next, fold this new determinant about the X axis and subtract. Note that the symbol ᴌ becomes the symbol F, and that the sum $\mathsf{L+⅃}$ becomes the sum $\mathsf{⌐+⅂}$.

$$\Delta = \begin{array}{|c|c|} \hline \mathsf{L+⅃} & \mathsf{ᴌ} \\ \hline \mathsf{⅂+⌐ \cdots ⌐-⅂} & \mathsf{⅂-F} \\ \hline \end{array} \qquad [14.4.2\text{-}2]$$

The lower left-hand subdeterminant is equal to zero. Then,

$$\Delta = \begin{array}{|c|c|} \hline \mathsf{L+⅃} & \mathsf{ᴌ} \\ \hline \mathsf{0} & \mathsf{⅂-F} \\ \hline \end{array} \qquad [14.4.2\text{-}3]$$

The determinant is upper triangularized. Consequently, the resulting determinant is given by the product of the diagonal entries.

$$\Delta = (\mathsf{L+⅃})(\mathsf{⅂-F}) \qquad [14.4.2\text{-}4]$$

---

*See Ref. [19].

■ **EXAMPLE 14.4.2-1** Folding a Six by Six Determinant
Given the following six by six determinant, reduce the order.

$$\Delta = \begin{vmatrix} M_1 & C_1 & C_2 & C_4 & C_5 & C_6 \\ C_7 & M_2 & C_8 & C_{10} & C_{11} & C_{12} \\ C_{13} & C_{14} & M_3 & C_{16} & C_{17} & C_{18} \\ C_{18} & C_{17} & C_{16} & M_3 & C_{14} & C_{13} \\ C_{12} & C_{11} & C_{10} & C_8 & M_2 & C_7 \\ C_6 & C_5 & C_4 & C_2 & C_1 & M_1 \end{vmatrix}$$

By inspection, the array is point symmetric. Hence the solution is given by eq. 14.4.2-3.

$$\Delta = \begin{vmatrix} M_1 + C_6 & C_1 + C_5 & C_2 + C_4 & C_4 & C_5 & C_6 \\ C_7 + C_{12} & M_2 + C_{11} & C_8 + C_{10} & C_{10} & C_{11} & C_{12} \\ C_{13} + C_{18} & C_{14} + C_{17} & M_3 + C_{16} & C_{16} & C_{17} & C_{18} \\ 0 & 0 & 0 & M_3 - C_{16} & C_{14} - C_{17} & C_{13} - C_{18} \\ 0 & 0 & 0 & C_8 - C_{10} & M_2 - C_{11} & C_7 - C_{12} \\ 0 & 0 & 0 & C_2 - C_4 & C_1 - C_5 & M_1 - C_6 \end{vmatrix}$$

14.4.3 Equivalent Symmetric Systems

It is not necessary to have a system that is a mirror image copy from right to left. Another, more adaptive, configuration is shown in Fig. 14.4.3-1. It should be pointed out that there are additional networks Z_2 and Z_3 (and these need not necessarily be equal to each other; one may even be equal to zero)

θ_1	θ_2	Inputs and IC
$J_1 s^2 + Z + Z_3 + Z_a$	$-Z$	$T_1 + Z_a \theta_a + IC_1$
$-Z$	$J_2 s^2 + Z + Z_2 + Z_b$	$T_2 + Z_b \theta_b + IC_2$

[14.4.3-1]

For symmetry, let

$$J_1 = J_2 = J \qquad \text{[14.4.3-2]}$$

$$Z_3 + Z_a = Z_2 + Z_b = Z_1 \qquad \text{[14.4.3-3]}$$

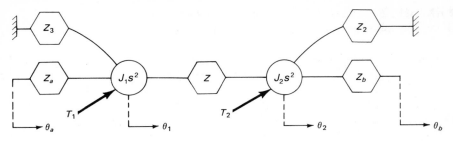

Figure 14.4.3-1   Criteria for maximum filtering in symmetric system.

The array is now symmetric and can be folded.

θ_1	$\theta_2 - \theta_1$	Inputs and IC
$J_1 s^2 + Z_1$	$-Z$	$T_1 + Z_a\theta_a + IC_1$
0	$J_2 s^2 + Z_1 + 2Z$	$T_2 + Z_b\theta_b + IC_2 - T_1 - Z_a\theta_a - IC_1$

$$[14.4.3\text{-}4]$$

Note that use has been made of the rules in Section 12.2.4, which include the input column and the headings (outputs).

The characteristic determinant is factored by inspection, using the product of the diagonal entries:

$$\Delta = \left(J_1 s^2 + Z_1\right)\left(J_2 s^2 + Z_1 + 2Z\right) \qquad [14.4.3\text{-}5]$$

Let the input θ_b be applied through Z_b to J_2, and let the system output be θ_1. Apply Cramer's rule to eq. 14.4.3-4.

$$\theta_1 = \frac{Z_b\theta_b Z}{\left(J_1 s^2 + Z_1\right)\left(J_2 s^2 + Z_1 + 2Z\right)} \qquad [14.4.3\text{-}6]$$

Assume all networks are of the form

$$Z_i = D_i s + K_i \qquad [14.4.3\text{-}7]$$

Then, after normalizing,

$$\frac{\theta_1}{\theta_b} = \frac{DD_b(s + \omega_k)(s + \omega_t)}{J^2\left(s^2 + 2\zeta_1\omega_1 s + \omega_1^2\right)\left(s^2 + 2\zeta_3\omega_3 s + \omega_3^2\right)} \qquad [14.4.3\text{-}8]$$

where

$$\omega_k = \frac{K}{D} \qquad\qquad \omega_t = \frac{K_b}{D_b} \qquad [14.4.3\text{-}9]$$

$$\omega_1 = \sqrt{\frac{K_1}{J}} \qquad\qquad \omega_3 = \sqrt{\frac{K_1 + 2K}{J}} \qquad [14.4.3\text{-}10]$$

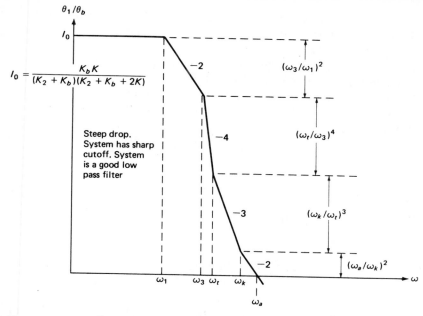

Figure 14.4.3-2   Generalized symmetric two mass system.

$$\zeta_1 = \frac{D_1}{2J\omega_1} \qquad \zeta_3 = \frac{D_1 + 2D}{2J\omega_3} \qquad [14.4.3\text{-}11]$$

$$K_1 = K_2 + K_b \qquad D_1 = D_2 + D_b \qquad [14.4.3\text{-}12]$$

If the system is to be a successful filter, the curve must drop sharply as soon as possible. This then requires that the frequencies in the denominator must occur as soon as possible. Referring to Fig. 14.4.3-2, note that the frequencies ω_1 and ω_3 ascend in that order. The frequencies ω_k and ω_t must be selected higher than ω_1 or ω_3, since they occur in the numerator, and hence will cause the frequency response curve to rise, the antithesis of filtering.

■ **EXAMPLE 14.4.3-2** Synthesis and Design of Two Mass Low Pass Filter Redesign the low pass filter of Ex. 9.4.4-1 for a more effective filter using a symmetric system as shown in Fig. 14.4.3-1. Retain the following from the original example:

$$\omega_1 = 600 \text{ rad/s} \qquad \omega_n = 80.7 \text{ rad/s}$$

$$J_1 = 10.0 \text{ g} \cdot \text{cm}^2 \qquad \zeta_1 = 0.6$$

$$\text{peak} = 4\% \quad (\text{or } R_r = 1.04)$$

Solution.   This problem will be approached in two stages. First, determine the design criteria and second, synthesize the system. In order to perform the first, determine the frequency response of the general symmetric system, optimize this for best low pass filter, and then specify the significant corner frequencies. In order to synthesize the system, relate the corner frequencies to the system parameters, forming sufficient equations to determine all of them.

The system array is given by eq. 14.4.3-1. Since the system is symmetric, it can be folded as shown in eq. 14.4.3-4, where the conditions for symmetry are given in eqs. 14.4.3-2 and 14.4.3-3. Use eqs. 14.4.3-9 through 14.4.3-12. Since $K_1 = K_2 + K_b$,

$$\omega_1 = \sqrt{\frac{K_2 + K_b}{J}} \tag{1}$$

or

$$K_b = J\omega_1^2 - K_2 \tag{2}$$

and

$$D_1 = D_2 + D_b = 2\zeta_1 J\omega_1 \tag{3}$$

or

$$D_b = 2\zeta_1 J\omega_1 - D_2 \tag{4}$$

Then

$$\omega_t = \frac{K_b}{D_b} = \frac{J\omega_1^2 - K_2}{2\zeta_1 J\omega_1 - D_2} \tag{5}$$

For effective filtering, it is desirable to make ω_t as high as possible. Then let

$$\boxed{K_2 = 0}$$

and

$$D_2 = \frac{3D_1}{4} \tag{6}$$

Substitute for D_1

$$D_2 = \frac{3(2\zeta_1 J\omega_1)}{4}$$

Apply eqs. (6) to eq. (5).

$$\omega_t = \frac{J\omega_1^2}{2\zeta_1 J\omega_1\left(1 - \frac{3}{4}\right)} = \frac{2\omega_1}{\zeta_1}$$

$$= \frac{2 \times 80.7}{0.6} = 272 \text{ rad/s} \tag{7}$$

Apply eq. (6) to eqs. (2) and (4).

$$K_b = 10 \times 80.7^2 = \boxed{6.5 \times 10^4 \, \text{dyn} \cdot \text{cm/rad}} \tag{8}$$

$$D_1 = 2 \times 0.6 \times 10 \times 80.7 = \boxed{970 \, \text{dyn} \cdot \text{cm} \cdot \text{s}} \tag{9}$$

Then,

$$D_2 = 970 \times \tfrac{3}{4} = \boxed{727 \, \text{dyn} \cdot \text{cm} \cdot \text{s}} \tag{10}$$

$$D_b = 970 - 727 = \boxed{243 \, \text{dyn} \cdot \text{cm} \cdot \text{s}} \tag{11}$$

From eqs. 14.4.3-9 through 14.4.3-12 and eq. (6)

$$\omega_3 = \sqrt{\frac{K_2 + K_b + 2K}{J}} = \sqrt{\frac{K_b + 2K}{J}} \tag{12}$$

For effective filtering, it is desirable to place ω_3 as close to ω_1 as possible, to impose the slope of -4 early. Then let

$$K = \tfrac{1}{2}K_b = \tfrac{1}{2}6.5 \times 10^4 = \boxed{3.25 \times 10^4 \, \text{dyn} \cdot \text{cm/rad}} \tag{13}$$

Then

$$\omega_3 = \sqrt{\frac{2K_b}{J}} = \omega_1\sqrt{2} = 115 \, \text{rad/s} \tag{14}$$

Also from eqs. 14.4.3-9 through 14.4.3-12,

$$\zeta_3 = \frac{D_1 + 2D}{2J\omega_3}$$

or

$$D = \tfrac{1}{2}(\zeta_3 2J\omega_3 - D_1) \tag{15}$$

For typical value, let $\zeta_3 = \tfrac{1}{2}$. Then eq. (15) becomes

$$D = \tfrac{1}{2}(\tfrac{1}{2} \times 2 \times 10 \times 115 - 970) = \boxed{90 \, \text{dyn} \cdot \text{cm} \cdot \text{s}} \tag{16}$$

Apply eqs. (13) and (16) to eqs. 14.4.3-9 through 14.4.3-12.

$$\omega_k = \frac{K}{D} = \frac{3.25 \times 10^4}{90} = 361 \, \text{rad/s} \tag{17}$$

All parameters and frequencies have been found. The frequency response curve can be drawn, using the numerical values thus found and by applying these to Fig. 14.4.3-2. The resulting frequency response curve is shown in Fig. 14.4.3-3. (After the asymptotes were drawn, the actual curve was sketched

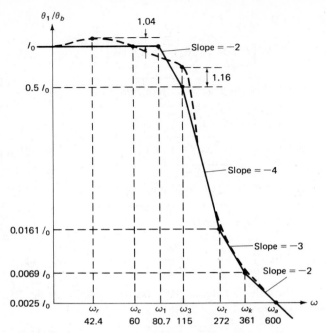

Figure 14.4.3-3   Frequency response for two mass filter.

using the values in Table 8.5.2-1.) Note that vibration frequencies of 600 rad/s or higher are reduced by a factor of 0.0025, which is considerably better than 0.018 as was the case for the original one mass filter. ∎

14.5 COMPATIBLE SYSTEMS AND A UNIVERSAL MODEL

14.5.1 Properties

If a system of any order can be expressed as two coupled subsystems A and B (not necessarily of the same order), the system is called compatible if the characteristic determinant can be factored by using a special form. This, in turn, requires the comparison of system parameters of like powers of s in what is called the compatibility equation.

Given some of the system parameters, or some of the system constraints, or a combination of some of both, the compatibility equation tests whether it is possible to find the remaining system parameters to make the system a compatible one. If the system passes this test, the technique shows how to design such a system.

The advantage of a compatible system is that it can be manipulated to satisfy many different requirements. For example, the system could be a resonator, a low pass filter, or a broadband system. In addition, the technique permits the involvement of a large number of parameters, all of which are determined one at a time in closed form.

It should be pointed out that the test for compatibility is an easy one to pass due to its flexibility in accepting so many terms. For example, given a fourth order system with as many as twelve parameters, as many as eleven may be given (or chosen arbitrarily) and the twelfth one could be found by using the compatibility equation. On the other hand, if only one parameter is known and there are sufficient system constraints, then the remaining eleven can be found one at a time in closed form.

The technique will be presented in three levels of advancement, each level providing more flexibility and more terms, and requiring more design effort.

14.5.2 First Level of Compatibility

The system array for a first level compatible system is given by

X_1	X_2	Input
$M + C$	$-gC$	F_1
$-cC$	$aM + bC$	F_2

[14.5.2-1]

where

$$M = \text{function of } s \text{ of order } m \qquad [14.5.2\text{-}2]$$

$$C = \text{function of } s \text{ of order } n \qquad [14.5.2\text{-}3]$$

$$a, b, c, g \text{ are constants} \qquad [14.5.2\text{-}4]$$

The characteristic determinant is given by

$$\Delta = (M + C)(aM + bC) - cCgC \qquad [14.5.2\text{-}5]$$

or

$$\frac{\Delta}{a} = M^2 + \frac{(b + a)}{a} MC + \frac{(b - cg)}{a} C^2 \qquad [14.5.2\text{-}6]$$

This is a quadratic in M and C that readily factors.

$$\frac{\Delta}{a} = (M + R_1 C)(M + R_2 C) \qquad [14.5.2\text{-}7]$$

where R_1 and R_2 are the negative of the roots of the quadratic,

$$R = \frac{b + a}{2a} \pm \sqrt{\left(\frac{b + a}{2a}\right)^2 - \frac{b - cg}{a}} \qquad [14.5.2\text{-}8]$$

$$= \frac{b + a}{2a} \pm \frac{1}{2a}\sqrt{(b - a)^2 + 4acg} \qquad [14.5.2\text{-}9]$$

■ **EXAMPLE 14.5.2-1** Factored High Order Compatible System
Construct a compatible system* and factor the characteristic determinant
given the following:

    (a) M and C are polynomials in s
    (b) M is one order higher than C
    (c) constant $g = 1$

The system array has six unknowns F through L as follows:

X_1	X_2	Inputs	
$s^4 + Fs^3 + Gs^2 + Hs + 120$	$-(s^3 + 3s^2 + 2s + 20)$	X_a	(d)
$-(Is^3 + Js^2 + Ls + 10)$	$2s^4 + 7s^3 + 23s^2 + 42s + 260$	X_b	

Solution. From (b) and the array (d), we conclude

$$M \text{ contains a term } s^4 \tag{1}$$

and

$$\boxed{a = 2} \tag{2}$$

Define the constant terms as follows:

$$K_m = \text{constant term of } M \tag{3}$$

$$K_c = \text{constant term of } C \tag{4}$$

Form the ratio of off-diagonal terms and use (c)

$$\frac{cC}{gC} = \frac{Is^3 + Js^2 + Ls + 10}{s^3 + 3s^2 + 2s + 20} = c \tag{5}$$

From eq. (5) we conclude

$$\boxed{c = \tfrac{1}{2}} \quad \boxed{I = 0.5} \quad \boxed{J = 1.5} \quad \boxed{L = 1} \tag{6}$$

and

$$\boxed{C = s^3 + 3s^2 + 2s + 20} \quad \boxed{K_c = 20} \tag{7}$$

Write the constant terms of entry 1-1, which is of the form $M + C$.

$$K_m + K_c = 120 = K_m + 20 \tag{8}$$

or

$$K_m = 100 \tag{9}$$

---

*It should be pointed out that not all systems can be made compatible. This one was selected by using tests to be demonstrated in a later section.

Apply eqs. (2), (7), and (9) to the constant terms of entry 2-2, which is of the form $aM + bC$.

$$aK_m + bK_c = 260 = 2 \times 100 + b20 \tag{10}$$

or

$$\boxed{b = 3} \tag{11}$$

Now use all the terms of entry 2-2.

$$2M + 3(s^3 + 3s^2 + 2s + 20) = 2s^4 + 7s^3 + 23s^2 + 42s + 260 \tag{12}$$

Solve for M.

$$\boxed{M = s^4 + 2s^3 + 7s^2 + 18s + 100} \tag{13}$$

Relate to entry 1-1.

$$\boxed{F = 2} \qquad \boxed{G = 7} \qquad \boxed{H = 18} \tag{14}$$

The parameters have been determined and can be checked against the terms in the array for a compatible system, eq. 14.5.2-1. Thus, the system is compatible. Consequently, the characteristic determinant can be factored into the form of eq. 14.5.2-7, where the roots are found by using eqs. (2) and (11) in eq. 14.5.2-9.

$$\boxed{R_1 = 2} \qquad \boxed{R_2 = \tfrac{1}{2}} \tag{15}$$

Then

$$\Delta = 2(s^4 + 4s^3 + 13s^2 + 22s + 140)(s^4 + 2.5s^3 + 8.5s^2 + 19s + 110) \tag{16}$$

14.5.3 Second Level of Compatibility

In order to provide more flexibility, the basic concept will be expanded; additional networks will be included in the system array. These may be in the form of a polynomial or one polynomial divided by another. The function M may be a polynomial or a quotient of polynomials. All polynomials may be of different orders, and any may have missing terms.

At this point we will eliminate the constants c and g by multiplying rows and columns according to the rules in Section 12.2.4. The following substitutions apply:

$$M = M_1 + Z_1, \qquad C = Z_3, \qquad c = g = 1 \qquad [14.5.3\text{-}1]$$

$$M_1, Z_1, \text{ and } Z_3 \text{ are functions of } s \qquad [14.5.3\text{-}2]$$

$$a \text{ and } b \text{ are constants} \qquad [14.5.3\text{-}3]$$

Apply eqs. 14.3.3-1 through 14.5.3-3 to eq. 14.5.2-1 (the array for a first level compatible system) to formulate the array for a second level compatible system as shown below:

X_1	X_2	Inputs
$M_1 + Z_1 + Z_3$	$-Z_3$	F_1
$-Z_3$	$a(M_1 + Z_1) + bZ_3$	F_2

[14.5.3-4]

In the typical array, entry 2-2 will seldom be in the form given in eq. 14.5.3-4. Instead it will be of the form:

X_1	X_2	Inputs
$M_1 + Z_1 + Z_3$	$-Z_3$	F_1
$-Z_3$	$M_2 + Z_2 + Z_3$	F_2

[14.5.3-5]

Since the array in eq. 14.5.3-4 was in compatible form (it could be factored by using the root equation), then in order for eq. 14.5.3-5 to qualify for compatibility, entry 2-2 of each array must be equal. Then for compatibility,

$$M_2 + Z_2 + Z_3 = a(M_1 + Z_1) + bZ_3 \qquad [14.5.3\text{-}6]$$

which implies

$$M_2 = aM_1 \qquad [14.5.3\text{-}7]$$

and

$$aZ_1 + bZ_3 = Z_2 + Z_3 \qquad [14.5.3\text{-}8]$$

or

$$\frac{Z_2 - aZ_1}{Z_3} \equiv b - 1 \qquad [14.5.3\text{-}9]$$

Equation 14.5.3-9 is called the *compatibility equation*. Note it is an identity. Hence, we may relate like powers of s. For example if

$$Z_i = I_i s^3 + J_i s^2 + D_i s + K_i \qquad [14.5.3\text{-}10]$$

then eq. 14.5.3-9 becomes

$$\frac{I_2 - aI_1}{I_3} = \frac{J_2 - aJ_1}{J_3} = \frac{D_2 - aD_1}{D_3} = \frac{K_2 - aK_1}{K_3} = b - 1 \quad [14.5.3\text{-}11]$$

Thus, if Z_i is of order n, then there are $n + 1$ equations. If the system satisfies the compatibility equation, the characteristic determinant factors as follows:

$$\Delta = a(M_1 + Z_1 + R_1 Z_3)(M_1 + Z_1 + R_2 Z_3) \qquad [14.5.3\text{-}12]$$

where

$$R = \frac{b + a}{2a} \pm \sqrt{\left(\frac{b + a}{2a}\right)^2 - \frac{b - 1}{a}} \qquad [14.5.3\text{-}13]$$

$$= \frac{b + a}{2a} \pm \frac{1}{2a}\sqrt{(b - a)^2 + 4a} \qquad [14.5.3\text{-}14]$$

On the other hand, given the roots, find the constants a and b. To do this, sum the roots in eq. 14.5.3-13.

$$R_1 + R_2 = (b + a)/a \qquad [14.5.3\text{-}15]$$

or

$$b = a(R_1 + R_2 - 1) \qquad [14.5.3\text{-}16]$$

Take the product of the roots.

$$R_1 R_2 = (b - 1)/a \qquad [14.5.3\text{-}17]$$

or

$$b = aR_1 R_2 + 1 \qquad [14.5.3\text{-}18]$$

Eliminate the constant b between eqs. 14.5.3-16 and 14.5.3-18.

$$a = \frac{1}{R_1 + R_2 - R_1 R_2 - 1} = \frac{1}{(R_1 - 1)(1 - R_2)} \qquad [14.5.3\text{-}19]$$

Unless otherwise stated, we will assume that all system elements may be either zero or positive, and finite.

$$0 \le M_i < \infty, \qquad 0 \le Z_i < \infty \qquad [14.5.3\text{-}20]$$

Using this constraint in eq. 14.5.3-7, we conclude

$$a \ge 0 \qquad [14.5.3\text{-}21]$$

The roots are employed in factoring the characteristic equation, for which the roots must be finite. Then

$$R < \infty \qquad [14.5.3\text{-}22]$$

Since the constant a appears in the denominator of the root equation, eq. 14.5.3-13, then

$$a \ne 0 \qquad [14.5.3\text{-}23]$$

which modifies eq. 14.5.3-21 to exclude 0.

$$a > 0 \qquad [14.5.3\text{-}24]$$

This implies of eq. 14.5.3-7 that

$$\begin{aligned} \text{if} \quad & M_1 \ne 0 \\ \text{then} \quad & M_2 \ne 0 \end{aligned} \qquad [14.5.3\text{-}25]$$

The limit on the constant a also implies limits on the roots.

$$\left.\begin{array}{c} R_1 \text{ and } R_2 \text{ are real} \\[4pt] R_1 \neq R_2 \\[20pt] R_1 > R_2 \end{array}\right\} \qquad [14.5.3\text{-}26]$$

Let us arbitrarily define

At this point it is appropriate to consider the possibility of a negative root. It is possible and does not conflict with the concept of compatibility. However caution must be exercised, since a negative root requires subtraction when factoring the characteristic determinant. This could possibly result in negative terms in the final result. This, in turn, implies either negative system elements or negative damping. Consequently, when using a negative root, it is advisable to test for the ultimate effect.

Consider an alternate form of the compatibility equation, eq. 14.5.3-9.

$$\frac{Z_2}{Z_3} - (b - 1) = \frac{aZ_1}{Z_3} \qquad [14.5.3\text{-}27]$$

If we apply the inequalities of eqs. 14.5.3-20 and 14.5.3-24, we conclude

$$\frac{Z_2}{Z_3} \geq (b - 1) \qquad [14.5.3\text{-}28]$$

Equation 14.5.3-28 is called the *compatibility inequality*. The inequality poses limits. Equality implies that at least one element of the system is equal to zero.

The system takes on special properties if

$$b = 1 \qquad [14.5.3\text{-}29]$$

From eq. 14.5.3-13, it is clear that when

$$\left.\begin{array}{c} b = 1 \\[12pt] R_1 = \dfrac{1 + a}{a} \\[16pt] R_2 = 0 \end{array}\right\} \qquad [14.5.3\text{-}30]$$

then

and

Also, the compatibility equation takes on a special form when

$$\left.\begin{array}{c} b = 1 \\[12pt] \dfrac{Z_2 - aZ_1}{Z_3} = 0 \\[20pt] Z_2 = aZ_1 \end{array}\right\} \qquad [14.5.3\text{-}31]$$

and

Then for all Z_3

■ **EXAMPLE 14.5.3-1** Compatible Mechanical System
Given the mechanical system shown in Fig. 14.5.3-1. All system parameters
are known except K_1. Perform the following:

(a) What is the limiting value* of K_2?
(b) If the limiting condition of part (a) is satisfied, determine K_1 for
compatibility.
(c) Write the characteristic determinant in factored form and check it
against the original (unfactored) determinant.
(d) Determine natural frequencies and damping ratios.

Given the following system parameters:

$$J_1 = 1.0, \qquad J_2 = 2.0 \tag{1}$$

$$Z_a = Ds + 0 = 20s, \qquad Z_b = Ds + 0 = 40s \tag{2}$$

$$\left. \begin{aligned} Z_1 &= D_1 s + K_1 = 10s + K_1 \\ Z_2 &= D_2 s + K_2 = 60s + 3800 \\ Z_3 &= D_3 s + K_3 = 20s + 600 \end{aligned} \right\} \tag{3}$$

Solution.   Formulate the system array

θ_1	θ_2	Inputs	
$J_1 s^2 + Z_1 + Z_3 + Z_a$	$-Z_3$	$T_1 + Z_a \theta_a$	(4)
$-Z_3$	$J_2 s^2 + Z_3 + Z_2 + Z_b$	$T_2 + Z_b \theta_b$	

The question arises concerning the treatment of the networks Z_a and Z_b.
Since they happen[†] to bear the same ratio as J_a and J_b, it seems provident
to let

$$M_1 = J_1 s^2 + Z_a = s^2 + 20s$$

$$M_2 = J_2 s^2 + Z_b = aM_1 \quad \text{where } a = 2 \tag{5}$$

Use the compatibility equation, eq. 14.5.3-9.

$$\frac{D_2 - aD_1}{D_3} = \frac{K_2 - aK_1}{K_3} = b - 1 \tag{6}$$

---

*Obviously there would be no problem if K_1 were given and if we were requested to determine K_2
for compatibility.

[†] There is no question that this problem was deliberately set up to make it easy to solve. But this is
seldom the case in practice and is treated in Problem 14.23.

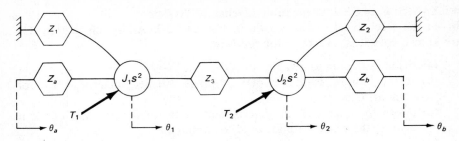

Figure 14.5.3-1   Compatible mechanical system.

Apply appropriate numerical values for D_i and a, and solve for b.

$$b = 3 \qquad (7)$$

Use the compatibility inequality, eq. 14.5.3-28.

$$\frac{K_2}{K_3} \geq b - 1 \qquad (8)$$

Use appropriate numerical values.

$$K_2 \geq 1200 \qquad (9)$$

Since the numerical value given for K_2 in eq. (3) is 3800, the inequality is satisfied and the system can be made compatible. Use numerical values in eq. (6) and solve for K_1.

$$K_1 = 1300 \qquad (10)$$

Use the numerical values of eqs. (2) and (9) in eq. 14.5.3-14 and find the roots.

$$R_1 = 2, \qquad R_2 = \tfrac{1}{2} \qquad (11)$$

Apply the appropriate numerical values to eq. 14.5.3-12 to formulate the factored form of the characteristic determinant. Collect terms.

$$\Delta = 2(s^2 + 70s + 2500)(s^2 + 40s + 1600) \qquad (12)$$

In order to check the result, multiply out

$$\Delta = 2s^4 + 220s^3 + 13{,}800s^2 + 424{,}000s + 8{,}000{,}000 \qquad (13)$$

Compare the result to that obtained for the original (unfactored) system array, eq. (4).

$$\Delta = \left(J_1 s^2 + Z_a + Z_1 + Z_3\right)\left(J_2 s^2 + Z_b + Z_2 + Z_3\right) - \left(Z_3\right)^2 \qquad (14)$$

Apply numerical values and multiply out. The identical function results.

Use eq. (12) to determine the natural frequencies and damping ratios.

$$\omega_1 = \sqrt{2500} = 50 \text{ rad/s} \qquad \omega_2 = \sqrt{1600} = 40 \text{ rad/s} \qquad (15)$$

$$\zeta_1 = \frac{70}{2 \times 50} = 0.7 \qquad \zeta_2 = \frac{40}{2 \times 40} = 0.5 \qquad (16)$$

■

■ **EXAMPLE 14.5.3-2** Design of Superheterodyne Interstage Amplifier
Refer to Ex. 12.3.1-2 (Superheterodyne Interstage Amplifier). Design the
system for the following requirements:

$$R_r = 10 \quad \text{(resonant peak)} \tag{1}$$

$$\omega_r \simeq \omega_2 = 4 \times 10^6 \text{ rad/s} \quad \text{(about 640 kHz)} \tag{2}$$

$$L_1 = 10^{-5} \text{ H} \tag{3}$$

Determine exact resonant frequency and bandwidth.

Solution. In Ex. 12.3.1-2, solve for I_2 using the system array given in that
example in eq. (6). Since so few constraints have been given, there is no doubt
that the system can be made compatible. Then we can assume that eventually
we will be able to factor the characteristic determinant, permitting us to write
the result as follows:

$$\frac{I_2}{E} = \frac{sZ_3}{a\left(L_1 s^2 + Z_1 + R_1 Z_3\right)\left(L_1 s^2 + Z_1 + R_2 Z_3\right)} \tag{4}$$

where

$$Z_i = D_i s + K_i = R_i s + \frac{1}{C_i} \tag{5}$$

For each quadratic term in eq. (4) we may determine the natural frequency
and damping ratio.

$$\frac{I_2}{E} = \frac{D_3 s (s + \omega_3)}{a L_1^2 \left(s^2 + 2\zeta_1 \omega_1 s + \omega_1^2\right)\left(s^2 + 2\zeta_2 \omega_2 s + \omega_2^2\right)} \tag{6}$$

where

$$\omega_1^2 = \frac{K_1 + R_1 K_3}{L_1} \qquad \omega_2^2 = \frac{K_1 + R_2 K_3}{L_1} \tag{7}$$

$$\zeta_1 = \frac{D_1 + R_1 D_3}{2 L_1 \omega_1} \qquad \zeta_2 = \frac{D_1 + R_2 D_3}{2 L_1 \omega_2} \tag{8}$$

The frequency response curve is shown on Fig. 14.5.3-2. It should be pointed
out that the damping ratios have a prominent role in this problem. It is
desirable to have a peak in the vicinity of ω_2 but none near ω_1. On the
strength of this statement, we may select the appropriate damping ratios with

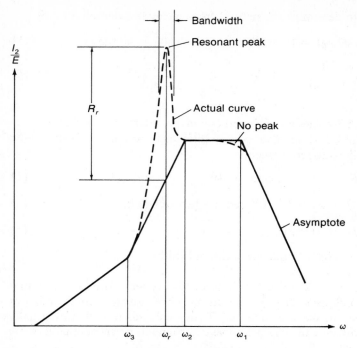

Figure 14.5.3-2 Frequency response of compatible resonator.

the help of either eq. 8.5.1-9 or Table 8.5.2-1.

$$\text{for } R_r = 10 \quad \zeta_2 = 0.05 \tag{9}$$

$$\text{for } R_r \approx 1 \quad \zeta_1 = 0.7 \tag{10}$$

Note that there is a large spread between the two required damping ratios.

$$\frac{\zeta_1}{\zeta_2} = \frac{0.7}{0.05} = 14 \tag{11}$$

Let us examine eqs. (7) and (8) to determine the parameter(s) that are to be responsible for such a large spread of damping ratios. Certainly the roots play an important part, but the relationship is neither linear nor obvious. However, we can be sure that it requires a large spread of the roots. Consulting eqs. 14.5.3-31, we are tempted to set $b = 1$. Also, for symmetry, we are tempted to set $a = 1$. However, let us deliberately make choices that are not optimum.*
Let

$$a = 2, \quad b = 1.25 \tag{12}$$

then

$$R_1 = 1.544, \quad R_2 = 0.0810 \tag{13}$$

---

*Better choices are used in Problem 14.30.

Since this is to be a resonator, it is desirable that the upper frequency ω_1 be close to the lower one ω_2. A fair* choice is to use a factor of 1.25. Then

$$\omega_1 \simeq 1.25\omega_2 = 1.25 \times 4 \times 10^6$$

$$= 5 \times 10^6 \text{ rad/s} \tag{14}$$

From eq. (6) we have

$$L_1\omega_1^2 = K_1 + R_1K_3 \tag{15}$$

and

$$L_1\omega_2^2 = K_1 + R_2K_3 \tag{16}$$

Equations (15) and (16) are two simple linear simultaneous equations in K_1 and K_3. Using appropriate numerical values, we find that

$$25 \times 10^7 = K_1 + 1.544K_3 \tag{17}$$

and

$$16 \times 10^7 = K_1 + 0.0810K_3 \tag{18}$$

Solving eqs. (17) and (18) simultaneously,

$$K_3 = 6.152 \times 10^7, \qquad C_3 = 1/K_3 = 0.01625 \ \mu\text{F} \tag{19}$$

$$K_1 = 15.5 \times 10^7, \qquad C_1 = 1/K_1 = 0.00645 \ \mu\text{F} \tag{20}$$

From eq. (8) we have

$$2L_1\omega_1\zeta_1 = D_1 + R_1D_3 \tag{21}$$

and

$$2L_1\omega_2\zeta_2 = D_1 + R_2D_3 \tag{22}$$

Apply numerical values.

$$70 = D_1 + 1.544D_3 \tag{23}$$

and

$$4 = D_1 + 0.081D_3 \tag{24}$$

Solve simultaneously for D_3 and D_1.

$$D_3 = 45.11 \tag{25}$$

$$D_1 = 0.35 \tag{26}$$

---

*Better choices are used in Problem 14.30.

Refer to the numerator of eq. (6).

$$\omega_3 = \frac{K_3}{D_3} = \frac{6.152 \times 10^7}{45.11} \tag{27}$$

$$= 1.36 \times 10^6 \text{ rad/s} \tag{28}$$

Since this frequency is lower than ω_2 the system will operate properly. Use the compatibility equation, eq. 14.5.3-11, for the K terms.

$$\frac{K_2 - aK_1}{K_3} = b - 1 \tag{30}$$

Apply the appropriate numerical values and solve for K_2.

$$K_2 = 32.54 \times 10^7, \qquad C_2 = 1/K_2 = 0.00307 \, \mu F \tag{31}$$

Use the compatibility equation for the D terms.

$$\frac{D_2 - aD_2}{D_3} = b - 1 \tag{32}$$

Apply numerical values and solve for D_2.

$$D_2 = 11.98 \tag{33}$$

Compare the original system array, eq. (5) of Ex. 12.3.1-2, to the one for this compatible system, eq. 14.5.3-5.

$$L_2 = aL_1 = 2 \times 10^{-5} \text{ H} \tag{34}$$

The resonant frequency ratio is found by using either eq. 8.5.1-8 or Table 8.5.2-1.

$$\gamma_r = 0.9974 \quad \text{for } \zeta_2 = 0.05 \tag{35}$$

Then the resonant frequency is given by

$$\omega_r = \gamma_r \omega_2 = 0.9974 \times 4 \times 10^6 \tag{36}$$

$$= 3.99 \times 10^6 \text{ rad/s} \tag{37}$$

which is close to ω_2 as required by eq. (2). For the bandwidth, refer to Fig. 8.7.4-1, eq. 8.7.4-2, and Table 8.7.4-1.

$$\begin{array}{l} \gamma_{b1} = 0.9461 \\ \gamma_{b2} = 1.0464 \end{array} \quad \text{for } \zeta_2 = 0.05 \tag{38}$$

The band limits for $\sqrt{2}$ attenuation are

$$\omega_{bi} = \gamma_{bi} \omega_2 \tag{39}$$

Then

$$\omega_{b1} = 0.9461 \times 4 \times 10^6 = 3.78 \times 10^6 \text{ rad/s} \tag{40}$$

$$\omega_{b2} = 1.0464 \times 4 \times 10^6 = 4.19 \times 10^6 \text{ rad/s} \tag{41}$$

∎

14.5.4 Third Level of Compatibility

The flexibility and complexity of compatible systems can be extended by including more and more network terms. For example, there could be terms involving connections to the ground, connections to power sources or supplies, entry networks through which noise enters the system, input networks through which the input is purposely processed or conditioned, intraconnection networks that tie in coupling effects, and so on. If each of these has means by which system constraints can be identified and employed, then each type of network will become an entity in the compatible system. Consequently, there can be a large number of terms. Ordinarily, a large number of terms complicates the system. However, the compatible system seems to thrive on large numbers of terms. In fact, the more terms, the easier it is to satisfy the compatibility equation. For this study, let us employ five networks.

For the third level compatible system, let

$$M = M_1 + Z_4 + Z_5 \qquad [14.5.4\text{-}1]$$

$$C = Z_3 \qquad [14.5.4\text{-}2]$$

$$M \text{ and } Z \text{ are functions of } s \qquad [14.5.4\text{-}3]$$

$$a \text{ and } b \text{ are constants} \qquad [14.5.4\text{-}4]$$

The system array for a third level compatible system is given by

X_1	X_2	Input	
$M_1 + Z_3 + Z_4 + Z_5$	$-Z_3$	$F_1 + Z_4 X_a$	$[14.5.4\text{-}5]$
$-Z_3$	$a(M_1 + Z_4 + Z_5) + bZ_3$	$F_2 + Z_6 X_b$	

While the array for the typical system is of the form

X_1	X_2	Input	
$M_1 + Z_3 + Z_4 + Z_5$	$-Z_3$	$F_1 + Z_4 X_a$	$[14.5.4\text{-}6]$
$-Z_3$	$M_2 + Z_3 + Z_6 + Z_7$	$F_2 + Z_6 X_b$	

System is compatible if

$$Z_6 + Z_7 + Z_3 = a(Z_4 + Z_5) + bZ_3 \qquad [14.5.4\text{-}7]$$

where

$$a = M_2/M_1 \qquad [14.5.4\text{-}8]$$

Equation 14.5.4-7 is the compatibility equation for third level compatible systems. It is clear that the introduction of additional terms does not complicate the technique, but instead makes it easier to satisfy the compatibility equation. The additional terms leave room to maneuver the sums of terms to help satisfy the equation.

Similarly to the approach in eq. 14.5.3-9, an alternate form of eq. 14.5.4-7 will prove fruitful.

$$\frac{Z_6 + Z_7 - a(Z_4 + Z_5)}{Z_3} \equiv b - 1 \qquad [14.5.4\text{-}9]$$

Similarly to the approach in eq. 14.5.3-11, eq. 14.5.4-9 may be treated as an identity, thus forming a number of equations. For example, if

$$Z_i = D_i s + K_i \qquad [14.5.4\text{-}10]$$

then eq. 14.5.4-9 becomes

$$\frac{D_6 + D_7 - a(D_4 + D_5)}{D_3} = \frac{K_6 + K_7 - a(K_4 + K_5)}{K_3} = b - 1 \quad [14.5.4\text{-}11]$$

Similarly to the logic employed in 14.5.3-11, if the networks are of order n and, since there are five parameters in each equation, plus the two quantities a and b, then

$$5(n + 1) + 2 = \text{number of terms}$$
$$4(n + 1) + 2 = \text{number of independent terms} \qquad [14.5.4\text{-}12]$$
$$(n + 1) = \text{number of terms to be found}$$

Once compatibility has been established, the characteristic determinant may be factored as follows:

$$\Delta = a(M_1 + Z_4 + Z_5 + R_1 Z_3)(M_1 + Z_4 + Z_5 + R_2 Z_3) \quad [14.5.4\text{-}13]$$

where the roots R_1 and R_2 are given by either eq. 14.5.3-13 or eq. 14.5.3-14. It should be pointed out that all the conditions covered in eqs. 14.5.3-15 through 14.5.3-26 apply to third level compatible systems, too.

Similarly to the treatment in eqs. 14.5.3-27 and 14.5.3-28, we may write a compatibility inequality for third level systems.

$$\frac{Z_4 + Z_5}{Z_3} \geq (b - 1) \qquad [14.5.4\text{-}14]$$

Similarly to the procedure that led to eqs. 14.5.3-31 the system has special properties when

then for all Z_3

$$\left.\begin{array}{c} b = 1 \\[1.2em] R_2 = 0 \\[1.2em] Z_6 + Z_7 = a(Z_4 + Z_5) \end{array}\right\} \qquad [14.5.4\text{-}15]$$

and

■ **EXAMPLE 14.5.4-1** Design Approach for Broadband System
Considering a general system (electrical, mechanical, fluidic, interdisciplinary, etc.) defined by the system array, eq. 14.5.4-6, formulate the design approach for a broadband system from input X_a to output X_1.

As a secondary objective, if possible, also provide a low pass filter with sharp cutoff to noise X_b. Given:

$$F_1 = F_2 = 0 \tag{1}$$

$$\omega_b = \text{bandwidth} \tag{2}$$

$$\epsilon = \text{allowable departure from flat curve} \tag{3}$$

$$M_1 = M_1 s^2 \tag{4}$$

$$M_2 = M_2 s^2 \tag{5}$$

$$Z_i = D_i s + K_i \tag{6}$$

$$\text{numerical value of any one element} \tag{7}$$

$$\text{constraints on networks} \tag{8}$$

$$\text{constraints on damping ratios} \tag{9}$$

Solution. Solve for X_1 by using Cramer's rule on the system array. Assume the system can be made compatible and factor the characteristic determinant.

$$X_1 = \frac{a(X_a Z_4)(M_1 s^2 + Z_4 + Z_5 + (b/a)Z_3) + Z_3 Z_6 X_b}{(M_1 s^2 + Z_4 + Z_5 + R_1 Z_3)(a)(M_1 s^2 + Z_4 + Z_5 + R_2 Z_3)} \tag{10}$$

Normalize

$$X_1 = \frac{a[X_a D_4(s + \omega_4)][s^2 + 2\zeta_a \omega_a s + \omega_a^2] + X_b D_3 D_6(s + \omega_3)(s + \omega_6)}{a(s^2 + 2\zeta_1 \omega_1 s + \omega_1^2)(s^2 + 2\zeta_2 \omega_2 s + \omega_2^2)} \tag{11}$$

where

$$\omega_a^2 = \frac{K_4 + K_5 + (b/a)K_3}{M_1} \tag{12}$$

$$\omega_1^2 = \frac{K_4 + K_5 + R_1 K_3}{M_1} \tag{13}$$

$$\omega_2^2 = \frac{K_4 + K_5 + R_2 K_3}{M_1} \tag{14}$$

$$2\zeta_a \omega_a = \frac{D_4 + D_5 + (b/a)D_3}{M_1} \tag{15}$$

$$2\zeta_1 \omega_1 = \frac{D_4 + D_5 + R_1 D_3}{M_1} \tag{16}$$

$$2\zeta_2 \omega_2 = \frac{D_4 + D_5 + R_2 D_3}{M_1} \tag{17}$$

and where

$$\omega_3 = \frac{K_3}{D_3} \tag{18}$$

$$\omega_4 = \frac{K_4}{D_4} \tag{19}$$

$$\omega_6 = \frac{K_6}{D_6} \tag{20}$$

Let

$$N_i = \frac{\omega_i}{\omega_0} \quad \text{frequency ratios} \tag{21}$$

where

$$\omega_0^2 = \frac{K_3}{M_1} \quad \text{scaling frequency*} \tag{22}$$

note

$$\frac{D_3}{M_1} = \frac{K_3}{M_1}\frac{D_3}{K_3} = \frac{\omega_0^2}{\omega_3} \tag{23}$$

Refer to Fig. 14.5.4-1, the frequency response curves. It is clear that the response to the *input* is *broadband*, which can be extended as shown by the propitious choice of frequencies. At the same time, without compromising the major objective of this design, it is clear from the figure that the system is a *low pass filter* to *noise*. So with cautious optimism, we may state that there is a possibility to achieve both objectives.

Apply eq. (22) to eq. (13).

$$\frac{K_4 + K_5}{K_3} = \frac{\omega_1^2}{\omega_0^2} - R_1 = N_1^2 - R_1 \tag{24}$$

Apply eq. (22) to eq. (14).

$$\frac{K_4 + K_5}{K_3} = \frac{\omega_2^2}{\omega_0^2} - R_2 = N_2^2 - R_2 \tag{25}$$

Since $K_4 + K_5$ cannot be negative, then

$$N_1^2 \geq R_1, \qquad N_2^2 \geq R_2 \tag{26}$$

Subtract eq. (14) from eq. (12) and apply eqs. (4), (21), and (23).

$$R_1 - 1 = \frac{\omega_a^2 - \omega_2^2}{\omega_0^2} = N_a^2 - N_2^2 \tag{27}$$

---

*This provides a rough scale, obtained by the ratio of the two most significant terms in the system.

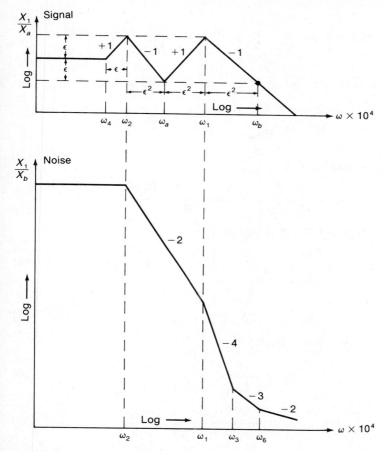

Figure 14.5.4-1   Compatible broadband system.

Similarly, subtract eq. (12) from eq. (13) and apply eqs. (4), (21), and (23).

$$R_2 - 1 = \frac{\omega_a^2 - \omega_1^2}{\omega_0^2} = N_a^2 - N_1^2 \tag{28}$$

Apply eqs. (27) and (28) to eq. 14.5.3-19.

$$a = \frac{1}{\left(N_a^2 - N_2^2\right)\left(N_1^2 - N_a^2\right)} \tag{29}$$

The scaling frequency ω_0 is arbitrary, but a good choice will facilitate the computations and design.
CHOICE #1

$$\omega_2 \leq \omega_0 \leq \omega_1 \tag{30}$$

CHOICE #2

Corresponding to $b = 1$ and $R_2 = 0$, set $R_2 = 0$ in eq. (28) and apply eq. (21).

$$\omega_0^2 = \omega_1^2 - \omega_a^2 \tag{31}$$

Once ω_0 has been chosen, use eq. (23) to determine K_3 and apply this to eq. (18) to determine D_3.

$$K_3 = M_1 \omega_0^2, \qquad D_3 = \frac{K_3}{\omega_3} \tag{32}$$

Subtract eq. (17) from eq. (15).

$$2\zeta_a \omega_a - 2\zeta_2 \omega_2 = \frac{(b/a - R_2)D_3}{M_1} \tag{33}$$

Substitute eqs. 14.5.3-15 and (23) into eq. (33).

$$R_1 - 1 = \frac{2\omega_3}{\omega_c^2}(\zeta_a \omega_a - \zeta_2 \omega_2) = 2N_3(\zeta_a N_a - \zeta_2 N_2) \tag{34}$$

Similarly,

$$R_2 - 1 = 2N_3(\zeta_a N_a - \zeta_1 N_1) \tag{35}$$

Equate eq. (27) to eq. (34), solve for N_3.

$$N_3 = \frac{N_a^2 - N_2^2}{2(\zeta_a N_a - \zeta_2 N_2)} \tag{36}$$

$$= \frac{N_a^2 - N_1^2}{2(\zeta_a N_a - \zeta_1 N_1)} \tag{37}$$

$$= \frac{N_1^2 - N_2^2}{2(\zeta_1 N_1 - \zeta_2 N_2)} \tag{38}$$

Note that the above set of equations has three unknowns, the three damping ratios. However, only two of the equations are independent, the third being derivable from the other two: There are three unknowns, but only two independent equations. It will be necessary (an asset rather than a liability) to provide a constraint, such as the minimum damping ratio. However, it is not obvious which will be minimum. Consequently, it will be necessary to try each one and solve for the other two, until the constraint is satisfied. Once this has been accomplished, then solve either eq. (24) or eq. (25) for K_4.

$$K_4 = K_3(N_1^2 - R_1) - K_5$$
$$K_4 = K_3(N_2^2 - R_2) - K_5 \tag{39}$$

Apply eqs. (18) and (22) to eq. (16).

$$\frac{D_4 + D_5}{D_3} = \frac{2\zeta_1 \omega_1 \omega_3}{\omega_c^2} - R_1 = 2\zeta_1 N_1 N_3 - R_1 \tag{40}$$

Apply eqs. (18) and (22) to eq. (17).

$$\frac{D_4 + D_5}{D_3} = \frac{2\zeta_2\omega_2\omega_3}{\omega_c^2} - R_2 = 2\zeta_2 N_2 N_3 - R_2 \tag{41}$$

Since $D_4 + D_5$ cannot be negative, then

$$2\zeta_1 N_1 N_3 > R_1, \qquad 2\zeta_2 N_2 N_3 > R_2 \tag{42}$$

If eq. (42) is satisfied, then proceed with the design. Solve either eq. (40) or eq. (41) for D_4.

$$D_4 = D_3(2\zeta_1 N_1 N_3 - R_1) - D_5$$
$$D_4 = D_3(2\zeta_2 N_2 N_3 - R_2) - D_5 \tag{43}$$

It should be pointed out that there are four unknowns, K_4, K_5, D_4, and D_5, but there are only three equations, (19), (39), and (43). Once again, we are entitled to make use of a constraint. Usually, a minimum value is implied by practical considerations. Otherwise, an element may be omitted (set = 0). Either way, the treatment is fairly direct. One important constraint is that frequency ω_4 is needed for the frequency response curve. Hence,

$$K_4 > 0, \qquad D_4 > 0 \tag{44}$$

The design strategy for D_5 and K_5 requires that, first, their minimum values be employed in eqs. (39) and (43). Use the computed values of D_4 and K_4 to compute a trial value of ω_4 by using eq. (19). Compare the trial value to the actual value. Then increase either D_5 or K_5 accordingly.

For the remaining terms, use the compatibility equation, eq. 14.5.4-9.

$$\frac{K_6 + K_7}{K_3} = a\frac{K_4 + K_5}{K_3} + b - 1 \tag{45}$$

$$\frac{D_6 + D_7}{D_3} = a\frac{D_4 + D_5}{D_3} + b - 1 \tag{46}$$

Substitute eqs. (24), 14.5.3-16, and (35) into eq. (45).

$$\frac{K_6 + K_7}{K_3} = a(N_1^2 - R_1) + a(R_1 + R_2 - 1) - 1$$

$$= aN_a^2 - 1 \tag{47}$$

Since $K_6 + K_7$ cannot be negative, then

$$aN_a^2 > 1 \tag{48}$$

If this test is satisfied, continue with the design. Solve for K_6 in eq. (47).

$$K_6 = K_3(aN_a^2 - 1) - K_7 \tag{49}$$

Similarly,

$$\frac{D_6 + D_7}{D_3} = a(2\zeta_1 N_1 N_3 - R_1) + a(R_1 + R_2 - 1) - 1 \tag{50}$$

$$= a2\zeta_a N_a N_3 - 1 \tag{51}$$

Since $D_6 + D_7$ cannot be negative,

$$a2\zeta_a N_a N_3 > 1 \tag{52}$$

If this test is satisfied, continue with the design. Solve for D_6 in eq. (50)

$$D_6 = D_3(2a\zeta_a N_a N_3 - 1) - D_7 \tag{53}$$

Once again, we have four unknowns, D_6, D_7, K_6, and K_7, but only three equations, (20), (49), and (53). Once again we are entitled to employ a constraint, as was done for eqs. (19), (39), and (43).

Recapping the design procedure, recall that it began with the identification of seven frequencies on the frequency response curves. A choice was made for the scaling frequency. There was one constraint on damping ratio and two on networks. With these eleven conditions, we have solved for eleven unknowns—not simultaneously, but one at a time. The system has met all requirements, and it has satisfied all inequalities. The design is complete.   ∎

SUGGESTED REFERENCE READING FOR CHAPTER 14

Modes of vibration [1], p. 225, 229, 306; [5], p. 164–169; [9], p. 103–112; [16], p. 96–108, 174–176; [17], chap. 12; [18], p. 257; [44], p. 170, 171, 196.
Computer methods [4], chap. 7; [47], p. 410–431.
Principal coordinates [4], p. 147; [5], p. 171; [16], p. 105, 135; [18], p. 263, 281, 307; [44], p. 170.
Transient vibration for undamped systems [1], p. 243–255; [17], chap. 14; [44], p. 306, 307.
Forced vibration in undamped systems [1], p. 256–260; [4], p. 129–132; [5], p. 175–179; [16], p. 114–116, 180–186; [17], chap. 15; [18], p. 269–278; [44], p. 171–178; [46], p. 194–198; [17], p. 401–410.
Damped high order systems [1], p. 270–288, 318–324; [4], p. 126–132; [5], p. 195–198, 233–238; [9], p. 119–133, 165–168; [16], p. 129–133; [17], p. 423–425, 447–456.
Optimum polynomials [20], p. 280–288; [26], p. 102–104.
Butterworth filter [20], p. 281.
ITAE criterion [20], p. 280–288; [26] p. 102–104.
Dynamic vibration absorber [1], p. 261–268; [4], p. 132; [5], p. 183–187; [9], p. 112–132; [17], p. 525–528; [44], p. 178–190.

PROBLEMS FOR CHAPTER 14

14.1. Using either an energy method or Lagrange's equations, show that the masses of an undamped system move either in phase or 180° out of phase.

14.2. Find the natural frequencies of a system whose free body diagram is given by Fig. P-12.19(a), where

$$M_1 = M_2 = M, \qquad K_1 = K_2 = K_3 = K, \qquad D_1 = 0$$

14.3. For Problem 14.2 let the initial condition be displacement with all other initial conditions and inputs equal to zero.
   (a) Show that the two natural frequencies are $\sqrt{K/M}$ and $\sqrt{3K/M}$.
   (b) If the initial displacements are equal and in the same direction, show that the system will vibrate in its lower frequency mode.
   (c) If the initial displacements are equal in magnitude but opposite in direction, show that the system will vibrate in its higher frequency mode.

14.4. Rework Problem 14.3 where $K_c \neq K_1$, but $K_1 = K_2$. Ans. $\omega_1 = \sqrt{K/M}$, $\omega_2 = \sqrt{(2K_c + K)/M}$.

14.5. Find the natural frequencies and sketch the frequency response for Problem 14.4 where

$$K_c = 400 \text{ N/m}, \qquad K_1 = K_2 = 100 \text{ N/m}, \qquad M_1 = M_2 = 100 \text{ kg}$$

14.6. Determine the transient response for Problem 14.5 when the inputs are zero and the only initial condition is an initial velocity $v_{10} = 0.1$ m/s.

14.7. In Ex. 14.2.4-1 (Dynamic Absorber for Reciprocating Pump), suppose the motor speed is equal to 52.9 rad/s and suppose that neither M_1 or K_1 can be altered.
   (a) If no dynamic absorber is used, explain the dilemma.
   (b) How does the use of an additional mass M_2 and spring K_2 alleviate the problem?
   (c) Why must the resonant frequencies be remote from the operating speed by at least 20%?

14.8. Redesign the dynamic absorber of Ex. 14.2.4-1 for 25%, 30%, 40%, and 50% remote location of the resonances. Compare results and list any conclusions.

14.9. *Design of Dynamic Absorber Using Limited Space.* A delicate instrument requires a small reciprocating pump, a miniature version of the system shown in Fig. 7.9.1-1. The unit is mounted on four springs $K_{1'}$, and is driven by a motor whose speed can vary by $\pm 30\%$ from its nominal speed of 600 rpm (10 Hz or 62.8 rad/s). Using the small space under the base (about $3\frac{1}{2}$ cm $\times$ $3\frac{1}{2}$ cm $\times$ 2 cm), design the dynamic absorber given the following:

$$F_1 = \text{acceleration force in pump} = 10^5 \text{ dyn}$$

$$M_1 = \text{mass of pump base and motor} = 100 \text{ g}$$

$$K_1 = \text{equivalent spring of four springs} = 2.8 \times 10^5 \text{ dyn/cm}$$

$$r_1 \text{ and } r_2 \text{ to be at least 40\% from } \omega_2$$

14.10. *Design of Dynamic Absorber for Zero Transmitted Force.* There are a number of devices that employ vibrators (such as hair clippers, shavers, sanders, saber saws, and so forth) for which vibration to the foundation is objectionable, but where the primary mass must move in order to accomplish its task. Hence, the technique in Section 14.2.4 is not applicable (since there, the *motion* of the primary mass was reduced to zero). In this problem, a different approach is required. Here, a third spring must be employed which connects the absorber mass to the foundation, where the sum of *forces* will be set equal to zero. The unknowns are K_1, K_2, K_3, M_2. Formulate a design procedure.

14.11. *Design Absorber for Hair Clipper.* A typical hair clipper consists of two sets of blades, each with a number of teeth. The blades are placed in sliding contact.

See Fig. P-14.11. One blade remains stationary, while the other translates. This produces a shearing action. The moving blade is actuated by a solenoid, which, through its rectification action, behaves like a frequency doubler (similar to the device in Ex. 9.3.1-1). To nullify the annoying vibration to the hand, a dynamic absorber is installed to vibrate in phase-opposition with the center of mass of the moving blade. The excitation frequency is 120 Hz. Design the complete system.
Given:

$$M_1 = \text{total moving mass} = M_b + \frac{M_c}{4} = 100 \text{ g}$$

$$F_1 = 3 \times 10^7 \text{ dyn (input force)}$$

$$X_2 = \text{allowable motion of absorber mass} = 1.5 \text{ cm}$$

$$X_1 = \text{required motion of primary mass} = 0.5 \text{ cm}$$

$$\omega_{2'} = \text{input frequency} = 120 \text{ Hz} = 754 \text{ rad/s}$$

$r_{1'}$ and $r_{2'}$ resonant frequency ratios to be
at least 20% remote from input frequency

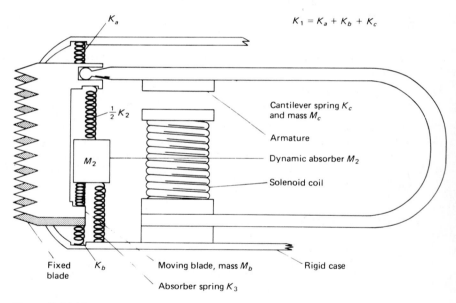

Figure P-14.11

14.12. Redesign the system of Ex. 14.2.4-2 (Band Eliminator for Alarm System) for 25%, 30%, 40%, and 50% remote location of resonances. List conclusions.

14.13. Refer to Ex. 12.3.1-2 (Superheterodyne Interstage Amplifier). List the requirements to make the system a proportional one. Design the system given the following:

$$R_r = 10 \quad \text{(resonant peak)}$$

$$\omega_r = 3.99 \times 10^6 \text{ rad/s} \quad \text{(about 640 kHz)}$$

$$L_1 = 10^{-5} \text{ H}$$

14.14. Similar to Problem 14.13, but applied to Ex. 12.3.1-1 (Motor and Pump), given the following:

$$J_1 = 0.5 \text{ kg} \cdot \text{m}^2, \qquad D_1 = 150 \text{ N} \cdot \text{m} \cdot \text{s}, \qquad K_1 = 45{,}000 \text{ N} \cdot \text{m}$$

14.15. In Fig. 12.3.1-1 (Motor and Pump), the two rotor masses are equal. The following parameters are given: $J = 0.5 \text{ kg} \cdot \text{m}^2$, $D = 150 \text{ N} \cdot \text{m} \cdot \text{s}$, and $K = 45{,}000 \text{ N} \cdot \text{m}$. Sketch frequency response.

14.16. (a) Plot the frequency response for the motor generator in Fig. P-14.16.
 (b) Indicate how this could have been done without factoring.
 (c) Find the transient response to a suddenly applied torque of 100 N · m.
 (d) Determine the maximum displacement of the spring coupling. Note that this is given by $\theta_2 - \theta_1$.

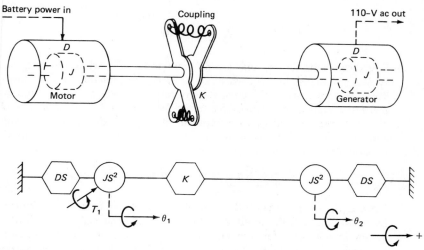

Figure P-14.16

14.17. For each of the following system arrays, make the appropriate tests for compatibility and solve for K.
 (a)

X_1	X_2	Inputs
$Ms^3 + 11s + 23$	$-(8s + 14)$	F_1
$-(8s + 14)$	$2Ms^3 + 30s + K$	F_2

 (b)

I_1	I_2	Inputs
$Ls^2 + 16s + 10$	$-(5s + 7)$	V_1
$-(5s + 7)$	$2Ls^2 + 37s + K$	V_2

14.18. Prove that all undamped systems of any order are compatible systems.

14.19. Prove that all systems with proportional damping of any order are compatible systems.

14.20. Prove that all symmetric systems of any order are compatible systems.

14.21. For a second level compatible system, arbitrarily select all the system parameters but K_2. Let $M_i = J_i s^2$.
(a) Determine K_2 for compatibility.
(b) Factor the characteristic determinant.
(c) Expand the original arbitrary system and compare to the factored form.

14.22. Similar to Problem 14.21, but applied to a third level compatible system.

14.23. *Arbitrary M Functions for Second Level.* The text problems have been careful to make the M functions proportional to one another. For this problem, let

$$M_i = J_i s^2 + L_i s + I_i$$

$$Z_i = D_i s + K_i$$

Select constants for M_1 that are different from M_2. Arbitrarily select all the system parameters but K_1, where any may be zero but finite and nonnegative. Test the system using the compatibility inequality, eq. 14.5.3-28. Find K_1 to make the system compatible and factor the characteristic determinant. *Hint*: You will need a strategy to manipulate the M_2 function to comply with the compatible model. To do this, first compute the constant $a = J_2/J_1$. Then define a manipulating network Z_b as follows:

$$Z_b = L_b s + I_b = (L_2 - aL_1)s + (I_2 - aI_1)$$

Since Z_b is not a real network, some of the terms may be negative. Subtract Z_b from M_2 and call the result M_2'. Add Z_b to Z_2 and call the result Z_2'. The system is now ready for compatibility. Observe the caution below eq. 14.5.3-26 for negative terms. Note that you have arbitirarily selected eleven parameters and have solved for one.

14.24. *All Entries of Array Are Quadratics.* Similar to Problem 14.23, but where

$$Z_i = H_i s^2 + D_i s + K_i$$

The order of networks Z_i is 2. Then there are two parameters to be found. Find K_1 and D_1.

14.25. *Eighth Order Second Level Arbitrary Compatible System.* Similar to Problem 14.23, but where

$$M_i = J_i s^4 + L_i s^3 + J_i s^2 + H_i s + G_i$$

$$Z_i = \text{same as that used in Problem 14.24}$$

14.26. *Arbitrary M Functions for Third Level.* Similar to Problem 14.23, but applied to third level. Note that you will select fifteen parameters and solve for one.

14.27. Given the circuit in Fig. P-14.27, formulate the system array in the form of eq. 14.5.4-6. What are the conditions for compatibility? *Hint:* Break up the circuit into two parts. Each part will become a quadratic form. Then tie the two together with appropriate coupling networks.

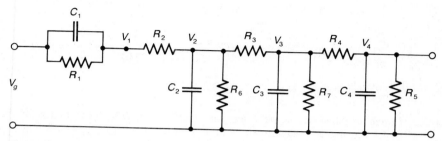

Figure P-14.27

14.28. *Transistor Amplifier With Various Input Networks.* Refer to the circuit in Fig. P-14.28, which is a variation of Fig. 12.7.5-1. Determine the frequency response for each of the following input networks:

(a) Pure resistance R_p.

(b) Pure capacitance C_1.

(c) Combination R_p-C_1 parallel network.

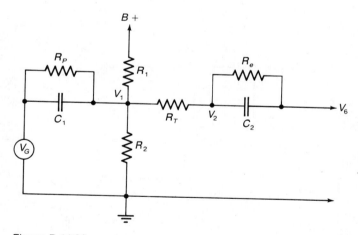

Figure P-14.28

14.29. *Compatible Superheterodyne Interstage Amplifier.* Refer to Ex. 12.3.1-2 (Superheterodyne Interstage Amplifier).

(a) Determine C_1 for compatibility.

(b) Write the characteristic determinant in factored form.

(c) Determine the natural frequencies and damping ratios.

(d) Sketch the frequency response and discuss.

Given:

$$D_1 = R_1 = 0.35 \ \Omega \qquad K_1 = \frac{1}{C_1} \quad \text{to be determined}$$

$$D_2 = R_2 = 12 \ \Omega \qquad K_2 = \frac{1}{C_2} = \frac{1}{0.00308 \ \mu F} = 32.5 \times 10^7 / F$$

$$D_3 = R_3 = 45.2 \ \Omega \qquad K_3 = \frac{1}{C_3} = \frac{1}{0.01625 \ \mu F} = 61.5 \times 10^7 / F$$

$$L_1 = 10^{-5} \ H \qquad L_2 = 2 \times 10^{-5} \ H$$

14.30. *Design of Superheterodyne Interstage Amplifier.* Redesign the system in Ex. 14.5.3-2 using IF transformers with $1:1$ ratios (thus $a = 1$) and assume that $b = 1$.

14.31. *Mechanical System with Arbitrary M Functions.* Rework Ex. 14.5.3-1 (Compatible Mechanical System), making the following changes in system parameters:

$$Z_a = 20s, \qquad Z_b = 50s + 200$$

Hint: See strategy for Problem 14.23.

14.32. *Design of Broadband Meter Movement.* A meter movement with input θ_a and noise θ_b is shown in Fig. P-14.32. Improve the performance of this device by changing the various components as needed. Design the instrument to meet the

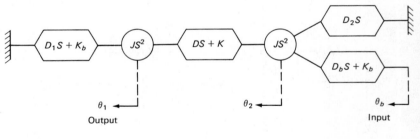

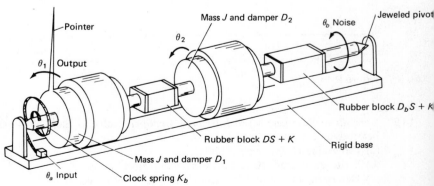

Figure P-14.32

following specifications: Given the bandwidth $\omega_b = 2.8 \times 10^2$ rad/s and the mass moment of inertia $J_1 = 10^{-2}$ dyn $\cdot$ s^2 $\cdot$ cm. Find all the system parameters for maximum bandwidth of the signal θ_a and for maximum filtering of noise θ_b. Assume that the noise can be represented by a random distribution of frequencies, predominantly in the range 1.5×10^2 to 10^3 rad/s. Note that the frequency range of the noise overlaps that of the signal. For stability, let all damping ratios be ≥ 0.5.

14.33. *Design of Transistor Amplifier.* Refer to Ex. 12.7.5-1 (Transistor Amplifier), where a transistor was modeled. Another model appears in Problem 14.28, which will help set up this problem in a form suitable for compatibility. Design the system for a bandwidth $= 1.20 \times 10^5$ rad/s (about 20,000 Hz) flat to within 3 dB, or $\sqrt{2}$. $M_1 = C_1 C_T$, where $C_1 = C_T = 0.118$ μF, or $M_1 = 1.4 \times 10^{-14}$. Design for broadband to input X_a but low pass to noise X_b. For stability, let damping ratios be ≥ 0.5. Use minimum number of elements. Constraint is $K_5 \geq 10^{-6}$.

Chapter **15**

Root Locus

The root locus method* is a graphical means with which to determine the transient properties of a system of any order. The method is tantamount to factoring the characteristic equation for each value of a parameter K that varies from zero to infinity, but without actually factoring.

The root locus method provides easy access to all of the transient properties—an asset to analysis, design, and synthesis. The method accomplishes this by plotting the path or locus of the roots without actually finding the roots themselves. This, in turn, is accomplished by following a set of rules that lead to a reasonably accurate plot in a relatively short time.

15.1 LOCUS OF ROOTS OF THE CHARACTERISTIC EQUATION

15.1.1 Information Provided by the Roots

In Chapters 12 through 14, methods were shown that either formulate or solve the differential equation of a system of any order. In general, the solution in the time domain is of the form

$$x(t) = C_1 e^{p_1 t} + C_2 e^{p_2 t} + \cdots + C_n e^{p_n t} \qquad [15.1.1\text{-}1]$$

---

*See references [59] and [60].

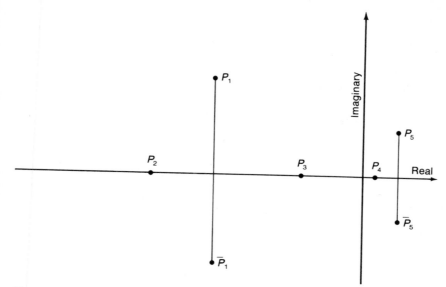

Figure 15.1.1-1   Roots in the complex plane.

where

$$p_i = \text{roots of the characteristic equation} \qquad [15.1.1\text{-}2]$$

In general, the roots can be complex numbers. From the theory of equations, it is known that if any roots are complex, they occur in conjugate pairs, while real roots may occur singly. If the real part of any root is positive, then the time domain solution for that particular term will grow without bound. Note that the remaining roots may be negative, thus having time domain solutions that approach limits, but if even one term is positive, the time domain solution grows without bound. Such a system is said to be unstable.

System stability can be determined by inspection of the roots. See Fig. 15.1.1-1. Roots P_1, $\overline{P}_1$ and P_5, $\overline{P}_5$ are conjugate pairs. Roots P_2, P_3, and P_4 are real. Roots P_4 and P_5, $\overline{P}_5$ have positive real parts, which means the solution will grow without bound, and the system is unstable.

In this chapter, we will consider only stable systems. Consequently, there is no need to construct the positive real half-plane, since any roots that lie there indicate an unstable system.

The task to be accomplished by the root locus method is to determine the closed loop properties of a system of any order from its open loop roots. From Chapter 13, we learned that the open loop is always in factored form, making it an easy task to determine the open loop roots.

Before we examine the root locus method, we will compute the closed loop roots and determine the transient information. This is not to be interpreted as a conflict with the ultimate goal. Rather, it provides two things: a preview of what the root locus will ultimately accomplish and a check on each bit of information as it is learned. This approach will help establish confidence in the root locus method and will help make comparisons. But in the end, the root locus will be used without the need to first find closed loop roots.

15.1.2 Influence of Damping Upon the Root Locus

In Fig. 15.1.1-1, the roots were static—they stayed in place, thus representing only one condition of the system. Now let us consider a system for which one parameter will be varied, but let us consider a system for which we know where all of its roots will lie, such as the second order system whose block diagram is modeled in Fig. 15.1.2-1. For this system, the damping ratio has been deliberately isolated so that we may study the effect of varying this particular parameter. We will track the roots as we vary the damping ratio from zero to infinity.

Use the closed loop rule, covered in Section 13.2.4, on Fig. 15.1.2-1.

$$\frac{X}{X_{in}}(s) = \frac{2\zeta\omega_n s}{s^2 + \omega_n^2 + 2\zeta\omega_n s} \qquad [15.1.2\text{-}1]$$

for which the characteristic equation is

$$s^2 + 2\zeta\omega_n s + \omega_n^2 = 0 \qquad [15.1.2\text{-}2]$$

or, solving for damping ratio ζ,

$$\frac{s^2 + \omega_n^2}{2\omega_n s} = -\zeta \qquad [15.1.2\text{-}3]$$

We will start the plot with $\zeta = 0$, in which case,

$$s^2 + \omega_n^2 = 0 \quad (\text{for } \zeta = 0) \qquad [15.1.2\text{-}4]$$

or

$$s = \pm j\omega_n \qquad [15.1.2\text{-}5]$$

The first pair of roots start on the imaginary axis. By the way, there is no point in considering damping ratio $\zeta < 0$ for this will produce roots in the positive real half-plane, indicating an unstable system. See Fig. 15.1.2-2. Since complex numbers occur in conjugate pairs, it will be convenient to perform all the computations in the upper imaginary half-plane and then make a mirror image copy for the conjugate in the lower imaginary half-plane.

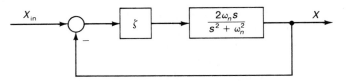

Figure 15.1.2-1 Second order system with isolated damping ratio.

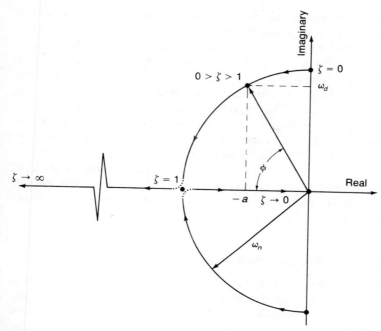

Figure 15.1.2-2   Root locus as damping ratio is varied.

The damping ratio ζ will be varied while holding the undamped natural frequency ω_n fixed. Note that for values of damping between zero and unity the coordinates of the point are

$$-a \quad \text{and} \quad \omega_d \qquad [15.1.2\text{-}6]$$

Refer to entry 6, Appendix 1,

$$a^2 + \omega_d^2 = \omega_n^2 \qquad [15.1.2\text{-}7]$$

Since ω_n is held fixed, then eq. 15.1.2-7 is the equation of a circle with radius r, where

$$r = \omega_n \qquad [15.1.2\text{-}8]$$

The radius vector r makes an angle ϕ with the negative real axis, where (in Appendix 1)

$$\cos \phi = a/\omega_n = \zeta \qquad [15.1.2\text{-}9]$$

$$\sin \phi = \omega_d/\omega_n = \beta \qquad [15.1.2\text{-}10]$$

$$\beta = \sqrt{1 - \zeta^2} \qquad [15.1.2\text{-}11]$$

As the damping ratio ζ increases from zero to unity, the two roots trace paths, or loci, in the form of circular arcs. At the point where the damping ratio is equal to unity, the two loci meet, indicating that the two roots are identical (repeated roots). If the damping ratio is increased further (over-damped) we know in advance that the roots will be real. From the theory of equations, we also know that the product of the roots is equal to the negative of the coefficient of the lowest power of s. Referring to the characteristic equation, eq. 15.1.2-2, such a term is ω_n^2. For this example, ω_n is held constant. Thus, the product of the roots is constant. Then, upon increasing the damping ratio, one root will increase, while the other decreases. The locus of each root will remain on the negative real axis, but one will approach $-\infty$, while the other approaches zero. (For this example, it is not important which root goes where. For the sake of the drawing, the upper root was selected as the one that approaches $-\infty$.) The point where the two loci join the negative real axis is called a *break-in point* and will be covered in detail in a subsequent section. The portions of the loci that lie on the negative real axis are called *segments* and will be covered in a subsequent section.

15.1.3 Influence of Other Parameters Upon Root Locus

Refer to Fig. 15.1.2-2. The general coordinates of a point in the complex plane are given in eq. 15.1.2-6. The damping ratio is related to angle ϕ by eq. 15.1.2-9. Consulting the figure and the two equations, we may draw the following conclusions:

(a) A small angle ϕ implies large damping ratio.
(b) Far from the real axis implies large damped frequency ω_d.
(c) A long radius vector implies high natural frequency ω_n.
(d) Far from the imaginary axis implies large decay coefficient a, which in turn implies short settling time.
(e) Close to the imaginary axis implies instability.

15.1.4 Roots of a Cubic System

Consider the system* modeled by a block diagram in Fig. 15.1.4-1. Its closed loop transfer function (refer to Section 13.2.4) is given by

$$\frac{\theta}{\theta_{\text{in}}}(s) = \frac{K}{s(s^2 + 570s + 115{,}000) + K} \qquad [15.1.4\text{-}1]$$

The characteristic equation becomes

$$s^3 + 570s^2 + 115{,}000s + K = 0 \qquad [15.1.4\text{-}2]$$

which is a cubic. Let us examine the locus of each of the three roots of this cubic as we vary K. For each value of K the characteristic equation can be

---

*This system is an electromechanical servo that will be derived in Chapter 16, Automatic Controls, Ex. 16.3.4-1.

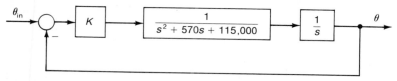

Figure 15.1.4-1   Block diagram of a cubic system.

factored as follows:

$$\Delta = (s + r_1)(s + r_2)(s + r_3) \qquad [15.1.4\text{-}3]$$

whose roots are

$$-r_1, -r_2, \text{ and } -r_3 \qquad [15.1.4\text{-}4]$$

and where two of the roots may be complex. There are various means by which the cubic can be factored and the roots obtained. The results are tabulated in Table 15.1.4-1 and are plotted in Fig. 15.1.4-2. From the figure, we note that one root locus remains on the negative real axis for all values of K. The other two start toward each other, each making an angle of about $-57°$ with the horizontal. This starting angle is called the *angle of departure* and will be covered in a subsequent section. When $K = 65.5 \times 10^6$, the locus crosses the imaginary axis. Hence, we conclude that this corresponds to the verge of instability—a value of K larger than this will result in an unstable system. Beyond this, the two loci approach straight line asymptotes. The subjects of stability and asymptotes will be discussed in subsequent sections.

We have completed the study of a special cubic system. Let us recap what has been learned.

In this study of a cubic system, the roots were determined first, and the loci plotted from the roots. From the plot, certain properties of the system were ascertained. However, the roots are not always easy to find, particularly

TABLE 15.1.4-1   ROOTS OF A CUBIC

Point	$K(10^6)$	r_1	r_2 and r_3	ζ	ω_n (rad / s)
0	0	0	$-285 \pm j184$	0.841	339
1	2.08	-20	$-275 \pm j168$	0.853	322
2	4.45	-50	$-260 \pm j146$	0.872	298
3	6.8	-100	$-235 \pm j113$	0.901	261
4	7.8	-150	$-210 \pm j89$	0.921	228
5	8.2	-200	$-185 \pm j82$	0.914	202
6	8.75	-250	$-160 \pm j97$	0.855	187
7	10.2	-300	$-135 \pm j126$	0.732	184
8	15.2	-370	$-100 \pm j176$	0.494	202
9	27.5	-450	$-60 \pm j240$	0.243	247
10	40.0	-500	$-35 \pm j281$	0.124	283
11	65.5	-570	$0 \pm j339$	0	339

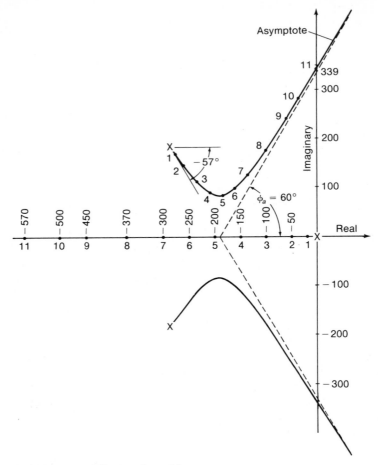

Figure 15.1.4-2 Roots of a cubic.

for design where numerical values and types of terms are not yet known. The objective is to construct the root locus plot *without* first knowing the roots. Paradoxical as it sounds, this is possible and is called the *root locus method*.

15.2 ROOT LOCUS METHOD

15.2.1 The Factor *K*

For the general system of any order, the characteristic equation may be written in the form;

$$C(s) = \frac{P(s)}{Z(s)} = -K \qquad [15.2.1\text{-}1]$$

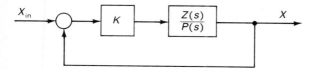

Figure 15.2.1-1   Block diagram representation of the Evans equation.

called the Evans equation,* where:

$P(s)$ and $Z(s)$ are polynomials that may be expressed as products of first order factors   [15.2.1-2]

K is a real positive number that varies from zero to infinity   [15.2.1-3]

It will prove instructive to create a block diagram representation of the Evans equation. Refer to Fig. 15.2.1-1. The open loop gain L is, by inspection,

$$L(s) = \frac{KZ(s)}{P(s)} \qquad [15.2.1-4]$$

and the closed loop becomes

$$\frac{X}{X_{in}}(s) = \frac{KZ(s)}{P(s) + KZ(s)} \qquad [15.2.1-5]$$

for which the characteristic equation is

$$P(s) + KZ(s) = 0 \qquad [15.2.1-6]$$

or

$$\frac{P(s)}{Z(s)} = -K \qquad [15.2.1-7]$$

which is identical to eq. 15.2.1-1. Hence, Fig. 15.2.1-1 may be considered an accurate representation of the Evans equation.

Considering eqs. 15.2.1-1 and 15.2.1-2, we may write the Evans equation in its factored form.

$$\frac{(s + P_1)(s + P_2) \cdots (s + P_n)}{(s + Z_1)(s + Z_2) \cdots (s + Z_m)} = -K \qquad [15.2.1-8]$$

Since we will be making a study of the system as we vary K from zero to infinity, let us consider the two limiting values of K in eqs. 15.2.1-1, 15.2.1-4, and 15.2.1-8. It is clear that when

$$K = 0, \infty \qquad [15.2.1-9]$$

---

*See references [59] and [60].

we may determine the roots

$$
\begin{array}{c}
-P_1, -P_2, \ldots, -P_n \\
-Z_1, -Z_2, \ldots, -Z_m
\end{array}
\qquad [15.2.1\text{-}10]
$$

but these roots are also the roots of the open loop $L(s)$ in eq. 15.2.1-4. Hence, these special roots, those corresponding to the limiting values of K (zero and infinity) will be called *open loop roots*.

It will subsequently prove convenient to distinguish between the open loop roots of the numerator and those of the denominator, as follows:

$$
\left.\begin{array}{l}
\text{when } K = 0 \\
\text{open loop roots } -P_1, -P_2, \ldots, -P_n \\
\text{will be labeled with an ex, X}
\end{array}\right\}
\qquad [15.2.1\text{-}11]
$$

$$
\left.\begin{array}{l}
\text{when } K = \infty \\
\text{open loop roots } -Z_1, -Z_2, \ldots, -Z_m \\
\text{will be labeled with a circle, O}
\end{array}\right\}
\qquad [15.2.1\text{-}12]
$$

Note that any (or all) open loop root(s) may be complex.

15.2.2 Vector Notation for Complex Factors

Refer to Fig. 15.2.2-1(a). Let $-\alpha$ be the location of an X (open loop root for $K = 0$), which in general is a complex number. This point may be expressed in either Cartesian or polar coordinates. Using the latter, draw a vector from the origin to the point, and label this vector $-\bar{\alpha}$. Now consider a general point s in the complex s plane. Draw a vector from the origin to this point and label it $\bar{s}$.

Given two vectors, we may add or subtract them vectorially. Let us subtract vector $-\bar{\alpha}$ from $\bar{s}$. The result is a vector that is drawn from point $-\alpha$ to the point s. We have subtracted vector $-\bar{\alpha}$ from $\bar{s}$. But this is the same as *adding* vector $+\bar{\alpha}$ *to* $\bar{s}$. Hence, the vector drawn from the open loop root X to the general point s represents a complex factor $(s + \alpha)$ in vector notation:

$$
\overline{s + \alpha} = \text{vector from X to } s
\qquad [15.2.2\text{-}1]
$$

or, more specifically,

$$
\left.\begin{array}{l}
\text{a complex factor } (s + \alpha) \\
\text{where } -\alpha = \text{open loop root} \\
\text{may be represented by a vector } \overline{s + \alpha} \\
\text{drawn from point } -\alpha \text{ to } s
\end{array}\right\}
\qquad [15.2.2\text{-}2]
$$

A similar approach applies to a O (open loop root when $K = \infty$), shown in Fig. 15.2.2-1(b) at the point $-\beta$.

$$
\overline{s + \beta} = \text{vector from O to } s
\qquad [15.2.2\text{-}3]
$$

$$
= \text{vector representation of} \qquad [15.2.2\text{-}4]
$$
$$
\text{the complex factor } (s + \beta)
$$

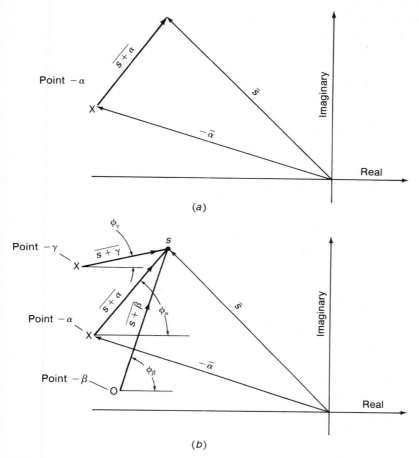

Figure 15.2.2-1   Vector notation for complex numbers. (a) Single quantity. (b) Three quantities.

Apply a similar approach to a third factor $(s + \gamma)$ for an X, as shown in Fig. 15.2.2-1(b).

$$\overline{s + \gamma} = \text{vector representation of} \qquad\qquad [15.2.2\text{-}5]$$
$$\text{the complex factor } (s + \gamma)$$

Note that each of the vectors associated with eqs. 15.2.2-2, 15.2.2-4, and 15.2.2-5 has a magnitude (equal to the length of the vector) and an angle (measured from the horizontal). This makes it possible to multiply and divide complex factors graphically. The rules are the same as those for complex numbers in polar form, as covered in Appendix 5: for a product, multiply magnitudes and add angles; for division, divide magnitudes and subtract angles.

■ **EXAMPLE 15.2.2-1** Products and Quotient of Complex Factors
Consider the complex quantity $W(s)$ given by the following:

$$W(s) = \frac{(s + \alpha)(s + \gamma)}{(s + \beta)} \tag{1}$$

where

$$s = \text{general point in complex plane as shown} \tag{2}$$
in Fig. 15.2.2-1(b)

$$-\alpha, -\gamma, -\beta \text{ are open loop roots as shown in} \tag{3}$$
Fig. 15.2.2-1(b)

Determine the complex quantity $W(s)$ in vector form.

Solution. $W(s)$ has a magnitude given by

$$|W| = \frac{|s + \alpha| \cdot |s + \gamma|}{|s + \beta|} \tag{4}$$

and an angle equal to

$$\phi_w = \phi_\alpha + \phi_\gamma - \phi_\beta \tag{5}$$

■

15.2.3 Angle Criterion

The Evans Equation was written in factored form in eq. 15.2.1-8, where the open loop roots were listed in eq. 15.2.1-10. Section 15.2.2 demonstrated how to represent complex factors in vector notation. Here, we will put all of this together.

Consider a general point s on the root locus in Fig. 15.2.3-1. Draw vectors from each open loop root to the point s. Since we know the point lies on the root locus, we know it will satisfy the Evans equation. To demonstrate this, draw a vector from each open loop root to the point s. Measure the length and angle of each vector. Then, using the rules for multiplying and dividing, the resulting complex number will have a magnitude and angle as follows:

$$K = \text{magnitude}, \qquad 180° = \text{angle} \qquad [15.2.3\text{-}1]$$

Using the graphical computational scheme described in Section 15.2.2, these same quantities are as follows:

$$|K| = \frac{|s + P_1| \cdot |s + P_2| \cdots |s + P_n|}{|s + Z_1| \cdot |s + Z_2| \cdots |s + Z_m|} \qquad [15.2.3\text{-}2]$$

$$\phi = \phi_{x1} = \phi_{x2} + \cdots + \phi_{xn} - \phi_{01} - \phi_{02} - \cdots - \phi_{0m}$$

$$= 180° \qquad [15.2.3\text{-}3]$$

The importance of these two equations, eqs. 15.2.3-2 and 15.2.3-3, is manifold.

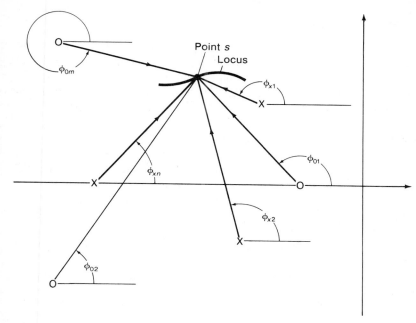

Figure 15.2.3-1   Evans equation in vector notation.

For one, they indicate how to compute the numerical value of the Evans equation. Second, they can be used as a test to see if a point does, indeed, satisfy the Evans equation. In particular, since we know that the result is a real number K that has a minus sign, then for all values of K, we need investigate only the angle. If the algebraic sum of the angles is

$$\phi = q180°$$
$$\text{where } q = \text{odd integer,}$$
$$\text{then the point } s \text{ lies on the root locus}$$

$$[15.2.3\text{-}4]$$

Equation 15.2.3-4 is called the *angle criterion*. It is fairly easy to apply, and yet it provides very important information. This will be the basis of a number of proofs to be covered subsequently. In particular, it will help accomplish the apparent paradox—to construct the root locus without first solving for the roots.

15.3 CONSTRUCTING THE ROOT LOCUS

15.3.1 Rule 1: The Evans Equation

Algebraically manipulate the system equation so that it is in the Evans form.

$$\frac{P(s)}{Z(s)} = -K \qquad [15.3.1\text{-}1]$$

15.3.2 Rule 2: Open Loop Roots

Once the system equation has been written in the Evans form, set $K = 0$ and determine the X roots. Set $K = \infty$ and determine the O roots. We will avoid the use of the terms "pole" and "zero," which are frequently used in the literature, since the root locus uses the reciprocal of the open loop function and confusion may arise.

■ **EXAMPLE 15.3.2-1** Open Loop Roots of a Cubic System
In Section 15.1.4, a cubic system* was analyzed. The roots were found for a number of values of one parameter. These roots were then used to plot the root locus.

In this section, we will ultimately plot the same root locus, but without knowing the roots first. This will be accomplished through the use of a number of examples that will divide the task into a number of simple steps. The first step, for this example, is to determine the open loop roots, a relatively simple problem.

The first task is to manipulate the characteristic equation of the original problem into the Evans form. The characteristic equation, given by eq. 15.1.4-2 is repeated below:

$$s^3 + 570s^2 + 115{,}000s + K = 0 \tag{1}$$

or in the Evans form,

$$s(s^2 + 570s + 115{,}000) = -K \tag{2}$$

The quadratic term factors readily and the equation becomes

$$s(s + 285 + j183)(s + 285 - j183) = -K \tag{3}$$

For $K = 0$ the X roots are 0, $-285 + j183$, and $-285 - j183$ (4)

For $K = \infty$ the O roots are all at infinity   (5)

■

15.3.3 Rule 3: Number of Paths; Where They Start and End

There will be a path, or locus, for each pair of open loop roots, where each locus starts at an X and ends at a O. For repeated roots, place two open loop symbols (either X or O, depending upon which type) at the same point. Two paths or loci will either start or end where the double root was located. For more than two repeated roots, place as many symbols as there are repeated roots. That many loci will either start or end at that point. While the loci will share the point indicated by a multiple root, the loci will most likely go along different routes beyond that point.

Keeping in mind that some open loop roots may exist at infinity, every locus will start at an X and end at a O, establishing the exact number of loci.

---

*See footnote on p. 562 .

15.3.4 Rule 4: Root Loci are Symmetrical About the Real Axis

All real roots lie on the real axis. All complex roots exist in conjugate pairs. Consequently, all points that are not on the real axis will have a mirror image, making the root loci symmetrical about the real axis.

15.3.5 Rule 5: Vector Notation

The Evans equation may be written in the form of a set of first order factors in the numerator and in the denominator. Each factor is of the form

$$(s + \alpha)$$

where $-\alpha$ is an open loop root

The factor may be represented by a vector drawn from the open loop root $-\alpha$ to any point s in the complex plane. When multiplying a set of such factors, multiply the magnitudes and add the angles. For division, divide the magnitudes and subtract the angles.

15.3.6 Rule 6: Angle Criterion

If a point lies on the root locus, it will satisfy the angle criterion, which states:

> The sum of all angles, constructed according to rule 5, will be equal to $q180°$, where $q =$ odd integer.

15.3.7 Rule 7: Sum of the Roots

From the theory of equations, we learn that the sum of the roots is equal to the negative of the coefficient of the second highest power of s in the characteristic polynomial. It should be pointed out that as K is varied, the roots will change, their sum will change, and the coefficient of the second highest power of s will change, but for any particular value of K the sum of the roots will obey the rule.

■ **EXAMPLE 15.3.7-1** Sum of Roots of Cubic System
Sum the open loop roots in Ex. 15.3.2-1. There are three roots as given by eq. (4) of the example. Sum these, separating real and imaginary parts and obtain

$$\sum \text{roots} = -570 + j0 \tag{1}$$

The coefficient of the second highest power of s is

$$570 \tag{2}$$

whose negative is

$$-570 \tag{3}$$

and the rule has been demonstrated.   ■

15.3.8 Rule 8: Segments on Real Axis

If a portion of the root locus lies on the real axis, it is called a segment. Segments were observed in Figs. 15.1.2-2 and 15.1.4-2. In both of those cases, the roots were known first, making it an easy task to determine the existence and the location of such segments. Now we wish to determine existence and location *without* first knowing the roots. Of course, we know the *open loop roots*, but at this time, we know nothing about the *closed loop roots* (for K between its limits of zero and infinity). Let us employ the rules listed up to this point to accomplish the task.

All we know for sure are the locations of the open loop roots. Let us test for the existence of roots on the real axis. Select an arbitrary point P on the real axis in the vicinity of an open loop root on the real axis. We will apply the angle criterion. To do this, draw a vector from each open loop root to the point P. See Fig. 15.3.8-1. Measure the angle for each vector. First let us examine the vectors from complex roots. Note that one will have an angle ϕ_1, while its conjugate will have an angle equal to $-\phi_1$. Consequently, when we add these angles they will sum to zero. Thus, in computing the total angle, only those open loop roots that lie on the real axis will make any contribution.

Let us concentrate upon the open loop roots on the real axis. Any open loop root to the left of the point P will have an angle equal to zero. Hence, these make no contribution to the total angle. Any open loop root to the right of point P will contribute either $\pm 180°$ depending upon whether it is an X or a O. But in summing the angles, $+180°$ will contribute the same as $-180°$. Then it won't matter which type of open loop root exists to the right of point P. If there is an even number of open loop roots, the sum is either zero or $360°$, again contributing nothing. However if there is an odd number, the sum is

$$q180° \hspace{3cm} [15.3.8\text{-}1]$$

and this satisfies the angle criterion, rule 6. Then the rule for existence of a segment reduces to:

> a segment lies on the real axis if there is an odd
> number of open loop roots to the right of it $\hspace{1cm}$ [15.3.8-2]

Let us apply this rule to Fig. 15.3.8-1. Starting at the origin, proceed along the real axis to $-\infty$. Until we reach the point $-a$, there are no roots to the right, hence no segment. As soon as we pass the point $-a$, the open loop root at $-a$ is on the right, hence there is a segment starting at $-a$ until we reach the point $-b$. As soon as we pass this point, there will be an even number of open loop roots, one at $-a$ and one at $-b$. Hence, the segment ends at $-b$. There will be no segment until we pass point $-d$. The open loop root there now changes the sum from even to odd. Hence, a segment starts at $-d$. Since there are no more open loop roots, the sum will remain odd, indicating that the segment runs all the way to $-\infty$. Notice that the two complex roots at $-c$ did not affect the location of the segment.

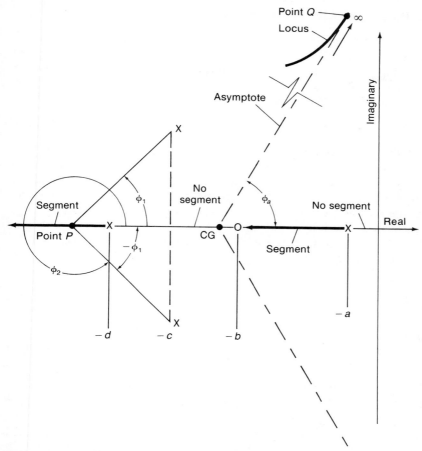

Figure 15.3.8-1   Segments and asymptotes.

■ **EXAMPLE 15.3.8-1** Segments of a Cubic System

For Ex. 15.3.2-1, there is only one open loop root on the real axis, at the origin, as determined in eq. (4). Starting at the origin, examine points on the real axis as we travel to $-\infty$. As soon as we pass the origin, the root at the origin counts as an odd number to the right, hence there is a segment that starts at the origin. Since no other roots are on the real axis, then the sum remains odd all the way to $-\infty$. Thus, there is a segment that runs from the origin to $-\infty$. Note that we have made this decision without first knowing where the closed loops lie.

Now that we have established the existence and the location of the segment, let us check this result against that in Fig. 15.1.4-2, when it was *necessary* to know the closed loop roots. The result checks. We are beginning to see that the root locus method *can* indeed provide closed loop information *without* first knowing the closed loop roots.   ■

15.3.9 Rule 9: Number and Angle of Asymptotes

Each branch of a root locus starts at an X and ends at a O, as stated in rule 3. However, if the number of Os is not equal to the number of Xs (including real and complex open loop roots) this implies that the missing ones must be located at infinity. If there are n Xs and m Os (and, usually, $n > m$), then the number of loci that end at infinity is

$$n - m = \text{number of branches that end at } \infty \qquad [15.3.9\text{-}1]$$

Each of these branches will approach infinity asymptotically. The task is to determine the angle ϕ_a each asymptote makes with the real axis. Refer to Fig. 15.3.8-1. Consider one asymptote at an angle ϕ_a and a point Q that lies on the locus where it approaches infinity (where the locus approaches the asymptote). Hence, the point Q is on both the locus and on the asymptote. Draw vectors from each open loop root to the point Q. In the limit, these vectors will all be parallel to the asymptote since the distance between open loop roots is finite while the distance to the point Q is infinite. Apply the angle criterion, rule 6. (Note all angles = ϕ_a.)

$$n\phi_a - m\phi_a = q180° \qquad [15.3.9\text{-}2]$$

or

$$\phi_a = \frac{q180°}{n - m} \qquad [15.3.9\text{-}3]$$

The treatment thus far involved one asymptote. Equation 15.3.9-1 indicated that there are $n - m$ branches that approach asymptotes. Then there are $n - m$ asymptotes. Equation 15.3.9-3 applies to each asymptote, but engaging different numerical values for q. It is a simple matter to systematically proceed through each value of q until all cases are encountered. For the case shown in Fig. 15.3.8-1, there are 4 Xs and 1 O. Then

$$n - m = 4 - 1 = 3 \qquad [15.3.9\text{-}4]$$

Then we can tabulate the results for various values of q in eq. 15.3.9-3, as follows:

q	$q180°/(n - m)$
1	60°
3	180°
5	300° = −60°
7	420° = +60° (repeats)

Thus, the asymptote make angles

$$60°, 180°, \text{ and } -60° \qquad [15.3.9\text{-}5]$$

15.3.10 Rule 10: CG of Asymptotes

Each of the asymptotes passes through a single point* CG which is the center of mass of all the open loop roots, taking Xs as unit positive masses and Os as unit negative masses. Since the root pattern is symmetrical about the real axis (rule 4), the CG will lie on the real axis. Hence, in computing the location of the CG, we need use only the real part of each open loop root.

Refer to Fig. 15.3.8-1. There are five open loop roots: Xs (positive)—one at $-a$, one at $-d$, and two (complex) at $-c$; Os (negative)—one at $-b$. Sum the products of each unit mass (including its algebraic sign) by its moment arm and divide the result by the total mass (obtained by summing masses algebraically).

$$CG = \frac{-a + b - 2c - d}{4 - 1} \qquad [15.3.10\text{-}1]$$

We now have the point through which each asymptote passes and we have the slopes, as found in Section 15.3.9. Draw the asymptotes, as shown on Fig. 15.3.8-1.

■ **EXAMPLE 15.3.10-1** Asymptotes of a Cubic System
For the system in Ex. 15.3.2-1, determine the number of asymptotes, their slopes, and CG.

The number of asymptotes is equal to

$$n - m = 3 - 0 = 3 \qquad (1)$$

The angle or slope of each asymptote becomes

$$\phi_a = \frac{q180°}{n - m} = 60°, 180°, -60° \qquad (2)$$

and the CG is located at

$$CG = \frac{(-285 - 285 - 0) - (0)}{3 - 0} = -190 \qquad (3)$$

Refer to Fig. 15.1.4-2, the root locus that was constructed *after* the closed loop roots were found. Note that there are three asymptotes that pass through the point -190 on the real axis; one has an angle $\phi_a = 60°$, one has an angle $= -60°$, and the third, which lies on the negative real axis, has an angle $= 180°$. Thus, the results found in this example check out exactly. Once again we have seen that the root locus method has provided closed loop information without first knowing the closed loop roots.   ■

---

*For proof, refer to reference [7], p. 249–250.

15.3.11 Rule 11: Breakaway and Break-In Points

In rule 4, it was established that the root loci appear in mirror images. Then it follows that if two open loop roots start on the real axis, and if increasing K eventually leads to complex roots, then the two root loci must leave the real axis in mirror images. The point where the two loci leave the real axis is called a *breakaway* point. Similarly, where the loci return to the real axis is called a *break-in* point. A break-in point appeared in Fig. 15.1.2-2 at the point $-\omega_n$ (when the variable $\zeta = 1$).

A heuristic proof will be offered. According to rule 3, each root locus starts at an X and ends at a O. If there are two Xs on a line segment (on the real axis), if the segment does not go to infinity, and if there are no Os on the segment, then each locus must leave the real axis, searching for a path to a O. But the loci pattern must be symmetrical, according to rule 4, requiring that the two loci leave the real axis at the same point, the breakaway point. Similarly, were the two loci to start at points not on the real axis (as is the case in Fig. 15.1.2-2), if they must return to the real axis, they must do so in symmetrical patterns. In the meantime, the value of K has been changing, reaching a maximum at the breakaway point and a minimum at the break-in point. Then the location (and existence) of breakaway and break-in points can be determined by setting

$$\frac{dK}{ds} = 0 \qquad\qquad [15.3.11\text{-}1]$$

and solve for s. If the value of s does not lie on a segment, as found by rule 8, then this establishes that there are no breakaway or break-in points.

■ **EXAMPLE 15.3.11-1** Breakaway and Break-in Points
Given the characteristic equation in the Evans form

$$\frac{s(s + a)}{s + b} = -K \qquad\qquad (1)$$

(a) Find the open loop roots.
(b) Determine the existence of segments, and if they exist, draw them.
(c) Determine existence of breakaway and break-in points and, if they exist, locate them.

Solution.   Corresponding to $K = 0$, there are two Xs, as follows:

$$\text{X at the origin, and at } -a \qquad\qquad (2)$$

Corresponding to $K = \infty$, there is one O, as follows:

$$\text{O at } -b \qquad\qquad (3)$$

There are segments on the real axis from the origin to $-a$ (odd number of open loop roots, the one at the origin). Between $-a$ and $-b$ there will be no segment. Beyond point $-b$, there are three open loop roots (odd number). Consequently, there is a segment from $-b$ to $-\infty$.

In order to determine both the existence and location of breakaway and break-in points, apply eq. 15.3.11-1 to eq. (1), collect terms, and simplify.

$$\frac{dK}{ds} = -\frac{s^2 + 2bs + ab}{(s + b)^2} = 0 \tag{4}$$

Solve for s

$$s = -b \pm \sqrt{b^2 - ab} \tag{5}$$

It can be shown geometrically that if

$$|b| > |a| \tag{6}$$

then

$$\begin{array}{l} \text{breakaway to the right of } -a \\ \text{break-in to the left of } -b \end{array} \tag{7}$$

Since there are segments in both areas, then the breakaway and break-in points exist and they are located by eq. (5). Refer to Fig. 15.3.11-1. The loci are shown dotted between the breakaway and break-in points since we have not completed the root locus study and don't yet know the shape of the curve.   ∎

■ **EXAMPLE 15.3.11-2** Breakaway and Break-in Points for a Cubic System Determine the existence of breakaway and/or break-in points for the cubic system in Ex. 15.3.2-1. If they exist, locate them.

Solution.   Apply eq. 15.3.11-1 to eq. (2) of Ex. 15.3.2-1.

$$-\frac{dK}{ds} = 3s + 1140s + 115{,}000 = 0 \tag{1}$$

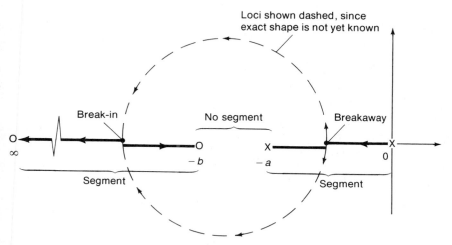

Figure 15.3.11-1   Breakaway and break-in points.

Solve for s

$$s = -190 \pm j46.9 \tag{2}$$

Since this is a complex number, it cannot be located on the real axis. Hence, there is neither a breakaway nor a break-in point. See Fig. 15.1.4-2, which verifies this result. ∎

15.3.12 Rule 12: Angles of Departure and Arrival

The angle at which the locus leaves a complex X is called the *angle of departure*. The angle at which the locus enters a complex O is called the *angle of arrival*. Refer to Fig. 15.3.12-1. Consider a point P on the root locus close to an X. Draw the vectors from each open loop root to the point P. Since P is on the root locus, we know that the algebraic sum of the angles will satisfy the angle criterion (rule 6). Let us follow the point P as it travels along the root locus toward the X. As long as P remains on the root locus, it will satisfy the angle criterion. In the limit, the vector drawn from the X to point P will become tangent to the root locus and, thus, will define the angle ϕ_d, the angle of departure. Also, in the limit, point P is at the X and we can measure the angle for each vector. Applying the angle criterion in the limit we have

$$\lim_{p \to x} \left(\sum \phi_{xi} - \sum \phi_{0i} \right) = q180° \qquad [15.3.12\text{-}1]$$

where

$$q = \text{odd integer} \qquad [15.3.12\text{-}2]$$

and

$$\sum \phi_{xi} \text{ includes angle } \phi_d \qquad [15.3.12\text{-}3]$$

Since all the angles but ϕ_d are known (can be measured) then eq. 15.3.12-1 provides a means to determine the angle ϕ_d.

A similar procedure applies for the angle of arrival ϕ_v at a O.

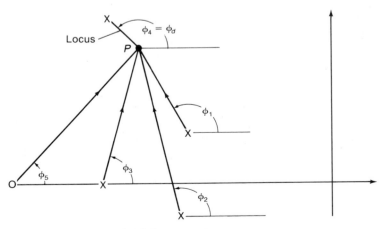

Figure 15.3.12-1   Angle of departure.

■ **EXAMPLE 15.3.12-1** Angle of Departure of a Cubic System
Determine the angle of departure for Ex. 15.3.2-1 (Cubic System).

Solution.   From the original example, we learned that the root locus plot has
three Xs, one at the origin and a complex conjugate pair at $-a \pm j\omega$. Let us
find the angle of departure from the X at the point $-a + j\omega$ (and copy it for
the symmetric image for its conjugate). See Fig. 15.3.12-2.
    In the limit, point P is at the X and

$$\lim_{p \to x} \phi_1 = 90° \tag{1}$$

$$\lim_{p \to x} \phi_2 = \tan^{-1} \frac{\omega}{-a} = 180° - \tan^{-1} \frac{\omega}{|a|} \tag{2}$$

Apply the angle criterion.

$$\phi_1 + \phi_2 + \phi_d = 180° \tag{3}$$

or

$$\phi_d = 180° - \phi_1 - \phi_2 \tag{4}$$

$$= 180° - 90° - \left[180° - \tan^{-1} \frac{\omega}{|a|}\right] \tag{5}$$

$$= -90° + \tan^{-1} \frac{\omega}{|a|} \tag{6}$$

Apply numerical values.

$$\tan^{-1} \frac{\omega}{|a|} = \tan^{-1} \frac{183}{285} = 32.7° \tag{7}$$

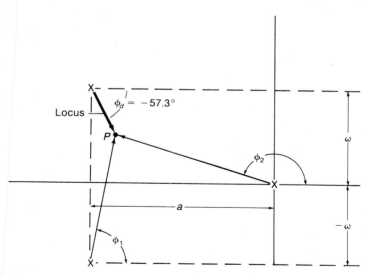

Figure 15.3.12-2   Angle of departure for a cubic system.

and

$$\phi_d = -90° + 32.7° = -57.3° \tag{8}$$

Refer to Fig. 15.1.4-1. The result is reasonably close to the angle found in the figure. Once again the root locus method has determined closed loop information without knowing the closed loop roots.   ■

15.3.13 Rule 13: Crossing the Imaginary Axis

If a closed loop root occurs in the positive real half-plane, the solution in the time domain will become unbounded, defining an unstable system. For this reason, we did not continue the root locus plot into that region. Consequently, the points where the root loci cross the imaginary axis represent points where the time domain solution will become unbounded. Such points define where the system is on the verge of instability.

See Fig. 15.1.4-2, where the loci cross the imaginary axis. The objective is to compute the value of K and to determine the coordinates where these crossings occur. At the point of crossing, the coordinates are

$$s = 0 + j\omega \qquad [15.3.13-1]$$

where ω is to be determined. This can be accomplished by substituting the coordinates of eq. 15.3.13-1 into the characteristic equation. Note that when this is done, the result is really two equations: one for real, one for imaginary coefficients. Thus, there are two equations, permitting us to solve for K and ω simultaneously.

■ **EXAMPLE 15.3.13-1** Crossing the Imaginary Axis for Cubic System
For the system in Ex. 15.3.2-1 (Cubic System), determine if the loci cross the imaginary axis. If so, determine ω and K at the crossings.

Solution.   The characteristic equation is given in eq. 15.1.4-2, repeated below:

$$\Delta = s^3 + 570s^2 + 115{,}000s + K = 0 \tag{1}$$

Apply eq. 15.3.13-1 to eq. (1).

$$-j\omega^3 - 570\omega^2 + j115{,}000\omega + K \equiv 0 + j0 \tag{2}$$

Equate coefficients of real and imaginary parts.

$$-j\omega^3 + j115{,}000\omega = j0 \tag{3}$$

$$-570\omega^2 + K = 0 \tag{4}$$

Solve for ω in eq. (3)

$$\omega = 0 \quad \text{and} \quad \omega = \sqrt{115{,}000} = 339 \tag{5}$$

The result at $\omega = 0$ is not an appropriate crossing, since this occurs at an open loop root, where $K = 0$. There would be a crossing only if $K < 0$, which is outside the limits for K. Hence, reject this point.

The result $\omega = 339$ is a valid crossing. Solve for K in eq. (4).

$$K = 570\omega^2 = 570 \times 115,000 = 65.5 \times 10^6 \tag{6}$$

Refer to Fig. 15.1.4-2. The result checks out exactly with that in the figure, as well as with that in Table 15.1.4-1, point 11. ∎

15.3.14 Rule 14: Trial and Error

By this time, a portion of the root locus plot is completed. It is possible to make some good educated guesses as to where the rest of the locus lies. Each point so selected must be tested with the angle criterion, which is a relatively simple matter of merely algebraically summing angles. By now, we have had considerable practice doing this.

∎ **EXAMPLE 15.3.14-1** Complete the Root Locus Plot for a Cubic System
Complete the root locus for the cubic system in Ex. 15.3.2-1.

Solution.   Summarize the results found thus far. In Ex. 15.3.2-1, the open loop roots were found to be

$$0 \text{ and } -285 \pm j183 \tag{1}$$

In Ex. 15.3.7-1, the sum of the open loop roots was found to be 570. Note that, for this particular problem, the coefficient of the second highest power of s in the characteristic equation, eq. 15.1.4-2, does not change with K. Hence, the sum of all roots will always be 570.

$$\sum \text{ roots} = 570 \tag{2}$$

Example 15.3.8-1 established the existence of a segment on the real axis from the origin to $-\infty$.

It was established in Ex. 15.3.10-1 that there are three asymptotes, where

$$\phi_a = 60°, 180°, \text{ and } -60° \tag{3}$$

$$CG = -190 \tag{4}$$

In Ex. 15.3.11-2, we learned that there are no breakaway or break-in points.
In Ex. 15.3.12-1, the angle of departure was shown to be

$$\phi_d = -57.3° \tag{5}$$

In Ex. 15.3.13-1, we found that the loci cross the imaginary axis at

$$\omega = 339 \tag{6}$$

for which

$$K + 65.5 \times 10^6 \tag{7}$$

We have a considerable amount of information, permitting us to make some good guesses. The first selection was chosen close to the open loop root. The first trial was made for point A on the 57.3° tangent. See Fig. 15.3.14-1. When the angle criterion was applied, the sum was off by nearly 10°. The second

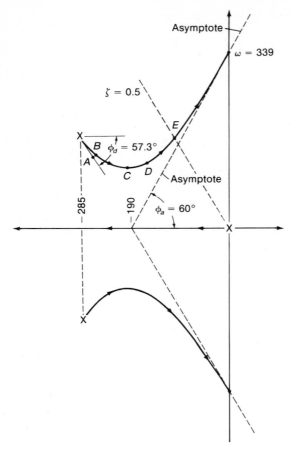

Figure 15.3.14-1 Root locus for cubic system.

trial, at point *B* was more successful. After several trials, the lowest point, *C*, was found. After that, points *D* and *E* were found quickly. Knowing the angle of departure, the asymptote, the crossing, and five points, the root locus was drawn. See Fig. 15.3.14-1. ∎

15.4 SYSTEM PROPERTIES DETERMINED BY THE ROOT LOCUS

15.4.1 Determination of *K*

In Section 15.2.3, where the angle criterion was derived, the magnitude *K* of the vector was found in eq. 15.2.3-2, repeated below:

$$|K| = \frac{|s + P_1| \cdot |s + P_2| \cdots |s + P_n|}{|s + Z_1| \cdot |s + Z_2| \cdots |s + Z_m|} \qquad [15.4.1\text{-}1]$$

Select a point that lies on the root locus and draw the vectors from each open loop root. Determine the length of each vector and perform the indicated arithmetic. This computation may either be done graphically (by physically measuring the magnitudes of each vector) or it can be done analytically (using the coordinates of each point involved).

■ **EXAMPLE 15.4.1-1** K for a Cubic
In Ex. 15.3.2-1, determine the maximum value of K for absolute stability, using:

    (a) Graphical means.
    (b) Analytical means.

Solution.   The point of interest is where the root locus crosses the imaginary axis. Draw vectors from each open loop root.

    (a) For graphical means, use the root locus that was constructed in Fig. 15.3.14-1. The appropriate vectors are shown on Fig. 15.4.1-1. Using a decimal scale, measure the lengths of each vector. The results are indicated on the

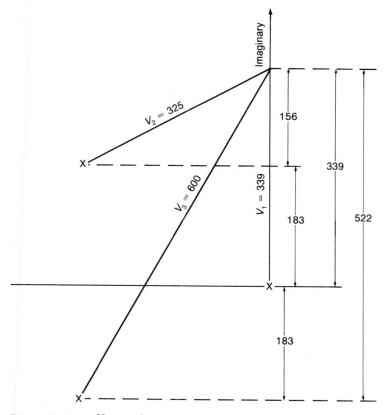

Figure 15.4.1-1   Vectors for a cubic.

figure: 339, 325, and 600. Multiply these.

$$K = 339 \times 325 \times 600 = 66.1 \times 10^6 \tag{1}$$

which is reasonably close to the value found in Table 15.1.4-1, entry 11 (where the roots were found first) and also is reasonably close to the value found in Ex. 15.3.13-1, eq. (6) (where the roots were unknown).

(b) For the analytical means, determine the coordinates of each point and compute the length of each vector. The appropriate dimensions are shown on Figs. 15.3.14-1 and 15.4.1-1.

$$V_1 = 339 \tag{2}$$

$$V_2 = \sqrt{285^2 + 156^2} = 324.9 \tag{3}$$

$$V_3 = \sqrt{285^2 + 522^2} = 594.7 \tag{4}$$

Upon multiplying these,

$$K = 339 \times 324.9 \times 594.7 = 65.5 \times 10^6 \tag{5}$$

which agrees with the two references above.   ■

15.4.2 Determination of Factors

There are several means by which the factors can be found from the root locus. Of course, we know the factors for the limiting cases where $K = 0, \infty$. The objective is to determine the factors (or the roots) for the cases between the limits.

Consider the case for which the numerical value of K is known on one branch locus and it is desirable to determine the factors (or roots) for each of the other branches. It will be necessary to hunt for the value of K by trial and error on each branch. Once the root has been located for the same value of K on each branch, the system factors can be obtained. To check, multiply the factors and compare to the characteristic equation.

Consider the case for which all but two factors (or roots) are known for one value of K. Multiply the known factors and form a known polynomial. Divide the characteristic equation by this polynomial and determine the remaining factors (which will appear as a quadratic term).

Consider the case for which the coefficient of the second highest power of s in the characteristic equation is easily found. Use rule 7, which states that the sum of the roots is equal to the negative of this coefficient. While extremely simple to use, keep in mind that when there is a complex root, it has a conjugate. In that case, the imaginary parts cancel. Hence, this rule is useful only for the real parts of complex roots.

In most practical cases, a combination of all three methods described above will be used.

■ **EXAMPLE 15.4.2-1** Factors of a Cubic Where the Root Locus Crosses the Imaginary Axis

Determine the factors of the cubic system in Ex. 15.3.2-1, where a locus crosses the imaginary axis.

Solution. The coordinates are known for two of the three roots, as determined in Ex. 15.3.13-1, eq. (5), where the zero root was rejected. The roots are as follows:

$$r_2 = 0 + j339 \tag{1}$$

$$r_3 = 0 - j339 \tag{2}$$

The coefficient of the second highest power of s does not change with K and can be found in Ex. 15.3.2-1, eq. (1). Then the sum of the roots becomes

$$\sum \text{roots} = -570 \tag{3}$$

Sum the roots.

$$r_1 + r_2 + r_3 = -570 \tag{4}$$

Apply eqs. (1) and (2) and solve for r_1.

$$r_1 = -570 - 0 = -570 \tag{5}$$

The corresponding factors are given by $s - r_i$:

$$\Delta = (s + 570)(s + 0 + j339)(s + 0 - j339) \tag{6}$$

■

■ **EXAMPLE 15.4.2-2** Factor When a Real Root is Known

For the cubic in Ex. 15.3.2-1, given one real root, $r_1 = -370$. Determine the following: remaining roots, K, undamped natural frequency ω_n, and the damping ratio ζ.

Solution. Divide the characteristic polynomial Δ by $(s + r_1)$, where Δ is given by eq. 15.1.4-2.

$$
\begin{array}{r}
s^2 + 200s + 41{,}000 \\
\hline
s + 370 \,\big|\, s^3 + 570s^2 + 115{,}000s + K \\
\underline{s^3 + 370s^2} \\
\end{array}
$$

$$
\begin{array}{r}
200s + 115{,}000s \\
\underline{200s + 74{,}000s} \\
\hline
41{,}000s + K \\
41{,}000s + 15.17 \times 10^6 \\
\hline
0 \qquad\qquad 0
\end{array}
\tag{1}
$$

From which

$$K = 15.17 \times 10^6 \tag{2}$$

The quotient is the quadratic

$$s^2 + 200s + 41{,}000 \tag{3}$$

from which

$$\omega_n = \sqrt{41{,}000} = 202.48 \tag{4}$$

$$\zeta = \frac{200}{2 \times 202.48} = 0.494 \tag{5}$$

$$a = \frac{200}{2} = 100 \tag{6}$$

$$\omega_d = \omega_n\sqrt{1 - \zeta^2}$$

$$= 202.48 \times \sqrt{1 - 0.494^2} = 176.1 \tag{7}$$

Then the remaining roots are

$$
\begin{aligned}
r_2 &= -100 + j176.1 \\
r_3 &= -100 - j176.1
\end{aligned} \tag{8}
$$

To check the results, sum the three roots.

$$\sum = -370 - 100 + j176.1 - 100 - j176.1$$

$$= -570 \quad \text{checks} \tag{9}$$

As an additional check, use entry 6 of Appendix 1.

$$\omega_n^2 = a^2 + \omega_d^2 \tag{10}$$

or

$$\omega_n = \sqrt{100^2 + 176.1^2} = 202.5 \quad \text{checks} \tag{11}$$

Note that all of the results of this example agree with the information in entry 8 of Table 15.1.4-1. ∎

15.4.3 System Stability

Where the root locus crosses the imaginary axis into the positive half-plane, the system will become unstable. Keep in mind that at the point of crossing the system goes from theoretically stable to absolutely unstable. Consequently, it is not desirable to allow the numerical value of K to come right up to the imaginary axis, but should remain a safe distance from it. An arbitrary safety margin is chosen as a factor equal to 3. Thus, if K_c is the value at the crossing, then for safety use

$$K \le K_c/3 \tag{15.4.3-1}$$

An alternate margin of safety can be stated in terms of the system damping ratio (or, approximately, as the damping ratio of the dominant factor). For safety, let

$$\zeta \geq 0.5 \qquad\qquad [15.4.3\text{-}2]$$

■ **EXAMPLE 15.4.3-1** Stability Criterion for a Cubic
Determine the value of K for a safe margin of stability for the cubic in Ex. 15.3.2-1.

Solution. In Ex. 15.3.13-1, the value of K_c when the root locus crossed the imaginary axis is given by eq. (6).

$$K_c = 65.5 \times 10^6 \qquad\qquad (1)$$

Use eq. 15.4.3-1.

$$K \leq (65.5 \times 10^6)/3 = 21.8 \times 10^6 \qquad\qquad (2)$$

In order to apply the damping ratio criterion, refer to entry 19 of Appendix 1.

$$\cos \phi = \zeta \qquad\qquad (3)$$

Use eq. 15.4.3-2 and solve for ϕ

$$\phi = \cos^{-1} 0.5 = 60° \qquad\qquad (4)$$

On Fig. 15.3.14-1 (the root locus plot for the cubic being used in this example), from the origin draw a vector making an angle of 60°. Any roots that lie on this vector have a damping ratio equal to 0.5. Where this vector crosses the root locus, determine the magnitude of K as shown in Section 15.4.1 (either graphically or analytically).

$$K = 15 \times 10^6 \qquad\qquad (5)$$

In order to comply with the inequality of eq. 15.4.3-2, the value of K must be equal to or less than that in eq. (5). Also, the value of K must be the lesser of the two values found, one in eq. (2), the other in eq. (5). Thus,

$$K \leq 15 \times 10^6 \qquad\qquad (6)$$

■

15.4.4 Transient Properties

In the previous three sections it was shown that the value of K, the undamped natural frequency ω_n, and the damping ratio ζ as well as the roots can be determined at any point on the root locus. Consequently, all of the transient properties can be determined for the dominant factor at any point on the root locus. Refer to Appendix 1 for definition of terms. Refer to Section 11.4 for the transient properties.

■ **EXAMPLE 15.4.4-1** Transient Properties of a Cubic
For the cubic in Ex. 15.3.2-1, determine the transient properties for a safe margin of stability.

Solution. Such a margin was found in Ex. 15.4.3-1 for which

$$K = 15 \times 10^6, \qquad \zeta = 0.5, \qquad \phi = 60° = 1.05 \text{ rad} \qquad (1)$$

Reading the coordinates of the point where the 60° vector crosses the root locus,

$$a = 100, \qquad \omega_d = 176 \qquad (2)$$

Using entries 5 and 6 of Appendix 1,

$$\omega_n = 202, \qquad \beta = 0.866 \qquad (3)$$

Assume the quadratic factor that contains the two complex roots is the dominant factor. For a step input, the system corresponds to entry 414 of Appendix 3, for which

$$n = 1 \qquad (4)$$

For the initial value of the transient, refer to Table 11.4.3-1, for $n = 1$

$$I_0 = 1 \qquad (5)$$

The rise time is given by eq. 11.4.4-3, for which $m = 1$.

$$T_r = (\pi - \phi)/\omega_d = (\pi - 1.05)/176 = 0.0119 \qquad (6)$$

Peak time is given by eq. 11.4.5-3, for which $m = 1$.

$$T_p = m\pi/\omega_d = (1 \times \pi)/176 = 0.0178 \qquad (7)$$

Overshoot is given by eq. 11.4.6-6, where $m = 1$, $n = 1$.

$$A_0 = e^{\zeta\pi/\beta} \qquad (8)$$

Apply appropriate numerical values.

$$A_0 = 0.163 \qquad (9)$$

The settling time will be found for a transient residue of 1.8%. Refer to Table 5.5.2-1,

$$N = 4 \qquad (10)$$

The time constant is

$$\tau = 1/a = 1/100 = 0.01 \qquad (11)$$

Settling time is

$$T_s = N\tau = 4 \times 0.01 = 0.04 \qquad (12)$$

∎

15.4.5 Design Using Transient Specifications

In the previous section, the transient properties were found for a given system in a known configuration. The same problem can be worked in reverse—given the transient properties (in the form of specifications), determine the system.

■ **EXAMPLE 15.4.5-1** Design Using Transient Specifications

Given the root loci for a group of systems, as shown in Fig. 15.4.5-1, select the most appropriate one and design it to meet the following specifications:

$$\zeta \geq 0.5 \tag{1}$$

$$T_s \leq 0.03 \quad \text{for } N = 4 \tag{2}$$

$$\omega_d \geq 100 \text{ rad/s} \tag{3}$$

$$\omega_n \leq 400 \text{ rad/s} \tag{4}$$

Solution.   Apply specification (1) to entry 6 of Appendix 1.

$$\phi \leq 60° \tag{5}$$

Apply specification (2) to eq. 11.4.7-2 and solve for *a*.

$$a \geq N/T_s = 4/0.03 = 133 \tag{6}$$

Specification (3) requires that the roots lie beyond 100 from the real axis.

Specification (4) requires that the roots lie within a circle of radius equal to 400.

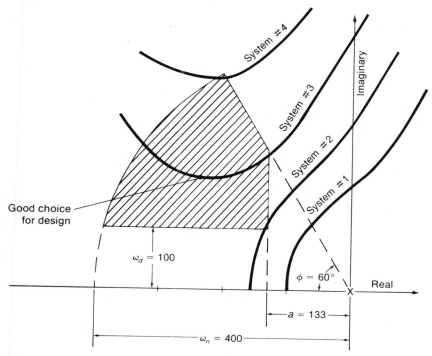

Figure 15.4.5-1   Transient specifications.

The four specified quantities form an area shown shaded. Any locus that passes through this closed area satisfies all four specifications. See Fig. 15.4.5-1. The root locus for System #1 does not qualify. Systems #2, #3, and #4 pass through the area, but systems #2 and #4 are so close to the boundary that manufacturing tolerances might thrust the system out of spec. System #3 passes well within the boundaries. Select a point that is somewhat in the middle, so that tolerances will not shift it out of spec. Now that a specific point has been selected, it is possible to determine all the system constants by employing the methods in the previous sections.

System #3 satisfies all four specifications (leaving considerable margin for errors and tolerances). The design is complete.   ■

15.4.6 Synthesis

The shape of the root locus can be altered by introducing Xs and Os. An X repels the curve while a O attracts it. With propitious choices and combinations, the curve can be molded into the desired shape.

■ **EXAMPLE 15.4.6-1** Repulsion and Attraction of the Root Locus
In the previous example, synthesize system #1 so that it satisfies the specifications.

Solution.   The specified area and the original system, system #1, are shown in Fig. 15.4.6-1. By locating an X on the real axis, the system is changed and is

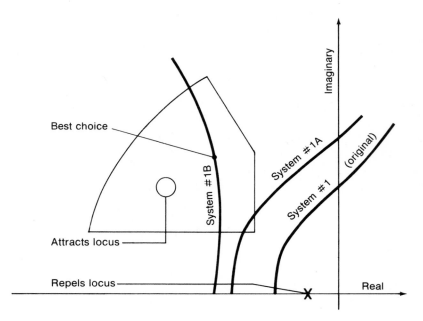

Figure 15.4.6-1 Synthesis of system to meet specifications.

called system #1A. The X repels the locus toward the desired area, but the curve barely remains within the boundary. It should be pointed out that the introduction of this open loop root has placed a first order term in the numerator of the characteristic equation, thus raising the order. This X has helped, but not enough, Now add a pair of complex Os to attract the locus deep into the area. This system is labeled system #1B. This offers a good solution, the best point being well within the specified region. The introduction of the pair of conjugate Os places a quadratic in the denominator of the characteristic equation, which makes the system more complicated, but doesn't change the order. ∎

SUGGESTED REFERENCE READING FOR CHAPTER 15

Damped high order systems [1], p. 270–288, 318–324; [4], p. 126–132; [5], p. 195–198, 233–238; [9], p. 119–133, 165–168; [16], p. 129–133; [17], p. 423–425, 447–456. Root locus [59]; [60]; [7], p. 249–250.

PROBLEMS FOR CHAPTER 15

15.1. In Section 15.1.2, the damping ratio ζ was isolated. Rework the problem so that the decay coefficient a is isolated, and vary a while holding ζ constant.

15.2. In Section 15.1.3, five conclusions are stated. Prove or verify each one.

15.3. Test the information listed in Table 15.1.4-1 in the following ways: Sum of roots equals negative of coefficient of second highest power of s; apply entry 6 of Appendix 1 (note that the real part of each pair of conjugate roots is equal to a and the imaginary part is equal to ω_d); apply entry 3 of Appendix 1; compute β and apply entry 4 of Appendix 1.

15.4. Test the angle criterion for each point on Fig. 15.1.4-2. Test the magnitude for each point.

15.5. For each of the following systems, determine if there are segments on the real axis and if so, find them:
(a) Eq. 15.1.2-1.
(b) $s^3 + 10s^2 + 100s + K = 0$
(c) $s^4 + 570s^3 + 115,000s^2 + 10^8 s + K = 0$
(d) $s^4 + 19s^3 + 80s^2 + (K - 100)s + K = 0$
(e) $\dfrac{s(s^2 + 4s + 13)(s + 1)}{s + 2} = -K$
(f) $\dfrac{s(s + 4)(s^2 + 6s + 13)}{s + 2} = -K$

15.6. For each system in Problem 15.5, construct the asymptotes.

15.7. For each system in Problem 15.5, determine if there are breakaway or break-in points. If so, find them.

15.8. For each system in Problem 15.5, determine the existence of angles of departure or arrival. If they exist, determine the appropriate angles.

15.9. For each system in Problem 15.5, determine if any of the root loci cross the imaginary axis. If so, determine the corresponding numerical values of K and ω.

15.10. For each system in Problem 15.5, complete the root locus plot.

15.11. For each system in Problem 15.5, if it crosses the imaginary axis, determine all the factors at that point.

15.12. For each system in Problem 15.5 that crosses the imaginary axis, determine the numerical value of K for a safe margin of stability.

15.13. For each system in Problem 15.5 whose locus crosses the line $\phi = 60°$ (where damping ratio $\zeta = 0.5$), determine the factors, the numerical value of K and the undamped natural frequency ω_n. Determine all the transient properties.

15.14. For each system in Problem 15.13, what value of K will halve the rise time?

15.15. For each system in Problem 15.13, what value of K will halve the overshoot?

15.16. For each system in Problem 15.13, what value of K will halve the settling time? (Note that neither the transient residue nor the numerical value of N is given.)

15.17. For each system in Problem 15.10, if there is an X on the real axis, place another one there, and replot the root locus. What conclusions can be drawn?

15.18. Similar to Problem 15.17, but for a O.

15.19. Using the method outlined in Ex. 15.4.2-2, determine all the roots (or factors) in Table 15.1.4-1. Why does this work for this system? What conclusions can be drawn?

Chapter **16**

Automatic Controls

This chapter considers the subject of the control of machine processes by means of feedback. As a result, the machine can function by itself, making the necessary adjustments in its operation. Systems are modeled by means of block diagrams. Each block contains the Laplace transfer function. A network of such blocks is employed to model the entire feedback control system, including the process that is being controlled.

Many of the properties of feedback control systems are determined using techniques that have been developed earlier in the text. As such, this provides a unified approach to the subject. Examples of these properties are: overshoot, time constant, settling time, steady state value, initial and final values, maximum values, frequency response, behavior as a filter or resonator, bandwidth, stability and means to optimize any of these properties.

16.1 DEFINITIONS AND NOTATIONS

16.1.1 Control System

A *control system* is an interconnection of subsystems which will provide a desired response to inputs. This is accomplished by means of a *controller*, a device that directs or controls the *process* or *plant*.

16.1.2 Open Loop

An open loop is a control system that uses a controller to drive or actuate a process or plant. In the open loop, the controller does *not* examine the results

Figure 16.1.2-1   Open loop (cascaded) control system.

of the process. The output does not return to the input. As such, the system is cascaded. See Fig. 16.1.2-1. As an example of an open loop control system, consider preparing to take a shower, setting the hot and cold water faucets at, say, their midpoints, and then stepping into the shower without testing the resulting temperature. Obviously, variations in the system components (water pressure, fluid resistance in valves, temperature of the hot and cold water supplies, and so forth) would make it unlikely that the temperature is correct. Note that these variables make it difficult to repeat the results each time. This is the nature of open loop or cascaded control systems.

16.1.3 Closed Loop System

A closed loop control system compares the output to the input, and if there is a difference, the system provides the appropriate correction force. As such, a closed loop control system tends to maintain a prescribed relationship between the output and the input. The closed loop control system is coupled. See Fig. 16.1.3-1.

The closed loop adds two more elements or components to the open loop system. These are a sensor (which examines the output) and a comparator (which compares the output, as seen by the sensor, to the input). The difference between the input and the output is referred to as the *error* ϵ. The larger the error, the harder the controller works to drive the process (plant) in such a manner as to reduce the error. Consequently, the *closed loop or feedback control system tends to minimize the error*.

In the case of the shower, the system could be converted to a closed loop control system by using a thermostat valve. This is a device that regulates the flow of hot water in order to maintain the preset temperature. As the various system components vary, the device automatically makes the correct adjustment. This is what the person who is taking the shower must do in the absence of such a device. The body sensing the temperature is really comparing the output (temperature of water) to the input (ideal temperature) and then makes

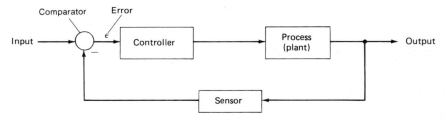

Figure 16.1.3-1   Closed loop (feedback) control system.

the necessary corrective means (adjust one or both of the valves). In this case, the bather is closing the loop.

16.1.4 Mathematical Model of Closed Loop Control System

The controller, the process, and the sensor may be modeled with blocks in which the Laplace transfer function of the corresponding operation may be written. Then, let

$$A(s) = \text{transfer function of the controller}$$

$$G(s) = \text{transfer function of the process or plant} \qquad [16.1.4\text{-}1]$$

$$H(s) = \text{transfer function of the sensor or feedback}$$

The mathematical model is shown in Fig. 16.1.4-1. Note that the system forms a closed loop. This makes it clear that the system is coupled.

Following the direction of flow (indicated by the arrows), from the input to the output is referred to as the *feed forward*. In the above diagram, this is equal to AG. Following the flow back to the comparator, this is referred to as *feedback*. In the above diagram, this is equal to H. Following the route around the entire loop is called the *open loop gain* and is given the symbol L. For the above diagram, note that in passing through the summing point, following the arrows, there is a change in sign. Consequently,

$$L = -AGH \qquad [16.1.4\text{-}2]$$

The difference between the input and output is called the *error* ϵ. For general purposes, consider an auxiliary input T_d located between the controller and the process (or plant). See Fig. 16.1.4-1(b). The auxiliary input will be discussed in Section 16.3.4.

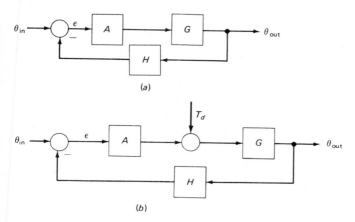

(a)

(b)

Figure 16.1.4-1 Generalized control system. (a) Basic feedback control system. (b) Control system with an auxiliary input.

16.2 CLOSED LOOP CONTROL

16.2.1 Primary Objective

The primary objective of a control system is that the output follow the input. There may be secondary goals involving the many properties of a dynamic system.

16.2.2 Single Loop

In order to achieve the primary objective with sufficient accuracy, a closed loop is required. Let us consider the simplest system, which consists of a single loop.

The student is advised to become familiar with the subject of block diagrams covered in Sections 13.1 and 13.2.

■ **EXAMPLE 16.2.2-1** Rolling Mill
In a foundry, hot slabs of steel are rolled into successively thinner slabs until sheets are formed. This is accomplished by squeezing the material between a pair of rollers. In the past, a constant pressure was applied to the rollers, assuming that this would produce a uniform thickness of the sheet. However, due to variations in the materials and in the hydraulic supply, the thickness of the finished product was found to vary. As a remedy, feedback control is employed today.

One method is shown in Fig. 16.2.2-1(a), where

$$M = 10^3 \text{ kg} \qquad D = 1.5 \times 10^5 \text{ N} \cdot \text{s/m}$$

$$K = 10^7 \text{ N/m} \qquad K_p = K_c K_m b/c = 1.5 \times 10^8 \text{ N} \cdot \text{m*}$$

If the material being rolled moves at a velocity V, and if the distance between the squeezing roller and the sensing roller is X, there will be a time delay α equal to X/V. Neglect the time delay for this example. It will be considered in Ex. 16.6.1-1.

Determine the block diagram for each subsystem. Draw the complete block diagram for the system and determine the closed loop transfer function.

Solution. A sensing roller measures the thickness Y of the finished product. This is mechanically compared to the preset (input) value Y_a (which is adjusted by means of a screw). Hence, feedback is unity.

$$H(s) = 1 \tag{1}$$

The difference or error ϵ is magnified by the linkage b/c and operates a valve which controls the flow rate Q from a pump. The valve is assumed to be

---

*This is the nominal value. For design purposes, it may be varied over a large range.

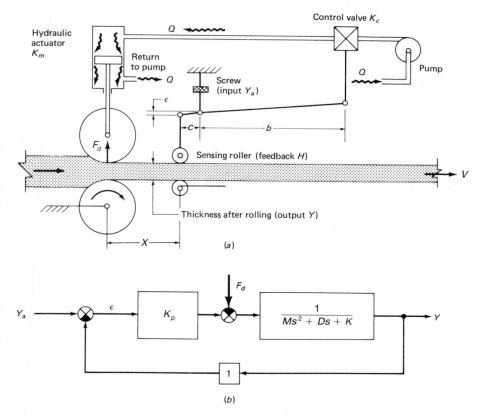

Figure 16.2.2-1   Rolling mill. (a) Actual system. (b) Block diagram.

linear, with constant of proportionality K_c. The resulting flow rate Q is

$$Q = \frac{K_c \epsilon b}{c} \tag{2}$$

Assuming the hydraulic actuator is not a positive displacement type, but is instead dependent upon flow rate, it produces a proportional force, where the constant of proportionality is K_m. Then the control force F_c is

$$F_c = K_m Q \tag{3}$$

The controller A is modeled by

$$A(s) = \frac{F_c}{\epsilon} = \frac{K_c K_m b}{c} = K_p \tag{4}$$

The plant is a reciprocal quadratic system (spring-mass system with damping) whose transfer function is

$$G(s) = \frac{1}{Ms^2 + Ds + K} \tag{5}$$

The material between the rollers exerts a force F_d against the hydraulic actuator. Applying all of the above to the generalized model shown in Fig. 16.1.4-1, the system block diagram may be constructed as shown in Fig. 16.2.2-1(b). In order to determine the closed loop transfer function, the forward gain (FG) is required. This is determined by tracing a path from the input Y_a to the output Y. Note that, in passing through the summing point, there is no change of sign. Consequently, FG becomes

$$\text{FG} = \frac{K_p}{Ms^2 + Ds + K} \tag{6}$$

The open loop gain L is found by tracing a path around the entire loop. Note that there is a change of sign in passing through the summing point.

$$L(s) = \frac{-K_p}{Ms^2 + Ds + K} \tag{7}$$

The closed loop transfer function is found by applying eqs. (6) an d(7) to eq. 13.2.4-6. This produces an improper fraction. Upon rationalizing, the result becomes

$$\frac{Y}{Y_a}(s) = \frac{K_p}{Ms^2 + Ds + K + K_p} \tag{8}$$

As an alternative means to determine the closed loop transfer function, note that the forward gain is a fraction. Hence, it will pay to use the fraction loop rule. Since the feedback is negative, using eq. 13.2.5-2, this requires that the numerator, K_p, be multiplied by the feedback (which is unity) and added to the denominator. The result is

$$\frac{Y}{Y_a}(s) = \frac{K_p}{Ms^2 + Ds + K + K_p} \tag{9}$$

which agrees with eq. (8).   ∎

16.2.3 Loop Within a Loop

Within any type of system (control or otherwise) there are occasions where the normal operation of a specific element involves a closed loop (and this may have nothing to do with a control function). Since the primary objective of a control system requires a loop (an overall one, not necessarily the previously mentioned one), then for such a system, there will be a loop within a loop.

■ **EXAMPLE 16.2.3-1** Hydraulic Servo

A control system in which the output follows the input is called a *servo*. If the quantity involved is displacement, the system is called a displacement servo. If velocity is used, the system is called a velocity servo. A displacement servo using hydraulic elements is shown in Fig. 16.2.3-1(a). It makes use of a reversing spool valve (like that discussed in Section 3.5.4 and shown in Fig.

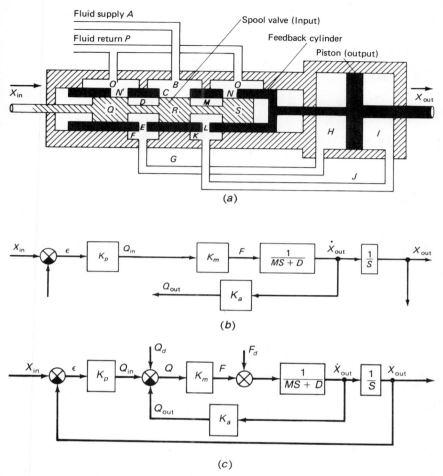

Figure 16.2.3-1 Hydraulic servo. (a) Actual system. (b) Partial block diagram. (c) Complete block diagram.

3.5.4-2) whose cylinder translates with the output (piston type hydraulic actuator). Displacement X_{in} of the spool valve is the input, while displacement X_{out} of the piston is the output.

The system is shown in the null or off position. Note that no fluid flows through the valve since spool plugs, Q, R, and S block the flow. Now apply an input displacement to the right. This is achieved by displacing the spool by amount X_{in}. Spool plug Q still blocks port N', but plug R uncovers port C and plug S uncovers port N. The former permits fluid to flow from supply A to fixed port B through feedback port C through spool notch D. This permits fluid to flow to feedback port E to fixed port F through pipe G to piston chamber H, causing the piston to move to the right. As it does, fluid is discharged from piston chamber I through pipe J, then following a route K, L, M, N, O to fluid return P. When the piston and feedback cylinder have traveled a distance X_{out} that is equal to the input displacement X_{in}, the system will have reached a new equilibrium or steady state condition. Thus, the output follows the input. Determine the block diagram for each subsystem. Draw the complete block diagram for the system and determine the following:

$$\frac{X_{out}}{X_{in}}, \quad \frac{\epsilon}{X_{in}}, \quad \frac{\epsilon}{Q_d}, \quad \frac{\epsilon}{F_d}$$

Solution. The flow rate Q is proportional to the valve port opening as given by eq. 13.1.7-3, but where the opening is equal to the relative displacement $X_{in} - X_{out}$. Then

$$Q_{in} = K_p (X_{in} - X_{out}) \tag{1}$$

The block diagram is essentially a summing point. The resulting flow rate Q_{in} travels to the piston, where it is multiplied by constant K_m and produces force F. The mass M [shown in black on Fig. 16.2.3-1(a)] has a damper D, but no spring. The function becomes

$$X_{out} = \frac{F}{Ms^2 + Ds} = \frac{F}{s(Ms + D)} \tag{2}$$

The displacement X_{out} is the system output.

As the piston moves at velocity $\dot{X}_{out}$, in order to obey the law of continuity, there is a flow rate Q_{out} given by

$$Q_{out} = K_a \dot{X}_{out} \tag{3}$$

where

$$K_a = \text{piston area} \tag{4}$$

Thus, it is clear that both the displacement and the velocity are required; the former for system feedback, the latter for continuity, eq. (3). Consequently, it will be necessary to break up eq. (2) into the product of two terms and to show it as the product of two blocks. See Fig. 16.2.3-1(b).

As fluid flows into the piston chamber at rate Q_{in}, there is also a flow out at rate Q_{out}. This will be shown by means of a second summing point. In addition to the normal input X_{in}, there are two additional inputs Q_d and F_d. Using the generalized model in Fig. 16.1.4-1 as a guide, the final block diagram is constructed as shown in Fig. 16.2.3-1(c), where:

$$K_m = 50 \text{ N} \cdot \text{s/m}^3 \qquad K_p = 72 \text{ m}^2/\text{s}*$$

$$K_a = 0.5 \text{ m}^2 \qquad\qquad D = 35 \text{ N} \cdot \text{s/m}$$

$$M = 3.5 \text{ kg}$$

Now that the block diagram has been constructed, we may proceed to determine the various output/input ratios requested in the example.

---

*This is a nominal value. For design purposes, it may be varied over a large range.

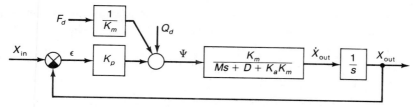

Figure 16.2.3-2   Hydraulic servo. Simplified block diagram.

In Ex. 13.2.7-1 the block diagram, containing a loop within a loop, was reduced to a single block in Fig. 13.2.7-3(d), repeated in Fig. 16.2.3-2. Note that the quantities required here were found in that example in eqs. (5), (9), (12), and (13), repeated below.

$$\frac{X_{out}}{X_{in}}(s) = \frac{K_p K_m}{Ms^2 + (D + K_a K_m)s + K_p K_m} \tag{5}$$

$$\frac{\epsilon}{X_{in}}(s) = \frac{s(Ms + D + K_a K_m)}{s(Ms + D + K_a K_m) + K_p K_m} \tag{9}$$

$$\frac{\epsilon}{Q_d}(s) = \frac{-K_m}{s(Ms + D + K_a K_m) + K_p K_m} \tag{12}$$

$$\frac{\epsilon}{F_d}(s) = \frac{-1}{s(Ms + D + K_a K_m) + K_p K_m} \tag{13}$$

■

16.3 ERROR

16.3.1 Output Should Follow Input

In Section 16.2.1 it was stated that the primary objective of a control system is that the output follow the input. Since the difference between the output and input is equal to the error ϵ, then inspection of the magnitude of the error will establish how well the control system is achieving its objective.

16.3.2 Error Due to Inability to Follow the Input

The system tries to follow the input, complying with its primary objective. However, it isn't always successful, depending upon how demanding the input happens to be. As a result, there is an error due to the inability of the system to follow the input.

To determine the error due to the input, refer to Fig. 16.1.4-1, the block diagram for a generalized control system. To determine the forward gain, trace a path from θ_{in} to ϵ. There is no sign change and the gain is equal to unity. The open loop gain L is equal to $-AGH$. Applying these relationships to the closed loop rule, eq. 13.2.4-6, we have

$$\frac{\epsilon}{\theta_{in}}(s) = \frac{1}{1 + AGH} \qquad [16.3.2\text{-}1]$$

The steady state response is found by making use of Section 4.4.4.

$$\epsilon_{ss} = \lim_{s \to 0} \frac{S\theta_{in}(s)}{1 + AGH} \qquad [16.3.2\text{-}2]$$

It should be pointed out that, to evaluate the steady state error, specific functions must be used in place of the symbols A, G, and H. Also, the form of the input must be known.

16.3.3 Disturbance

A disturbance is an outside influence that affects the control system. The disturbance is usually unknown, both in magnitude and direction, making it unpredictable. Examples of disturbances are wind loads, bumps in the road, particles falling in or on the machine, vibration, noise, stray electric and magnetic fields, eddies in fluids, leaks into or out of pressure lines, heat flow into or out of the system, bubbles and dirt, flaws in the working substance, and so on.

The disturbance tends to divert the system away from the input, thus preventing the system from achieving its primary objective. Therefore, the disturbance is an error source, an unwelcome random input. While great pains may be taken to minimize disturbances, they still remain, unpredictable and uncorrectable. Consequently, the designer of a control system must make estimates of possible disturbances and create means to minimize their diverse effects upon the system.

16.3.4 Error Due to a Disturbance

To determine the error due to a disturbance, refer to Fig. 16.1.4-1(b). The disturbance may appear at any point in the loop, a typical location being between the controller and the process (or plant). For the forward gain, trace a path from the disturbance T_d to the error ϵ. Note that there is a sign change and that the forward gain is equal to $-GH$. The open loop gain L is the same as that found in Section 16.3.2. Apply the closed loop rule and multiply both sides of the equation by T_d.

$$\epsilon(s) = \frac{-GHT_d(s)}{1 + AGH} \qquad [16.3.4\text{-}1]$$

Similar to that used in eq. 16.3.2-2, the steady state response is given by

$$\epsilon_{ss} = \lim_{s \to 0} \left[\frac{-sGHT_d(s)}{1 + AGH} \right] \qquad [16.3.4\text{-}2]$$

Similar to the statements made for eq. 16.3.2-2, in order to evaluate the steady state error, the specific functions A, G, and H must be known. Also, it will be necessary to assume the form and the maximum numerical value of the disturbance. It may be necessary to make a set of such assumptions and choose the worst case.

■ **EXAMPLE 16.3.4-1** Electromechanical Displacement Servo
An angular displacement servo using electromechanical elements is shown in Fig. 16.3.4-1(a). The input device is a remotely located pot (or potentiometer, as shown in Fig. 3.3.2-1). The voltage on the wiper of this pot is compared to that of a second pot mounted on the output shaft. The difference between the

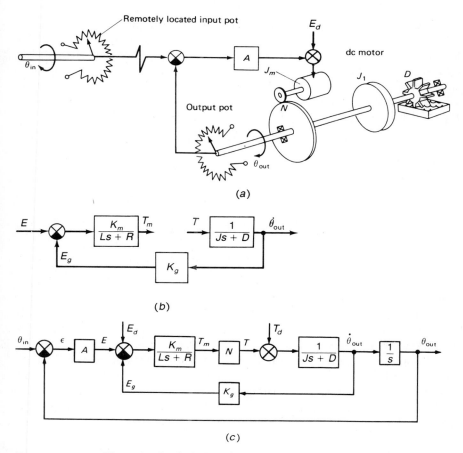

Figure 16.3.4-1   Electromechanical displacement servo. (a) Actual system. (b) Blocks for motor and load. (c) Complete system.

two pot voltages is amplified and this is fed to a dc motor with fixed field. The motor uses a gear train with a speed ratio (reduction) $1/N$ to drive a load (such as an antenna) with moment of inertia J and damper D.

In addition to the normal input θ_{in}, there are two disturbances, E_d electrical noise that enters the system after the amplifier A before the signal enters the motor, and T_d a mechanical disturbance in the form of a wind load at the antenna.

Draw the block diagram, find the closed loop transfer function, and determine the error due to the normal input and due to each of the disturbances.

Solution.   Assume that pot voltage is synonymous with shaft angle. Hence, the difference represents the error ϵ. The block diagram of the motor is shown in Fig. 13.1.6-1(b). The transfer function of the process (or plant) is given by eq. 13.1.5-3, where

$$\omega = \dot{\theta} \tag{1}$$

The torque T_m of the motor is multiplied by the gear ratio N and this drives the load or process, closing the motor loop. The output of the motor loop is shaft velocity $\dot{\theta}_{out}$, which, upon multiplication by $1/s$ (the equivalent of an integration) produces output shaft angle θ_{out}. This, in turn, is fed back and is compared to the input angle θ_{in}. Each of the disturbances is appropriately located. The diagram can be completed with the help of the generalized model, Fig. 16.1.4-1(b). The result is shown in Fig. 16.3.4-1(c), where:

$$J = J_1 + J_m N^2 \qquad\qquad D = 7\,\text{N}\cdot\text{m}\cdot\text{s}$$

$$J_1 = 0.05\,\text{kg}\cdot\text{m}^2 \qquad\quad K_m = 8\,\text{N}\cdot\text{m/A}$$

$$J_m = 5\times10^{-6}\,\text{kg}\cdot\text{m}^2 \qquad K_g = 1.0\,\text{V/rad/s} \tag{1a}$$

$$N = 100 \qquad\qquad A = 400\,\text{V/rad*}$$

$$R = 50\,\Omega \qquad\qquad L = 0.1\,\text{H}$$

From Fig. 16.3.4-1 it is clear that the system consists of a loop within a loop. To reduce the inner loop to a single block, it will be necessary to shift the disturbances T_d and E_d. See Fig. 16.3.4-2(a). The inner loop now can readily be reduced to a single block. Inserting the result in the original diagram, the result appears in Fig. 16.3.4-2(b).

We are now ready to determine the various closed loop terms. First the closed loop transfer function, from the input θ_{in} to the output θ_{out}. This path has the form of a fraction, inviting the use of the fraction loop rule. The feedback is equal to -1 and the numerator is equal to AK_mN. This product is subtracted algebraically from the denominator. Multiplying the terms to form

---

*This is the nominal value. For design purposes, it may be varied over a large range.

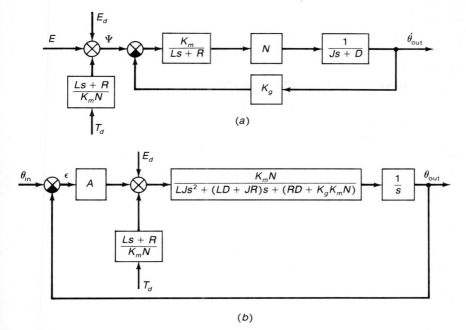

Figure 16.3.4-2 Reduction of block diagram for electromechanical servo. (a) Inner loop. (b) Complete system.

a polynomial in the denominator, we have

$$\frac{\theta_{out}}{\theta_{in}}(s) = \frac{AK_mN}{\Delta} \qquad (2)$$

where

$$\Delta(s) = LJs^3 + (LD + JR)s^2 + (RD + K_gK_mN)s + AK_mN \qquad (3)$$

or, in normalized form,

$$\frac{\Delta(s)}{LJ} = s^3 + \frac{(LD + JR)}{LJ}s^2$$

$$+ \frac{(RD + K_gK_mN)}{LJ}s + \frac{AK_mN}{LJ} \qquad (3a)$$

Apply numerical values.

$$\frac{\Delta(s)}{LJ} = s^3 + 570s^2 + 115,000s + K \qquad (3b)$$

where

$$K = \frac{AK_mN}{LJ} = 8 \times 10^4 A \qquad (3c)$$

Note that the characteristic polynomial $\Delta(s)$ is a cubic. Hence, this is a third order system.

The path from the input θ_{in} to the error ϵ has a forward gain FG that is equal to $+1$. Since this is not a fraction, the closed loop rule is used. After rationalizing the resulting improper fraction, we have

$$\frac{\epsilon}{\theta_{in}}(s) = \frac{s\left[LJs^2 + (LD + JR)s + (RD + K_gK_mN)\right]}{\Delta} = \frac{s(Q)}{\Delta} \qquad (4)$$

where Q is used to represent the numerator.

To find the error ϵ due to the electrical disturbance E_d, the forward gain is a fraction. Using the fraction loop rule, we have

$$\frac{\epsilon}{E_d}(s) = \frac{K_mN}{\Delta} \qquad (5)$$

To find the error due to the mechanical disturbance T_d, multiply the quantity in eq. (5) by $(Ls + R)/K_mN$.

$$\frac{\epsilon}{T_d}(s) = \frac{Ls + R}{\Delta} \qquad (6)$$

Now, let us examine the steady state errors for eqs. (4), (5), and (6). For this, we need to know the form(s) of the error source. Each will be treated separately.

Input

$$\epsilon_{ss} = \lim_{s \to 0} s\left[\frac{\epsilon}{\theta_{in}}(s)\theta_{in}(s)\right] \qquad (7)$$

Four input forms are likely to occur: impulse, step, ramp, and parabolic (constant acceleration).

Impulse input: $\theta_{in}(s) = \theta_i$.

$$\epsilon_{ss} = \lim_{s \to 0} s\left[\frac{s(Q)\theta_i}{\Delta}\right] = 0 \qquad (8)$$

Step input: $\theta_{in}(s) = \theta_s/s$.

$$\epsilon_{ss} = \lim_{s \to 0} s\left[\frac{s(Q)\theta_s/s}{\Delta}\right] = 0 \qquad (9)$$

Ramp input: $\theta_{in}(s) = \theta_r/s^2$.

$$\epsilon_{ss} = \lim_{s \to 0} s\left[\frac{s(Q)\theta_r/s^2}{\Delta}\right] = \frac{(RD + K_gK_mN)\theta_r}{AK_mN} \qquad (10)$$

Parabolic input: $\theta_{in}(s) = 2\theta_p/s^3$.

$$\epsilon_{ss} = \lim_{s \to 0} s\left[\frac{s(Q)2\theta_p/s^3}{\Delta}\right] = \infty \qquad (11)$$

Electrical Disturbance

Two forms of this disturbance are likely: impulse and step.

Impulse electrical disturbance: $E_d(s) = E_i$.

$$\epsilon_{ss} = \lim_{s \to 0} s \left[\frac{K_m N E_i}{\Delta} \right] = 0 \tag{12}$$

Step electrical disturbance: $E_d(s) = E_s/s$.

$$\epsilon_{ss} = \lim_{s \to 0} s \left[\frac{K_m N E_s/s}{\Delta} \right] = \frac{E_s}{A} \tag{13}$$

Torque Disturbance

Two forms of this disturbance are likely: impulse and step.

Impulse torque disturbance: $T_d(s) = T_i$.

$$\epsilon_{ss} = \lim_{s \to 0} s \left[\frac{(Ls + R)}{\Delta} T_i \right] = 0 \tag{14}$$

Step torque disturbance: $T_d(s) = T_s/s$.

$$\epsilon_{ss} = \lim_{s \to 0} s \left[\frac{(Ls + R)}{\Delta} T_s/s \right] = \frac{R T_s}{A K_m N} \tag{15}$$

Conclusions

Equations (8) and (9) indicate that the system has no difficulty following impulse or step inputs. It has a little difficulty following a ramp input, according to eq. (10), since there is an error, but this error can be reduced if

$$A \gg 1 \tag{16}$$

Equation (11) indicates that the system cannot follow a parabolic input at all. This means that the user of the system will have to be so informed, restricting the system to applications where a parabolic input will not occur. Equations (12) and (14) indicate that the system is capable of rejecting or ignoring either impulse disturbance. Equations (13) and (15) indicate that the system has a little difficulty with step disturbances, causing some error, but this can be reduced in the same manner as that in eq. (16).   ∎

16.4 FEEDBACK

16.4.1 Types of Feedback

Many simple control systems use unity feedback. However, there are cases for which additional terms in the feedback are needed for specific performance requirements. Some commonly used terms are as follows:

  unity

  unity plus derivative

  unity plus integral

  polynomial in s

16.4.2 Unity Plus Derivative Feedback

Perhaps one of the most common types is unity plus derivative feedback. Servo motors are commercially available that have a built-in tachometer. This provides derivative feedback. The system may also have a position readout to provide unity feedback. The combination provides unity plus derivative feedback.

In high performance aircraft and rockets, there is a need for an attitude and a rate gyro. The attitude device measures the angle the craft makes with the vertical. This is accomplished by the nature of a free gyro, which remains fixed in space as the vehicle rolls and pitches. In this way, the attitude gyro measures the relative angle between its own fixed position and the vehicle's changing one. This is unity feedback. The rate gyro is an instrument that measures the vehicle's rate of turn with respect to fixed space. This information is used in several places. One, to coordinate the bank angle with the rate of turn so that the net force (gravity plus centripetal acceleration) is perpendicular to the air foil.

A more sophisticated installation requires roll rate information for stability when strong crosswinds tend to turn the aircraft about its roll axis. In that case, both the attitude angle and the roll rate are fed back.

For very sophisticated vehicles, there may be three axes of angle measurement and three angular velocities. Each axis does a different job, and each may involve unity plus derivative feedback.

Some systems have very little internal damping. For stability, damping must be provided somewhere. Some systems place damping in the controller, others place it in the feedback. This is used in conjunction with position feedback, making a combination of unity plus derivative feedback.

Thus, it is clear that unity plus derivative feedback deserves a thorough investigation.

■ **EXAMPLE 16.4.2-1** Aircraft with Roll Stabilization

A high speed aircraft with roll stabilization is shown in Fig. 16.4.2-1(a). The aircraft is approximated as a one mass torsional system. Two gyros are employed, one sensing the roll angle θ, the other sensing roll rate $\dot{\theta}$. A torque T_c can be applied about the roll (longitudinal) axis of the aircraft by tilting two ailerons (movable aerodynamic surfaces at the trailing edge of each wing) in opposite directions. Model the system, draw the block diagram, and determine the following: open loop again L, closed loop transfer function θ/θ_{in}, error due to generalized input $\theta_{in}(s)$, error due to generalized disturbance $T_d(s)$, and the steady state errors.

Solution.   The roll angle gyro reads the difference between the vehicle roll angle θ and the required angular input θ_{in}. (An angle may be required to bank the vehicle while making a turn.) Since the angle is measured directly, this is tantamount to unity feedback. The roll rate gyro produces a signal that is proportional to $\dot{\theta}$ where the constant of proportionality is D_g. This is equiva-

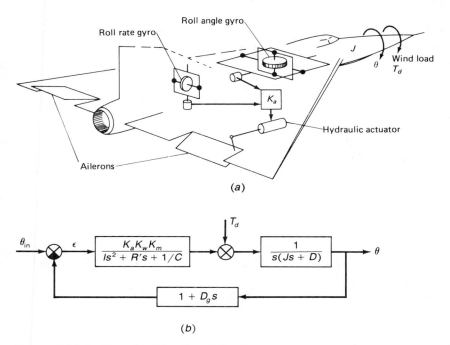

Figure 16.4.2-1 Aircraft with roll stabilization. (a) Actual system. (b) Block diagram.

lent to derivative feedback. The total feedback is equal to the sum. Then

$$H(s) = 1 + D_g s \tag{1}$$

The system has unity plus derivative feedback.

The controller A consists of an amplifier K_a and a pair of hydraulic actuators (only one of which is shown in the figure). The amplifier K_a produces a pressure P that is proportional to the error ϵ. Then

$$P = K_a \epsilon \tag{2}$$

Since this is a high speed aircraft, we cannot neglect the dynamic response of the hydraulic actuator. However, we can assume that the torque T_c produced by the ailerons is instantaneous (once they move). In order to model the hydraulic actuator, assume that it can be represented by a series fluid circuit whose transfer function is given by eq. 13.1.7-2, repeated below.

$$\frac{W}{P}(s) = \frac{1}{Is^2 + R's + 1/C} \tag{3}$$

Assuming that the hydraulic actuator is a positive displacement type, the displacement (tilt) of each aileron is proportional to the volume W of the fluid, where the constant of proportionality is K_w. Assume that the torque T_c applied to the aircraft is proportional to the displacement of the ailerons (when they are tilted in opposite directions), where the constant of proportionality is K_m. Then

$$T_c = W K_w K_m \tag{4}$$

$$= \frac{K_a K_w K_m \epsilon}{I s^2 + R' s + 1/C} \tag{5}$$

The torque T_c will be referred to as the control torque. This and the disturbance torque T_d (wind load) are both applied to the aircraft structure. The process (plant) is the aircraft structure whose moment of inertia is J and whose air damping is D. The transfer function is given by

$$G(s) = \frac{1}{J s^2 + D s} = \frac{1}{s(J s + D)} \tag{6}$$

Apply eqs. (1), (5), and (6) to the general model in Fig. 16.1.4-1. The resulting block diagram is shown in Fig. 16.4.2-1(b).

The following numerical values apply.

$J = 3 \times 10^5$ kg $\cdot$ m^2 $D = 15 \times 10^5$ N $\cdot$ m $\cdot$ s

$D_g = 0.05$ V $\cdot$ s/rad* $C = 0.015$ m^5/N

$R' = 0.8$ N $\cdot$ s/m^5 $I = 0.01$ N $\cdot$ s^2/m^5

$K_p{}^* = K_a K_w K_m = 2 \times 10^9$ N $\cdot$ m $\cdot$ N/m^5

Refer to Fig. 16.4.2-1(b). The open loop gain is

$$L(s) = \frac{-K_a K_w K_m (1 + D_g s)}{s(J s + D)(I s^2 + R' s + 1/C)} \tag{7}$$

The closed loop transfer function is

$$\frac{\theta}{\theta_{in}}(s) = \frac{K_a K_w K_m}{s(J s + D)(I s^2 + R' s + 1/C) + K_a K_w K_m (1 + D_g s)} \tag{8}$$

The error due to a generalized input is

$$\frac{\epsilon}{\theta_{in}}(s) = \frac{s(J s + D)(I s^2 + R' s + 1/C)}{s(J s + D)(I s^2 + R' s + 1/C) + K_a K_w K_m (1 + D_g s)} \tag{9}$$

---

*These are nominal values. They may be varied over a large range for design purposes.

The error due to a generalized disturbance is

$$\frac{\epsilon}{T_d}(s) = \frac{-(1 + D_g s)(Is^2 + R's + 1/C)}{s(Js + D)(Is^2 + R's + 1/C) + K_a K_w K_m(1 + D_g s)} \quad (10)$$

Note that, upon expanding the characteristic polynomial, it is an unfactored quartic. This is a fourth order system. Leave the denominator as shown. It will not pay to expand it, particularly since the next task is to find the steady state errors. To do this, apply the final value theorem as given in Section 4.4.4.

The steady state error due to the generalized input is given by

$$\epsilon_{ss} = \lim_{s \to 0} \left\{ s \left[\frac{\epsilon}{\theta_{in}}(s) \right] \theta_{in}(s) \right\} \quad (11)$$

$$= \lim_{s \to 0} \frac{s^2 D \theta_{in}(s)/C}{K_a K_w K_m} \quad (12)$$

Since there is an s^2 term in the numerator, obviously an impulse and a step input will have a zero steady state error. A ramp input would have the following response:

$$\theta_{in}(s) = V/s^2 \quad (13)$$

then

$$\epsilon_{ss} = \frac{DV}{K_a K_w K_m C} \quad (14)$$

The steady state error due to the generalized disturbance is given by

$$\epsilon_{ss} = \lim_{s \to 0} \left\{ s \left[\frac{\epsilon}{T_d}(s) \right] T_d(s) \right\} \quad (15)$$

$$= \lim_{s \to 0} \frac{-s T_d(s)/C}{K_a K_w K_m} \quad (16)$$

Although the disturbance is unknown, we may assume that it could have the form of either an impulse or a step. (A ramp disturbance is highly unlikely, since this implies that the disturbance grows without bound.) Since there is an s in the numerator of eq. (16), we can immediately see that an impulse disturbance will have a zero steady state error. Thus, we need study only the step disturbance for this system. For a step disturbance

$$T_d(s) = \frac{U_d}{s} \quad (17)$$

and

$$\epsilon_{ss} = \frac{U_d}{K_a K_w K_m C} \quad (18)$$

16.5 STABILITY OF CONTROL SYSTEMS

16.5.1 Definition of Stability

A system is *stable* if its output remains *bounded* for a small input or disturbance. A system is *unstable* if a small input or disturbance produces an *unbounded* output. These cases are shown in Fig. 16.5.1-1. A cone resting on its base returns if tilted through a small angle. Hence, the cone in this position is *stable*. If the cone is precariously balanced on its apex, the slightest disturbance will cause it to move away from that initial position. Only the floor limits the motion. In this position, the cone is *unstable*.

Now lay the cone on its side. If the surface is level, then the cone will remain in any position. This case lies between the first two. On the one hand the motion is not unbounded, but on the other it does not approach a limit. This case is referred to as *neutral stability*. Another example of neutral stability is an undamped pendulum. Subject to a disturbance, the pendulum will oscillate indefinitely, the output does not become unbounded, but it does not approach a limit.

The considerations in the previous two paragraphs may be termed *absolute stability*. It is insufficient to merely establish absolute stability. Some systems are more stable than others. For example, the broad based cone in Fig. 16.5.1-1(d) is more stable than the one shown in Fig. 16.5.1-1(e). Hence, the notion of *relative stability*, or as it will be termed in Section 16.5.6, the margin of stability.

The differential equation of a linear system of nth order will have the form

$$\left(P^n + A_1 P^{n-1} + \cdots + A_{n-1} P + A_n \right) x = f(t) \qquad [16.5.1-1]$$

The solution will be of the form

$$x(t) = C_1 e^{r_1 t} + C_2 e^{r_2 t} + \cdots + C_n e^{r_n t} \qquad [16.5.1-2]$$

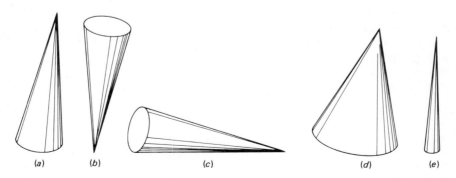

Figure 16.5.1-1   Stability of cone. (a) Stable. (b) Unstable. (c) Neutral stability. (d) High relative stability. (e) Low relative stability.

where

$$r_i = \text{root of the characteristic polynomial}$$

$$P = \text{derivative operator} \qquad [16.5.1-3]$$

$$A_i, C_i = \text{constants}$$

In general, the roots r_i will be complex. The real part will describe a decay envelope, while the imaginary part will provide an oscillation. If any *one* of the real terms is *positive*, then the solution will grow *unbounded* with time. Consequently, if the real part of any root is *positive*, then the system is *unstable*.

16.5.2 Instability Identified by Either Negative or Missing Terms

The closed loop transfer function for the general system of nth order is of the form

$$\frac{\theta}{\theta_{in}}(s) = \frac{\text{numerator}(s)}{A_n s^n + A_{n-1} s^{n-1} + \cdots + A_1 s + A_0} \qquad [16.5.2-1]$$

For the ensuing discussion, the numerator will have little consequence. Let us concentrate upon the denominator. If there is but one negative term, or if any term of the polynomial in s is missing, from the theory of equations, this implies that one of the roots is positive, resulting in an *unstable* system.

16.5.3 Instability for Positive Feedback or Positive Open Loop Gain

If the control system has positive feedback, the controller drives the system in the *same* direction as the error. But this increases the error. This is the case of the cone balanced on its apex in Fig. 16.5.1-1(b). The error increases without bound and the system is *unstable*.

The open loop gain L is given by the product of the feed forward and the feedback. In the previous paragraph it was established that positive feedback results in an unstable system. So, to begin with, we must consider only negative feedback. If the open loop gain is positive (and we have established that feedback must be negative) then this implies that there is a change of sign in the feed forward loop. The feedback being negative provides an error signal that would tend to drive the system in the direction that would correct the error. Hence, a change in sign will divert the system away, resulting in a growing error. Thus, a positive open loop gain results in an *unstable* system.

16.5.4 Stability Using Root Locus

The root locus method plots the roots of the *closed* loop transfer function starting with two special cases where

$$K = 1, \qquad K = \infty \qquad\qquad [16.5.4\text{-}1]$$

But these two conditions are the same as determining the roots of the *open* loop. Of course, a great deal of effort must be expended to determine the roots for all the cases of K between the two limits. In this vein, a number of innovative techniques have been created to assist in making the plot. Chapter 15 covers a number of these techniques.

■ **EXAMPLE 16.5.4-1** Stability of Electromechanical Servo Using the Root Locus

Refer to Ex. 16.3.4-1. In that example, eq. (3b) provides the characteristic equation. This particular system was analyzed throughout Chapter 15 (Root Locus). Consequently, some of the parameters were found in that chapter. In Ex. 15.3.13-1, it was determined that the root locus crossed the imaginary axis (where the system is on the verge of instability). At this point it was found that

$$K = 65.5 \times 10^6 = K_c \qquad\qquad (1)$$

$$\zeta = 0 \qquad\qquad (2)$$

Refer to Ex. 16.3.4-1, eq. (3c), repeated below.

$$K = K_c = \frac{AK_mN}{LJ} = 8 \times 10^4A \qquad\qquad (3)$$

Apply the numerical value of K_c and solve for A.

$$A = 819 \qquad\qquad (4)$$

In Section 15.4.3, the criteria for a safe margin of stability is given by eqs. 15.4.3-1 and 15.4.3-2, repeated below.

$$K \le K_c/3 \qquad\qquad (5)$$

$$\zeta \ge 0.5 \qquad\qquad (6)$$

In Ex. 15.4.3-1, using these two criteria, it was found, respectively,

$$K \le 21.8 \times 10 \qquad\qquad (7)$$

$$K \le 15.0 \times 10 \qquad\qquad (8)$$

The lower value satisfies both inequalities. Then

$$K = 15 \times 10^6 \qquad\qquad (9)$$

for which

$$\zeta = 0.5 \qquad\qquad (10)$$

Use this value of K in eq. (3c) to determine the corresponding value of A for a very stable system.

$$A = 188 \tag{11}$$

We have the two extreme values of A, corresponding to a system on the verge of instability and to the very stable system. Select a value between these limits.

$$A = 400 \tag{12}$$

Using eq. (3c), determine the corresponding value of K.

$$K = 32 \times 10^6 \tag{13}$$

Using the root locus plot, determine the location of this value of K and note the (complex) roots:

$$-49.9 \pm j256 = -a \pm j\omega_d \tag{14}$$

From these, compute ω_n and ζ.

$$\omega_n = 261, \qquad \zeta = 0.19 \tag{15}$$

Now that all the numerical values are available, the rest of the problem may be simulated with TUTSIM. The resulting plots appear in Fig. 16.5.4-1, where the transient response is plotted for each value of A. Note that for

$$A = 819, \qquad \zeta = 0 \tag{16}$$

the oscillation does not die out, as expected for zero damping. It is clear that the system is on the verge of instability.

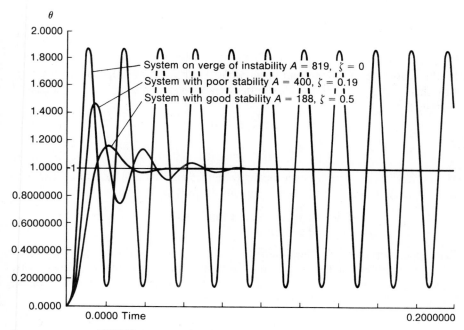

Figure 16.5.4-1   TUTSIM plots for electromechanical servo.

Consider the opposite extreme.

$$A = 188, \qquad \zeta = 0.5 \tag{17}$$

The oscillation dies out rapidly, as expected for this optimum value of damping.

Finally, consider the intermediate value,

$$A = 400, \qquad \zeta = 0.19 \tag{18}$$

Here, the oscillation dies out slowly, demonstrating that the system has only a small margin of stability.

It was pointed out that the root locus also provides transient properties. For these, refer to Ex. 16.6.2-2.   ∎

16.5.5 Stability Using the Open Loop Gain

The closed loop transfer function is given by eq. 13.2.4-6, repeated below.

$$\frac{out}{in} = \frac{FG}{1 - L} \qquad [16.5.5\text{-}1]$$

where   FG = forward gain
L = open loop gain (including algebraic sign)

It was established in Section 16.5.4 that the open loop gain must be negative. Including the necessary sign, eq. 16.5.5-1 becomes

$$\frac{out}{in} = \frac{FG}{1 + L} \qquad [16.5.5\text{-}2]$$

In the *closed* loop, the system is *unstable* if the function approaches infinity. But this is the same as the denominator approaching zero: For an unstable system,

$$1 + L = 0 \qquad [16.5.5\text{-}3]$$

or

$$L = -1 \qquad [16.5.5\text{-}4]$$

Hence, eq. 16.5.5-4 provides a means to detect an unstable system by examining the *open* loop. The following conclusion may be drawn:

$$\left. \begin{array}{l} \textit{closed } \text{loop approaching infinity is the same as the} \\ \textit{open } \text{loop approaching } -1 \end{array} \right\} \quad [16.5.5\text{-}5]$$

The open loop gain is a function of the Laplace operator s that may be plotted on the complex plane or on the complex frequency plane. Either way, *instability* is established if $L = -1$ or if L has a magnitude $= 1$ with a phase angle $\pm 180°$. Thus, the system is *unstable* if

$$L = 1 \underline{/\pm 180°} \qquad [16.5.5\text{-}6]$$

Keep in mind that the minus sign associated with the summing point has already been accounted for. Consequently, we are now discussing the function L itself: the *positive* blocks that make up the loop, exclusive of the summing point. If this, itself, is equal to -1, the system is unstable.

Now the function L itself is always in factored form, while the closed loop transfer function seldom is. Equation 16.5.5-5 offers two ways to determine stability, one using the closed loop, the other using the open loop. Since the open loop L is always in factored form, it offers the more convenient technique. In performing the computation, since L is a function of s, set

$$s = j\omega \qquad\qquad [16.5.5\text{-}7]$$

Here, too, a labor saving task is available. The frequency methods in Chapter 8 perform this computation for five factors, the most commonly encountered in the subject of automatic control.

■ **EXAMPLE 16.5.5-1** Stability of Undamped System
Let us ascertain the stability of an undamped system using several methods. Since an undamped system does not approach a limit, we anticipate that it will be unstable. An undamped system is shown in Fig. 16.5.5-1. The closed loop transfer function is

$$\frac{X_{\text{out}}}{X_{\text{in}}}(s) = \frac{A}{MS^2 + A} = \frac{A/M}{S^2 + A/M} \qquad (1)$$

Upon examination of eq. (1), we note that one term is missing from the denominator. This is an immediate clue to instability.

Next, consider the response of the closed loop to any finite input. Since any input may be used, let us apply a sinusoidal input and examine the frequency response. Fig. 8.4.7-1 provides the response for all reciprocal quadratics. The system given by eq. (1) belongs to this class. Fig. 8.4.7-1 indicates that for zero damping ratio, the output is unbounded, thus branding this an unstable system. Note that the frequency at which this occurs is equal to $\sqrt{A/M}$.

Next, consider the open loop plot. The open loop gain $L(s)$ is given by

$$L(s) = \frac{-A}{Ms^2} \qquad (2)$$

Recall that the negative sign (as far as the frequency plot is concerned) has been accounted for in formulating the stability criterion. Hence, let us plot L

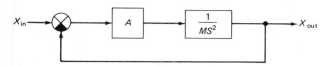

Figure 16.5.5-1   Undamped system.

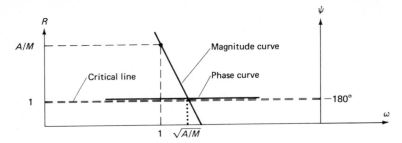

Figure 16.5.5-2   Unstable system.

itself, in which case

$$L(j\omega) = \frac{A}{Ms^2} \tag{3}$$

In constructing the plot, it will prove convenient to set up the scales so that magnitude = 1 lines up with phase = $-180°$. In this way, the two simultaneous conditions required in eq. 16.5.5-6 can be observed by inspection. The line joining magnitude = 1 with phase = $-180°$ is called the *critical line*. The magnitude and phase curves are shown on one plot in Fig. 16.5.5-2. The phase curve lies entirely on the critical line, waiting for the magnitude curve to cross. Of course, this happens, demonstrating that the system is unstable. Incidentally, the frequency at which the curves cross on the critical line is equal to $\sqrt{A/M}$, the same frequency found before.

Here we see the advantage of lining up the scales and using the critical line. By inspection, we see that the curves cross together, meaning that magnitude is equal to unity at the same frequency for which the phase is equal to $-180°$. This pair of simultaneous conditions meets the conditions for instability indicated in eq. 16.5.5-6.

For the fourth proof, consider the root locus method. The root locus for the typical second order system with all values of damping is shown in Fig. 15.1.2-2. The root locus plot holds ω_n fixed while ζ is varied. The locus crosses the imaginary axis, indicating instability when $\zeta = 0$, at which point $K = \omega_n$. In turn,

$$\omega_n = \sqrt{A/M} \tag{4}$$

This reveals the same information as all the other methods have done. All results are consistent. ∎

16.5.6 Margin of Stability

In the previous section, the two curves attained their respective critical values at the same time. The magnitude became unity at exactly the same frequency where the phase became equal to $-180°$. Now the question arises, "What if the two curves to not cross the critical line simultaneously, but instead cross at points that are near one another?"

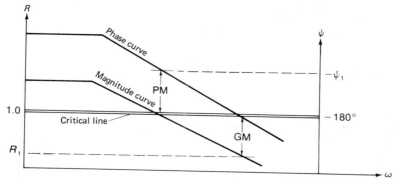

Figure 16.5.6-1   Margin of stability.

This concept is shown in Fig. 16.5.6-1. The two curves do *not* cross the *critical* line together. Examine each one where it alone crosses, and determine the distance the other curve *missed* the line. The amount of miss, or more precisely, the vertical distance the opposite curve is away from the *critical* line is defined as the *margin of stability*.

When the magnitude curve crosses the critical line, the phase curve should lie above it. The *distance* from the *critical* line *up* to the *phase curve* is called *phase margin PM*. When the phase curve crosses the critical line, the distance from the *critical* line *down* to the *magnitude curve* is called *gain margin GM*.

Refer to Fig. 16.5.6-1. The PM is measured on the (linear) phase scale in degrees by subtracting the angle ψ_1 from 180°. If the phase curve lies *above* the critical line (as shown) then the PM is given a positive sign. The GM is measured on the (log) magnitude scale as a ratio.

$$GM = 1.0/R_1$$

How much of a margin is suitable? Mathematically speaking, if PM > 0° the system is stable. However, a small change in numerical values of parameters due to manufacturing tolerances or small internal phase shifts within components could conceivably thrust the system into the unstable region. In addition to this, even if the accumulation of manufacturing tolerances does not cast the system as an unstable one, if the margin of stability is insufficient, the actual system will chatter, shimmy, or hum. Although the output does *not* approach infinity, these disturbances will make the system *unusable* for practical purposes.

Considering practical experience with feedback control systems, as a rule of thumb, three conditions (all three) must be met for the system to be *stable*:

$$
\left.
\begin{aligned}
&\text{PM} \geq 45° \\
&\text{GM} \geq 3 \quad \text{(about 10 dB)} \\
&\text{at point of crossing, magnitude curve} \\
&\text{must have slope more negative than } -\tfrac{1}{2}
\end{aligned}
\right\}
\qquad [16.5.6\text{-}1]
$$

■ **EXAMPLE 16.5.6-1** Margin of Stability

Determine the margin of stability of the following open loop

$$L(s) = \frac{200}{(s + 2)(s + 5)(s + 10)} \qquad (1)$$

Since the expression is in the form of products of factors, sketch the asymptotes (Bode plots) using methods shown in Sections 8.4 and 8.6. The composites are shown in Fig. 16.5.6-2. For this example, it can be shown that the asymptotes provide fair accuracy. If this were not the case, then the actual curves would be required. Note that the magnitude curve crosses the critical line first, a necessary condition for stability. At the point where the magnitude curve crosses the critical line, a vertical line is drawn until it strikes the phase curve. Reading horizontally to the phase scale, the reading is equal to $-130°$, which is $50°$ above $-180°$. Hence, the PM (phase margin) is equal to

$$PM = 50° > 45° \quad OK \qquad (2)$$

Follow the phase curve until it crosses the critical line. At that point, drop vertically until it strikes the magnitude curve. The reading on the magnitude scale is 0.21. Hence the GM (gain margin) is equal to

$$GM = \frac{1}{0.21} = 4.8 > 3 \quad OK \qquad (3)$$

When the magnitude curve crosses the critical line, the curve has a slope $= -1$, or

$$slope = -1 < -\tfrac{1}{2} \quad OK \qquad (4)$$

The system has met all three requirements for stability.   ■

■ **EXAMPLE 16.5.6-2** Stability of Rolling Mill

Refer to Ex. 16.2.2-1 (Rolling Mill). Determine the phase margin for the following values of K_p:

(a) 5×10^8
(b) 1.5×10^8
(c) 0.44×10^8

Solution.   The open loop gain is given by eq. (7) of Ex. 16.2.2-1. Delete the minus sign, since this was accounted for in establishing the stability criteria.

$$L = \frac{K_p}{Ms^2 + Ds + K} \qquad (1)$$

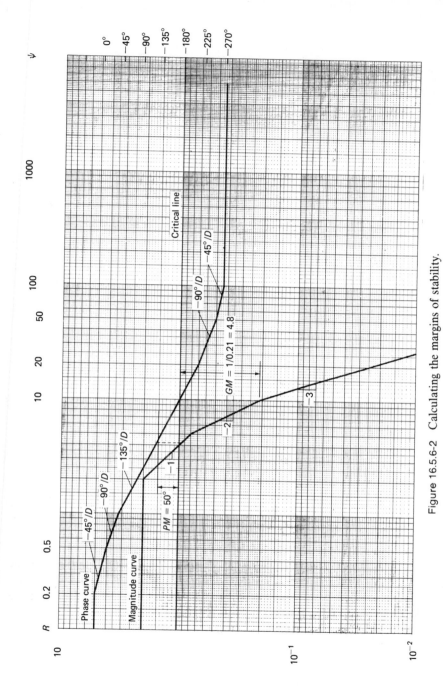

Figure 16.5.6-2 Calculating the margins of stability.

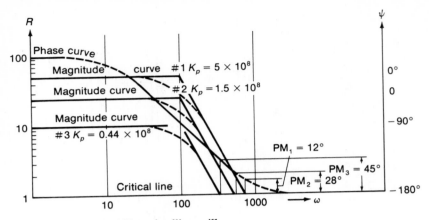

Figure 16.5.6-3   Stability of rolling mill.

Normalize.

$$L = \frac{K_p/M}{s^2 + 2as + \omega_0^2} \tag{2}$$

$$M = 10^3 \text{ kg}, \qquad D = 1.5 \times 10^5 \text{ N} \cdot \text{s/m}, \qquad K = 10^7 \text{ N/m} \tag{3}$$

$$\omega_0 = \sqrt{K/M} = \sqrt{\frac{10^7}{10^3}} = 100 \text{ rad/s} \tag{4}$$

$$\zeta_0 = \frac{D/M}{2\omega_0} = \frac{1.5 \times 10^5/10^3}{2 \times 100} = 0.75 \tag{5}$$

$$I_0 = \lim_{s \to 0} L(s) = \frac{K_p}{K} = \frac{K_p}{10^7} \tag{6}$$

The open loop stability curves are plotted in Fig. 16.5.6-3. Since the point of interest is close to the corner frequency of the phase curve, asymptotes will not suffice. The actual curves must be plotted with the help of Tables 8.5.2-1 and 8.5.2-2, using $\zeta_0 = 0.75$. The phase margin PM is read graphically on Fig. 16.5.6-3, which made use of the open loop plot.

The closed loop properties are found by the use of Ex. 16.2.2-1, eq. (8), where

$$\omega_n = \sqrt{\frac{K + K_p}{M}} \tag{7}$$

$$\zeta = \frac{D}{2M\omega_n} \tag{8}$$

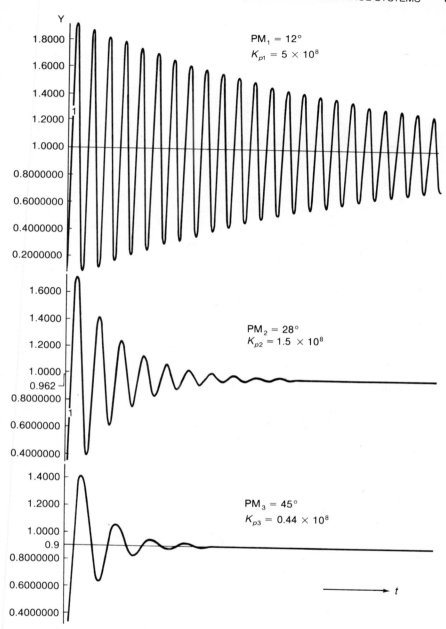

Figure 16.5.6-4   Response to unit step input for rolling mill.

The information is tabulated as follows:

Curve #	K_p	PM	ω_n	ζ	
1	5×10^8	12°	714	0.105	
2	1.5×10^8	28°	400	0.188	(7)
3	0.44×10^8	45°	232	0.323	

The rolling mill was programmed using TUTSIM for each of the above values of K_p. The results are shown in Fig. 16.5.6-4. It is clear that a phase margin less than 45° results in excessive oscillation before the system settles. For the rolling mill, this means the rolling operation will produce sheets whose thickness varies. Of course, such material cannot be used and thus represents waste. For $PM_1 = 12°$, there will be an enormous amount of waste, while for $PM_3 = 45°$, there will be only a small amount of waste. The computer runs make it clear why the criterion of 45° phase margin was established.   ■

16.6 ADDITIONAL SYSTEM PROPERTIES

16.6.1 Time Delay

A discrete type of time delay* occurs when the signal must physically travel slowly, particularly if the signal is physically carried. Examples of this are:

(a) rolling mill where the material being pressed travels some distance before it is measured; see Fig. 16.2.2-1   [16.6.1-1]

(b) tank filling system where the fluid travels some distance before its effect is measured; see Fig. 5.1.3-2   [16.6.1-2]

This type of time delay can be idealized and modeled by the impulse delay function, entry 156, Appendix 3.

$$G_d(s) = e^{-\alpha s}$$ [16.6.1-3]

where

$$\alpha = \text{delay time}$$ [16.6.1-4]

The delay time is equal to the time required for the signal to travel from the point of action to the point of measurement.

$$\alpha = X/V$$ [16.6.1-5]

---

*See reference [26], p. 238–241.

where

$$X = \text{distance traveled} \qquad [16.6.1\text{-}6]$$

$$V = \text{average velocity} \qquad [16.6.1\text{-}7]$$

A time delay described by eq. 16.6.1-3 can be treated as a block in the conventional manner. Its frequency response is computed in the same manner as all functions in Chapter 8; let

$$s = j\omega \qquad [16.6.1\text{-}7]$$

Apply this to eq. 16.6.1-3

$$G_d(j\omega) = e^{-j\omega\alpha} \qquad [16.6.1\text{-}8]$$

Using the polar form for a complex number, this can be expressed in terms of a magnitude and phase where

$$R = \text{magnitude} = 1 \qquad [16.6.1\text{-}9]$$

$$\psi = \text{phase} = -\alpha\omega \text{ rad} \qquad [16.6.1\text{-}10]$$

Since α has units of seconds and ω has units of rad/sec, the phase ψ is given in radians. Note that the time delay* results in a phase lag, but no change in magnitude. The phase lag can become excessive if

$$\left. \begin{array}{ll} X & \text{is very large} \\ V & \text{is very small} \\ \omega & \text{is very high} \end{array} \right\} \qquad [16.6.1\text{-}11]$$

■ **EXAMPLE 16.6.1-1** Time Delay in Rolling Mill
Refer to Ex. 16.2.2-1 (Rolling Mill). Given the following:

$$X = 0.1 \text{ m} \qquad\qquad V = 100 \text{ m/s} \qquad D = 1.5 \times 10^5 \text{ N} \cdot \text{s/m}$$

$$K_p = 0.44 \times 10^8 \text{ N} \cdot \text{m} \qquad M = 10^3 \text{ kg} \qquad K = 10^7 \text{ N/m}$$

Draw the block diagram and determine stability using the numerical values given above.

Solution. The delay occurs between the sensing roller and the actuator. The sensing roller is the comparator. Hence, the delay block is located between the summing point and K_p. The block diagram is shown in Fig. 16.6.1-1.
Compute the value of the delay time using eq. 16.6.1-4.

$$\alpha = X/V = 0.1/100 = 0.001 \text{ s} \qquad\qquad (1)$$

---

*For root locus treatment of time delay, see reference 40, p. 335–341.

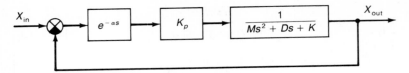

Figure 16.6.1-1   Rolling mill with time delay.

The phase lag is given by eq. 16.6.1-10.

$$\psi = -\alpha\omega \times 57.3° \qquad\qquad (2)$$
$$= -0.57° \quad \text{at} \quad \omega = 10 \text{ rad/s}$$
$$= -5.7° \quad \text{at} \quad \omega = 100 \text{ rad/s} \qquad\qquad (3)$$
$$= -57.3° \quad \text{at} \quad \omega = 1000 \text{ rad/s}$$

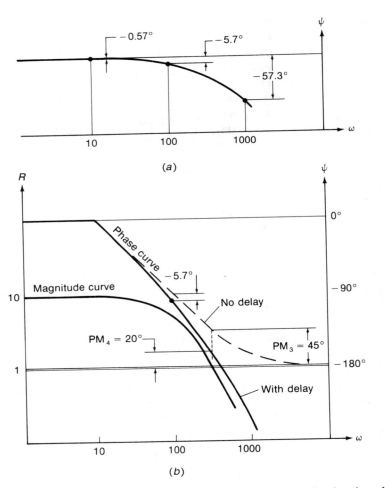

Figure 16.6.1-2   Rolling mill with time delay. (a) Phase plot for time delay function. (b) Composite.

The open loop gain in Fig. 16.6.1-1 is

$$L = \left(\frac{K_p}{Ms^2 + Ds + K} \right)(e^{-as}) \tag{4}$$

Normalize and apply numerical values.

$$L = \left(\frac{7.5 \times 10^4}{s^2 + 150s + 10^4} \right)(e^{-0.001s}) \tag{5}$$

The quadratic provides the natural frequency and damping ratio for the open loop.

$$\omega_{n0} = 100 \text{ rad/s} \tag{6}$$

$$\zeta_0 = 0.75 \tag{7}$$

The phase plot for the time delay function is shown in Fig. 16.6.1-2(a). Note that at frequencies above 1000 rad/s the phase lag is excessive. Since the time delay function does not alter the magnitude, the magnitude curve need not be drawn.

The composite is shown in Fig. 16.6.1-2(b). From the curve it can be seen that with a time delay

$$PM_4 = 20° \tag{8}$$

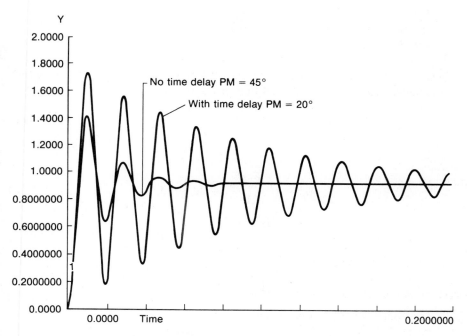

Figure 16.6.1-3   TUTSIM plots for rolling mill with time delay.

For comparison, the curve with*out* a time delay is shown dotted, for which

$$PM_3 = 45°   \tag{9}$$

as determined in Ex. 16.5.6-2.

This problem was simulated using TUTSIM for a unit step input, with and without a time delay. The results are shown in Fig. 16.6.1-3. Note that for the time delay, where the phase margin is far below the allowable value, the oscillation dies out very slowly, demonstrating poor stability. Compare this to the case without a time delay, where the system is very stable. ∎

16.6.2 Transient Properties

The transient solution to a differential equation is the complete solution, which can be obtained by several methods. One is to use the extensive table of Laplace transförm pairs in Appendix 3. However, many times, the complete solution is unnecessary. Instead, there are times when just the properties are required. In most cases, the properties can be determined without solving the differential equation. The problem becomes difficult when there are a number of terms acting simultaneously. In that case, the dominant term (or the one assumed to be dominant) will be used to estimate the entire system performance.

∎ **EXAMPLE 16.6.2-1** Transient Properties of Hydraulic Servo
Refer to Ex. 16.2.3-1 (Hydraulic Servo). Determine the following:

(a) Overshoot due to a step input with magnitude = 0.1 m.
(b) Settling time for 5% transient residue.
(c) Rise time.

Solution.   The out/in ratio is given by eq. (5) of the original example. Normalize and apply numerical values given in the example.

$$\frac{X_{out}}{X_{in}}(s) = \frac{C}{s^2 + 2\zeta\omega_n s + \omega_n^2}   \tag{1}$$

where

$$\omega_n = \sqrt{\frac{K_p K_m}{M}} = 32.1 \text{ rad/s}   \tag{2}$$

$$\zeta = \frac{(D + K_a K_m)}{M} = 0.267   \tag{3}$$

$$C = \frac{K_p K_m}{M} = 1029   \tag{4}$$

The response to a step input where

$$X_{in}(s) = \frac{U}{s} = \frac{0.1}{s}   \tag{5}$$

is found by applying eq. (5) to eq. (1) and by normalizing.

$$\frac{X_{out}\omega_n^2}{CU}(s) = \frac{\omega_n^2}{s(s^2 + 2as + \omega_n^2)} \tag{6}$$

This corresponds to entry 414 of Appendix 3. Referring to the response in the time domain, note

$$n\phi = \phi \quad \text{or} \quad n = 1 \tag{7}$$

For overshoot, use eq. 11.4.6-6, where

$$\begin{aligned} m &= 1 & n &= 1 \\ \zeta &= 0.267 & \beta &= \sqrt{1 - \zeta^2} = 0.9637 \end{aligned} \tag{8}$$

Apply numerical values to eq. 11.4.6-6

$$A_0 = e^{-0.267\pi/0.9637} = 0.419 \tag{9}$$

To use this number, replace the right-hand side of eq. (6) with A_0.

$$\left(\frac{X_{out}\omega_n^2}{CU}\right)_{os} = A_0 = 0.419 \tag{10}$$

or

$$X_{out} = \frac{A_0 CU}{\omega_n^2} \tag{11}$$

$$= \frac{0.419 \times 1029 \times 0.1}{32.1^2} = 0.0418 \text{ m} \tag{12}$$

For 5% transient residue, according to Table 5.5.2-1,

$$N = 3 \tag{13}$$

$$T_s = \frac{N}{\zeta\omega_n} \tag{14}$$

$$= \frac{3}{0.267 \times 32.1} = 0.35 \text{ s} \tag{15}$$

Dimensionless rise time is given by eq. 11.4.4-5, where:

$$\begin{aligned} m &= 1, & n &= 1, & \zeta &= 0.267 \\ \beta &= 0.9637, & \phi &= \sin^{-1}\beta = 1.302 \end{aligned} \tag{16}$$

Solve for T_r and apply numerical values

$$T_r = \frac{\pi - 1 \times 1.302}{0.9637 \times 32.1} = 0.0595 \text{ s} \tag{17}$$

∎

■ **EXAMPLE 16.6.2-2** Transient Properties of Electromechanical Servo
Refer to Ex. 16.3.4-1. Equation (3b) indicates that the system is an unfactored cubic. Consequently, the transient properties will require more than the methods of Chapter 11. Fortunately, much of the computational effort has been accomplished throughout Chapter 15. In particular, refer to Ex. 15.4.4-1,

where the rise time T_r, peak time T_p, overshoot A_0, and settling time T_s (for 1.8% transient residue) have been found in eqs. (6), (7), (9), and (12), repeated below.

$$T_r = 0.0119 \tag{1}$$

$$T_p = 0.0178 \tag{2}$$

$$A_0 = 0.163 \tag{3}$$

$$T_s = 0.04 \tag{4}$$

This system was also simulated on the computer using TUTSIM in Ex. 16.5.4-1, where the results were shown on Fig. 16.5.4-1. Considering only the very stable system, for which

$$\zeta = 0.5 \tag{5}$$

the same transient properties can be measured on the figure, as follows:

$$T_r = 0.012 \tag{6}$$

$$T_p = 0.02 \tag{7}$$

$$A_0 = \theta_{max} - 1.00 = 1.16 - 1.00 = 0.16 \tag{8}$$

$$T_s = 0.04 \tag{9}$$

all of which show reasonable agreement with their respective counterparts in eqs. (1) through (4).   ∎

■ **EXAMPLE 16.6.2-3** Transient Properties of Roll Stabilization System for Aircraft

Refer to Ex. 16.4.2-1 (Aircraft with Roll Stabilization). The denominators of eqs. (8), (9), and (10) indicate that the system is an unfactored quartic (fourth order). In that example, it was pointed out that the system employs a rate gyro D_g to provide system damping via the feedback term. It was also mentioned that both the rate gyro constant D_g and the controller gain K_p can be varied.

This system was simulated on the computer using TUTSIM for a unit step input. The results are shown on Fig. 16.6.2-1. In part (a) of the figure, three values were selected for the controller gain K_p. It is apparent that the system is stable for all three. At the same time, we note that the overshoot did not vary much, but that the rise time T_r varied considerably. For the nominal gain,

$$K_p = 2 \times 10^9 \tag{1}$$

measure the rise time,

$$T_r = 0.23 \text{ s} \tag{2}$$

Assume there is a dominant factor and that it is a reciprocal quadratic with a step input, of the form of entry 414 of Appendix 3. Correspondingly apply eq. 11.4.4-3 and determine the damping ratio ζ by iteration.

$$\zeta \simeq 0.4 \tag{3}$$

The system damping ratio is close to the optimum value of 0.5, explaining why the system was fairly stable.

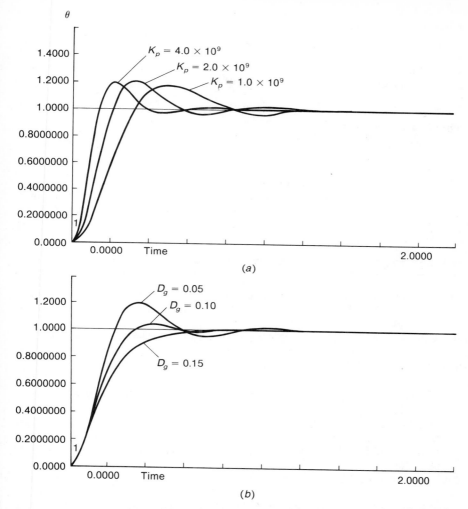

Figure 16.6.2-1 TUTSIM plots for roll stabilization system. (a) Effect of varying controller gain. (b) Effect of varying rate gyro constant.

The period of oscillation can be determined by measuring the time for a full cycle (crossings on the steady state line).

$$\lambda = 0.35 \text{ s} \tag{4}$$

or the damped frequency is

$$\omega_d = \frac{2\pi \text{ rad}}{0.35 \text{ s}} \simeq 18 \text{ rad/s} \tag{5}$$

In part (b) of Fig. 16.6.2-1, it is clear that the system is stable for all three values of rate gyro constant. For the nominal value

$$D_g = 0.05 \tag{6}$$

measure the overshoot.

$$A_0 = \theta_{max} - 1.00 = 1.20 - 1.00 = 0.20 \qquad (7)$$

Using the same assumption of the existence and form of a dominant factor, apply eq. 11.4.6-6 and solve for damping ratio ζ by iteration.

$$\zeta \approx 0.4 \qquad (8)$$

which agrees with that found in eq. (3).

Measure the period.

$$\lambda \approx 0.35 \text{ s} \qquad (9)$$

which agrees with that found in eq. (4).

The stability determined in this example may be verified by making a log-log plot of the open loop. The transient properties may be verified by making a root locus plot. ∎

SUGGESTED REFERENCE READING FOR CHAPTER 16

Modeling of control elements and systems [7], chap. 3; [17], p. 273–279, 495–502, 559–562; [26], chap. 2; [40], chaps. 3–6; [42], chaps. 2, 7; [57], chap. 16; [47], p. 494–513.

Block diagram algebra [7], p. 71–72; [26], p. 43; [40], p. 194–199; [42], p. 130–133; [46], p. 308–313; [47], p. 483–488, 537.

Transient response [7], chap. 6; [17], p. 631–635, 683–691; [26], chap. 9; [47], p. 509–525.

Control system error [17], p. 631–637; [26], p. 73–74, 94–97; [42], p. 91–94; [47], p. 511, 512.

Stability [7], chap. 12, p. 485–489; [17], p. 394, chap. 21; [26], chaps. 5, 8; [40], chaps. 10, 15; [42], p. 94, 203, 209, 211, 268, 342; [46], p. 275–277, 398–400; [47], p. 479.

Types of control [7], chap. 4; [40], p. 61, 73, 347; [42], chap. 8; [46], chap. 15; [47], p. 487–509.

Root locus method [7], chap. 7; [17], chap. 21; [26], chap. 6; [40], chap. 11; [42], chap. 12.

Bode diagram [7], p. 400–405, 451, 452; [26], chap. 7; [42], chap. 11.

Nyquist diagram [7], p. 439–448, 475–477; [26], p. 216–229; [40], p. 387, 407, 416, 442; [42], chap. 10.

Time delay [26], p. 238–241; [40], p. 335–341.

PROBLEMS FOR CHAPTER 16

16.1. The forward gain FG and the feedback H are listed below for a number of systems. For each one, determine the closed loop transfer function.

        FG H

(a) $\dfrac{as + b}{cs + d}$ -1

(b) $\dfrac{As + K}{Ms^2 + Ds + K}$ -1

(c) $\dfrac{(A + 1)K}{(B + 1)(Ms + D)}$ $-C$

(d) $\dfrac{AR}{s^2(Ls^2 + Rs + 1/C)} \quad - 1$

(e) $\dfrac{KN}{s(Ls + R)(Js + D)} \quad - (1 + Rs)$

16.2. Similar to Problem 16.1, but where the headings FG and H are interchanged.

16.3. Given the function

$$\frac{\epsilon}{\theta_{in}}(s) = \frac{s^n(Rs + 1/C)}{(Js^2 + Ds + K)(Ls + R) + K/C}$$

determine the steady state error ϵ_{ss} due to each of the following inputs: doublet, impulse, step, ramp, and parabola, where $n = 2$.

16.4. Similar to Problem 16.3, but where $n = 1$.

16.5. Similar to Problem 16.3, but where $n = 0$.

16.6. What conclusions and suggestions can be made from the results of Problems 16.3 through 16.5?

16.7. The output/input ratio for a certain system is modeled with entry 415 of the Table of Laplace Transform Pairs in Appendix 3. Determine the overshoot OS due to each of the following unit inputs: doublet, impulse, step, ramp, and parabola, where $\zeta = 0.5$ and $\omega_n = 10$.

16.8. Determine the rise time for each input in Problem 16.7.

16.9. For each input in Problem 16.7, determine the settling time T_s for transient residue of 1.8%. How does the input affect the settling time?

16.10. *Instability Due to Negative or Missing Terms.* For each of the following output/input ratios, determine which are absolutely unstable and explain.

(a) $\dfrac{Ms^2 + K}{D}$

(b) $\dfrac{Ls^2 + 1/C}{Rs + 1}$

(c) $\dfrac{LJs^4 + RDs^3 + Ks^2}{C}$

(d) $\dfrac{s^2(Ls^2 + Rs + 1/C)}{Js - 1}$

(e) $\dfrac{As^2 + B}{Is^2 - Rs - 1/C}$

(f) $\dfrac{s^2 + 1}{s^3(Ms^2 + Ds + K)}$

(g) $\dfrac{Fs - G}{(As^3 + B)(Cs + D) + Es^2}$

(h) $\dfrac{1}{(As^3 + B)(Cs + D)}$

(i) $\dfrac{s - 1}{(As^2 + Bs + C)(Ds + E)}$

(j) $\dfrac{1}{As^4 + Bs^3 + Cs + D}$

16.11. *System Properties Using TUTSIM.* For each function in Problem 8.28, determine the following:
(a) Signal flow or block diagram.

(b) Model structure (identify blocks by number).

(c) Model parameters and ranges (numerical data).

(d) Plotblocks and ranges* (types of outputs, running data, number of cycles of oscillation, and numerical ranges of time and variables.)

(e) Timing data (time interval and final time* to terminate the run).

(f) Changes to promote a particular aspect or system property, such as gain or damping ratio.

16.12. *Margin of Stability Using Log-Log Plot.* For each function in Problem 8.28 determine the following:

(a) Open loop gain $L(s)$.

(b) I_0, I_e, and I_c for frequencies $\omega = 0.1, 1.0$, and 10.

(c) Corner frequencies.

(d) Open loop damping ratios for quadratic factors.

(e) Construct the asymptotes only.

(f) Determine PM and GM.

(g) Construct actual curves.

(h) Determine PM and GM for part (g).

(i) Determine the gain that results in PM = 0.[†]

(j) Determine where the magnitude curve crosses the critical line for part (i).[†]

16.13. *Transient Properties Using Laplace Tables.* Using the table of Laplace transform pairs in Appendix 3, partial fractions, and techniques in Chapter 11, determine the following transient properties[‡] for each function in Problem 8.28 for a unit step input:

(a) Dominant factor and defend your choice.

(b) ω_n, ω_d, ζ, ϕ, β, and decay coefficient a.

(c) Settling time for 5% transient residue.

(d) Rise time.

(e) Peak time and overshoot.

16.14. *System Properties Using the Root Locus.* For each function in Problem 8.28, determine the following:

(a) The Evans equation.

(b) Number of paths.

(c) Open loop roots.

(d) Sum of the roots.

(e) Segments on the real axis.

(f) Number of asymptotes.

(g) Angles for each asymptote.

(h) Point through which asymptotes pass.

(i) Breakaway and break-in points.

(j) Angles of departure and arrival.

---

*If poorly chosen, data is not spread out efficiently. It will appear either truncated or too compressed.

[†]Compare results to those in Problem 16.14, part (k).

[‡]Compare results to those in Problem 16.14, parts (n) through (p).

(k) Points where loci cross the imaginary axis, and corresponding values* of K and ω.

(l) Trial points as required.

(m) Construct the Root Locus.

(n) Dominant factor and explain reasons for your choice.

(o) ω_n, ω_d, ζ, ϕ, and a.

(p) Each of the transient properties listed in Problem 16.13.[†]

16.15. *Properties of Control Systems.* For the system in Fig P-13.5(a), determine the following:

(a) Closed loop transfer function (algebraic form—no numbers).

(b) Error due to general input (algebraic form).

(c) Error due to general disturbance. If there is more than one disturbance, write the expression for each one (algebraic form).

(d) Steady state error due to specific inputs (numerical values) as follows:
(1) impulse input = ± 1.0.
(2) step input = ± 1.0.

(e) Steady state error due to specific disturbances (numerical values) as follows:
(1) impulse disturbance[‡] = ± 1.0.
(2) step disturbance[‡] = ± 1.0.

(f) Open loop gain (algebraic form)

(g) PM using log-log method outlined in Problem 16.12, part (h).

(h) K_p for which PM = 0, using log-log method. Also find frequency at which magnitude curve crosses critical line. Compare results to that in part (j).

(i) Transient properties using Laplace tables, outlined in Problem 16.13.

(j) Numerical values of K and ω where loci cross the imaginary axis, using the root locus method, outlined in Problem 16.14, part (k).

(k) Transient properties using the root locus method. Compare to part (m).

(l) Stability using TUTSIM, outlined in Problem 16.11.

(m) Transient properties using TUTSIM.

16.16. *Rolling Mill.* Apply the format of Problem 16.15 [parts (a) through (m)] to Ex. 16.2.2-1 (Rolling Mill), using the following specific inputs and disturbances:

(d) Inputs:
(1) impulse input = 0.01 m · s.
(2) step input = 0.01 m.

(e) Disturbances:
(1) impulse disturbance = 100 N · s.
(2) step disturbance = 10^6 N.

16.17. *Rolling Mill With Time Delay.* Apply the format of Problem 16.15 [parts (a) through (m)] to Ex. 16.6.1-1 (Time Delay[¶] in Rolling Mill), using the specific inputs and disturbances given in Problem 16.16.

---

*Compare results to those in Problem 16.12, parts (i) and (j).

[†]Compare results to those in Problem 16.13, parts (a) through (e).

[‡]Note that the disturbance is unknown.

[¶]For root locus treatment of time delay, see reference [40], p. 335–341.

16.18. *Hydraulic Servo.* Apply the format of Problem 16.15 [parts (a) through (m)] to Ex. 16.2.3-1 (Hydraulic Servo), using the following specific inputs and disturbances:

    (d) Inputs:

        (1) step input = 0.1 m.

        (2) ramp input = 3.6 m/s.

        (3) parabolic input = 0.05 m/s^2.

    (e) Disturbances:

        (1) impulse flow disturbance Q_d = 1.5 m^3.

        (2) step flow disturbance Q_d = 0.15 m^3/s.

        (3) impulse force disturbance F_d = 75 N · s

        (4) step force disturbance F_d = 7.5 N.

16.19. *Hydraulic Servo With Time Delay.* Similar to Problem 16.18, but where the time delay* is not negligible, and where X = 0.175 m and V = 300 m/s.

16.20. *Electromechanical Servo.* Apply the format of Problem 16.15 [parts (a) through (m)] to Ex. 16.3.4-1 (Electromechanical Servo), using the following specific inputs and disturbances:

    (d) Inputs:

        (1) step input = 1.5 rad.

        (2) ramp input = 0.5 rad/s.

        (3) parabola = 0.01 rad/s^2.

    (e) Disturbances:

        (1) impulse electrical disturbance E_d = 36 V · s.

        (2) step electrical disturbance E_d = 3.6 V.

        (3) impulse torque disturbance T_d = 160 N · m · s.

        (4) step torque disturbance T_d = 16 N · m.

16.21. Discuss the most likely magnitude of a time delay in the electromechanical servo.

16.22. *Aircraft Roll System.* Apply the format of Problem 16.15 [parts (a) through (m)] to Ex. 16.4.2-1 (Aircraft with Roll Stabilization), using the following specific inputs and disturbances:

    (d) Inputs:

        (1) impulse input = 0.1 rad · s.

        (2) step input = 0.1 rad.

        (3) ramp input = 0.3 rad/s.

        (4) parabolic input = 0.5 rad/s^2.

    (e) Disturbances:

        (1) impulse disturbance = 5 × 10^6 N · m · s.

        (2) step disturbance = 10^6 N · m.

16.23. *Aircraft Roll System with Time Delay.* Similar to Problem 16.22, but where time delay* is not negligible, and where X = 0.07 m and V = 200 m/s.

16.24. *Servo Positioner for Control Valve.* The control valve shown in Fig. 11.4.7-2 suffered from poor accuracy, since the pressure and velocity of the fluid affected the valve opening Y. An order of improvement was achieved by using a

---

*See the last footnote on page 635.

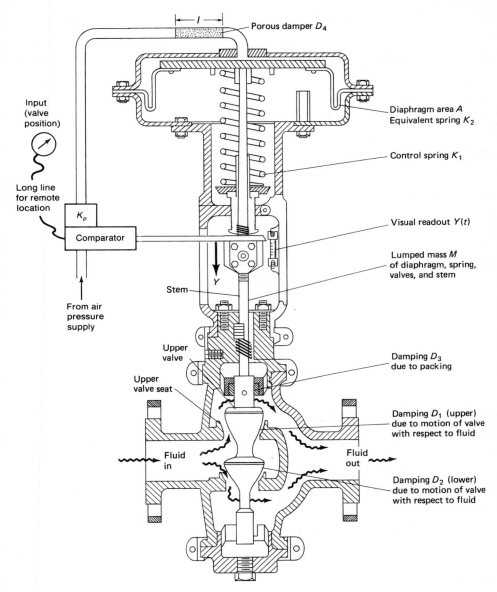

Figure P-16.24

balanced configuration, like that in Fig. P-11.34, where these effects were almost
balanced out. In either case, the valve had limited accuracy because it used an
open loop—the actual valve position was *not* fed back. Consider an improved
valve that employs a position servo in Fig. P-16.24.

The actual valve displacement Y is fed to a comparator that compares this
to the required displacement Y_{in} from a remote station. The error ϵ is amplified

by a pneumatic device K_p, which transmits air pressure to the diaphragm. Model each component of the system; draw the block diagram; determine the closed loop transfer function, ω_n, ζ, settling time for 5% transient residue, and PM. Use the numerical values of Problem 11.34, plus the following:

$$M_v = 1.4 \text{ kg} \qquad K_p = 100 \text{ N/m}^3$$

$$D_4 = 113 \text{ N} \cdot \text{s/m} \qquad K_1 = 9.44 \times 10^4 \text{ N/m}$$

16.25. *Pressure Regulator.* A pressure regulator is an automatically operated valve that places a controlled restriction in the flow line in order to maintain a specific pressure even though line loads and pressure sources vary. A typical configuration is shown in Fig. P-16.25.

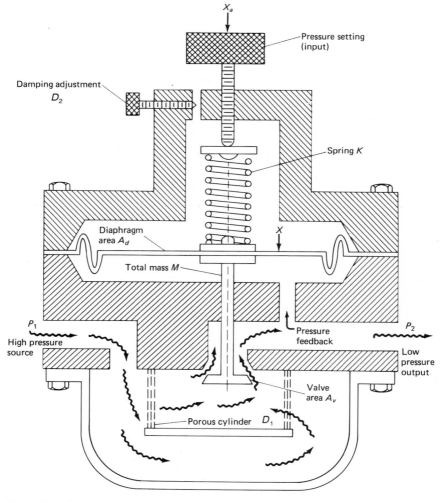

Figure P-16.25

Output pressure P_2 acts against the diaphragm, exerting a force. A preset spring K also acts against the diaphragm. Hence, the diaphragm is the comparator. The difference, or error ϵ, results in movement of the diaphragm and, with it, there is movement of the valve. This causes a fluid resistance, resulting in a pressure drop, attempting to make the output pressure equal to the preset value.

Model each element of the system and draw the block diagram, assuming the fluid resistance is proportional to displacement Y.

16.26. *Temperature Control for Room Heater.* A central heating system for a hotel employs individual thermostatic controls for each room. See Fig. P-16.26. The heat source is warm air at temperature T. A temperature sensor in the room reads room temperature. This is compared to a preset value, and the error ϵ is fed to an amplifier K_p. This is applied to a motor K_m, which supplies a torque to open or close the valve-baffle system, which in turn regulates the amount of warm air admitted to the room. There is a heat loss (disturbance) Q_d.

Assume that the flow rate $\dot{W}$ of warm air is proportional to the baffle opening θ. This in turn is a function of the baffle dynamics, which consists of a mass J and a damper D. Assume that the heat transfer is modeled by eq. 3.4.1-1, but where conduction is neglected. Model each element of the system and draw the block diagram.

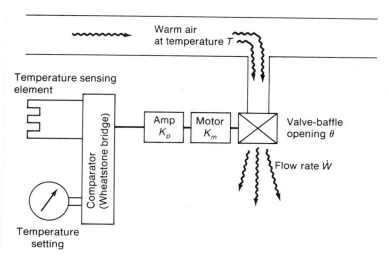

Figure P-16.26

16.27. *Water Level Regulator.* Refer to Fig. 5.1.3-2 (Water Tower). As shown, the system uses an on-off switch, causing the water level to vary from minimum to maximum continuously. An improved system makes use of a motor speed control, as shown in Fig. P-16.27. In an emergency, such as a fire, the tank can deliver water at a rate of 0.4 m³/s. But under normal conditions, the average pump flow rate is 0.004 m³/s, for which the pump runs at half-speed. The pump then runs either faster or slower as needed to maintain the level in the tank.

Model each element, draw the block diagram, and determine system stability. [*Suggestion:* The motor-pump subsystem is a flow-supply parallel fluid circuit, like that in Fig. 3.6.2-1(f).]

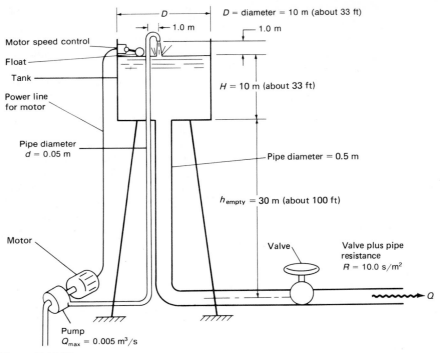

Motor speed control

Float

Tank

Power line
for motor

Pipe diameter
$d = 0.05$ m

Motor

Pump
$Q_{max} = 0.005$ m^3/s

D

$\leftarrow$ 1.0 m

D = diameter = 10 m (about 33 ft)

1.0 m

$H = 10$ m (about 33 ft)

Pipe diameter = 0.5 m

$h_{empty} = 30$ m (about 100 ft)

Valve

Valve plus pipe
resistance
$R = 10.0$ s/m^2

Q

Figure P-16.27

16.28. *Water Level Regulator with Time Delay.* Similar to Problem 16.27, but where the time delay is not negligible. Model the system, draw the block diagram, and determine PM. [*Suggestion*: The difficulty is in determining the distance X. As a rule of thumb, consider the fluid that continues to travel after the motor (or flow) stops. Due to the inertance of the fluid, a certain amount will continue to travel, with sufficient kinetic energy to rise to the top of the U section. After that, the fluid will continue by action of gravity. When the motor starts again, the fluid will be required to traverse this path until it reaches the liquid level. Note that most of the water will remain in the pipe. Consequently, X will be much less than 40 m.]

16.29. *Robot Hand.* A simple robot hand is essentially a two-fingered "claw." See Fig. P-16.29(a). The fingers move in parallel planes. Pressure sensitive elements provide pressure feedback. The opening of the hand is also fed back. A rotary version of the robot hand is shown in Fig. 16.29(b). In this case, a pulse torquer, like that in Fig. P-11.36, may be employed to give "instant" response. The pulse torquer drives one finger directly using pulse width modulation for high linearity. A precision "synchro" reads the angular displacement. The combination of these two devices makes it possible to have high accuracy and fast response.

A third possibility makes use of a hydraulic drive like that in Fig. 16.2.3-1.

Select the scheme to actuate the hand or finger and defend your choice. Lay out the specific device that will act as the comparator. Choose the appropriate amplifier. Model the system, draw the block diagram, and indicate the means to determine response time and system stability.

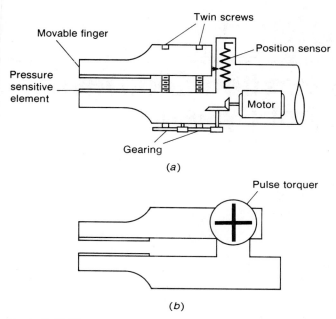

Figure P-16.29

16.30. *Creative Engineering*. Make a detailed layout of each of the following control systems:

(a) Robot wrist device where the motor or actuator is in the robot arm and not in the hand (where there is little room).

(b) Multifingered hand, where the fingers are all driven by a single motor with springs that enable the fingers to adapt to the shape of the object being held.

(c) Robot transport that enables the robot to move over rough terrain, or a mechanism that adapts to the terrain.

(d) A machine that will hunt and find metal objects that may be buried up to 0.05 m below the surface.

(e) A heat-seeking robot.

(f) Mobile robot that avoids objects in its path.

(h) Power steering for a small vehicle.

(i) Power brakes for a small vehicle.

(j) Programmable robot earth mover, bulldozer, and so on.

(k) A tracker that finds and tracks a celestial object.

(l) Force-balance accelerometer. Instead of measuring the amount of deflection of a cantilever beam (open loop technique in current use today), use either a magnetic pulse-width force or an electrostatic force to balance the force due to acceleration.

(m) Programmable device that "writes" using either a pencil or an engraving tool.

(n) Contour milling machine. Like copying a key, this device follows a master contour. But unlike the key maker (which is open loop), this device uses a closed loop with high accuracy.

(o) Any idea of your own.

Chapter **17**

Compensation and Design of Control Systems

17.1 GENERAL TERMS

17.1.1 General Model

For design purposes, it will prove fruitful to define the general closed loop system as follows:

$$\text{control function} = A(s) \tag{17.1.1-1}$$

$$\text{process} = \frac{G(s)}{s^m F(s)} \tag{17.1.1-2}$$

$$\text{feedback} = H(s) \tag{17.1.1-3}$$

The block diagram is shown in Fig. 17.1.1-1.
   The closed loop transfer function is

$$\frac{\theta_{\text{out}}}{\theta_{\text{in}}}(s) = \frac{AG}{s^m F + AGH} \tag{17.1.1-4}$$

17.1.2 Limits

There are a number of occasions for taking limits. It will be helpful to summarize some appropriate rules. The limit of a product (or quotient) is equal to the product (or quotient) of the limits.

$$\lim AB = (\lim A)(\lim B) \tag{17.1.2-1}$$

$$\lim \frac{A}{B} = \frac{\lim A}{\lim B} \tag{17.1.2-2}$$

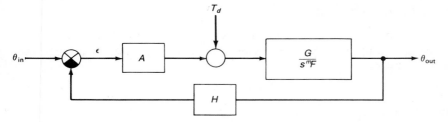

Figure 17.1.1-1   Generalized model for design.

When taking the limit as s approaches zero, any term multiplied by s or power of s that is added to a constant will vanish. For example,

$$\lim_{s \to 0} \frac{(K_d s + K_p)(K_a s + 1)}{(\tau s + 1)(K_b s + 1)} = K_p \qquad [17.1.2\text{-}3]$$

Take the limit of the time delay function of eq. 16.6.1-3.

$$\lim_{s \to 0} e^{-as} = 1 \qquad [17.1.2\text{-}4]$$

When taking the limit of powers of s divided by powers of s, strictly speaking, it is mathematically improper to cancel one s with another, for this implies division by s, which will be zero in the limit. To be more precise, use L'Hôpital's rule and continue to take derivatives of the numerator and of the denominator until there is either a limit or an unbounded result. For example:

$$\lim_{s \to 0} \frac{s^m}{s^n} = \lim_{s \to 0} \frac{m s^{m-1}}{n s^{n-1}} = \cdots = \lim_{s \to 0} \frac{m!}{n!} s^{m-n} = \begin{cases} 0 & \text{when } m > n \\ 1 & \text{when } m = n \\ \infty & \text{when } m < n \end{cases}$$

$$[17.1.2\text{-}5]$$

The process, as defined in eq. 17.1.1-2, deliberately has excluded any exponential terms from either the function $F(s)$ or $G(s)$. However, it is conceivable to have exponentials in the other system functions $A(s)$ and $H(s)$. Nevertheless, most of the time all four functions will be devoid of exponentials, in which case, their limits will reduce to constants, as follows:

$$\lim_{s \to 0} A(s) = A_0 \quad \text{(no exponentials)}$$

$$\lim_{s \to 0} F(s) = F_0$$

$$\lim_{s \to 0} G(s) = G_0 \qquad [17.1.2\text{-}6]$$

$$\lim_{s \to 0} H(s) = H_0 \quad \text{(no exponentials)}$$

17.1.3 Steady State Errors

The input, in general, may be expressed as a jump function with magnitude = U. The unit jump functions are listed in Appendix 3, entries 141 through 148. Then,

$$\theta_{in}(s) = Us^u \qquad [17.1.3\text{-}1]$$

Similarly, a disturbance may be expressed as

$$T_d(s) = Vs^v \qquad [17.1.3\text{-}2]$$

For the general system in Fig. 17.1.1-1, the error due to the input is

$$\frac{\epsilon_i}{\theta_{in}}(s) = \frac{s^m F}{s^m F + AGH} \qquad [17.1.3\text{-}3]$$

Apply eq. 17.1.3-1 and solve for the error.

$$\epsilon_i(s) = \frac{s^m F U s^u}{s^m F + AGH} \qquad [17.1.3\text{-}4]$$

For the steady state value, multiply by s and take the limit.

$$\epsilon_{ssi} = \lim_{s \to 0} \frac{UF s^{m+1+u}}{s^m F + AGH} \qquad [17.1.3\text{-}5]$$

If there are no exponentials in both A and H, we may take advantage of eq. 17.1.2-6. Equation 17.1.3-5 becomes

$$\epsilon_{ssi} = \lim_{s \to 0} \frac{UF_0 s^{m+1+u}}{s^m F_0 + A_0 G_0 H_0} \qquad [17.1.3\text{-}6]$$

$$= \begin{cases} 0 & \text{when } m + 1 + u > 0 \\ \text{finite} & \text{when } m + 1 + u = 0 \\ \infty & \text{when } m + 1 + u < 0 \end{cases} \qquad [17.1.3\text{-}7]$$

Equation 17.1.3-7 can be used as a design tool. Given the allowable error and the type of input (or, more exactly, the value of the exponent u), the numerical value of the exponent m and the product $A_0 G_0 H_0$ can be determined. For example, if the input is a parabola (constant acceleration), then $u = -3$. For a finite error, let $m = 2$.

For design, we need consider only the case of finite error in eq. 17.1.3-7. Solve for A_0.

$$A_0 = \frac{UF_0 - \epsilon_{allow} F_0}{\epsilon_{allow} G_0 H_0} \qquad \text{when } m = 0, u = -1 \qquad [17.1.3\text{-}8]$$

$$A_0 = \frac{UF_0}{\epsilon_{allow} G_0 H_0} \qquad \text{when } m > 0, m = 1 - u \qquad [17.1.3\text{-}9]$$

A similar approach applies to a disturbance. Referring to Fig. 17.1.1-1, the error due to a disturbance is

$$\frac{\epsilon_d}{T_d}(s) = \frac{GH}{s^m F + AGH} \qquad [17.1.3\text{-}10]$$

Apply eq. 17.1.3-2, solve for error, multiply by s, and take the limit for steady state.

$$\epsilon_{ssd} = \lim_{s \to 0} \frac{-VGHs^{1+v}}{s^m F + AGH} \qquad [17.1.3\text{-}11]$$

Similarly to the previous case, if there are no exponentials in A and H, eq. 17.1.3-11 becomes

$$\epsilon_{ssd} = \lim_{s \to 0} \frac{-VG_0 H_0 s^{1+v}}{s^m F_0 + A_0 G_0 H_0} \qquad [17.1.3\text{-}12]$$

$$= \begin{cases} 0 & \text{when } v + 1 > 0 \\ \text{finite} & \text{when } v + 1 = 0 \\ \infty & \text{when } v + 1 < 0 \end{cases} \qquad [17.1.3\text{-}13]$$

Considering only the finite case for design, solve for A_0 in eq. 17.1.3-13.

$$A_0 = \frac{-VG_0 H_0 + F_0 \epsilon_{\text{allow}}}{G_0 H_0 \epsilon_{\text{allow}}} \qquad \text{when } m = 0, v = -1 \quad [17.1.3\text{-}14]$$

$$A_0 = \frac{V}{\epsilon_{\text{allow}}} \qquad \text{when } m > 0, v = -1 \quad [17.1.3\text{-}15]$$

17.1.4 Stability Using Log-Log Plots

Refer to Fig. 17.1.1-1. The open loop gain is

$$L(s) = \frac{AGH}{s^m F} \qquad [17.1.4\text{-}1]$$

If there are any exponential terms in either A or H, they should be lumped with the exponent m. The log-log plot can be formulated as shown in Chapter 8. If the functions A, F, G, and H are in factored form, use the plots in Figs. 8.4.4-1 through 8.4.7-1. If any or all of these is unfactored, use the method in Section 8.6.6.

The composite intercept I_c is computed by the method shown in eqs. 8.6.2-2 through 8.6.2-4. These are adapted for the general model in Fig. 17.1.1-1 as follows:

$$I_c = I_0 I_e \qquad [17.1.4\text{-}2]$$

where

$$I_0 = \lim_{s \to 0} \frac{AGH}{F} \qquad [17.1.4\text{-}3]$$

$$I_e = \frac{1}{(\omega_e)^m} \quad \text{at frequency } \omega_e \qquad [17.1.4\text{-}4]$$

Both terms, I_0 and I_e, must be computed at the same frequency.

Refer to Section 16.6.1 for treatment of time delays.

Gain and phase margins are computed as shown in Section 16.5.6.

17.1.5 Stability Using Root Locus

The general procedure is shown in Section 16.5.4. Details are covered in Section 15.3.13 (Rule 13: Crossing the Imaginary Axis). The points where the loci cross the imaginary axis are found by setting

$$s = 0 + j\omega \qquad [17.1.5\text{-}1]$$

This will produce two equations, one real, one imaginary, permitting the solution of two unknowns, K and ω. This determines absolute stability. It is then necessary to set boundaries (the equivalent of a margin of stability) to avoid instability.

Incidentally, it is possible to determine the value of K that represents the boundary between stable and unstable without drawing the root locus. The loci do help, however, in determining the setting of a margin of stability. But more important, the root locus is very useful in determining transient properties of the dominant factor.

17.2 PROPORTIONAL CONTROL

17.2.1 Description

If the control function $A(s)$ in Fig. 17.1.1-1 is a constant K_p, the output signal is proportional to the error ϵ and the system has proportional control. This is a fairly popular system in practice because of its simplicity. It rates average in all of its characteristics, not outstanding in any one. It has fair steady state errors, fair stability, and fair transient response.

To meet performance specifications, K_p must be selected.

For finite steady state errors, eqs. 17.1.3-6 and 17.1.3-12 indicate that, to minimize the error, the constant K_p must be large.

For stability, eq. 17.1.4-3 indicates that the constant K_p must be small.

Needless to say, the two conditions—steady state errors and stability—are mutually exclusive.

Since the only parameter that can be adjusted in proportional control is the constant K_p, it is difficult to adjust the damping, a requirement to meet transient specifications.

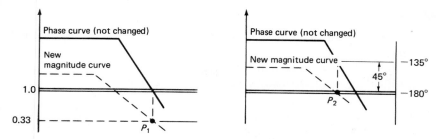

Figure 17.2.2-1   Control system design using gain adjustment.

17.2.2 Design for Stability Using Gain Adjustment

The design technique involves adjusting the numerical value of the constant K_p. From eq. 17.1.4-3, it is apparent that this will raise or lower the magnitude curve without affecting the phase curves. Consequently, this provides a direct means to produce exactly the margins of stability that are required. Decide where the magnitude curve must cross the critical line and move the curve up or down, like a rigid body, accordingly. Since there are two margins involved, this procedure must be exercised twice. See Fig. 17.2.2-1.

First, to satisfy the gain margin, find where the *phase curve* (note that the phase curve will *not* be altered in the process of satisfying either margin) crosses the *critical line*. Drop a distance equal to 3 (note that this is a log scale, and that a distance of 3 means a *ratio* of 3). The point P_1 is such a point, and since the critical line is at unity, this point is at $\frac{1}{3}$ or 0.333. Since this point is uniquely determined, and since the slope of the magnitude curve is unchanged, the new magnitude curve (adjusted for suitable gain margin) can be drawn. Secondly, to satisfy the phase margin, locate a point on the *phase curve* that is 45° *above* the critical line. Draw a vertical to the critical line, point P_2. Translate the magnitude curve until it passes through this point. Now there are two locations for the magnitude curve, one passing through point P_1 to satisfy GM and one passing through point P_2 to satisfy PM. The lower curve prevails since it will produce the case for which both margins are satisfied.

■ **EXAMPLE 17.2.2-1** Hydraulic Servo with Proportional Control
Refer to Ex. 16.2.3-1 (Hydraulic Servo).
    Specifications:

$$\epsilon_{ss} \leq 0.002 \text{ m}, \qquad \text{PM} = 45°, \qquad \text{ramp input} = 3.6 \text{ m/s} = X_i$$

$$Q_d \text{ step} = 0.15 \text{ m}^3/\text{s}^\dagger, \qquad F_d \text{ step} = 7.5 \text{ N}^\dagger$$

[†]Refer to eqs. (12) and (13) of Ex. 16.2.3-1.

Design the system* for each of the following conditions:

(1) Determine all quantities using parameters as given in the original example.
(2) Adjust K_p to meet stability specification only.
(3) Adjust K_p to meet steady state specification only.

Solution.   The steady state errors due to ramp input X_i and step disturbances Q_d and F_d are determined by applying the final value theorem of Section 4.4.4 to eqs. (9), (12) and (13) of Ex. 16.2.3-1.

$$\epsilon_{ssi} = \left(\frac{D + K_a K_m}{K_p K_m} \right) X_i \tag{1}$$

$$\epsilon_{ssq} = \frac{Q_d}{K_p}, \qquad \epsilon_{ssf} = \frac{F_d}{K_p K_m} \tag{2}$$

Refer to Fig. 16.2.3-2. Open loop gain is

$$L(s) = \frac{K_p K_m}{s(Ms + D + K_a K_m)} = I_c \frac{a}{s(s + a)} \tag{3}$$

where

$$a = \frac{D + K_a K_m}{M} = \frac{35 + 0.5 \times 50}{3.5} = 17.1 \tag{4}$$

$$I_c = I_0 I_e \qquad I_e = 1 \text{ at } \omega = 1 \tag{5}$$

$$I_0 = \frac{K_p K_m}{D + K_a K_m} = \frac{50 K_p}{35 + 0.5 \times 50} = \frac{5 K_p}{6} \tag{6}$$

The stability plot for open loop is shown on Fig. 17.2.2-2. Note: actual phase curve must be used.
    Part (1), use the given value of K_p.

$$K_{p1} = 72 \tag{7}$$

then

$$I_{c1} = 60, \qquad PM_1 = 31° \tag{8}$$

$$\epsilon_{ssi} = 0.06, \qquad \epsilon_{ssq} = 0.0021, \qquad \epsilon_{ssf} = 0.0021 \tag{9}$$

    Part (2), note that PM_1 is too low. To *increase* the stability, *lower* the magnitude curve as shown in Fig. 17.2.2-2, and determine the corresponding value of K_p.

$$K_{p2} = I_0 6/5 = 19.5, \qquad PM_2 = 45° \tag{10}$$

---

*For block diagram, refer to Fig. 16.2.3-2.

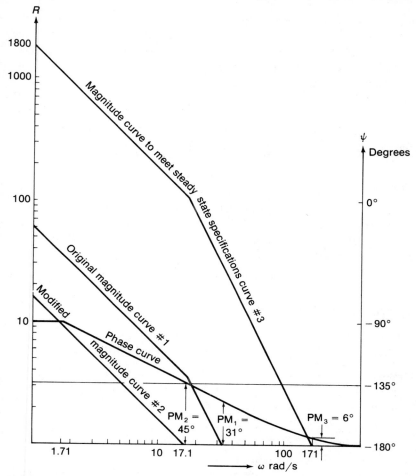

Figure 17.2.2-2   Open loop stability for hydraulic servo.

Part (3), note that steady state errors in eq. (9) are too large. Adjust the proportional gain K_p to meet the steady state specifications.* Solve eqs. (1) and (2) for K_p and apply numerical values. The largest value of K_p applies.

$$K_{p3} = 2160 \qquad (11)$$

For this,

$$I_{c3} = 1800, \qquad PM_3 \simeq 6° \qquad (12)$$

In part (1), the system did not meet either the steady state or the stability specifications. In part (2), the system was adjusted to meet the stability

*Refer to eqs. (12) and (13) of Ex. 16.2.3-1.

specification (there was no need to test for steady state specifications since the system gain was *lowered*, making the errors even larger) at the expense of the steady state specifications. In part (3), the system was adjusted to meet the steady state specifications (and corresponding to that, the PM = 6°) at the expense of stability. Here we have experienced the noted fact that steady state and stability requirements are mutually exclusive.   ∎

17.3 PROPORTIONAL PLUS DERIVATIVE CONTROL

17.3.1 Description

When there is insufficient damping in the (plant) process $G(s)$, it becomes necessary to introduce it in the controller. However, it can be shown that pure derivative control has poor steady state error performance. Then use a combination of proportional and derivative controls. In Fig. 17.1.1-1,

$$A(s) = K_d s + K_p \qquad [17.3.1\text{-}1]$$

where K_d and K_p are to be selected.

17.3.2 Design Using Proportional Plus Derivative Control

The introduction of the derivative term K_d causes the magnitude curve to change its slope by +1 at a frequency ω_a, and to add a phase shift of 45°/decade starting at frequency $\omega_a/10$. See Fig. 17.3.2-1, where

$$\omega_a = K_p/K_d \qquad [17.3.2\text{-}1]$$

Needless to say, the choice of frequency ω_a requires more art than mathematics. If this corner frequency is chosen small, the phase advance starts early, which is desirable. But that will also cause the magnitude curve to bend up (change slope by +1) early, causing it to cross the critical line late, which is not desirable. The advantage to the technique lies in the fact that the phase advance starts at $\omega_a/10$, while the change in the magnitude curve starts at frequency ω_a. Consequently, there can be a phase advance before the magnitude curve is affected. If the corner frequency is chosen properly, then

$$PM_2 > PM_1 \qquad [17.3.2\text{-}2]$$

From eq. 17.3.1-1, the definition of proportional plus derivative control, it is clear that this provides a factor of the form shown in Fig. 8.4.4-1. Such a factor is also called a (phase) lead network since it provides a phase advance (lead), which helps make a system stable. The manner in which the lead network functions is shown in Fig. 17.3.2-1.

Referring to eqs. 17.1.3-6 and 17.1.3-12, it is clear that the introduction of the derivative term $K_d s$ does not change the steady state errors. Referring to eq. 17.1.4-3, it is clear that the derivative term does not change the composite

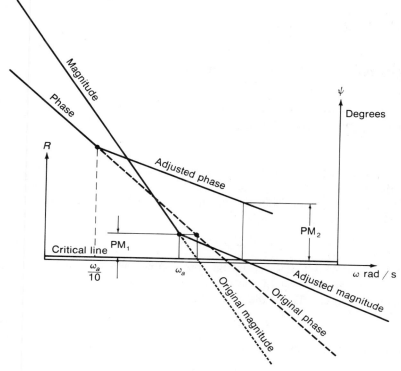

Figure 17.3.2-1   Proportional plus derivative control.

intercept I_c. Holding these quantities fixed permits the designer to concentrate upon using the derivative to improve system stability.

It should be pointed out that there is a range of numerical values for ω_a that will provide stability. This gives the designer considerable latitude, making it possible to meet one more specification, perhaps a transient property such as rise time, settling time, peak time, or overshoot. If the closed loop transfer function is of second order, these properties can be determined by the Laplace transform methods in Section 11.4; if it is of higher order, use the root locus method in Section 15.4. These appear as exercises in the Problem section of this chapter.

■ **EXAMPLE 17.3.2-1** Hydraulic Servo with Proportional Plus Derivative Control
Refer to Ex. 17.2.2-1, where proportional control was used, but the system could not meet all specifications. Using derivative plus proportional control, redesign the system to meet all specifications.

Solution.   The block diagram is constructed by replacing K_p with $K_d s + K_p$ in Fig. 16.2.3-2. The result is shown in Fig. 17.3.2-2.

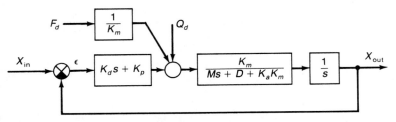

Figure 17.3.2-2   Hydraulic servo with proportional plus derivative control.

Referring to eqs. 17.1.3-6 and 17.1.3-12, it is clear that the introduction of the derivative term $K_d s$ does not change the steady state errors. Then use the value of K_{p3} in the original example that met the steady state specifications. See eq. (11) of the original example.

$$K_p = K_{p3} = 2160 \tag{1}$$

Referring to eq. 17.1.4-3, it is clear that the composite intercept I_c is unchanged by the introduction of the derivative term $K_d s$. Consequently, I_c remains as it was in eq. (12) of the original example.

$$I_c = I_{c3} = 1800 \tag{2}$$

The open loop gain is determined from the block diagram on Fig. 17.3.2-2.

$$L(s) = \frac{(K_d s + K_p)K_m}{s(Ms + D + K_a K_m)} \tag{3}$$

Normalize

$$L(s) = \frac{K_d K_m}{M} \frac{(s + \omega_a)}{s(s + \omega_b)} \tag{4}$$

where

$$\omega_a = \frac{K_p}{K_d} \tag{5}$$

$$\omega_b = \frac{D + K_a K_m}{M} \tag{6}$$

Start the open loop plot with curve #3 of Fig. 17.2.2-2. For this, $PM_3 = 6°$. See Fig. 17.3.2-3. Now, include the proportional plus derivative term $s + \omega_a$. This is a first order factor that will change the slope of the magnitude curve from -2 to -1. It will also change the slope of the phase curve from $-45°$/decade to zero, until the next break point alters that. But in the meantime, this offers a straightforward means to select the frequency ω_a. Select it where the actual phase curve is 45° above the critical line. This requires that

$$\omega_a/10 \simeq 24 \text{ rad/s} \tag{7}$$

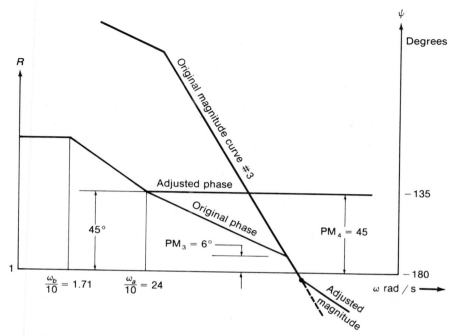

Figure 17.3.2-3   Hydraulic servo with proportional plus derivative control.

or

$$\omega_a = 24 \times 10 = 240 \text{ rad/s} \tag{8}$$

Note that the magnitude curve crosses the critical line at a frequency slightly below 180 rad/s. Consequently, the change in its slope does not alter the result. Then

$$\text{PM}_4 = 45° \tag{9}$$

To conclude the design, note that from eq. (5),

$$K_d = K_p/\omega_a = 2160/240 = 9.0 \tag{10}$$

■

17.4 PROPORTIONAL PLUS INTEGRAL CONTROL

17.4.1 Description

It can be shown that integral control by itself has the advantage of small steady state errors, but it has poor stability. Combining integral with proportional control provides the designer with two constants to adjust, permitting the optimum mixture. In Fig. 17.1.1-1,

$$A(s) = K_p\left(K_i + \frac{1}{s}\right) \qquad [17.4.1\text{-}1]$$

where K_p and K_i are to be selected.

17.4.2 Design Approach

Write eq. 17.4.1-1 as a proper fraction

$$A(s) = \frac{K_p(K_i s + 1)}{s} \qquad [17.4.2\text{-}1]$$

or

$$A(s) = \frac{K_p K_i(s + \omega_i)}{s} \qquad [17.4.2\text{-}2]$$

where

$$\omega_i = \frac{1}{K_i} \qquad [17.4.2\text{-}3]$$

The steady state error cannot be evaluated by using eq. 17.1.3-6 since this assumed there were no exponential terms in the control function. Instead apply eq. 17.4.2-1 to eq. 17.1.3-5.

$$\epsilon_{ssi} = \lim_{s \to 0} \frac{UFs^{m+1+u}}{s^m F + \dfrac{K_p(K_i s + 1)GH}{s}} \qquad [17.4.2\text{-}4]$$

Rationalize the improper fraction

$$\epsilon_{ssi} = \lim_{s \to 0} \frac{UFs^{m+2+u}}{s^{m+1}F + K_p(K_i + 1)GH} \qquad [17.4.2\text{-}5]$$

Note that this derivation has accounted for the exponential in the control function $A(s)$. Assuming that there is no exponential in the feedback $H(s)$, we may apply the limits listed in eq. 17.1.2-6.

$$\epsilon_{ssi} = \lim_{s \to 0} \frac{UF_0 s^{m+2+u}}{s^{m+1}F_0 + K_p G_0 H_0} \qquad [17.4.2\text{-}6]$$

$$= \text{finite} \quad \text{when } m + 2 + u = 0 \qquad [17.4.2\text{-}7]$$

Thus, we see firsthand, that proportional plus integral control reduces system error, not merely by a factor, but by a power of s. This scheme enables the system to have an input with a higher power of s in the denominator. For example, a system that originally had an infinite steady state error due to a ramp input, will now have a finite error.

Consider the behavior of this control for stability. The composite intercept will have an exponential that is one power higher in the denominator. Then, eqs. 17.1.4-2 and 17.1.4-3 will remain unchanged, while eq. 17.1.4-4 will become

$$I_e = 1/(\omega_e)^{m+1} \qquad [17.4.2\text{-}8]$$

On the other hand, the form in eq. 17.4.2-2 indicates a first order lead term in the numerator. The pure integration due to the s in the denominator starts the open loop stability plot with a phase lag of $-90°$, which will impose a problem with stability. However, at the same time, the lead network in the numerator will permit the designer to advance the phase by $+90°$. Noting the form of the function in eq. 17.4.2-2, we may anticipate that the design technique will be very similar to that for proportional plus derivative control. Hence, we may employ the same strategy to select the corner frequency ω_i as was done for ω_a.

It should be pointed out that, although there will be a certain amount of difficulty in selecting the numerical value of ω_i, there is still a range that will provide stability. This gives the designer considerable latitude, making it possible to meet one more specification, perhaps a transient property such as rise time, settling time, peak time, or overshoot. If the closed loop transfer function is of second order, these properties can be determined by the Laplace transform methods in Section 11.4. If the system is of higher order, use the root locus method in Section 15.4. These appear as exercises in the Problem section of this chapter.

■ **EXAMPLE 17.4.2-1** Aircraft With Roll Stabilization and Proportional Plus Integral Control

Refer to Ex. 16.4.2-1 (Aircraft with Roll Stabilization). Consider this a high performance aircraft, for which the aircraft is requested to acquire the bank angle θ_{in} in the shortest time (for a sudden turn). Keep in mind that when any aircraft makes a turn, it must bank (tilt the wings) so that the total acceleration vector is perpendicular to the airfoil. The fastest way to acquire any angle is via a step function. But this would tear the wings off an airplane. Hence, there is a limit imposed on angular acceleration. Design the system using the parameters of the original example, but with the following specifications:

$$\theta_{bank} = 1.0 \text{ rad} \quad \text{maximum bank angle} \tag{1}$$

$$t_m = \text{minimum time} \tag{2}$$

$$\ddot{\theta}_{max} = \pm 0.5 \text{ rad/s}^2 \quad \text{maximum angular acceleration} \tag{3}$$

$$\epsilon_{ss} = \pm 0.0125 \text{ rad} \quad (\text{about } 0.7°) \tag{4}$$

Solution. The block diagram is constructed by replacing K_p in Fig. 16.4.2-1(b) with eq. 17.4.2-1. The result is shown in Fig. 17.4.2-1. The open loop gain is

$$L(s) = \frac{K_p(K_i s + 1)(D_g s + 1)}{s^2(Js + D)(Is^2 + R's + 1/C)} \tag{5}$$

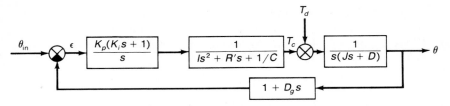

Figure 17.4.2-1   Aircraft roll system with proportional plus integral control.

The numerical value of K_p will be determined so that the system meets the steady state error specification. The steady state error due to a generalized input is

$$\epsilon_{ssi} = \lim_{s \to 0} \frac{s^3 D \theta_{in}(s)}{K_p C} \tag{6}$$

To determine the form and magnitude of the input, consider the restrictions upon achieving the bank angle. The shortest time will be achieved by rolling with angular acceleration $+\ddot{\theta}_{max}$ for an angle equal to $\theta_{bank}/2$ and following this by rolling at $-\ddot{\theta}_{max}$ for the second half of θ_{bank}. For each half of the roll motion, we have

$$\theta = \tfrac{1}{2}\ddot{\theta}t^2 \tag{7}$$

where

$$\theta = \theta_{bank}/2 = 0.5 \text{ rad} \tag{8}$$

$$\ddot{\theta} = \ddot{\theta}_{max} = 0.5 \text{ rad/s}^2 \tag{9}$$

Solve for t.

$$t = \sqrt{2 \times 1/0.5} = 2 \text{ s} \tag{10}$$

and

$$t_m = 2t = 2 \times 2 = 4 \text{ s} \tag{11}$$

Assume the aircraft is traveling at a speed

$$V = 500 \text{ m/s} \quad \text{(about 1115 mph)}$$

Then preparation for the turn should start at distance

$$X = Vt_m = 500 \times 4 = 2000 \text{ m} \quad \left(\text{about } 1\tfrac{1}{4} \text{ miles}\right) \tag{12}$$

no trivial amount, which explains why there is a stringent requirement in the specifications. We are now ready to apply numerical values to eq. (6). The input is a constant acceleration (parabola) with magnitude 0.5.

$$\theta_{in}(s) = \frac{2U}{s^3} = \frac{2 \times 0.5}{s^3} \tag{13}$$

Apply numerical values, including eq. (4) and solve for K_p

$$K_p = \frac{UD}{C\epsilon_{ss}} = \frac{15 \times 10^5 \times 2 \times 0.5}{0.015 \times 0.0125} = 8 \times 10^9 \tag{14}$$

There is only one more term K_i to find. This will be determined for stability. Apply numerical values to the open loop gain in eq. (5).

$$L(s) = K_0 \frac{(s + \omega_i)(s + \omega_g)}{s^2(s + \omega_j)(s^2 + 2\zeta_0\omega_0 s + \omega_0^2)} \tag{15}$$

where

$$\omega_i = 1/K_i \qquad \omega_j = D/J = 5 \text{ rad/s}$$

$$\omega_g = 1/D_g = 1/0.05 = 20$$

$$\omega_0 = \sqrt{1/(CI)} = \sqrt{1/(0.015 \times 0.0125)} = 82 \tag{16}$$

$$\zeta_0 = R'/2I\omega_0 = 0.8/(2 \times 0.01 \times 82) \simeq 0.49$$

$$I_0 = K_p C/D = 8 \times 10^9 \times 0.015/(15 \times 10^5) = 80$$

$$K_0 = K_p K_i D_g/JI$$

The open loop stability plot appears in Fig. 17.4.2-2. For the first trial, let $\omega_i = \omega_j = 5$.

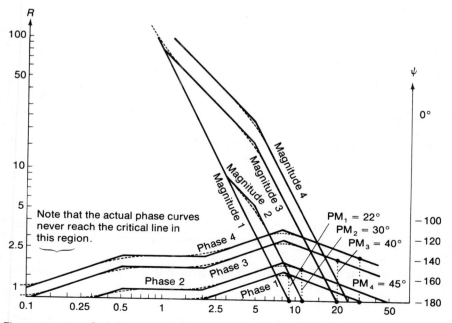

Figure 17.4.2-2   Stability of aircraft system with proportional plus integral control.

Note that two terms cancel. For the first trial,

$$PM_1 \simeq 22° \tag{17}$$

After a few trials,

$$\omega_i \simeq 0.5 \quad \text{and} \quad K_i = 2.0 \tag{18}$$

and

$$PM_4 \simeq 45° \tag{19}$$

The design is complete.   ■

17.5 LEAD-LAG CONTROL

17.5.1 Description

It was pointed out that derivative control produced a phase lead, while integral control produced phase lag. They had almost opposite effects upon stability. If the two are combined, the result is referred to as *lead-lag* control.

The objective of lead-lag control is to improve the system stability without altering any steady state property. The lead-lag network consists of a pair of first order terms, one in the numerator, the other in the denominator. In Fig. 17.1.1-1,

$$A(s) = \frac{K_p(K_a s + 1)}{(K_b s + 1)} \qquad [17.5.1\text{-}1]$$

where K_p, K_a and K_b are to be selected.

17.5.2 Design Approach, Phase Curve on Critical Line

The lead-lag network will be used to stabilize a system that was previously shown to be unstable. Refer to Ex. 16.5.5-1 (Stability of Undamped System). The system is clearly unstable, since the phase curve lies on the critical line and the magnitude curve must cross it. See Fig. 16.5.5-2. The magnitude crosses the critical line at a frequency equal to $\sqrt{A/M}$, and the phase margin is

$$PM_1 = 0. \qquad [17.5.2\text{-}1]$$

This information is repeated on Fig. 17.5.2-1, where the original magnitude curve is shown dotted, and where the original phase curve lies on the critical line.

Now consider the effect of adding a lead-lag network. Normalize eq. 17.5.1-1.

$$A(s) = K_0 I_c \frac{(s+a)}{(s+b)} \qquad [17.5.2\text{-}2]$$

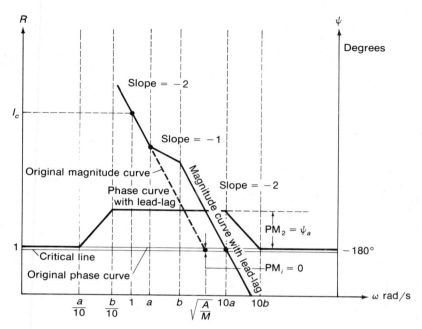

Figure 17.5.2-1   Lead-lag control, phase curve on critical line.

where

$$a = 1/K_a$$
$$b = 1/K_b$$
$$I_c = K_p$$
$$K_0 = K_a/K_b = b/a$$

[17.5.2-3]

The block diagram is shown in Fig. 17.5.2-2. The parameters are chosen so that frequency a is lower than frequency b. Since a is in the numerator, then at a frequency equal to $a/10$ the phase curve will rise and will level off at a frequency equal to $b/10$. At a frequency equal to $10a$ the curve will descend and finally level off (having returned to the original phase angle before the influence of either frequency a or b had any effect) at a frequency equal to $10b$. The final effect of the lead-lag control was to temporarily raise the phase curve and then return it. This may be compared to raising the bridge temporarily when a tall ship passes. In the interval where the phase curve has

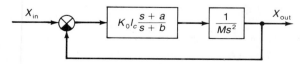

Figure 17.5.2-2   Undamped system with lead-lag control.

TABLE 17.5.2-1   PHASE SHIFT FOR LEAD-LAG CONTROL

b/a	1	2.06	4.63	10	21.5	46.6	100	> 100
ψ	0°	15°	30°	45°	60°	75°	90°	90°

been raised, the phase angle has been shifted by an angle

$$\psi_a = \text{phase shift} \qquad [17.5.2\text{-}4]$$

At the same time, the magnitude curve has also been shifted. Consequently, instead of crossing the critical line at frequency $\sqrt{A/M}$, it crosses at a frequency somewhat higher. But at this point, the phase curve has been shifted by an angle ψ_a and has not yet returned to its original value. Measuring the distance up, to determine the phase margin PM_2, it is clear that

$$PM_2 = \psi_a \qquad [17.5.2\text{-}5]$$

The lead-lag network has changed the phase margin from zero to ψ_a and this has been accomplished because the lead-lag network produced the corresponding phase shift. The amount of phase shift is a function of log-log geometry, where

$$\psi_a = \left(\frac{1}{2} \log_{10} \frac{b}{a} \right) 90° \qquad [17.5.2\text{-}6]$$

Equation 17.5.2-6 is tabulated in Table 17.5.2-1.

Note that, had frequency a been chosen somewhat higher, the phase margin would have been the same. However, had a been chosen lower, the phase margin would have been less than ψ_a. Thus, there is a range of numerical values for a that will provide stability. This gives the designer considerable latitude, making it possible to meet one more specification, perhaps a transient property such as rise time, settling time, peak time, or overshoot. If the closed loop transfer function is of second order, these properties can be determined by the Laplace transform methods in Section 11.4. If it is of higher order, use the root locus method in Section 15.4. These appear as exercises in the Problem section of this chapter.

17.5.3 Design Approach, General System

Consider a general system for which the region of interest is shown on Fig. 17.5.3-1. Assume asymptotes will provide sufficient accuracy. The original magnitude and phase curves are shown dotted. The original phase margin is

$$PM_1 < 45° \qquad [17.5.3\text{-}1]$$

Now apply the lead-lag network. On the adjusted phase curve (shown in solid lines), the first corner occurs at frequency $a/10$. If the original slope was

$$\text{original slope} = X°/\text{decade}.$$

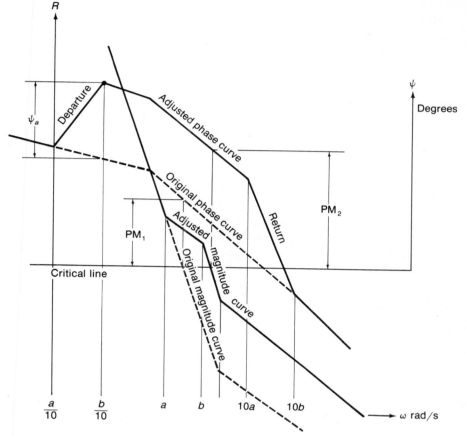

Figure 17.5.3-1   Lead-lag control, general system.

the new slope at frequencies above $a/10$ will be

$$\text{new slope} = (45° + X°)/\text{decade} \qquad [17.5.3\text{-}2]$$

summed algebraically. The adjusted phase curve will depart from the original one until frequency $b/10$. Now the adjusted phase curve will remain parallel to the original curve, even around corners remaining a distance equal to the phase shift,

$$\text{phase shift} = \psi_a \qquad [17.5.3\text{-}3]$$

This will continue until frequency $10a$. Here, the slope of the adjusted phase curve is

$$\text{return slope} = (-45° + X°)/\text{decade} \qquad [17.5.3\text{-}4]$$

and will continue until the phase curve returns to the original curve.

If the magnitude curve didn't change, it would be a simple matter to determine the expected phase margin. However, the magnitude curve does change. At frequency a the adjusted magnitude curve (shown in solid lines) bends upward, having changed its slope so that

$$\text{new slope} = \text{original slope} + 1 \qquad [17.5.3\text{-}5]$$

The adjusted magnitude curve will depart from the original curve until frequency b, after which, the adjusted curve will follow parallel to the original magnitude curve. If the departure occurs before the adjusted magnitude curve crosses the critical line, then some of the phase lead that was gained by the lead-lag network will be lost. Consequently the phase margin becomes

$$PM_2 \leq (PM_1 + \psi_a) \qquad [17.5.3\text{-}6]$$

Due to the lost phase shift, it may become necessary to repeat the operation, making new choices of the frequencies a and b.

When the final choices produce the correct phase margin, the design may be completed by computing the numerical values of K_a and K_b with the use of eq. 17.5.2-3.

■ **EXAMPLE 17.5.3-1** Electromechanical Servo with Lead-Lag Control
Refer to Ex. 16.3.4-1 (Electromechanical Servo). Design the system to meet the following specifications:

$$PM = 45°$$

$$|\epsilon_{ss}| \leq 0.001 \text{ rad} \quad \text{due to any one error source}$$

$$U_i = \text{ramp input} = 0.5 \text{ rad/s} \qquad (1)$$

$$E_d = \text{step electrical disturbance} = 3.6 \text{ V}$$

$$T_d = \text{step torque disturbance} = 16 \text{ N} \cdot \text{m}$$

Solution.   The block diagram is constructed by replacing A in Fig. 16.3.4-2(b) with eq. 17.5.1-1. The result is shown in Fig. 17.5.3-2.

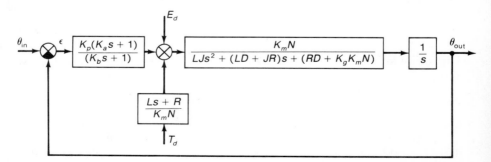

Figure 17.5.3-2   Electromechanical servo with lead-lag control.

The steady state errors are the same as those in the original example. The steady state error due to a ramp input is given by eq. (10) of the original example.

$$\epsilon_{ssi} = \frac{(RD + K_g K_m N)U_i}{K_p K_m N} \tag{2}$$

Solve for K_p and apply numerical values.

$$K_p \geq 720 \tag{3}$$

The steady state error due to a step electrical disturbance is given by eq. (13).

$$\epsilon_{sse} = E_d / K_p \tag{4}$$

Solve for K_p and apply numerical values.

$$K_p \geq 3600 \tag{5}$$

The steady state error due to a step torque disturbance is given by eq. (15).

$$\epsilon_{sst} = \frac{RT_d}{K_p K_m N} \tag{6}$$

Solve for K_p and apply numerical values.

$$K_p \geq 1000 \tag{7}$$

The largest value of K_p rules. Then,

$$K_p = 3600 \tag{8}$$

Refer to Fig. 17.5.3-2. The open loop gain is

$$L(s) = \frac{(K_a s + 1) K_p K_m N}{(K_b s + 1)s \left[LJs^2 + (LD + JR)s + RD + K_g K_m N \right]} \tag{9}$$

Normalize.

$$L(s) = \frac{K_0(s + \omega_a)}{s(s + \omega_b)(s^2 + 2\zeta_0 \omega_0 s + \omega_0^2)} \tag{10}$$

where

$$\omega_a = \frac{1}{K_a}, \qquad \omega_b = \frac{1}{K_b} \tag{11}$$

$$\left. \begin{array}{l} \omega_0 = \sqrt{\dfrac{RD + K_g K_m N}{LJ}} = 339 \\[2mm] \zeta_0 = \dfrac{RJ + LD}{2LJ\omega_0} = 0.84 \\[2mm] I_0 = \dfrac{K_p K_m N}{RD + K_g K_m N} = 0.6957 K_p \\[2mm] K_0 = \dfrac{K_a K_p K_m N}{K_b LJ} \end{array} \right\} \tag{12}$$

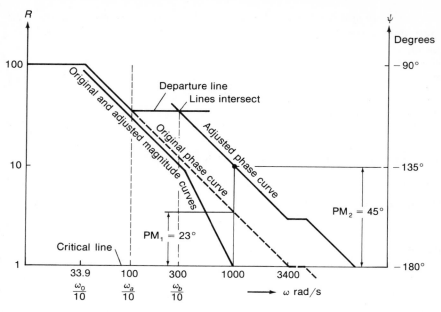

Figure 17.5.3-3   Stability plot for electromechanical servo with lead-lag control.

Using the appropriate numerical quantities, the open loop stability plot is drawn. See Fig. 17.5.3-3. First, draw the asymptotic curves for magnitude and phase (shown dotted). The magnitude curve crosses the critical line at a frequency

$$\omega_{\text{cross}} = 1000 \text{ rad/s} \tag{13}$$

and the phase margin is

$$PM_1 = 23° \quad \text{which is insufficient} \tag{14}$$

We are now ready to include the lead-lag network. For the first trial, attempt to select the corner frequency so that it won't affect the magnitude curve until it crosses the critical line. Then let

$$\omega_a = 1000 \text{ rad/s} \tag{15}$$

which provides the corner for the phase curve at

$$\omega_a/10 = 100 \text{ rad/s} \tag{16}$$

which locates the first corner frequency.* Now, if the adjusted magnitude curve will not be altered until *after* it crosses the critical line, then we know exactly where it will cross—the same place it already crossed in eq. (13). From

---

*Where the adjusted phase curve starts to depart from the original phase curve.

the point of crossing, draw a vertical line upward, corresponding to a phase angle

$$\psi = 45° \tag{17}$$

The adjusted phase curve must pass through this point. Also, in this region, the adjusted phase curve will have passed frequency $\omega_b/10$ and thus will have the same slope as the original phase curve. We have a point and a slope. Continue the adjusted phase curve upward until it intersects the departure line that started at frequency $\omega_a/10$. Where the two lines intersect defines frequency $\omega_b/10$, which is

$$\omega_b/10 = 300 \tag{18}$$

or

$$\omega_b = 3000 \tag{19}$$

Normally, several trials would be necessary. However, in this case, the geometric proportions made it possible to select the optimum location of the corner frequencies. In this case, the choice did not affect or change the point where the magnitude curve crossed the critical line. The appropriate corner frequencies have been found. To conclude the design, use eq. (11).

$$K_a = 1/1000 = 1.0 \times 10^{-3} \tag{20}$$

$$K_b = 1/3000 = 3.33 \times 10^{-3} \tag{21}$$

All specifications have been met. All parameters have been found. The design is complete. ∎

SUGGESTED REFERENCE READING FOR CHAPTER 17

Compensation [7], chap. 12; [17], p. 691–696; [26], chap. 10; [40], p. 61, 73, 199, 287, 347, 357; [42], chap. 13; [46], p. 321–322; [47], p. 525–531.
Design of control systems [7], p. 318–319; [17], p. 691–710; [26], chap. 10; [40], chaps. 11–13, 16; [47], p. 531–536.

PROBLEMS FOR CHAPTER 17

17.1. *Design of Rolling Mill with Proportional Control.* The rolling mill is shown in Fig. P-17.1, where the following information applies:
  System parameters:

$$M = 10^3 \text{ kg} \qquad D = 1.5 \times 10^5 \text{ N} \cdot \text{s/m}$$

$$K = 10^7 \text{ N/m} \qquad K_p = K_c K_m b/c = 1.5 \times 10^8 *$$

---

*This is the nominal value. It may be varied over a large range for design purposes.

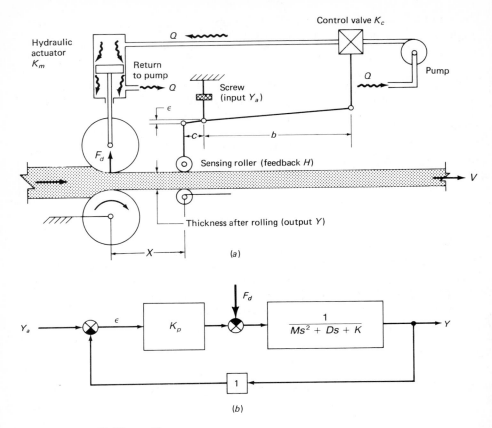

Figure P-17.1   Rolling mill.

Normal error sources:

$$Y_a = \begin{cases} \text{impulse input} = 0.01 \text{ m} \cdot \text{s} \\ \text{step input} = 0.01 \text{ m} \end{cases}$$

$$F_d = \begin{cases} \text{impulse disturbance} = 100 \text{ N} \cdot \text{s} \\ \text{step disturbance} = 10^7 \text{ N} \end{cases}$$

Special error sources: none.

Time delay:

$$X = 0.1 \text{ m}, \qquad V = 100 \text{ m/s}$$

Steady state specifications:

$$|\epsilon_{ss}| \le 0.001 \text{ m} \quad \text{for any one error source}$$

Stability specifications:

either $\text{PM} \ge 45°$ or $\zeta \ge 0.5$ (depending upon method)

Transient specifications:

$$T_s \leq 0.035 \text{ s} \quad \text{for } 1.8\% \text{ transient residue.}$$

$$|OS| \leq |5\epsilon_{ss}| \quad \text{for step input} = 0.01 \text{ m}$$

Using proportional control, design the system for each of the following:
(a) Only the steady state specs for normal error sources.
(b) Only the steady state specs for special error sources.
(c) Only the stability specs.
(d) Only the stability specs with a time delay.
(e) Why is it not possible to meet the transient specs?
(f) Why is it not possible to meet the steady state and stability specs simultaneously?

17.2. *Design of Rolling Mill with Proportional Plus Derivative Control.* Using proportional plus derivative control and the information in Problem 17.1, design the system to meet all of the following requirements:
    steady state specs for normal error sources;
    stability specs.

17.3. *Design of Rolling Mill with Proportional Plus Integral Control.* Using proportional plus integral control, meet all the specifications in Problem 17.2.

17.4. *Design of Rolling Mill with Lead-Lag Control.* Using lead-lag control, meet all specifications in Problem 17.2.

17.5. *Design of Rolling Mill with Transient Requirements.* Using any control(s) of your choice and the information in Problem 17.1, design the system to meet all of the following requirements:
    steady state specs for normal error sources;
    stability specs;
    transient specs.

17.6. *Design of Rolling Mill with Time Delay.* Including the time delay, use any control(s) of your choice to meet all of the following requirements:
    steady state specs for normal error sources;
    stability specs.

17.7. *Design of Hydraulic Servo with Proportional Control.* The hydraulic servo is shown in Fig. P-17.7, where the following information applies:
    System parameters:

$$K_m = 50 \text{ N} \cdot \text{s}/\text{m}^3 \qquad K_p = 72 \text{ m}^2/\text{s}*$$

$$K_a = 0.5 \text{ m}^2 \qquad\qquad D = 35 \text{ N} \cdot \text{s}/\text{m}$$

$$M = 3.5 \text{ kg}$$

---

*This is the nominal value. For design purposes, it may be varied over a large range.

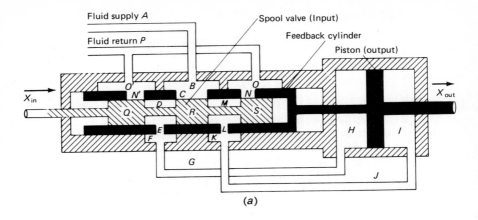

(a)

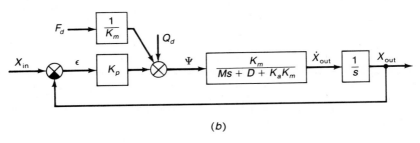

(b)

Figure P-17.7   Hydraulic servo.

Normal error sources:

$$X_{in} = \begin{cases} \text{step input} = 0.1 \text{ m} \\ \text{ramp input} = 3.6 \text{ m/s} \end{cases}$$

$$Q_d = \begin{cases} \text{impulse flow disturbance} = 1.5 \text{ m}^3 \\ \text{step flow disturbance} = 0.15 \text{ m}^3/\text{s} \end{cases}$$

$$F_d = \begin{cases} \text{impulse force disturbance} = 75 \text{ N} \cdot \text{s} \\ \text{step force disturbance} = 7.5 \text{ N} \end{cases}$$

Special error sources:

$$X_{in} = \text{parabolic input} = 0.05 \text{ m/s}^2$$

Time delay:

$$X = 0.175 \text{ m}, \qquad V = 300 \text{ m/s}$$

Steady state specifications:

$$|\epsilon_{ss}| \leq 0.002 \text{ m} \quad \text{for any one error source}$$

Stability specifications:

$$\text{PM} \geq 45 \quad \text{or} \quad \zeta \geq 0.5 \quad (\text{depending upon method})$$

Transient specifications:

$$T_s \leq 0.35 \text{ s} \quad \text{for } 1.8\% \text{ transient residue}$$

$$|OS| < |5\epsilon_{ss}| \quad \text{for step input} = 0.1 \text{ m}$$

Using proportional control, do parts (a), (c), and (f) of Problem 17.1.

17.8. *Design of Hydraulic Servo with Proportional Plus Derivative Control.* Using proportional plus derivative control and the information in Problem 17.7, design the system to meet all of the following requirements:
    steady state specs for normal error sources;
    stability specs.

17.9. *Design of Hydraulic Servo with Proportional Plus Integral Control.* Using proportional plus integral control, meet all the specifications in Problem 17.8.

17.10. *Design of Hydraulic Servo with Lead-Lag Control.* Using lead-lag control, meet all the specifications in Problem 17.8.

17.11. *Design of Hydraulic Servo with Transient Requirements.* Using any control(s) of your choice and the information in Problem 17.7, design the system to meet all of the following requirements:
    steady state specs for normal error sources;
    stability specs;
    transient specs.

17.12. *Design of Hydraulic Servo with Added Error Source.* Using any control(s) of your choice and the information in Problem 17.7, design the system to meet all of the following requirements:
    steady state specs for normal error sources;
    steady state specs for special error sources;
    stability specs.

17.13. *Design of Hydraulic Servo with Demanding Requirements.* Using any control(s) of your choice and the information in Problem 17.7, design the system to meet all of the following requirements:
    steady state specs for normal error sources;
    steady state specs for special error sources;
    stability specs;
    transient specs.

17.14. *Design of Hydraulic Servo with Time Delay.* Including a time delay, use any control(s) of your choice to meet all of the following requirements:
    steady state specs for normal error sources;
    stability specs.

17.15. *Design of Electromechanical Servo with Proportional Control.* The electromechanical servo is shown in Fig. P-17.15, where the following information applies:

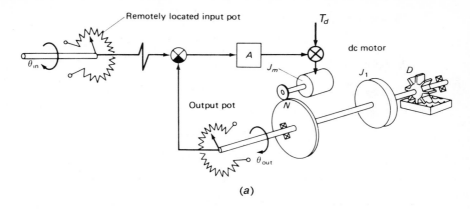

(a)

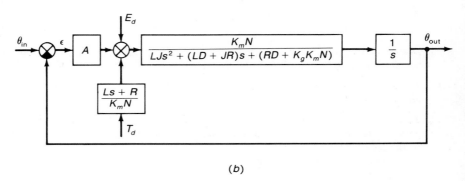

(b)

Figure P-17.15   Electromechanical servo.

System parameters:

$$J = J_1 + J_m \; N^2 \qquad\qquad D = 7.0 \; \text{N} \cdot \text{m} \cdot \text{s}$$

$$J_1 = 0.05 \; \text{kg} \cdot \text{m}^2 \qquad\quad K_m = 8.0 \; \text{N} \cdot \text{m/A}$$

$$J_m = 5 \times 10^{-6} \; \text{kg} \cdot \text{m}^2 \qquad K_g = 1.0 \; \text{V/rad/s}$$

$$N = 100 \qquad\qquad\qquad A = 400 \; \text{V/rad*}$$

$$R = 50 \; \Omega \qquad\qquad\qquad L = 0.1 \; \text{H}$$

Normal error sources:

$$\theta_{in} = \begin{cases} \text{step input} = 1.5 \; \text{rad} \\ \text{ramp input} = 0.5 \; \text{rad/s} \end{cases}$$

$$E_d = \begin{cases} \text{impulse electrical disturbance} = 36 \; \text{v} \cdot \text{s} \\ \text{step electrical disturbance} = 1.15 \; \text{V} \end{cases}$$

$$T_d = \begin{cases} \text{impulse torque disturbance} = 160 \; \text{N} \cdot \text{m} \cdot \text{s} \\ \text{step torque disturbance} = 16 \; \text{N} \cdot \text{m} \end{cases}$$

---

*This is the nominal value. It may be varied over a large range for design purposes.

Special error sources:

$$\theta_{in} = \text{parabolic input} = 4.01 \text{ rad/s}^2$$

Time delay: negligible

Steady state specifications:

$$|\epsilon_{ss}| \leq 0.001 \text{ rad} \quad \text{for any one error source}$$

Stability specifications:

$$\text{PM} \geq 45° \quad \text{or} \quad \zeta \geq 0.5 \quad (\text{depending upon method})$$

Transient specifications:

$$T_s \leq 0.35 \text{ s} \quad \text{for } 1.8\% \text{ transient residue}$$

$$|\text{OS}| \leq |5\epsilon_{ss}| \quad \text{for step input} = 1.5 \text{ rad}$$

Using proportional control, design the system to meet only the stability specification.

17.16. *Design of Electromechanical Servo with Proportional Plus Derivative Control.* Using proportional plus derivative control and the information in Problem 17.15, design the system to meet all of the following requirements:
steady state specs for normal error sources;
stability specs.

17.17. *Design of Electromechanical Servo with Proportional Plus Integral Control.* Using proportional plus integral control, meet all the specifications in Problem 17.16.

17.18. *Design of Electromechanical Servo with Lead-Lag Control.* Using lead-lag control, meet all the specifications in Problem 17.16.

17.19. *Design of Electromechanical Servo with Added Error Source.* Using any control(s) of your choice and the information in Problem 17.15, design the system to meet all of the following requirements:
steady state specs for normal error sources;
steady state specs for special error sources;
stability specs.

17.20. *Design of Electromechanical Servo with Transient Requirements.* Using any control(s) of your choice and the information in Problem 17.15, design the system to meet all of the following requirements:
steady state specs for normal error sources;
stability specs;
transient specs.

17.21. *Design of Electromechanical Servo with Demanding Requirements.* Using any control(s) of your choice and the information in Problem 17.15, design the system to meet all of the following requirements:
steady state specs for normal error sources;
steady state specs for special error sources;
stability specs;
transient specs.

17.22. *Design of Aircraft Roll System with Proportional Control and Unity Plus Derivative Feedback.* The aircraft roll system is shown in Fig. P-17.22, where the

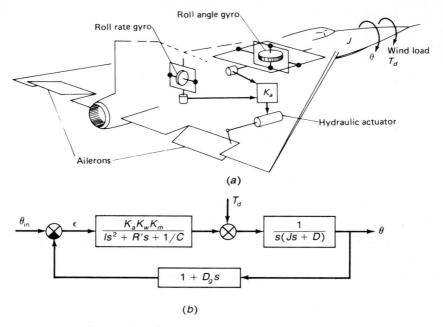

(a)

(b)

Figure P-17.22 Aircraft roll system.

following information applies:
    System parameters:

$$J = 3 \times 10^5 \text{ kg} \cdot \text{m}^2 \qquad D = 15 \times 10^5 \text{ N} \cdot \text{m} \cdot \text{s}$$

$$D_g = 0.05 \text{ V} \cdot \text{s/rad*} \qquad C = 0.015 \text{ m}^5/\text{N}$$

$$R' = 0.8 \text{ N} \cdot \text{s/m}^5 \qquad I = 0.01 \text{ N} \cdot \text{s}^2/\text{m}^5$$

$$^\dagger K_p = K_a K_w K_m = 2 \times 10^9 \text{ N} \cdot \text{m} \cdot \text{N/m}^5$$

Normal error sources:

$$\theta_{in} = \begin{cases} \text{impulse input} = 0.1 \text{ rad} \cdot \text{s} \\ \text{step input} = 0.1 \text{ rad} \\ \text{ramp input} = 0.3 \text{ rad/s} \end{cases}$$

$$T_d = \begin{cases} \text{impulse disturbance} = 5 \times 10^6 \text{ N} \cdot \text{m} \cdot \text{s} \\ \text{step disturbance} = 10^7 \text{ N} \cdot \text{m} \end{cases}$$

Special error sources:

$$\text{parabolic input} = 0.5 \text{ rad/s}^2$$

---

*This is the nominal value, which may be varied when using unity plus derivative feedback.

†This is the nominal value, which may be varied for all types of control.

Time delay:

$$X = 0.07 \text{ m}, \qquad V = 200 \text{ m/s}$$

Steady state specifications:

$$|\epsilon_{ss}| \leq 0.0125 \text{ rad} \quad \text{for any one error source}$$

Stability specifications:

$$\text{PM} \geq 45° \quad \text{or} \quad \zeta \geq 0.5 \quad \text{(depending upon method)}$$

Transient specifications:

$$T_s \leq 0.03 \text{ s} \quad \text{for } 1.8\% \text{ transient residue}$$

$$|\text{OS}| \leq |5\epsilon_{ss}| \quad \text{for step input} = 0.1 \text{ rad}$$

Using proportional control and unity plus derivative feedback, design the system to meet all of the following requirements:
    steady state specs for normal error sources;
    stability specs.

17.23. *Design of Aircraft Roll System with Proportional Plus Derivative Control.* Using proportional plus derivative control (the rate gyro D_g is not varied, but is retained at its nominal value), meet all of the following requirements:
    steady state spec for normal error sources;
    stability spec.

17.24. *Design of Aircraft Roll System with Lead-Lag Control.* Using lead-lag control, meet the specifications in Problem 17.22.

17.25. *Design of Aircraft Roll System with Added Error Source.* Using proportional plus integral control and the information in Problem 17.22, design the system to meet all of the following requirements:
    steady state specs for normal error sources;
    steady state specs for special error source;
    stability specs.

17.26. *Design of Aircraft Roll System with Transient Requirements.* Using any control(s) of your choice (including unity plus derivative feedback) and the information in Problem 17.22, design the system to meet all of the following requirements:
    steady state specs for normal error sources;
    stability specs;
    transient specs.

17.27. *Design of Aircraft Roll System with Demanding Requirements.* Using any control(s) of your choice (including unity plus derivative feedback) and the information in Problem 17.22, design the system to meet all of the following requirements:
    steady state specs for normal error sources;
    steady state specs for special error source;
    stability specs;
    transient specs.

17.28. *Design of Aircraft Roll System with Time Delay.* Using any control(s) of your choice (including unity plus derivative feedback) and the information in Problem 17.22, including a time delay, design the system to meet all of the following requirements:

> stead state specs for normal error sources;
>
> stability specs.

17.29. *Design of Aircraft Roll System with Time Delay and Demanding Requirements.* Including a time delay and all the information in Problem 17.22, design to meet all the specifications for all conditions, special error sources, and so on. Use any control(s) of your choice (including unity plus derivative feedback).

References

1. A. H. Church, *Mechanical Vibrations* 2nd ed. (New York: Wiley, 1963).
2. T. Baumeister and L. Marks, *Standard Handbook for Mechanical Engineers* 9th ed. (New York: McGraw-Hill, 1985).
3. M. F. Spotts, *Design of Machine Elements* (Englewood Cliffs, N.J.: Prentice-Hall, 1961).
4. C. M. Haberman, *Vibration Analysis* (Columbus, Ohio: Merrill, 1968).
5. W. T. Thomson, *Vibration Theory and Applications* 2nd ed. (Englewood Cliffs, N.J.: Prentice-Hall, 1968).
6. I. Cochin, *Analysis and Design of the Gyroscope* (New York: Wiley, 1963).
7. F. H. Raven, *Automatic Control Engineering* 4th ed. (New York: McGraw-Hill, 1987).
8. L. Shearer, A. Murphy, and H. Richardson, *Introduction to System Analysis* (Reading, Mass.: Addison-Wesley, 1967).
9. J. Den Hartog, *Mechanical Vibrations* (New York: McGraw-Hill, 1956).
10. R. Roark and W. Young, *Formulas for Stress and Strain* 6th ed. (New York: McGraw-Hill, 1989).
11. J. John and W. Haberman, *Introduction to Fluid Mechanics* 3rd ed. (Englewood Cliffs, N.J.: Prentice-Hall, 1986).
12. E. O. Doebelin, *System Dynamics Modeling and Response* (Columbus, Ohio: Merrill, 1972).
13. L. Prandtl, *Essentials of Fluid Dynamics* (New York: Macmillan (Hafner), 1952).
14. J. F. Blackburn, *Fluid Power Control* (New York: Wiley, 1960).
15. S. Crandall, D. Karnopp, E. Kurtz, and D. Pridmore-Brown, *Dynamics of Mechanical and Electro Mechanical Systems* (New York: McGraw-Hill, 1968).
16. L. Meirovitch, *Elements of Vibration Analysis* 2nd ed. (New York: McGraw-Hill, 1985).
17. R. H. Cannon, Jr., *Dynamics of Physical Systems* (New York: McGraw-Hill, 1967).
18. S. Timoshenko and D. Young, *Advanced Dynamics* (New York: McGraw-Hill, 1948).

19. I. Cochin, "Connectivity Theory and Applications," *Transactions of the Eighteenth Conference of Army Mathematicians*, May 24, 1972, p. 117–143.

20. D. Graham and R. Lathrop, "Synthesis of Optimum Response," *Trans. AIEE*, Nov. 1953, p. 273–288.

21. M. Marcus and H. Minc, *Introduction to Linear Algebra* (New York: Macmillan, 1965).

22. H. Marganau and C. Murphy, *Mathematics of Physics and Chemistry* (New York: Van Nostrand, 1964).

23. V. Fadeeva, *Computational Methods of Linear Algebra* (New York: Dover, 1959).

24. R. Schwarz and B. Friedland, *Linear Systems* (New York: McGraw-Hill, 1965).

25. H. Palm, *Modeling Analysis and Control of Dynamic Systems* (New York: John Wiley, 1983).

26. R. Dorf, *Modern Control Systems* 5th ed. (Reading, Mass.: Addison-Wesley, 1989).

27. I. Cochin, "Analysis and Synthesis of Inertial Navigation Systems in Universal Terms," Ph.D. thesis, Cooper Union, 1969.

28. C. Broxmeyer, *Inertial Navigation Systems* (New York: McGraw-Hill, 1964).

29. C. Wilson and W. Michels, *Mechanism Design-Oriented Kinematics* (Chicago: American Technical Society, 1969).

30. R. W. Hamming, *Numerical Methods for Engineers and Scientists* (New York: McGraw-Hill, 1962).

31. W. E. Milne, *Numerical Solution of Differential Equations* (New York: Wiley, 1953).

32. F. B. Hildebrand, *Introduction to Numerical Analysis* (New York: McGraw-Hill, 1956).

33. G. Korn and T. Korn, *Electronic Analog Computers* (New York: McGraw-Hill, 1956).

34. K. Ogata, *Automatic Control*, 2nd ed. (Englewood Cliffs, N.J.: Prentice-Hall, 1984).

35. A. Deutschman, W. Michels, and C. Wilson, *Machine Design, Theory, and Practice* (New York: Macmillan, 1975).

36. W. F. Stoeker, *Design of Thermal Systems* 3rd ed. (New York: McGraw-Hill, 1989).

37. F. B. Hildebrande, *Advanced Calculus for Applications* 2nd ed. (Englewood Cliffs, N.J.: Prentice-Hall, 1976).

38. J. E. Shigley, *Dynamic Analysis of Machines* (New York: McGraw-Hill, 1961).

39. D. T. Greenwood, *Principles of Dynamics* 2nd ed. (Englewood Cliffs, N.J.: Prentice-Hall, 1987).

40. J. W. Brewer, *Control Systems* (Englewood Cliffs, N.J.: Prentice-Hall, 1974).

41. D. Auslander, Y. Takahashi, and M. Rabins, *Introduction to Systems and Control* (New York: McGraw-Hill, 1974).

42. H. Harrison and J. Bollinger, *Introduction to Automatic Controls* 2nd ed. (New York: Harper & Row (Intext), 1970).

43. J. Reswick and C. Taft, *Introduction to Dynamic Systems* (Englewood Cliffs, N.J.: Prentice-Hall, 1967).

44. M. Phelan, *Dynamics of Machinery* (New York: McGraw-Hill, 1967).

45. F. Kreith, *Principles of Heat Transfer* 4th ed. (New York: Harper & Row, 1984).

46. N. Beachley and H. Harrison, *Introduction to System Analysis* (New York: Harper & Row, 1978).

47. K. Ogata, *System Dynamics* (Englewood Cliffs, N.J.: Prentice-Hall, 1978).

48. C. Hsu, "Micromotors," *California Engineer*, vol. 24, 1988, p. 82–90.

49. K. E. Peterson, "Fabrication of an Integrated, Planar Silicon Ink Jet Structure," *IEEE Transactions on Electron Devices*, vol. ED-26, Dec. 1979, p. 1918.

50. K. E. Peterson, "Silicon Torsional Scanning Mirror," *IBM Research and Development*, vol. 24, 1981, p. 631.

51. K. E. Peterson, "Silicon as a Mechanical Material," *IEEE Transactions, Proceedings*, vol. 70, May 1982, p. 1122.

52. J. B. Angell, "Silicon Micromechanical Devices," *Scientific American*, April 1983, p. 44–55.

53. P. L. Chen, "Integrated Silicon Microbeam PI-FET Accelerometer," *IEEE Transactions on Electron Devices*, vol. ED-1, Jan. 1982, p. 27.

54. Y. S. Lee and K. D. Wise, "A Batch-Fabricated Silicon Capacitive Pressure Transducer with Low Temperature Sensitivity," *IEEE Transactions on Electron Devices*, vol. ED-1, Jan. 1982, p. 42.

55. M. Esashi, "Fabrication of Catheter-Tip and Sidewall Miniature Pressure Sensors," *IEEE Transactions on Electron Devices*, vol. ED-1, Jan. 1982, p. 57.

56. S. C. Kim and K. D. Wise, "Temperature Sensitivity in Silicon Piezoresistive Pressure Transducer," *IEEE Transactions on Electron Devices*, vol. ED-30, July 1983, p. 802.

57. W. A. Little, "Microminiature Refrigeration," *Review on Scientific Instruments*, May 1984, p. 592.

58. J. Gosch, "Miniature Components Carved from Silicon May be Used as Microsensors and Valves," *Electronics*, May 1984, p. 82.

59. W. R. Evans, "Graphical Analysis of Control Systems," *Trans. AIEE*, vol. 68, 1949, p. 765–777.

60. W. R. Evans, *Control System Dynamics* (New York: McGraw-Hill, 1954).

61. I. Cochin, "Gyroscope," in *Encyclopedia of Physics* 3rd ed. (New York: Van Nostrand Reinhold, 1985).

62. I. Cochin, "Inertial Guidance," in *Encyclopedia of Physics* 3rd ed. (New York: Van Nostrand Reinhold, 1985).

63. W. W. Hagerty and H. J. Plass, Jr., *Engineering Mechanics* (Princeton: D. Van Nostrand, 1967).

64. W. E. Newell and R. A. Wickstrom, "The Tunister for Microcircuits," *IEEE Transactions on Electron Devices*, vol. ED-16, no. 9, 1969, p. 781–787.

65. L. S. Fan, Y. C. Tal, and R. S. Muller, "Integrated Movable Micromechanical Structures for Sensors and Actuators," *IEEE Transactions on Electron Devices*, vol. 35, no. 6, June 1988, p. 724–730.

Appendix 1

Useful Formulae

ω_n = undamped natural frequency (rad/s)

$\omega_d = \omega$ = damped natural frequency (rad/s)

ζ = damping ratio

β = damped frequency ratio

τ = time constant (s)

λ = period of sinusoid (s)

γ = frequency ratio

ϕ = phase angle (rad)

a = decay coefficient (1/s)

r = first order term (rad/s)

SYSTEM ELEMENTS AND UNITS

M = mass (kg or N · s^2/m)

D = damper (N · s/m)

K = spring (N/m)

J = mass moment of inertia (kg · m or N · m · s^2)

D_t = torsional damper (N · m · s/rad)

K_t = torsional spring (N · m/rad)

C = electrical capacitance (F or farad)

R = electrical resistance (Ω or ohm)

L = electrical inductance (H or henry)

---

*Note: In Appendices 1 and 3, whenever the symbol ω appears without a subscript, subscript d will be understood. If no units are shown, the term is dimensionless.

$$1/C_f = \text{reciprocal fluid capacitance } (\text{N}/\text{m}^5)$$
$$R_f' = \text{fluid resistance } (\text{N} \cdot \text{s}/\text{m}^5)$$
$$I = \text{inertance } (\text{N} \cdot \text{s}^2/\text{m}^5)$$

VIBRATION FORMULAE*

1. $\omega_n^2 = K/M = K_t/J = 1/LC = 1/C_f I$
2. $\zeta = D/D_c = D/2M\omega_n = D/2J\omega_n = R/2L\omega_n = R_f/2I\omega_n$
3. $a = \zeta\omega_n = D/2M = D_t/2J = R/2L = R_f/2I$
4. $\omega = \omega_d = \omega_n\beta$
5. $\beta = \sqrt{1 - \zeta^2}, \ \beta^2 + \zeta^2 = 1$
6. $\omega_n^2 = a^2 + \omega^2$
7. $r^2 = (c - a)^2 + \omega^2$
8. $\tau = 1/a = 1/\zeta\omega_n$
9. $\lambda = 2\pi/\omega$
10. $\gamma = \omega/\omega_n$
11. e^{-at}, decay envelope
12. $s^2 + 2as + \omega_n^2$, underdamped quadratic $(1/\text{s}^2)$

PHASE ANGLES

13. $\phi = \sin^{-1}\beta = \cos^{-1}\zeta = \tan^{-1}\omega/a$
14. $\phi_r = \sin^{-1}\omega/r$
15. $\sin\phi = \beta = \omega/\omega_n$
16. $\sin 2\phi = 2\beta\zeta = 2a\omega/\omega_n^2$
17. $\sin 3\phi = \beta(4\zeta^2 - 1)$
18. $\sin 4\phi = 4\zeta\beta(2\zeta^2 - 1)$
19. $\cos\phi = a/\omega_n = \zeta$
20. $\cos 2\phi = 2\zeta^2 - 1 = (a^2 - \omega^2)/\omega_n^2$
21. $\cos 3\phi = \zeta(4\zeta^2 - 3)$
22. $\cos 4\phi = 1 + 8\zeta^2(\zeta^2 - 1)$
23. $\tan\phi = \omega/a = \beta/\zeta$
24. $\tan(\pi - \phi) = \omega/-a$
25. $\tan 2\phi = \dfrac{2a\omega}{a^2 - \omega^2} = \dfrac{2\beta\zeta}{2\zeta^2 - 1}$
26. $\sin\phi_r = \omega/r$
27. $\cos\phi_r = (c - a)/r$
28. $\tan\phi_r = \omega/(c - a)$

---

*Note: See footnotes on page 679.

EULER'S IDENTITIES

29. $e^{jx} = \cos x + j \sin x$

30. $e^{-jx} = \cos x - j \sin x$

31. $\cos x = \dfrac{e^{jx} + e^{-jx}}{2}$

32. $\sin x = \dfrac{e^{jx} - e^{-jx}}{2j}$

33. $e^x = \cosh x + \sinh x$

34. $e^{-x} = \cosh x - \sinh x$

35. $\cosh x = \dfrac{e^x + e^{-x}}{2}$

36. $\sinh x = \dfrac{e^x - e^{-x}}{2}$

Appendix 2

Modeling of System Elements

Table 2.1.3-1 MASSES AND MOMENTS OF INERTIA OF RIGID BODIES

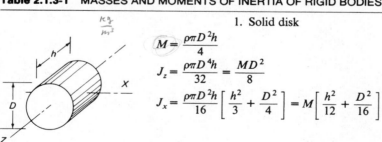

1. Solid disk

$$M = \frac{\rho \pi D^2 h}{4}$$

$$J_z = \frac{\rho \pi D^4 h}{32} = \frac{MD^2}{8}$$

$$J_x = \frac{\rho \pi D^2 h}{16}\left[\frac{h^2}{3} + \frac{D^2}{4} \right] = M\left[\frac{h^2}{12} + \frac{D^2}{16} \right]$$

2. Tube

$$M = \frac{\rho \pi h(D^2 - d^2)}{4}$$

$$J_z = \frac{\rho \pi h(D^4 - d^4)}{32} = \frac{M(D^2 + d^2)}{8}$$

$$J_x = \frac{\rho \pi h}{16}(D^2 - d^2)\left[\frac{h^2}{3} + \frac{D^2}{4} + \frac{d^2}{4} \right]$$

$$= M\left[\frac{h^2}{12} + \frac{D^2}{16} + \frac{d^2}{16} \right]$$

3. Rectangular solid

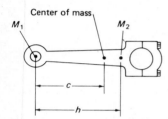

Center of mass

$$M = \rho Lbh$$

$$J_z = \frac{\rho Lbh}{12}(L^2 + h^2 + 12r^2)$$

$$= M\left[\frac{L^2}{12} + \frac{h^2}{12} + r^2\right]$$

$$J_c = \frac{\rho Lbh(L^2 + h^2)}{12}$$

$$= \frac{M[L^2 + h^2]}{12}$$

4. Connecting rod

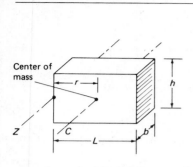

Center of mass

M_1 M_2

$$M_1 = M\left(1 - \frac{c}{h}\right)$$

$$M_2 = \frac{Mc}{h}$$

$$h = \frac{J_1}{Mc}$$

$J_1 =$ moment of inertia about M_1

Table 2.1.4-1   MASSES AND MOMENTS OF INERTIA OF ELASTIC BODIES

1. Cantilever beam

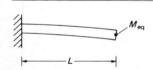

M_{eq}

$$M_{eq} \approx \frac{M}{4}$$

2. Simply supported beam

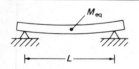

M_{eq}

$$M_{eq} = \tfrac{1}{2}M$$

3. Coiled spring

M_{eq}

$$M_{eq} = \frac{M}{3}$$

4. Bellows

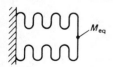

M_{eq}

$$M_{eq} = \frac{M}{3}$$

5. Shaft in torsion

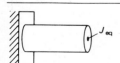

J_{eq}

$$J_{eq} = \frac{J}{3}$$

Table 2.1.4-1   (cont.)

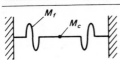

6. Diaphragm

$$M_{eq} = M_c + \frac{M_f}{4}$$

7. Clock spring
$M = \rho b h L$

$$J_{eq} \approx \frac{M d^2}{16}$$

L = total length of spiral

Table 2.1.5-1   FLUID MASSES

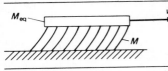

1. Translating surfaces

$$M_{eq} = \frac{M}{3}$$

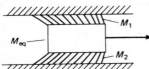

2. Translating body

$$M_{eq} = \frac{M_1 + M_2}{3}$$

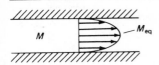

3. Laminar flow in pipe

$$M_{eq} = \frac{4M}{3}$$

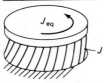

4. Rotating parallel surfaces

$$J_{eq} = \frac{J}{3}$$

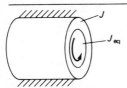

5. Rotating concentric cylinders

$$J_{eq} = \frac{J}{3}$$

Table 2.2.3-1   MECHANICAL SPRINGS (TRANSLATORY)

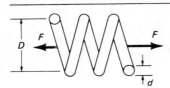

1. Coiled spring (round wire)

$$K_{eq} = \frac{d^4 G}{64 R^3 N}$$

where

G = modulus in shear

N = number of coils

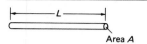

2. Rod or wire in tension

$$K_{eq} = \frac{EA}{L}$$

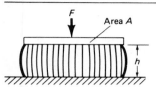

3. Rubber in compression

$$K_{eq} = \frac{1 + F/AE}{h/AE}$$

where

A = area of plate

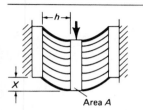

4. Rubber in double shear

$$K_{eq} = \frac{2AG}{h}$$

where

A = area of plate providing shear surface

5. Pendulum

$$K_{eq} = \frac{gM}{L}$$

where

g = acceleration due to gravity

Table 2.2.3-2   DISK SPRINGS

1. Uniform load, edges simply supported

$$K = \frac{64\pi Z(1 + \nu)}{a^2(5 + \nu)}$$

$$\sigma_{max} = \frac{3Pa^2(3 + \nu)}{8t^2}$$

$$Z = \frac{Et^3}{12(1 - \nu^2)}$$

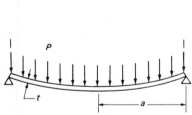

a = radius of disk

2. Uniform load, edges clamped

$$K = \frac{64\pi Z}{a^2}$$

$$\sigma = \frac{3P(1 + \nu)a^2}{8t^2}$$

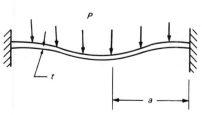

Table 2.2.3-2   (cont.)

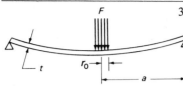

3. Concentrated load, edges simply supported

$$K = \frac{16\pi Z(1 + \nu)}{a^2(3 + \nu)}$$

$$\sigma_{max} = \frac{3F}{2\pi t^2}\left[(1 + \nu)\ln\frac{a}{r_0} + 1\right]$$

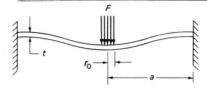

4. Concentrated load, edges clamped

$$K = \frac{16\pi Z}{a^2}$$

$$\sigma_{max} = \frac{3F}{2\pi t^2}(1 + \nu)\ln\frac{a}{r_0}$$

Table 2.2.3-3   EQUIVALENT SPRINGS FOR BEAMS

1. Cantilever

$$K_{eq} = \frac{3EI}{L^3}$$

where

$$I = \frac{bh^3}{12}$$

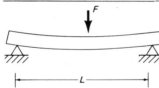

2. Simply supported

$$K_{eq} = \frac{48EI}{L^3}$$

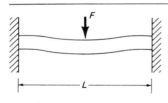

3. Both ends fixed or clamped

$$K_{eq} = \frac{192EI}{L^3}$$

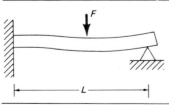

4. Cantilever with pinned end

$$K_{eq} = \frac{110EI}{L^3}$$

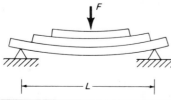

5. Leaf spring

$$K_{eq} = \sum K_i$$

where

K_i = spring constant for each leaf

Table 2.2.3-4   TORSION SPRINGS

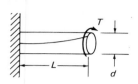

1. Round rod
$$K_{eq} = \frac{\pi d^4 G}{32L}$$
where
$$G = \text{modulus in shear}$$

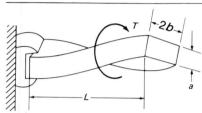

2. Rectangular bar
$$K_{eq} = \frac{CG}{L}$$

$$C = ab^3 \left[\frac{16}{3} - 3.36\frac{b}{a}\left(1 - \frac{b^4}{12a^4}\right) \right]$$

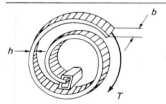

3. Coiled spring in torsion (round wire)
$$K_{eq} = \frac{Ed^4}{64DN}$$
where
$$E = \text{Young's modulus}$$
$$N = \text{number of coils}$$

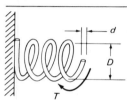

4. Square wire coiled spring
$$K_{eq} = \frac{Ea^4}{12\pi DN}$$

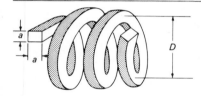

5. Clock spring
$$K_{eq} = \frac{EI}{L}$$
or
$$K_{eq} = \frac{Ebh^3}{12L}$$

Table 2.2.4-1   FLUID SPRINGS

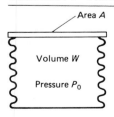

1. Air spring, bellows
$$K = \frac{A^2 P_0}{W}$$

Table 2.2.4-1 (cont.)

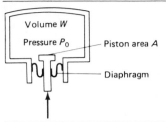

2. Air spring, piston

$$K = \frac{A^2 P_0}{W}$$

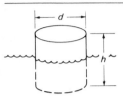

3. Floating cylinder

$$K = \frac{\pi d^2 \rho_f g}{4}$$

ρ_f = mass density of fluid

g = acceleration due to gravity

Table 2.3.3-1 FLUID DAMPERS USING VISCOUS SHEAR

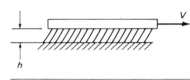

1. Translating surfaces

$$D_{eq} = \frac{\mu A}{h}$$

where

A = wetted surface area

μ = coefficient of viscosity (absolute)

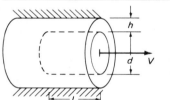

2. Translating cylinder

$$D_{eq} = \frac{\mu A}{h}$$

where

A = wetted surface area

$= \pi L d$

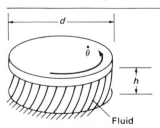

3. Rotating disks

$$D_{eq} = \frac{\pi \mu d^4}{16h}$$

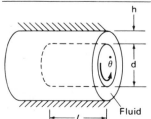

4. Rotating cylinders

$$D_{eq} = \frac{\pi \mu d^3 L}{4h}$$

Table 2.3.4-1   DAMPING DUE TO FLOW THROUGH RESTRICTION

1. Porous material*

$$R = \frac{4C_1 L}{\pi d^2 \rho g}$$

C_1 = coefficient of fluid resistance

$$D_{eq} = \frac{A_1^2 4 C_1 L}{\pi d^2}$$

2. Pipe* (Laminar flow)

$$R = \frac{128 \mu L}{\pi d^4 \rho g}$$

$$D_{eq} = \frac{A_1^2 128 \mu L}{\pi d^4}$$

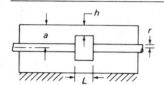

3. Dashpot

$$D_{eq} = \frac{6 \pi \mu L}{h^3} \left[\left(a - \frac{h}{2} \right)^2 - r^2 \right] \left[\frac{a^2 - r^2}{a - h/2} - h \right]$$

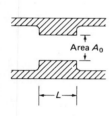

4. Orifice

$$Q = C_d A_0 \sqrt{\frac{2 \Delta p}{\rho}}$$

A_0 = area of orifice
C_d = discharge coefficient
ρ = fluid density
Δp = pressure drop through orifice

$$N_r = \frac{\rho Q d}{A_0 \mu}$$

Table 2.3.5-1   DAMPING DUE TO DRAG IN A VISCOUS MEDIUM

1. Sphere[†]
$$D_{eq} = 3 \pi \mu d$$

2. Disk[†]
$$D_{eq} = 30 \mu d$$

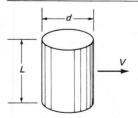

3. Cylinder[†]
$$D_{eq} = 3 \mu L$$

*A_1 = area of piston in Fig. 2.6.2-2. μ = absolute (dynamic) viscosity. See Appendix 6.
[†]See footnote on page 690.

Table 2.3.5-1   (cont.)

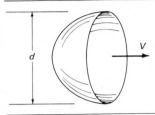

4. Cup or parachute[†]
    $D_{eq} = 100\mu d$

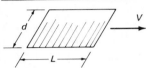

5. Skin friction[†]
    $D_{eq} = 100\mu L^2 d$

Table 2.3.6-1   MECHANICAL DAMPERS

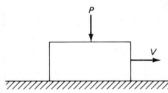

1. Coulomb friction

    $$D_{eq} = \frac{4\mu P}{\pi\omega X_m}$$

    μ = coefficient of dry friction
    P = normal force
    ω = input frequency
    X_m = amplitude of oscillation
        of mass relative to base

2. Velocity squared

    $$D_{eq} = \frac{4\omega C_d X_m}{3\pi}$$

    C_d = drag coefficient

3. Structural damping

    $$D_{eq} = \frac{C}{\pi\omega}$$

    $$C = \frac{E}{X_m{}^2}$$

    E = energy lost per cycle

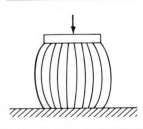

4. Rubber pad

    $$D_{eq} = \frac{E}{\pi\omega X_m{}^2}$$

    Rubber acts like a spring
    damper network Z where

    $Z(s) = DS + K$

[†]Fair approximation valid for low Reynold's number, N_r

$$N_r = \frac{\rho V d}{\mu}$$

Table 2.4.1-1   MAGNETIC DAMPERS AND SPRINGS

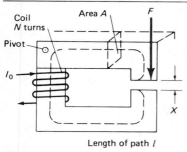

Coil N turns
Pivot
Area A
F
I_0
X
Length of path l

1. Small air gap

$$L = \frac{\mu N^2 A}{l(1 + \mu X/l)}$$

$$F = \frac{A}{2}\left[\frac{\mu N I_0}{l(1 + \mu X/l)}\right]^2$$

$$K_{eq} = -AN^2 I_0^2 \left[\frac{\mu}{l(1 + \mu X/l)}\right]^3$$

$$D_{eq} = \frac{I_0^2 A^2}{R}\left[\frac{\mu N}{l(1 + \mu X/l)}\right]^4$$

where $\mu = \mu_{core}/\mu_0$

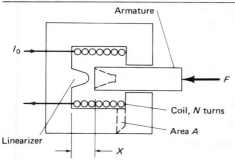

Armature
I_0
F
Coil, N turns
Area A
Linearizer
X

2. Solenoid
   All relations are identical to those for either small air gap (Item 1) or large air gap (Item 4) depending upon the stroke. The protrusion and corresponding well in the armature tends to linearize the system.

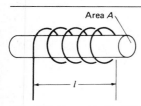

Area A
l

3. Open Core
   $$L = \mu N^2 A / l$$

   where $\mu = \mu_{core}/\mu_0$

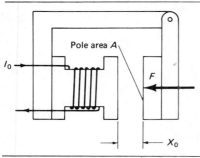

Pole area A
I_0
F
X_0

4. Large air gap

$$F = \frac{AN^2 I_0^2 \mu_0}{2X_0^2}$$

$$K_m = \frac{-AN^2 I_0^2 \mu_0}{X_0^3}$$

$$D_m = \frac{A^2 I_0^2 N^4 \mu_0}{R_c X_0^4}$$

$$L = \frac{N^2 A \mu_0}{X_0}$$

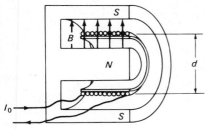

S
B
N
d
I_0
S

5. Voice coil

$$F = \pi d N B I_0$$
$$V = \pi d N B \dot{X}$$
$$K = 0$$
$$D_m = \frac{(\pi d N B)^2}{10^8 R_c}$$

Table 2.4.1-1 (cont.)

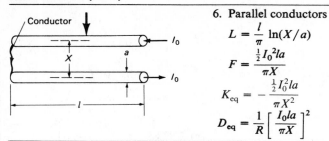

6. Parallel conductors

$$L = \frac{l}{\pi} \ln(X/a)$$

$$F = \frac{\frac{1}{2}I_0^2 la}{\pi X}$$

$$K_{eq} = -\frac{\frac{1}{2}I_0^2 la}{\pi X^2}$$

$$D_{eq} = \frac{1}{R}\left[\frac{I_0 la}{\pi X}\right]^2$$

units in mks system

F = force (N)	I_0 = nominal current (A)
μ = permeability	V = voltage (V)
L = inductance (H)	R_c = coil resistance (Ω)
	B = mag. field (Wb/m^2)
	l = length (m)

Table 2.4.2-1 ELECTROSTATIC DAMPERS AND SPRINGS

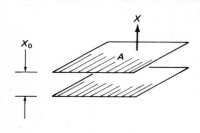

1. Parallel planes

$$C = \frac{\epsilon A}{X}$$

$$F_0 = \frac{V_0^2 \epsilon A}{2X_0^2}$$

$$K_{eq} = -\frac{V_0^2 \epsilon A}{X_0^3}$$

$$D_{eq} = \frac{(V_0 \epsilon A)^2 R}{X_0^4}$$

C = capactance (farad)

ϵ = permittivity

See Appendix 6

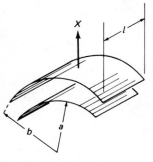

2. Concentric cylinders

$$C = \frac{2\pi \epsilon l}{\ln(b/a)}$$

$$F_0 = \frac{V_0^2 \pi \epsilon l}{b(\ln X_0/a)^2}$$

$$K_{eq} = -\frac{V_0^2 2\pi \epsilon l}{b^2(\ln X_0/a)^3}$$

$$D_{eq} = \frac{4(V_0 \pi \epsilon l)^2 R}{b^2(\ln X_0/a)^4}$$

$X_0 = b - a$

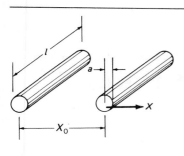

3. Parallel rods

$$C = \frac{\pi \epsilon l}{\ln\left(\dfrac{b + \sqrt{b^2 - 4a^2}}{2a}\right)}$$

$$F_0 = \frac{V_0^2 \pi \epsilon l}{2(\ln X_0/a)^2}$$

$$K_{eq} = \frac{-V_0^2 \pi \epsilon l}{(\ln X_0/a)^3}$$

$$D_{eq} = \frac{2(V_0 \pi \epsilon l)^2 R}{(\ln X_0/a)^4}$$

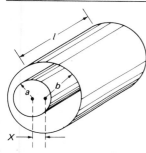

4. Eccentric cylinders

$$C = \frac{2\pi \epsilon i}{\cosh^{-1}\left(\dfrac{a^2 + b^2 - X^2}{2ab}\right)}$$

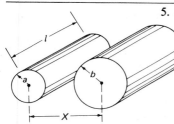

5. Parallel cylinders

$$C = \frac{2\pi \epsilon l}{\cosh^{-1}\left(\dfrac{a^2 + b^2 - X^2}{2ab}\right)}$$

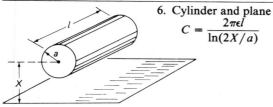

6. Cylinder and plane

$$C = \frac{2\pi \epsilon l}{\ln(2X/a)}$$

7. Concentric spheres

$$C = \frac{4\pi \epsilon ab}{b - a}$$

Appendix 3

Laplace Transform Pairs

For definition of terms and symbols, see Appendix 1. The entry numbers in the following table of Laplace transform pairs are arranged as follows:

hundreds place: family or type of system;

tens place: group within a family;

units place: ascending numerical value indicates derivative, descending value indicates an integration.

Note that certain entries have been deliberately omitted to provide exercises for the student.

	Laplace domain	Time domain
100	*General Operators*	
110	Basics	
111	F	$\int_0^\infty e^{-st} f(t)\, dt$ Definition
112	$F + G$	$f(t) + g(t)$ Sum
113	aF	$af(t)$ Multiplication by constant
114	$F(s + a)$	$e^{-at} f(t)$ Shifting theorem
115	$F(s)G(s)$	$\int_0^t f(t - \tau)g(\tau)\, d\tau$ Convolution theorem
		$= \int_0^t f(\tau)g(t - \tau)\, d\tau$
116	$\lim_{s \to 0} sF$	$\lim_{t \to \infty} f(t)$ Final value theorem
117	$\lim_{s \to \infty} sF$	$\lim_{t \to 0} f(t)$ Initial value theorem

	Laplace domain	Time domain
120	Integrals and Derivatives in the Time Domain	
122	$\dfrac{F}{s^n}$	$\displaystyle\int\int^n\int_{0^+}^t f(t)\,dt^n$
123	$\dfrac{F}{s^2}$	$\displaystyle\int_{0^+}^t\int_{0^+}^t f(t)\,dt^2$
124	$\dfrac{F}{s}$	$\displaystyle\int_{0^+}^t f(t)\,dt$
125	F	$f(t)$ (also listed as entry 111)
126	$sF - f(0^-)$	$df(t)/dt$
127	$s^2F - sf(0^-) - \dot{f}(0^-)$	$d^2f(t)/dt^2$
128	$s^nF - \displaystyle\sum_{k=1}^n s^{n-k}\dfrac{d^{k-1}f(0^-)}{dt^{k-1}}$	$\dfrac{d^nf(t)}{dt^n}$
130	Integrals and Derivatives in the Laplace Domain	
134	$\displaystyle\int_s^\infty F\,ds$	$\dfrac{1}{t}f(t)$
135	F	$f(t)$ (also listed as entry 111)
136	$\dfrac{dF}{ds}$	$-tf(t)$
137	$\dfrac{d^2F}{ds^2}$	$t^2f(t)$
138	$\dfrac{d^nF}{ds^n}$	$(-1)^n t^n f(t)$
140	Jump Functions	
141	$\dfrac{1}{s^{n+1}}$	$\dfrac{t^n}{n!}$ $n = 0, 1, 2; \ldots$
142	$\dfrac{1}{s^4}$	$\dfrac{t^3}{3!}$
143	$\dfrac{1}{s^3}$	$\dfrac{t^2}{2!}$ parabolic function
144	$\dfrac{1}{s^2}$	t ramp function
145	$\dfrac{1}{s}$	1 step function
146	1	$\delta(t)$ impulse function
147	s	$\delta'(t)$ doublet impulse
148	s^2	$\delta''(t)$ triplet impulse

	Laplace domain	Time domain
150	Delayed Functions	
151	$e^{-\alpha s}F(s)$	$f(t - \alpha)$ Function delayed by time α (also listed as entry 115)
152	$\dfrac{e^{-\alpha s}}{s^2}$	$(t - \alpha)v(t - \alpha)$ Delayed ramp
153	$\dfrac{1 - e^{-\alpha s}}{s^2}$	$vt - \{(t - \alpha)v(t - \alpha)\}$ Ramp step
154	$\dfrac{e^{-\alpha s}}{s}$	$U(t - \alpha)$ Delayed step
155	$\dfrac{1 - e^{-\alpha s}}{s}$	$U(t) - U(t - \alpha)$ Rectangular pulse
156	$e^{-\alpha s}$	$\delta(t - \alpha)$ Delayed impulse
157	$se^{-\alpha s}$	$\delta'(t - \alpha)$ Delayed doublet

200	*First Order Factors*	
210	Single Factor	
211	$\dfrac{a^5}{s^5(s + a)}$	$\dfrac{a^4t^4}{4!} - \dfrac{a^3t^3}{3!} + \dfrac{a^2t^2}{2!} - at + 1 - e^{-at}$
212	$\dfrac{a^4}{s^4(s + a)}$	$\dfrac{a^3t^3}{3!} - \dfrac{a^2t^2}{2!} + at - 1 + e^{-at}$
213	$\dfrac{a^3}{s^3(s + a)}$	$\dfrac{a^2t^2}{2!} - at + 1 - e^{-at}$

	Laplace domain	Time domain
214	$\dfrac{a^2}{s^2(s+a)}$	$at - 1 + e^{-at}$
215	$\dfrac{a}{s(s+a)}$	$1 - e^{-at}$
216	$\dfrac{1}{s+a}$	e^{-at}
217	$\dfrac{s}{a(s+a)}$	$\dfrac{\delta}{a} - e^{-at}$
218	$\dfrac{s^2}{a^2(s+a)}$	$\dfrac{\delta'}{a^2} - \dfrac{\delta}{a} + e^{-at}$
219	$\dfrac{s^3}{a^3(s+a)}$	$\dfrac{\delta''}{a^3} - \dfrac{\delta'}{a^2} + \dfrac{\delta}{a} - e^{-at}$

220　Two First Order Factors
(If the damping ratio ζ of a quadratic is greater than unity, the quadratic *must* be factored into two first order factors). $a \neq b$.

222	$\dfrac{a^3 b^3}{s^3(s+a)(s+b)}$	$\dfrac{a^2 b^2 t^2}{2!} - \dfrac{(b^2 - a^2)abt}{b-a} + \dfrac{b^3 - a^3}{b-a}$
		$-\left[\dfrac{b^3 e^{-at} - a^3 e^{-bt}}{b-a}\right]$
223	$\dfrac{a^2 b^2}{s^2(s+a)(s+b)}$	$abt - \dfrac{b^2 - a^2}{b-a} + \left[\dfrac{b^2 e^{-at} - a^2 e^{-bt}}{b-a}\right]$
224	$\dfrac{ab}{s(s+a)(s+b)}$	$1 - \left[\dfrac{be^{-at} - ae^{-bt}}{b-a}\right]$
225	$\dfrac{1}{(s+a)(s+b)}$	$\dfrac{e^{-at} - e^{-bt}}{b-a}$
226	$\dfrac{s}{(s+a)(s+b)}$	$-\left[\dfrac{ae^{-at} - be^{-bt}}{b-a}\right]$
227	$\dfrac{s^2}{(s+a)(s+b)}$	$\delta + \left[\dfrac{a^2 e^{-at} - b^2 e^{-bt}}{b-a}\right]$
228	$\dfrac{s^3}{(s+a)(s+b)}$	$\delta' - \dfrac{(b^2 - a^2)\delta}{b-a} - \left[\dfrac{a^3 e^{-at} + b^3 e^{-bt}}{b-a}\right]$
229	$\dfrac{s^4}{(s+a)(s+b)}$	$\delta'' - \dfrac{(b^2 - a^2)\delta'}{b-a} + \dfrac{(b^3 - a^3)\delta}{b-a}$
		$+ \left[\dfrac{a^4 e^{-at} - b^4 e^{-bt}}{b-a}\right]$

	Laplace domain	Time domain

240   Three First Order Factors
The cubic can be factored into first and quadratic factors. (If the damping ratio ζ of the quadratic is greater than unity, then it *must* be factored. In that case, there will be three first order factors.) $a \neq b \neq c$.

243
$$\frac{abc}{s(s + a)(s + b)(s + c)}$$
$$1 - \frac{bce^{-at}}{(b - a)(c - a)} - \frac{ace^{-bt}}{(a - b)(c - b)}$$
$$- \frac{abe^{-ct}}{(a - c)(b - c)}$$

244
$$\frac{1}{(s + a)(s + b)(s + c)}$$
$$\frac{e^{-at}}{(b - a)(c - a)} + \frac{e^{-bt}}{(a - b)(c - b)}$$
$$+ \frac{e^{-ct}}{(a - c)(b - c)}$$

245
$$\frac{s}{(s + a)(s + b)(s + c)}$$
$$- \frac{ae^{-at}}{(b - a)(c - a)} - \frac{be^{-bt}}{(a - b)(c - b)}$$
$$- \frac{ce^{-ct}}{(a - c)(b - c)}$$

246
$$\frac{s^2}{(s + a)(s + b)(s + c)}$$
$$\frac{a^2 e^{-at}}{(b - a)(c - a)} + \frac{b^2 e^{-bt}}{(a - b)(c - b)}$$
$$+ \frac{c^2 e^{-ct}}{(a - c)(b - c)}$$

247
$$\frac{s^3}{(s + a)(s + b)(s + c)}$$
$$- \delta - \frac{a^3 e^{-at}}{(b - a)(c - a)} - \frac{b^3 e^{-bt}}{(a - b)(c - b)}$$
$$- \frac{c^3 e^{-ct}}{(a - c)(b - c)}$$

260   Two Repeated First Order Factors
Note: If the damping ratio ζ is equal to unity, the quadratic is a perfect square, yielding two identical or repeated factors.

261
$$\frac{a^5}{s^4(s + a)^2}$$
$$\frac{a^3 t^3}{3!} - a^2 t^2 + 3at - 4 + (at + 4)e^{-at}$$

262
$$\frac{a^4}{s^3(s + a)^2}$$
$$\frac{a^2 t^2}{2} - 2at + 3 - (at + 3)e^{-at}$$

263
$$\frac{a^3}{s^2(s + a)^2}$$
$$at - 2 + (at + 2)e^{-at}$$

264
$$\frac{a^2}{s(s + a)^2}$$
$$1 - (at + 1)e^{-at}$$

	Laplace domain	Time domain
265	$\dfrac{a}{(s+a)^2}$	ate^{-at}
266	$\dfrac{s}{(s+a)^2}$	$-(at-1)e^{-at}$
267	$\dfrac{s^2}{a(s+a)^2}$	$\dfrac{\delta}{a} + (at-2)e^{-at}$
268	$\dfrac{s^3}{a^2(s+a)^2}$	$\dfrac{\delta'}{a^2} - \dfrac{2\delta}{a} - (at-3)e^{-at}$
269	$\dfrac{s^4}{a^3(s+a)^2}$	$\dfrac{\delta''}{a^3} - \dfrac{2\delta'}{a^2} + \dfrac{3\delta}{a} + (at-4)e^{-at}$
280	*N* Repeated First Order Factors	
285	$\dfrac{a^n}{(s+a)^{n+1}}$	$\dfrac{a^n t^n e^{-at}}{n!} \quad n > -1$
286	$\dfrac{s}{(s+a)^{n+1}}$	$\dfrac{(n-at)t^{n-1}e^{-at}}{n!} \quad n > 0$
300	*Undamped Quadratics*	
310	Single Quadratic	
311	$\dfrac{\omega^5}{s^4(s^2+\omega^2)}$	$\dfrac{\omega^3 t^3}{3!} - \omega t + \sin \omega t$
312	$\dfrac{\omega^4}{s^3(s^2+\omega^2)}$	$\dfrac{\omega^2 t^2}{2!} - 1 + \cos \omega t$
313	$\dfrac{\omega^3}{s^2(s^2+\omega^2)}$	$\omega t - \sin \omega t$
314	$\dfrac{\omega^2}{s(s^2+\omega^2)}$	$1 - \cos \omega t$
315	$\dfrac{\omega}{s^2+\omega^2}$	$\sin \omega t$
316	$\dfrac{s}{s^2+\omega^2}$	$\cos \omega t$
317	$\dfrac{s^2}{\omega(s^2+\omega^2)}$	$\dfrac{\delta}{\omega} - \sin \omega t$
318	$\dfrac{s^3}{\omega^2(s^2+\omega^2)}$	$\dfrac{\delta'}{\omega^2} - \cos \omega t$
319	$\dfrac{s^4}{\omega^3(s^2+\omega^2)}$	$\dfrac{\delta''}{\omega^3} - \dfrac{\delta}{\omega} + \sin \omega t$

Laplace domain	Time domain

320   Two Undamped Quadratics, $\alpha \neq \omega$

321    $\dfrac{\alpha^3 \omega^3}{s^2(s^2 + \omega^2)(s^2 + \alpha^2)}$

$t - \left[\dfrac{\alpha^3 \sin \omega t - \omega^3 \sin \alpha t}{\alpha^2 - \omega^2} \right]$

322    $\dfrac{\alpha^2 \omega^2}{s(s^2 + \omega^2)(s^2 + \alpha^2)}$

$1 - \left[\dfrac{\alpha^2 \cos \omega t - \omega^2 \cos \alpha t}{\alpha^2 - \omega^2} \right]$

323    $\dfrac{\alpha \omega}{(s^2 + \omega^2)(s^2 + \alpha^2)}$

$\dfrac{\alpha \sin \omega t - \omega \sin \alpha t}{\alpha^2 - \omega^2}$

324    $\dfrac{s}{(s^2 + \omega^2)(s^2 + \alpha^2)}$

$\dfrac{\cos \omega t - \cos \alpha t}{\alpha^2 - \omega^2}$

325    $\dfrac{s^2}{(s^2 + \omega^2)(s^2 + \alpha^2)}$

$-\left[\dfrac{\omega \sin \omega t - \alpha \sin \alpha t}{\alpha^2 - \omega^2} \right]$

326    $\dfrac{s^3}{(s^2 + \omega^2)(s^2 + \alpha^2)}$

$-\left[\dfrac{\omega^2 \cos \omega t - \alpha^2 \cos \alpha t}{\alpha^2 - \omega^2} \right]$

327    $\dfrac{s^4}{(s^2 + \omega^2)(s^2 + \alpha^2)}$

$\delta + \left[\dfrac{\omega^3 \sin \omega t - \alpha^3 \sin \alpha t}{\alpha^2 - \omega^2} \right]$

328    $\dfrac{s^5}{(s^2 + \omega^2)(s^2 + \alpha^2)}$

$\delta' + \left[\dfrac{\omega^4 \cos \omega t - \alpha^4 \cos \alpha t}{\alpha^2 - \omega^2} \right]$

329    $\dfrac{s^6}{(s^2 + \omega^2)(s^2 + \alpha^2)}$

$\delta'' + \dfrac{(\omega^4 - \alpha^4)\delta}{\alpha^2 - \omega^2}$

$- \left[\dfrac{\omega^5 \sin \omega t - \alpha^5 \sin \alpha t}{\alpha^2 - \omega^2} \right]$

330   Repeated Undamped Quadratics

332    $\dfrac{2\omega^5}{s^2(s^2 + \omega^2)^2}$

$2\omega t + \omega t \cos \omega t - 3 \sin \omega t$

333    $\dfrac{2\omega^4}{s(s^2 + \omega^2)^2}$

$2 - \omega t \sin \omega t - 2 \cos \omega t$

334    $\dfrac{2\omega^3}{(s^2 + \omega^2)^2}$

$-\omega t \cos \omega t + \sin \omega t$

335    $\dfrac{2\omega^2 s}{(s^2 + \omega^2)^2}$

$\omega t \sin \omega t$

	Laplace domain	Time domain
336	$\dfrac{2\omega s^2}{\left(s^2 + \omega^2\right)^2}$	$\omega t \cos \omega t + \sin \omega t$
337	$\dfrac{2s^3}{\left(s^2 + \omega^2\right)^2}$	$-\omega t \sin \omega t + 2 \cos \omega t$
338	$\dfrac{2s^4}{\omega\left(s^2 + \omega^2\right)^2}$	$\dfrac{2\delta}{\omega} + \omega t \cos \omega t - 3 \sin \omega t$
340	Undamped Modulation	
341	$\dfrac{4\omega\alpha}{s^2\left\{s^2 + (\omega + \alpha)^2\right\}\left\{s^2 + (\omega - \alpha)^2\right\}}$	$-\dfrac{4\omega\alpha t}{\left(\omega^2 - \alpha^2\right)^2} + \dfrac{\sin(\omega + \alpha)t}{(\omega + \alpha)^3}$ $-\dfrac{\sin(\omega - \alpha)t}{(\omega - \alpha)^3}$
342	$\dfrac{4\omega\alpha}{s\left\{s^2 + (\omega + \alpha)^2\right\}\left\{s^2 + (\omega - \alpha)^2\right\}}$	$-\dfrac{4\omega\alpha}{\left(\omega^2 - \alpha^2\right)^2} + \dfrac{\cos(\omega + \alpha)t}{(\omega + \alpha)^2}$ $-\dfrac{\cos(\omega - \alpha)t}{(\omega - \alpha)^2}$
343	$\dfrac{4\omega\alpha}{\left\{s^2 + (\omega + \alpha)^2\right\}\left\{s^2 + (\omega - \alpha)^2\right\}}$	$-\dfrac{\sin(\omega + \alpha)t}{(\omega + \alpha)}$ $+\dfrac{\sin(\omega - \alpha)t}{(\omega - \alpha)}$
344	$\dfrac{4\omega\alpha s}{\left\{s^2 + (\omega + \alpha)^2\right\}\left\{s^2 + (\omega - \alpha)^2\right\}}$	$-\cos(\omega + \alpha)t + \cos(\omega - \alpha)t$
345	$\dfrac{4\omega\alpha s^2}{\left\{s^2 + (\omega + \alpha)^2\right\}\left\{s^2 + (\omega - \alpha)^2\right\}}$	$(\omega + \alpha)\sin(\omega + \alpha)t$ $-(\omega - \alpha)\sin(\omega - \alpha)t$
346	$\dfrac{4\omega\alpha s^3}{\left\{s^2 + (\omega + \alpha)^2\right\}\left\{s^2 + (\omega - \alpha)^2\right\}}$	$(\omega + \alpha)^2 \cos(\omega + \alpha)t$ $-(\omega - \alpha)^2 \cos(\omega - \alpha)t$
347	$\dfrac{4\omega\alpha s^4}{\left\{s^2 + (\omega + \alpha)^2\right\}\left\{s^2 + (\omega - \alpha)^2\right\}}$	$4\omega\alpha\delta - (\omega + \alpha)^3 \sin(\omega + \alpha)t$ $+(\omega - \alpha)^3 \sin(\omega - \alpha)t$

	Laplace domain	Time domain
348	$$\dfrac{4\omega\alpha s^5}{\left\{s^2 + (\omega + \alpha)^2\right\}\left\{s^2 + (\omega - \alpha)^2\right\}}$$	$4\omega\alpha\delta' - (\omega + \alpha)^4 \cos(\omega + \alpha)t$ $+ (\omega - \alpha)^4 \cos(\omega - \alpha)t$

400 *Underdamped Quadratics*
Note: The 400 family can be used only if the damping ratio ζ is greater than zero but less than unity. For other values of ζ, see groups 220, 260, and 310. The term a is a special constant; $a = \zeta\omega_n$. See Appendix 1 for definitions of symbols.

410 Numerator is Single Term

412	$$\dfrac{\omega_n^4}{s^3\left(s^2 + 2as + \omega_n^2\right)}$$	$\dfrac{\omega_n^2 t^2}{2} - 2at + (4\zeta^2 - 1) - \dfrac{\omega_n}{\omega_d}e^{-at}\sin(\omega_d t + 3\phi)$
413	$$\dfrac{\omega_n^3}{s^2\left(s^2 + 2as + \omega_n^2\right)}$$	$\omega_n t - 2\zeta + \dfrac{\omega_n}{\omega_d}e^{-at}\sin(\omega_d t + 2\phi)$
414	$$\dfrac{\omega_n^2}{s\left(s^2 + 2as + \omega_n^2\right)}$$	$1 - \dfrac{\omega_n}{\omega_d}e^{-at}\sin(\omega_d t + \phi)$
415	$$\dfrac{\omega_n}{s^2 + 2as + \omega_n^2}$$	$\dfrac{\omega_n}{\omega_d}e^{-at}\sin\omega_d t$
416	$$\dfrac{s}{s^2 + 2as + \omega_n^2}$$	$-\dfrac{\omega_n}{\omega_d}e^{-at}\sin(\omega_d t - \phi)$
417	$$\dfrac{s^2}{\omega_n\left(s^2 + 2as + \omega_n^2\right)}$$	$\dfrac{\delta}{\omega_n} + \dfrac{\omega_n}{\omega_d}e^{-at}\sin(\omega_d t - 2\phi)$
418	$$\dfrac{s^3}{\omega_n^2\left(s^2 + 2as + \omega_n^2\right)}$$	$\dfrac{\delta'}{\omega_n^2} - \dfrac{2\zeta\delta}{\omega_n} - \dfrac{\omega_n}{\omega_d}e^{-at}\sin(\omega_d t - 3\phi)$
419	$$\dfrac{s^4}{\omega_n^3\left(s^2 + 2as + \omega_n^2\right)}$$	$\dfrac{\delta''}{\omega_n^3} - \dfrac{2\zeta\delta'}{\omega_n^2} + (4\zeta^2 - 1)\dfrac{\delta}{\omega_n} + \dfrac{\omega_n}{\omega_d}e^{-at}\sin(\omega_d t - 4\phi)$

420 Numerator Contains First Two Terms of Quadratic
(See notes for entry 400.)

422	$$\dfrac{\omega_n^2(s + 2a)}{s^2\left(s^2 + 2as + \omega_n^2\right)}$$	$2at - (4\zeta^2 - 1) + \dfrac{\omega_n}{\omega_d}e^{-at}\sin(\omega_d t + 3\phi)$
423	$$\dfrac{\omega_n(s + 2a)}{s\left(s^2 + 2as + \omega_n^2\right)}$$	$2\zeta - \dfrac{\omega_n}{\omega_d}e^{-at}\sin(\omega_d t + 2\phi)$
424	$$\dfrac{s + 2a}{s^2 + 2as + \omega_n^2}$$	$\dfrac{\omega_n}{\omega_d}e^{-at}\sin(\omega_d t + \phi)$

	Laplace domain	Time domain
425	$\dfrac{s(s + 2a)}{\omega_n\left(s^2 + 2as + \omega_n^2\right)}$	$\dfrac{\delta}{\omega_n} - \dfrac{\omega_n}{\omega_d}e^{-at}\sin\omega_d t$
426	$\dfrac{s^2(s + 2a)}{\omega_n^2\left(s^2 + 2as + \omega_n^2\right)}$	$\dfrac{\delta'}{\omega_n^2} + \dfrac{\omega_n}{\omega_d}e^{-at}\sin(\omega_d t - \phi)$
427	$\dfrac{s^3(s + 2a)}{\omega_n^3\left(s^2 + 2as + \omega_n^2\right)}$	$\dfrac{\delta''}{\omega_n^3} - \dfrac{\delta}{\omega_n} - \dfrac{\omega_n}{\omega_d}e^{-at}\sin(\omega_d t - 2\phi)$

430   Numerator Contains First and Last Terms of Quadratic
(See notes for entry 400.)

432	$\dfrac{\omega_n^4\left(s^2 + \omega_n^2\right)}{2as^4\left(s^2 + 2as + \omega_n^2\right)}$	$\dfrac{\omega_n^4 t^3}{(2a)3!} - \dfrac{\omega_n^2 t^2}{2} + 2at - (4\zeta^2 - 1)$ $+ \dfrac{\omega_n}{\omega_d}e^{-at}\sin(\omega_d t + 3\phi)$
433	$\dfrac{\omega_n^3\left(s^2 + \omega_n^2\right)}{2as^3\left(s^2 + 2as + \omega_n^2\right)}$	$\dfrac{\omega_n^3 t^2}{4a} - \omega_n t + \dfrac{2a}{\omega_n} - \dfrac{\omega_n}{\omega_d}e^{-at}\sin(\omega_d t + 2\phi)$
434	$\dfrac{\omega_n^2\left(s^2 + \omega_n^2\right)}{2as^2\left(s^2 + 2as + \omega_n^2\right)}$	$\dfrac{\omega_n^2 t}{2a} - 1 + \dfrac{\omega_n}{\omega_d}e^{-at}\sin(\omega_d t + \phi)$
435	$\dfrac{\omega_n\left(s^2 + \omega_n^2\right)}{2as\left(s^2 + 2as + \omega_n^2\right)}$	$\dfrac{\omega_n}{2a} - \dfrac{\omega_n}{\omega_d}e^{-at}\sin\omega_d t$
436	$\dfrac{s^2 + \omega_n^2}{2a\left(s^2 + 2as + \omega_n^2\right)}$	$\dfrac{\delta}{2a} + \dfrac{\omega_n}{\omega_d}e^{-at}\sin(\omega_d t - \phi)$
437	$\dfrac{s\left(s^2 + \omega_n^2\right)}{2a\omega_n\left(s^2 + 2as + \omega_n^2\right)}$	$\dfrac{\delta'}{2a\omega_n} - \dfrac{\delta}{\omega_n} - \dfrac{\omega_n}{\omega_d}e^{-at}\sin(\omega_d t - 2\phi)$
438	$\dfrac{s^2\left(s^2 + \omega_n^2\right)}{2a\omega_n^2\left(s^2 + 2as + \omega_n^2\right)}$	$\dfrac{\delta''}{2a\omega_n^2} - \dfrac{\delta'}{\omega_n^2} + \dfrac{2\zeta\delta}{\omega_n} + \dfrac{\omega_n}{\omega_d}e^{-at}\sin(\omega_d t - 3\phi)$

440   Numerator Contains Last Two Terms of Quadratic
(See notes for entry 400.)

442	$\dfrac{\omega_n^4\left(2as + \omega_n^2\right)}{s^5\left(s^2 + 2as + \omega_n^2\right)}$	$\dfrac{\omega_n^4 t^4}{4!} - \dfrac{\omega_n^2 t^2}{2!} + 2at - (4\zeta^2 - 1)$ $+ \dfrac{\omega_n}{\omega_d}e^{-at}\sin(\omega_d t + 3\phi)$
443	$\dfrac{\omega_n^3\left(2as + \omega_n^2\right)}{s^4\left(s^2 + 2as + \omega_n^2\right)}$	$\dfrac{\omega_n^3 t^3}{3!} - \omega_n t + 2\zeta - \dfrac{\omega_n}{\omega_d}e^{-at}\sin(\omega_d t + 2\phi)$

	Laplace domain	Time domain
444	$\dfrac{\omega_n^2(2as + \omega_n^2)}{s^3(s^2 + 2as + \omega_n^2)}$	$\dfrac{\omega_n^2 t^2}{2!} - 1 + \dfrac{\omega_n}{\omega_d} e^{-at} \sin(\omega_d t + \phi)$
445	$\dfrac{\omega_n(2as + \omega_n^2)}{s^2(s^2 + 2as + \omega_n^2)}$	$\omega_n t - \dfrac{\omega_n}{\omega_d} e^{-at} \sin \omega_d t$
446	$\dfrac{2as + \omega_n^2}{s(s^2 + 2as + \omega_n^2)}$	$1 + \dfrac{\omega_n}{\omega_d} e^{-at} \sin(\omega_d t - \phi)$
447	$\dfrac{2as + \omega_n^2}{\omega_n(s^2 + 2as + \omega_n^2)}$	$-\dfrac{\omega_n}{\omega_d} e^{-at} \sin(\omega_d t - 2\phi)$
448	$\dfrac{s(2as + \omega_n^2)}{\omega_n^2(s^2 + 2as + \omega_n^2)}$	$\dfrac{2\zeta\delta}{\omega_n} + \dfrac{\omega_n}{\omega_d} e^{-at} \sin(\omega_d t - 3\phi)$
449	$\dfrac{s^2(2as + \omega_n^2)}{\omega_n^3(s^2 + 2as + \omega_n^2)}$	$\dfrac{2\zeta\delta'}{\omega_n^2} - \dfrac{(4\zeta^2 - 1)\delta}{\omega_n} - \dfrac{\omega_n}{\omega_d} e^{-at} \sin(\omega_d t - 4\phi)$

460   Numerator Contains s Plus Half-Damping
(See notes for entry 400.)

Note that, generally, damping produces the term $2a$. Thus, the appearance of the term a represents half-damping. See Appendix 1 for definition of symbols.

462	$\dfrac{\omega_n^3(s + a)}{s^3(s^2 + 2as + \omega_n^2)}$	$\dfrac{a\omega_n t^2}{2} - (2\zeta^2 - 1)\omega_n t + \zeta(4\zeta^2 - 3)$ $-e^{-at} \cos(\omega t + 3\phi)$
463	$\dfrac{\omega_n^2(s + a)}{s^2(s^2 + 2as + \omega_n^2)}$	$at - (2\zeta^2 - 1) + e^{-at} \cos(\omega t + 2\phi)$
464	$\dfrac{\omega_n(s + a)}{s(s^2 + 2as + \omega_n^2)}$	$\zeta - e^{-at} \cos(\omega t + \phi)$
465	$\dfrac{s + a}{s^2 + 2as + \omega_n^2}$	$e^{-at} \cos \omega t$
466	$\dfrac{s(s + a)}{\omega_n(s^2 + 2as + \omega_n^2)}$	$\dfrac{\delta}{\omega_n} - e^{-at} \cos(\omega t - \phi)$
467	$\dfrac{s^2(s + a)}{\omega_n^2(s^2 + 2as + \omega_n^2)}$	$\dfrac{\delta'}{\omega_n^2} + \dfrac{\zeta\delta}{\omega_n} + e^{-at} \cos(\omega t - 2\phi)$
468	$\dfrac{s^3(s + a)}{\omega_n^3(s^2 + 2as + \omega_n^2)}$	$\dfrac{\delta''}{\omega_n^2} + \dfrac{\zeta\delta'}{\omega_n^2} + \dfrac{(2\zeta^2 - 1)\delta}{\omega_n} - e^{-at} \cos(\omega t - 3\phi)$

	Laplace domain	Time domain
470	Half-Damping Plus ω_n^2 in Numerator (See notes for entry 400.) See note for entry 460. See Appendix 1 for definition of symbols.	

| 472 | $\dfrac{\omega_n^3\left(as + \omega_n^2\right)}{s^4\left(s^2 + 2as + \omega_n^2\right)}$ | $\dfrac{\omega_n^3 t^3}{3!} - \dfrac{a\omega_n t^2}{2} + (2\zeta^2 - 1)\omega_n t - \zeta(4\zeta^2 - 3)$ $+ e^{-at}\cos(\omega t + 3\phi)$ |

| 473 | $\dfrac{\omega_n^2\left(as + \omega_n^2\right)}{s^3\left(s^2 + 2as + \omega_n^2\right)}$ | $\dfrac{\omega_n^2 t^2}{2!} - at + (2\zeta^2 - 1) - e^{-at}\cos(\omega t + 2\phi)$ |

| 474 | $\dfrac{\omega_n\left(as + \omega_n^2\right)}{s^2\left(s^2 + 2as + \omega_n^2\right)}$ | $\omega_n t - \zeta + e^{-at}\cos(\omega t + \phi)$ |

| 475 | $\dfrac{as + \omega_n^2}{s\left(s^2 + 2as + \omega_n^2\right)}$ | $1 - e^{-at}\cos\omega t$ |

| 476 | $\dfrac{as + \omega_n^2}{\omega_n\left(s^2 + 2as + \omega_n^2\right)}$ | $e^{-at}\cos(\omega t - \phi)$ |

| 477 | $\dfrac{s\left(as + \omega_n^2\right)}{\omega_n^2\left(s^2 + 2as + \omega_n^2\right)}$ | $-\dfrac{\zeta\delta}{\omega_n} - e^{-at}\cos(\omega t - 2\phi)$ |

| 478 | $\dfrac{s^2\left(as + \omega_n^2\right)}{\omega_n^3\left(s^2 + 2as + \omega_n^2\right)}$ | $-\dfrac{\zeta\delta'}{\omega_n^2} + \dfrac{(2\zeta^2 - 1)\delta}{\omega_n} + e^{-at}\cos(\omega t - 3\phi)$ |

| 490 | General First Degree Factor in Numerator *Note*: The constant c is not related to any other constant in the system. For the special constant a, see notes for entry 400. | |

| 493 | $\dfrac{\omega_n^2(s + c)}{s^2\left(s^2 + 2as + \omega_n^2\right)}$ | $ct + \left[2\zeta^2 - 1 + 2\zeta\dfrac{(c - a)}{\omega}\right]$ $+ \dfrac{r}{\omega_d}e^{-at}\sin(\omega t + \phi_r + 2\phi)$ |

| 494 | $\dfrac{\omega_n(s + c)}{s\left(s^2 + 2as + \omega_n^2\right)}$ | $\dfrac{c}{\omega_n} - \dfrac{r}{\omega_d}e^{-at}\sin(\omega t + \phi_r + \phi)$ |

| 495 | $\dfrac{s + c}{s^2 + 2as + \omega_n^2}$ | $\dfrac{r}{\omega_d}e^{-at}\sin(\omega t + \phi_r)$ |

| 496 | $\dfrac{s(s + c)}{\omega_n\left(s^2 + 2as + \omega_n^2\right)}$ | $\dfrac{\delta}{\omega_n} - \dfrac{r}{\omega_d}e^{-at}\sin(\omega t + \phi_r - \phi)$ |

| 497 | $\dfrac{s^2(s + c)}{\omega_n^2\left(s^2 + 2as + \omega_n^2\right)}$ | $\dfrac{\delta'}{\omega_n^2} + \left[\dfrac{(c - a)}{\omega} - \zeta\right]$ $\dfrac{\delta}{\omega_n} + \dfrac{r}{\omega_d}e^{-at}\sin(\omega t + \phi_r - 2\phi)$ |

	Laplace domain	Time domain
500	*Underdamped Cubics*	

Note: The 500 family can be used only if the damping ratio ζ of the quadratic factor is greater than zero but less than unity (see notes for entry 400). For overdamped cubics, see group 240. $\omega = \omega_d$

510   Numerator is Single Term

513
$$\frac{\omega_n^2 r^2}{s^2(s + c)(s^2 + 2as + \omega_n^2)}$$
$$\frac{c^2 t}{\omega_n^2} + \frac{c^2 + 2(a^2 - \omega^2)}{\omega_n^2} - \frac{\omega_n^2}{c^2} + \frac{\omega_n^2 e^{-ct}}{c^2}$$
$$+ \frac{r}{\omega_d} e^{-at} \sin(\omega t - \phi_r + 2\phi)$$

514
$$\frac{\omega_n r^2}{s(s + c)(s^2 + 2as + \omega_n^2)}$$
$$\frac{r^2}{c\omega_n} - \frac{\omega_n e^{-ct}}{c} - \frac{r}{\omega_d} e^{-at} \sin(\omega t - \phi_r + \phi)$$

515
$$\frac{r^2}{(s + c)(s^2 + 2as + \omega_n^2)}$$
$$e^{-ct} + \frac{r}{\omega_d} e^{-at} \sin(\omega t - \phi_r)$$

516
$$\frac{r^2 s}{\omega_n(s + c)(s^2 + 2as + \omega_n^2)}$$
$$-\frac{ce^{-ct}}{\omega_n} - \frac{r}{\omega_d} e^{-at} \sin(\omega t - \phi_r - \phi)$$

517
$$\frac{r^2 s^2}{\omega_n^2(s + c)(s^2 + 2as + \omega_n^2)}$$
$$\frac{c^2 e^{-ct}}{\omega_n^2} + \frac{r}{\omega_d} e^{-at} \sin(\omega t - \phi_r - 2\phi)$$

518
$$\frac{r^2 s^3}{\omega_n^3(s + c)(s^2 + 2as + \omega_n^2)}$$
$$\frac{(c - \omega_n)^2 \delta}{\omega_n^3} - \frac{c^3 e^{-ct}}{\omega_n^3}$$
$$- \frac{r}{\omega_d} e^{-at} \sin(\omega t - \phi_r - 3\phi)$$

520   Numerator Contains First Two Terms of Quadratic
(See notes for entry 400.)

524
$$\frac{r^2(s + 2a)}{\omega_n(s + c)(s^2 + 2as + \omega_n^2)}$$
$$\frac{2r^2}{c\omega_n} - \frac{(\omega_n^2 + r^2)e^{-ct}}{c\omega_n}$$
$$- \frac{r}{\omega_d} e^{-at} \sin(\omega t - \phi_r + \phi)$$

525
$$\frac{r^2 s(s + 2a)}{\omega_n^2(s + c)(s^2 + 2as + \omega_n^2)}$$
$$2e^{-ct} + \frac{r}{\omega_d} e^{-at} \sin(\omega t - \phi_r)$$

526
$$\frac{r^2 s^2(s + 2a)}{\omega_n^3(s + c)(s^2 + 2as + \omega_n^2)}$$
$$\frac{r^2 \delta}{\omega_n^3} - \frac{c(\omega_n^2 + r^2)e^{-ct}}{\omega_n^3}$$
$$- \frac{r}{\omega_d} e^{-at} \sin(\omega t - \phi_r - \phi)$$

Laplace domain	Time domain

530 **Numerator Contains First and Last Terms of Quadratic**
(See notes for entry 400.)

534
$$\frac{\omega_n r^2\left(s^2 + \omega_n^2\right)}{2as^2(s + c)\left(s^2 + 2as + \omega_n^2\right)}$$

$$\frac{\omega_n r^2 t}{2ac} - \frac{r^2\left(\omega_n^2 - 2ac\right)}{2ac^2\omega_n}$$

$$+ \frac{\omega_n\left(r^2 - 2ac\right)e^{-ct}}{2ac^2}$$

$$- \frac{r}{\omega_d}e^{-at}\sin(\omega t - \phi_r + \phi)$$

535
$$\frac{r^2\left(s^2 + \omega_n^2\right)}{2as(s + c)\left(s^2 + 2as + \omega_n^2\right)}$$

$$\frac{r^2}{2ac} - \frac{\left(r^2 - 2ac\right)e^{-ct}}{2ac}$$

$$+ \frac{r}{\omega_d}e^{-at}\sin(\omega t - \phi_r)$$

536
$$\frac{r^2\left(s^2 + \omega_n^2\right)}{2a\omega_n(s + c)\left(s^2 + 2as + \omega_n^2\right)}$$

$$\frac{\left(r^2 - 2ac\right)e^{-ct}}{2a\omega_n}$$

$$- \frac{r}{\omega_d}e^{-at}\sin(\omega t - \phi_r - \phi)$$

540 **Numerator Contains Last Two Terms of Quadratic**
(See notes for entry 400.)

544
$$\frac{\omega_n r^2\left(2as + \omega_n^2\right)}{s^3(s + c)\left(s^2 + 2as + \omega_n^2\right)}$$

$$\frac{\omega_n r^2 t^2}{2c} - \frac{\omega_n r^2 t}{c^2} + \frac{r^2\left(\omega_n^2 + c^2\right)}{c^3\omega_n}$$

$$- \frac{\omega_n\left(r^2 + c^2\right)e^{-ct}}{c^3}$$

$$- \frac{r}{\omega_d}e^{-at}\sin(\omega t - \phi_r + \phi)$$

545
$$\frac{r^2\left(2as + \omega_n^2\right)}{s^2(s + c)\left(s^2 + 2as + \omega_n^2\right)}$$

$$\frac{r^2 t}{c} - \frac{r^2}{c^2} + \frac{\left(r^2 + c^2\right)e^{-ct}}{c^2}$$

$$+ \frac{r}{\omega_d}e^{-at}\sin(\omega t - \phi_r)$$

546
$$\frac{r^2\left(2as + \omega_n^2\right)}{\omega_n s(s + c)\left(s^2 + 2as + \omega_n^2\right)}$$

$$\frac{r^2}{c\omega_n} - \frac{\left(r^2 + c^2\right)e^{-ct}}{c\omega_n}$$

$$- \frac{r}{\omega_d}e^{-at}\sin(\omega t - \phi_r - \phi)$$

Appendix **4**

Conversions

Pressure or Stress

$1 \text{ psi} = 1 \text{ lb/in.}^2 = 6895 \text{ N/m}^2 = 6.895 \times 10^4 \text{ dyn/cm}^2$

$\text{N/m}^2 = \text{Newtons per meter squared} = 1.45 \times 10^{-4} \text{ psi}$

$\quad = 10 \text{ dyn/cm}^2 = 10^{-7} \text{ dyn/}\mu\text{m}^2$

Mass Density

$1 \text{ g/cm}^3 = 1 \text{ gram per cubic centimeter} = 0.91 \times 10^{-4} \text{ lb} \cdot \text{s}^2/\text{in.}^4$

$\quad = \text{dyn} \cdot \text{s}^2/\text{cm}^4 = 10^3 \text{ kg/m}^3$

Velocity

$1 \text{ m/s} = 100 \text{ cm/s} = 39.4 \text{ in./s} = 3.28 \text{ ft/s} = 2.23 \text{ mph}$

Viscosity

$1 \text{ poise} = 1 \text{ dyn} \cdot \text{s/cm}^2 = 0.1 \text{ N} \cdot \text{s/m}^2 = 100 \text{ centipoises}$

$\text{lb} \cdot \text{s/ft}^2 = 479 \text{ poises} = 47.9 \text{ N} \cdot \text{s/m}^2$

Mass

$1 \text{ kilogram} = 1 \text{ kg} = 1000 \text{ g} = 2.248 \text{ lb} = 1 \text{ N} \cdot \text{s}^2/\text{m} = 1000 \text{ dyn} \cdot \text{s}^2/\text{cm}$

$1 \text{ g} = 1 \text{ dyn} \cdot \text{s}^2/\text{cm}$

Length

$1 \text{ meter} = 1 \text{ m} = 100 \text{ cm} = 10^6 \ \mu\text{m} = 39.37 \text{ in.} = 3.281 \text{ ft.} = 0.621 \times 10^{-3} \text{ mile}$

$1 \text{ micron} = 1 \ \mu\text{m} = 10^{-6} \text{ m} = 10^{-4} \text{ cm} = 39.37 \times 10^{-6} \text{in.}$

Force

1 newton = 1 N = 10^5 dyn = 0.2248 lb

Torque

1 N $\cdot$ m = 8.850 lb $\cdot$ in.

Voltage

1 volt = 1 V = 1 N $\cdot$ m/C = 8.850 lb $\cdot$ in./C.

   C = coulomb

Translatory Spring Constant K

1 N/m = 5.71 $\times$ 10^{-3} lb/in. = 10^3 dyn/cm

Translatory Damper, D

1 N $\cdot$ s/m = 10^3 dyn $\cdot$ s/cm = 5.71 $\times$ 10^{-3} lb $\cdot$ s/in.

Moment of Inertia, J

1 kg $\cdot$ m^2 = N $\cdot$ m $\cdot$ s^2 = 10^7 dyn $\cdot$ cm $\cdot$ s^2 = 8.85 lb $\cdot$ in. $\cdot$ s^2

Torsional Spring, K_t

1 N $\cdot$ m/rad = 10^7 dyn $\cdot$ cm/rad = 8.85 lb $\cdot$ in./rad

Torsional Damper, D_t

1 N $\cdot$ m $\cdot$ s/rad = 10^7 dyn $\cdot$ cm $\cdot$ s/rad = 8.85 lb $\cdot$ in. $\cdot$ s/rad

Angular Velocity, $\dot{\theta}$

1 rad/s = 9.549 rpm = 57.3°/s

Power, P

1 watt = W = N $\cdot$ m/s = 8.85 lb $\cdot$ in./s = 1.341 $\times$ 10^{-3} hp

   1 hp = 745.7 W

Heat Flow, q

1 N $\cdot$ m = joule = J = 2.389 $\times$ 10^{-4} kcal = 10^7 dyn $\cdot$ cm
          = 8.85 lb $\cdot$ in. = 9.481 $\times$ 10^{-4} Btu

Thermal Conductivity, k

1 W/m $\cdot$ °C = 0.578 Btu/h $\cdot$ ft $\cdot$ °F

Appendix 5

Brief Review of Complex Numbers

A complex quantity is represented on a two axis plot, referred to as the Argand plane, or the complex plane. The complex quantity has two parts, a real and an imaginary part. The real part is shown on the horizontal or real axis. The imaginary part j is shown on the vertical axis. For example, the complex number $a + jb$ may be shown on the complex plane as shown in Fig. A5.1-1(a).

The above representation is also referred to as the rectangular or Cartesian form of a complex number. The same information may be shown in polar form. In polar form, a vector is drawn from the origin to the point $a + jb$. The magnitude of this vector (the hypotenuse of the triangle whose sides are a and b) is the magnitude of the complex number, and the angle this vector makes with the real axis is the phase angle of the complex number. See Fig. A5.1-1(b). Summarizing, a complex number in both the rectangular and polar notations is given below.

$$a + jb = Re^{j\psi} = \text{complex quantity} \qquad [\text{A5.1-1}]$$

where

$$R = \sqrt{a^2 + b^2} = \text{magnitude}$$

$$\psi = \tan^{-1} \frac{b}{a} = \text{phase angle} \qquad [\text{A5.1-2}]$$

The choice of symbols makes it immediately apparent how the complex plane will provide the frequency response. Obviously, the polar form provides the frequency response directly. There is another advantage to using the polar form. If the solution to the differential equation is given in factored form, each term may be transformed individually to the complex plane, and the results can then be multiplied. The rules for algebra of complex numbers in the complex plane make use of the interchange between

710

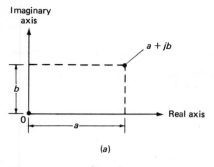

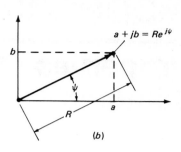

Figure A5.1-1   Complex numbers and the Argand plane. (a) Rectangular notation. (b) Polar notation.

the two forms of the complex number. Given two complex numbers,

$$A_1 + jB_1 = R_1 e^{j\psi_1}$$
$$A_2 + jB_2 = R_2 e^{j\psi_2}$$

[A5.1-3]

The sum of the two numbers is best performed in rectangular form

$$\text{sum} = A_1 + A_2 + j(B_1 + B_2)$$

[A5.1-4]

The product is best performed in the polar form

$$\text{product} = R_1 R_2 e^{j(\psi_1 + \psi_2)}$$

[A5.1-5]

$$\text{quotient} = \frac{R_1}{R_2} e^{j(\psi_1 - \psi_2)}$$

[A5.1-6]

Appendix 6

Properties of Materials

Material	Density g / cm³	E 10¹² dyn / cm²	G	Yield strength* 10⁹ dyn / cm²	ν
Cast iron	7.2	1.7	0.64	12.6	
Steel, low carbon	7.8	2.1	0.83	4.2	0.3
Steel, high carbon	7.8	2.1	0.83	5.8	0.3
Steel, hardened and tempered	7.8	2.1	0.83	9.7	0.3
Stainless steel	7.9	2.0	0.76	7.0	
Aluminum	2.7	0.7	0.26	1.7	
Brass	8.5	1.03	0.39	1.7	
Magnesium alloy	1.8	0.41	0.17	1.7	
Copper, annealed	8.9	1.2	0.44	0.7	
Silicon	2.3	1.9	0.6	4.2	0.18
Glass	2.5	0.7	0.2	1.3	

*Applies for tension and compression. For the yield strength in shear use $\frac{2}{3}$ of this value.

PROPERTIES OF WATER

Temperature		Density ρ		Absolute viscosity μ	
°F	°C	10^{-2} lb / in.^{3}	g / cm^3	10^{-7} lb · s / in.^{2}	10^{-3} N · s / m^2
32	0	3.612	0.999	3.75	2.60
40	4.44	3.613	1.000	3.23	2.24
60	15.56	3.609	0.9992	2.36	1.74
80	26.67	3.601	0.9968	1.80	1.22
100	37.78	3.588	0.9932	1.42	0.989
120	48.89	3.571	0.9886	1.17	0.821
160	71.11	3.530	0.9772	0.838	0.582

PROPERTIES OF AIR

Temperature		Density ρ		Absolute viscosity μ	
°F	°C	10^{-5} lb / in.^{3}	10^{-3} g / cm^3	10^{-9} lb · s / in.^{2}	10^{-5} N · s / m^2
− 100	− 73.33	6.045	1.673	1.92	1.33
− 40	− 40.0	5.468	1.513	2.19	1.51
0	− 17.78	4.991	1.381	2.36	1.63
40	4.44	4.593	1.271	2.52	1.74
60	15.56	4.416	1.222	2.60	1.80
80	26.67	4.252	1.177	2.68	1.85
100	37.78	4.099	1.134	2.76	1.90
120	48.89	3.958	1.095	2.83	1.95
160	71.11	3.703	1.025	2.97	2.05

PROPERTIES OF OIL

Temperature		SAE number Absolute viscosity μ (10^{-3} N · s / m^2)						
°F	°C	10	20	30	40	50	60	70
60	15.56	172	310	600	—	—	—	—
80	26.67	82.7	121	228	345	655	—	—
100	37.78	48.3	69.0	103	159	276	372	552
120	48.89	24.1	41.4	55.2	82.7	138	176	241
160	71.11	11.0	13.8	20.0	26.9	44.8	60.0	75.9

ELECTRICAL RESISTANCE

Material	Resistivity ρ ($\Omega \cdot m$)
Silver	1.6×10^{-8}
Copper	1.72×10^{-8}
Aluminum	2.83×10^{-8}
Insulators	10^{6} to 10^{16}
Semiconductors	10^{-5} to 10^{5}

MAGNETIC PROPERTIES

Material	Permeability μ (H / m)
Free space	1.257×10^{-6}
Air	1.257×10^{-6}
Iron	6.53×10^{-3}
Steel	8.80×10^{-3}
Permalloy	1.257×10^{-1}

DIELECTRIC PROPERTIES

Material	Permittivity ϵ 10^{-11} F / m	Allowable voltage gradient V / l 10^{6} V / m
Air	0.885	1
Ether	4.0	5
Free space	0.8854	1
Pyrex	3.54	1000
Mica	4.69	1000
Mylar	4.16	100
Quartz	3.54	1000
Pure water	71	100
Titanium dioxide	150	4

THERMODYNAMIC PROPERTIES OF GASES

Gas	Gas constant R ft · lb / lb · °R	Specific heat C_p	C_v	Adiabatic coefficient K
Air	53.35	0.24	0.17	1.4
CO_2	35.11	0.21	0.17	1.3
H_2	766.53	3.4	2.4	1.4
N_2	55.16	0.25	0.18	1.4
O_2	48.29	0.22	0.16	1.4

SPECIFIC HEAT

Material	C_p (Btu / lb · °F)
Water	1.00
Alcohol	0.54
Aluminum	0.22
Steel	0.11
Lead	0.03
Wood	0.62
Asbestos	0.19
Glass	0.19

COEFFICIENT OF THERMAL EXPANSION

Material	α (10^{-6} / °C)
Sodium	77.0
Aluminum	25.0
Copper	16.8
Nickel	12.8
Steel	7.0
Silicon	2.3
Glass	0.55

THERMAL CONDUCTIVITY k (W / cm · °C)

Insulators	k	Conductors	k
Asphalt	0.014	Aluminum	2.36
Brick	0.013	Copper	4.7
Glass	0.014	Silver	5.0
Wood	0.002	Steel	0.97
Asbestos	0.0017	Silicon	1.57
Fibre glass	0.0007		
Gases		Liquids	
Air	0.0007	Water	0.013
CO_2	0.0003	Oil	0.003

Appendix 7

Programming System Problems in TUTSIM

Chapters 4, 5, 6, 7, and 16 contain examples of systems that are solved using TUTSIM, the software accompanying this book. This software may be used with any IBM compatible personal computer. The programs in TUTSIM for each solved example in the text are developed specifically for those examples. They are also available as TUTSIM disk files stored on the included disk and can be called for study if desired. For any of these filed programs it is possible to answer the question: "What if I change the value of one or more of the parameters?" This question can easily be answered by the TUTSIM software. For any of the filed examples it is possible to make such parameter studies by changing the appropriate parameters. The technique for doing this will be discussed later. File names for each text example are stated here for the reader's convenience.

Example number	System name	File name
4.6.3-1	First Order System	FIRST.SIM
4.6.3-2	Second Order System	SECOND.SIM
5.1.3-3	Water Tank with Nonlinearity	TANK.SIM
5.4.1-1	Nonlinear Annealing Oven	ANNEAL.SIM
6.3.4-2	Spring Scale (Linear)	SPRING.SIM
6.3.4-3	Spring Scale (Velocity Squared Damping)	VELSQR.SIM
6.3.4-4	Spring Scale (Coulomb Damping)	COULMB.SIM
7.2.3-1	Dial Movement (Sinusoidal Input)	DIAL.SIM
7.3.3-1	Forced Vibration (Second Order)	MKDSYS.SIM
7.9.1-1	Startup Transient in Pump	STARTUP.SIM
16.5.4-1	Electromechanical Servo	SERVO.SIM
16.5.6-2	Rolling Mill	MILL.SIM
16.6.2-3	Roll Stabilization of Aircraft	ROLCON.SIM

Note: The suffix SIM means simulation.

Two separate sets of instructions for the use of TUTSIM are presented here. They are (1) the procedure for using the TUTSIM files, and (2) the procedure for creation of new TUTSIM programs, such as those requested in some of the text problems. To do either of these tasks requires the availability of an IBM compatible computer with DOS 3 or higher. Boot up the computer with DOS. Then insert the accompanying disk into slot A. After the A: > appears on the screen, type TUTSIMHR, or TUTSIMHR/D = 2,* and press Return. A message appears on the screen for which you are given choices of where your input is to come from: from the keyboard (K) or from a file (F). You respond by typing K or F and press Return.

If you have typed F, you will get another message requesting the file name. You may respond to this by typing one of the file names in the list above. For example you may type FIRST.SIM followed by Return. You will then see COMMAND: on the screen, to which you respond by any of the following:

SD   simulation to display (graphical output)

SN   simulation to numerical results

SNP   simulation to numerical results on the printer

SN:10   numerical output to screen, every 10th point

SNP:10   numerical output to printer, every 10th point

CS   change structure of the model

CP   change parameters of the model

CB   change plotblocks and ranges of the model

CT   change timing numbers

L   lists the model data on the screen

LP   lists the model data on the printer

CL   clear the display screen

If you wish to reproduce the graphical output given in the text examples, call the desired file and after COMMAND: type SD. The graphical output appears on the screen. Numerical scale values are obtained by pressing one of the F keys. F1 is typed if the scale numbers are desired for output Y1, already specified in the model program. F5 activates the time scale. F10 permits the entry of a title at the top of the graph. The command HC copies the graphical information to the printer. HC means hard copy. [*Note*: it is necessary to type GRAPHICS after DOS is loaded into the computer in order to copy graphical results.]

If you wish to change one of the parameters, type CP followed by a comma, and the new parameter value. Simulation can then be repeated with the new parameter value. For example, to change the damping constant in Ex. 4.6.3-1 from 4 to 2, type after COMMAND: the following: 2,.5. The new parameter for block 2 is 0.5, changed from 0.25, the original value of $1/D$. Type Return and produce a new simulation run by typing SD. The new results are displayed along with the previous results obtained by the SD command. If only the new results are desired, then type CL followed by Return before the new simulation is run.

To create a new TUTSIM program, you need the system differential equations, the initial values of the dependent variables, and an operational block diagram for the system, such as that shown in Fig. 5.1.3-3(a). Start the TUTSIM diagram by indicating

---

*Typing TUTSIMHR/D = 2 produces graphical output like those in the text examples. Typing TUTSIMHR only produces a graphical frame that fills the entire screen and permits color.

as input to Block 1, an integrator (INT), the highest derivative of the dependent variable (in this case $\dot{H}$). The output of Block 1 is H. Note that the initial condition for H is prescribed as the parameter for the INT block. The output H is raised to the power n in Block 2. Here n is the parameter for Block 2. This operation is followed by division by the constant AR, by means of the attenuator Block 3, ATT. The output of Block 3 is returned as an input to Block 1, with the sign changed.

A TUTSIM program is entered on the keyboard by typing K (keyboard) after calling TUTSIMHR or TUTSIMHR/D = 2. You will see the prompt MODEL STRUCTURE with the words FORMAT:Block Number, Type, Input, Optional Comment on the screen, below which is a colon : followed by a flashing cursor. Here you enter the structure data as printed in the text following Fig. 5.1.3-3. Block 1 is an integrator whose input is negative and comes from Block 3. Type the following:

:1,INT, – 3

Similarly Block 2 is a power block (PWR) whose input comes from Block 1. Type the following:

:2,PWR,1

Finally Block 3 is an attenuator block (ATT) whose input comes from Block 2. Type the following:

:3,ATT,2

This completes what is called the model structure. Press Return twice and get the following message:

Model Parameters
Format:Block number,Parameters   (in our case only one)

After the colon with the flashing cursor type the following:

1,40   (initial value of H)
2,1.25   (the power to which H is raised)
3,7850   (the value of AR)

This completes the entry of the model parameters. Press Return twice and get the prompt:

Plotblocks and Ranges
Format:Block Nbr,Plot minimum,Plot maximum

followed by the message Horz: after which is the flashing cursor. Horz is the horizontal axis, in our case time. For Ex. 5.1.3-3 the time axis starts at 0 and ends at 20000. To enter this from the keyboard, type the following after Horz:0,0,20000. The first zero is the time "block", the second zero is the initial time for the plot, and the 20000 is the final time on the plot. Press Return and get Y1: after which is the flashing cursor. After the colon type 1,0,50. Here 1 signifies the block number whose output is to be plotted (in this case H). Press Return again and get Y2:. This is for a second output if desired. In this example no second output is to be found. We bypass Y2: by pressing Return, and getting Y3: for a possible third output. Again bypass this by pressing Return and get Y4:. Again bypass by pressing Return. Now the message on the screen is:

Timing Data

Format:Delta time,Final time.

In response to this, type 300,20000 and press Return. The numbers 300 and 20000 are the time interval for numerical integration and the final time. After entering these numbers, press Return and get Command:. Here we enter the letters depicting the kind of output desired. If we want a graphical output, we type SD, followed by Return. The graph of H versus time is displayed. If numerical output is desired, type SN. (Refer to the list of commands given earlier in this appendix.)

If you want to change any of the parameters, such as the exponent of H, type CL (clear the screen) followed by Return. The word Command: appears on the screen. After the colon type CP (Change parameters). Then type 2,1. The parameter of Block 2 has been changed to 1. Press Return, then SD to get the new output.

Scale values may be displayed on the graph by pressing one of the F keys. Pressing F1 causes scale numbers to appear on the vertical axis (0 to 50 in our case for H). Keys F2, F3, and F4 would display scales for outputs Y2, Y3, and Y4, respectively. Our example does not have these. Pressing the key F5 causes the minimum and maximum time values to appear on the horizontal axis. Key F10 is pressed if you want to enter a short title at the top of the grid. Type the desired title and press Return. The graph as it appears on the screen can be printed by typing HC (hard copy), followed by Return.

If you want to compare the outputs for $n = 1.25$ with that for $n = 1$, first run the simulation for $n = 1.25$. Do not clear the screen by typing CL in this case. Instead, type CP after the Command prompt and change the parameter of Block 2 to 1, then press Return and type SD once more, after which the new output is plotted on the same grid.

Only three blocks, INT, PWR, and ATT, were used in this example. Other blocks are used in other simulations. In Ex. 4.6.3-1 the blocks used are CON (constant), GAI (gain), SUM (sum of inputs), and INT. The model structure and model parameters are shown in that example, as well as plotblocks and ranges, and timing data. The commands SD, SN, and SNP are also discussed.

Other block names appear in Ex. 5.4.1-1. They are ABS (absolute value) and MUL (multiplication). Refer to that example for the model data. Another useful block occurs in Ex. 6.3.4-4. This is REL (Relay), which is explained there. Two new blocks appear in Ex. 7.2.3-1. They are TIM (time, a unit ramp) and SIN (sine of the input in radians). Note that some blocks do not require parameters, such as SIN, TIM, and SUM. Others require more than one parameter, such as REL. The COS (cosine) block is used in Ex. 7.3.3-1. Example 7.9.1-1 uses the EXP (exponential function) block.

Chapter 16 (Automatic Controls) utilizes additional blocks, namely FIO (first order system) and SEO (second order system). These are used only for linear first and second order systems. For example, the differential equation in Section 3.7.1 is

$$D\dot{y} + Ky = F$$

$$\text{IC:} \quad y(0) = y_0$$

which can be written in the Laplace domain as

$$Y(s) = \frac{F(s) + Dy_0}{Ds + K} = \frac{1}{K} \frac{F(s) + Dy_0}{\tau s + 1}$$

where τ is the time constant, $\tau = D/K$.

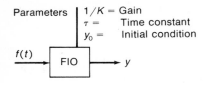

Figure A7.1   First order system.

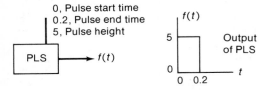

Figure A7.2   Pulse block.

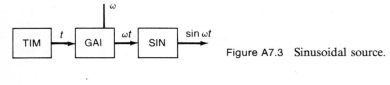

Figure A7.3   Sinusoidal source.

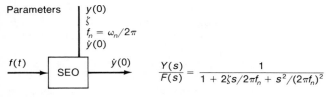

Figure A7.4   Second order system.

This system can be modeled as shown in Fig. A7.1.

The blocks that generate $f(t)$ are not shown here. If $f(t)$ were a pulse of magnitude 5 and duration 0.2, starting at $t = 0$, the PLS block could be used to provide the pulse input to the FIO block above. The diagram of the PLS block is shown in Fig. A7.2. Note that CON (constant) could be used if the input $f(x) = $ constant. The value of the constant is entered as a parameter of CON. If the input to FIO is a unit ramp, then TIM may be used to supply the input to the FIO block. If the input to FIO is sinusoidal, the input to FIO is generated by the system shown in Fig. A7.3. Figure A7.4 shows the SEO (second order system). Note the parameters required to use this block.

Index

ISBN 0-06-041314-X

90000

9 780060 413149